# Geographies for Advanced Study

The British Isles: A Geographical and Economic Survey
Central Europe
**Concepts in Climatology**
Eastern Europe
Geography of Population
Geomorphology
The Glaciations of Wales and Adjoining Regions
An Historical Geography of South Africa
An Historical Geography of Western Europe before 1800
Human Geography
Land, People and Economy in Malaya
Malaya, Indonesia, Borneo and the Philippines
North America
The Polar World
A Regional Geography of Western Europe
The Scandinavian World
The Soviet Union
Statistical Methods and the Geographer
The Tropical World
Urban Essays: Studies in the Geography of Wales
Urban Geography
West Africa
The Western Mediterranean World

# Concepts in Climatology

## P. R. Crowe

Professor of Geography
University of Manchester

**Longman**

LONGMAN GROUP LIMITED
London
*Associated companies, branches and representatives*
*throughout the world*

© Longman Group Limited 1971

ISBN   0 582   48148   1

*Printed in Great Britain by*
*William Clowes & Sons Limited*
*London, Colchester and Beccles*

# Contents

# Preface

The behaviour of the atmosphere is of immediate concern to all mankind, both directly as it affects each individual during his daily round of activities and indirectly via the resource-potential of distant lands with which he has established social or economic contacts. Yet the air-ocean is a vast continuum which is beyond man's comprehension unless he can break it down either regionally (as to its geographical differentiation), stratigraphically (as to its vertical characteristics) or mechanically (as to the various processes which occur within it). It presents a number of fascinating but still only partially-answered questions.

This book is about ideas; about theories and speculations concerning all three of these interrelated aspects of the atmosphere. Hence its title. Such ideas are rarely either completely right or completely wrong. They have undergone an evolution which is often a matter of subtle changes of emphasis. Hence a comparatively historical approach is just as appropriate as the more mathematical-physical techniques employed by meteorological specialists.

All the major countries of the world spend large sums each year in acquiring factual information about the weather and in supporting professional staff to make the day-to-day forecasts which are of the greatest practical importance to seamen, airmen, farmers, transport concerns, sportsmen and others. But the information thus gathered calls for analysis on a wider canvas; this is the task of the climatologist. The book illustrates numerous devices by which such an approach can be facilitated and is the first climatology text to employ systematically the great mass of upper-air data which has become available during the last 15 years on a world-wide scale. Indeed facts about the weather from surface stations, upper-air stations and, still more recently, from satellites, are now being fed to us at a dizzily accelerating pace. We are in real danger of being overwhelmed unless we can strengthen the conceptual framework into which they must be fitted. Computers may assist in marshalling the facts but they cannot be expected to do our thinking for us.

The work was already in the hands of the publisher when the question of metrication became an urgent issue. It was inevitable that the solution should be in the nature of a compromise. Values are frequently presented in duplicate scales but complete metrication appeared to be absurd in a field where much of the basic data—and the literature derived from it—is expressed for all time in non-metric terms. Conversion is never difficult (hence Appendix 3) and acquiring the habit of making conversions is surely as important to the climatologist as having at least some mastery of more than one language.

It is a pleasure to acknowledge a longstanding indebtedness to the late Henry V. Janau, schoolmaster, who first stimulated my interest in climatology, and to Miss Elaine Austin under whose benign but alert guidance that interest was ripened at the Meteorological Office during the Second World War. Since that time the staff of the office Library, now at Bracknell, have proved unfailing in their courtesy. The debt to university colleagues is invariably as deep as it is difficult to specify. The diagrams were prepared by Miss Anne Lowcock and Mrs Margaret Steele typed the final copy of the text. The faults are all unquestionably my own.

P. R. CROWE

*University of Manchester*
*January 1971*

# Maps and Diagrams

xi

# Acknowledgements

We are grateful to the following for permission to reproduce copyright material:

Cambridge University Press for extracts from *Manual of Meteorology* by Sir Napier Shaw; Macmillan & Co. Ltd., for an extract from *A Popular Treatise on the Winds* by W. Ferrel.

# Abbreviations

| | |
|---|---|
| A | average |
| cal | calorie |
| CCL | convective condensation level |
| cm | centimetre |
| $D_1$, $D_9$ | deciles (lower and upper) |
| DALR | dry adiabatic lapse rate |
| $E$ | evaporation |
| EPT | equivalent potential temperature |
| ft | feet |
| g, gr | gram |
| hr | hour |
| ICAN | International Commission for Air Navigation |
| in | inch |
| kg | kilogram |
| km | kilometre |
| kw | kilowatt |
| LCL | lifting condensation level |
| m | metre |
| $M$ | median |
| max | maximum |
| mb | millibar |
| min | minimum or minute |
| mm | millimetre |
| M.O. | Meteorological office |
| MR | mixing ratio |
| mw | milliwatt |
| obs | observations |
| $P$ | precipitation (in tables) |
| PE | potential evapotranspiration |
| PT | potential temperature (in tables) |
| $Q_1$, $Q_3$ | quartiles (lower and upper) |
| RH | relative humidity |

| SALR | saturated adiabatic lapse rate |
|------|-------------------------------|
| *sd* | saturation deficit |
| SD | standard deviation—usually $\sigma$ |
| sec | second |
| SFC | surface observation (in tables) |
| SPT | saturation potential temperature |
| $T$ | temperature |
| $T_c$ | tropical continental air |
| $T_m$ | tropical maritime air |
| tr. coef. | transmission coefficient |

# 1
# Insolation or solar income

It is generally recognised by biologists that all food chains, no matter how long or complex they may be, lead back initially to the absorption and utilisation of solar energy by the plant cell. All life is thus parasitic upon the sun or, as a Japanese scientist has put it, we live 'by consuming the sun's body' (Fukui, 1954). Nor is this all; the power which stirs the waters of the oceans, impels the movement of the wind and abrades and transports the sediments of the continents is derived from the same source. Furthermore, geological evidence suggests that these processes have continued, apparently with little modification, during the past 500 million years and probably for very much longer. The fossil fuels, coal, oil and lignite, are properly regarded as entrapped solar energy which may be made available, through oxidation, to move the wheels of modern industry. It requires a greater effort of the imagination to grasp the fact that all the sedimentary rocks, tens of thousands of feet of them covering millions of square miles, are mute evidence of solar work done in the past. The imagination may well boggle at the magnitude of energy, mass and time involved in such a picture of the world about us.

## Earth and sun

Science begins with measurement and some of the magnitudes involved are comparatively well known.

Mean solar distance $R$ 92 870 000 miles (149 450 000 km)
Sun's semidiameter $r_s$ 432 700 miles (695 300 km)
Earth's equatorial semidiameter $r_e$ 3963 miles (6378 km)

Given these figures we can make some quite elementary and yet stimulating inferences. Thus a table of natural tangents (or sines) will show us that, as seen from the earth, the sun's diameter subtends an angle of $2 \times 0° 16'$ or $0° 32'$ (approximately). In simple household terms this means that the sun 'looks as big as' a shilling seen from a distance of 8 feet.[1] Alternatively, by comparing the area of the sun's disc ($\pi r_s^2$) with the area of the dome of the sky, that is the celestial hemisphere at solar distance ($2\pi R^2$), we can show that

[1] A new five-penny piece at 2.4 metres.

the sun occupies no more than 1/92 000th of the visible sky. The photographer at high altitudes is familiar with the problem of the sharp, deep shadows cast by such a comparatively small lamp. It is only because of atmospheric diffusion that we who live nearer sea-level can find our way about in comfort in the shade.

Transferring our viewpoint, in imagination, to the sun's surface, a similar calculation has even more impressive results. With a diameter only 1/109th that of the sun, the illuminated earth would subtend an angle of only about $17\frac{1}{2}$ seconds of arc. It would look as big as a shilling at 265 metres or a pin-head at 15 metres. The real significance of this fact becomes apparent when we consider it too in terms of area. The earth covers only about 1/2210 millionth of the sun's celestial sphere; its total income of solar radiation is therefore only this infinitesimally small proportion of the sun's total output. Apart from the similarly minuscule proportions intercepted by the other planets and their satellites, the remainder passes outwards into space for objects and destinations quite unknown. Yet it is this minute proportion of solar energy, continued through the ages, that has had the momentous consequences outlined in our opening paragraph. Grounds enough for scientific humility!

We leave it to the astrophysicist to explain how this vast atomic furnace converts 800 million tons of hydrogen into helium every second, using carbon and nitrogen nuclei as catalysts (see Gamow, 1967, 123–4), and to assure us that it is capable of maintaining a mean surface temperature of about 5750°C for at least 10 000 million years and possibly for very much longer. What is of more immediate concern to the climatologist is that it is possible to estimate the mean radiative energy delivered per unit area, normal to the rays, at the outer limits of the earth's atmosphere. This is known as the 'solar constant'.

Solar constant = 1.94 gram-calories per square centimetre per minute.

Recent research at very high altitudes suggests that this estimate may be a trifle low, but since the extra heat is absorbed in the extreme outer layers of the atmosphere the revision has apparently no climatological significance.

Indeed, for geographical purposes, the units of heat, time and area in this statement may appear to be distressingly small. Let us remind ourselves, however, that, theoretically, such radiation could raise the temperature of a 1-centimetre layer of distilled water 1.94°C per minute, or a layer a foot deep nearly 7°F per hour. Fully devoted to evaporation it would be capable of removing nearly two millimetres or 0.075 inches of water per hour. Yet both of these processes are notoriously great consumers of heat. Applied uniformly to a column of air a kilometre deep the same energy would raise its temperature by 3.9°C (7°F) per hour.

It is when we consider this solar input in terms of terrestrial areas that the figures become really impressive. Thus the solar constant, which may be expressed alternatively as 1.35 kw/m², becomes 13 500 kw/hectare (5 470 kw/acre or 3 500 000 kw/mile²). In engineering terms this is the same as 18 000 hp/hectare (7300 hp per acre or 4.7 million hp/mile²). Of course, much of this energy is lost in transit through the atmosphere or reflected and dispersed when it reaches the earth's surface. Nevertheless, this is the measure of the earth's solar endowment.

## The geometry of insolation

The distribution of solar energy over the earth's surface is a function of certain properties of the sphere and of certain features of planetary motion which, although complex in detail, are simple enough in principle. We are entitled, at first, to ignore the presence of the atmosphere so as to simplify the presentation but we are not entitled to forget that we have done so.

If the earth were a flat disc facing the sun the solar constant would be a measure of the intensity of radiation upon each point of it but in fact, at any given moment of time, we are dealing with an illuminated hemisphere which has *twice* the area of a disc of corresponding radius. The intercepted radiation is spread over this wider area most unequally, but the distribution satisfies a simple mathematical law—the intensity varies as the sine of the angle of incidence or as the cosine of the zenith angle of the sun (Fig. 1*a*). Figure 1*b*, which is plotted directly from trigonometrical tables, shows clearly that intensity varies only slightly for angles of incidence above 60°, that it is halved at 30° and that thereafter it falls off almost rectilinearly by about 5 per cent of the possible maximum for every reduction in angle of incidence of 3°. The first of these three features has often been insufficiently stressed in the past and the comparatively rare occasions, even within the Tropics, when the sun is 'vertically overhead' have been permitted to usurp a climatological status to which they are by no means entitled.

In Fig. 2 essentially the same facts are looked at from a less familiar angle. It is shown that, at any given moment of time, one quarter of the intercepted beam is received at an angle of incidence of over 60°, one half has an angle of incidence of over 45°, and threequarters of the beam is intercepted at angles above 30°. The areas illuminated under each of these conditions can be shown to amount to 13.4, 29.3 and 50 per cent of the illuminated hemisphere, half of which is thus receiving rays at angles below 30°. Expressed in terms of the total area of the globe this means that, at any one moment of time, half of the solar income received by the earth is being paid into 14.6 per cent or barely one seventh of its surface. The other half is spread unevenly over 35.4 per cent of the surface and the half of the globe which remains in shadow receives nothing.

It will be shown hereafter that the general effect of the presence of the atmosphere on solar *income* (apart from some minor refraction phenomena) is to exaggerate still further these inequalities of distribution. The truth of this fact is nevertheless usually obscured by the very fundamental part the atmosphere plays in the back-radiation or *expenditure* side of the account.

The above remarks would be true of an immobile earth and we have brought them into touch with reality by the use of the phrase 'at any given moment of time'. It is now necessary to take account of planetary and rotational motion. The earth moves round the sun in a near-circular or slightly elliptical path in very slightly less than 365¼ days.

Perihelion    2 January    solar distance 146.9 million kilometres
                                          (91.3 million miles)
Aphelion    3 July    solar distance 152.1 million kilometres
                                          (94.5 million miles)

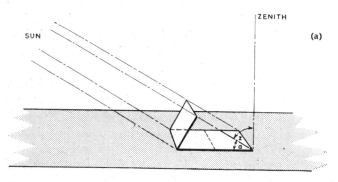

α = ANGLE OF INCIDENCE = 30°
Z = ZENITH ANGLE = 60°

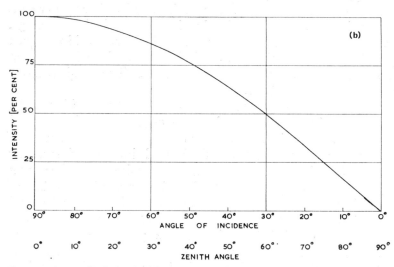

Fig. 1a. A pencil of rays incident at an angle of 30°. It is evident that the area illuminated is twice that of the cross-section of the beam. The intensity of illumination at the surface is thus one half of that normal to the beam. (Sine 30° = 0.500). The zenith angle (60°) is the complement of the angle of incidence (Cosine 60° = 0.500). Fig. 1b. The sine curve from 90° to 0°. Climatologists have frequently exaggerated the significance of the 'vertical sun' at midday in tropical latitudes. It is evident that within 25° of the vertical the difference in solar intensity is less than 10 per cent. (Sine 65° = 0.906)

It spins meanwhile, once in 24 hours (15° of arc per hour), on an axis set at an angle of 66°33′ to the plane of its orbit (the ecliptic). That axis of rotation, it must be emphasised, remains set in space, pointing in the north to within a degree of Polaris and in the south to an unmarked position in the heavens about 28° below the foot of the Southern Cross. Each pole is thus inclined towards the sun and away from the sun alternately once per year as the annual planetary motion proceeds.

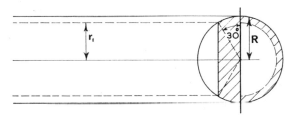

$\dfrac{r_1}{R} = \cos 30° \therefore r_1 = 0.866R$

Proportion of beam $= (0.866)^2 = 0.75$
Proportion of hemisphere illuminated $= (1 - \sin 30°)$
$= 1 - 0.5 = \underline{0.5}$

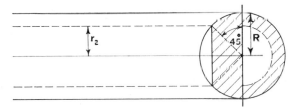

$\dfrac{r_2}{R} = \cos 45° \therefore r_2 = 0.707R$

Proportion of beam $= (0.707)^2 = 0.5$
Proportion of hemisphere illuminated $= (1 - \sin 45°)$
$= 1 - 0.707 = \underline{0.293}$

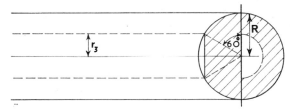

$\dfrac{r_3}{R} = \cos 60° \therefore r_3\, 0.5R$

Proportion of beam $= (0.5)^2 = 0.25$
Proportion of hemisphere illuminated $= (1 - \sin 60°)$
$= 1 - 0.866 = \underline{0.134}$

FIG. 2. The distribution of total insolation on the illuminated hemisphere

Once the nature of this motion is fully comprehended the climatologist can afford to forget Copernicus and think of the world in geocentric or Ptolemaic terms. As the earth rolls daily from west to east, the sun and all the stars apparently move from east to west along planes set parallel to the celestial equator, the plane of the earth's equator projected into space. The stars are indeed set in their courses but the sun exhibits a degree of waywardness, almost imperceptible from day to day but clearly recognisable over longer periods of time.[1] Not only does it move across the starry backdrop, a fact we can most clearly recognise by looking in diametrically the *opposite* direction at night (akin to judging the turning of a car through the back window), but it does so along a plane set at an angle of 23°27′ to the paths of all the other stars and hence to the plane of the celestial equator.

Obliquity of the ecliptic: 23°27′.
Declination of the sun: the angular distance of the sun north (+) or south(−) of the celestial equator at any moment of time.

This angle will be recognised at once as the complement of the angle between axis and orbit quoted above and the seasonal behaviour of the sun, with all its momentous consequences, is thus a direct result of the earth's planetary motion. To the earthbound observer the sun's declination (judged most clearly apart from its daily rotation by changes in its angular height above the horizon at midday) increases to a maximum of +23°27′ on 22 June and decreases to a minimum of −23°27′ on 22 December (Fig. 4). These are the 'solstices' and the moments between them on 21 March and 23 September when the declination is zero are known as the 'equinoxes'. At the equinox therefore the sun lies directly upon the celestial equator and the northern and southern hemispheres of the earth are equally illuminated. Solar income is then distributed over the earth, according to recognised 'differentials' based simply upon latitude.

It is easy, however, to be misled by too static a view of the balance between the hemispheres at the equinoxes. Seen dynamically they are the *periods of most rapid change* of solar declination and the consequences that ensue therefrom depend on the direction of that change. The traditional association of stormy weather with the equinoxes is not entirely mythical—air exchange between the hemispheres is most active at this period (see Fig. 59).

The seasonal variation in the apparent path of the sun, together with the variation in the length of day which it produces, is most readily illustrated by a simple type of diagram which the student should learn to draw for himself for any given latitude. To accommodate the three dimensions it is drawn in perspective, as if seen from outside (Fig. 3), but to *interpret* it the reader must make the conscious effort to project himself into the position of the observer at 'O'. In Fig. 3 the observer is presumed to be in the northern hemisphere. The plane N, E, S, W, is then the observer's horizon, a plane tangent to the surface

[1] The inconstant moon is, of course, proverbial but let us say once and for all that its influence on climate or meteorology is negligible.

of the earth upon which he is standing, and the points named represent the positions of the cardinal points upon it. $Z$ is the zenith of the point of observation and $n$ is its nadir.

For any desired point of observation set up the angle $P\,O\,N$ equal to its latitude $(\theta)$, the simplest near approximation to latitude being given by the angular elevation of Polaris $(P)$. Draw $C\,O\,L$ at right angles to $O\,P$ and sketch in the ellipse $C\,E\,L\,W$ to represent the plane of the celestial equator. This plane thus cuts the horizon at the cardinal points $E$ and $W$ and reaches its highest angular elevation above the observer $O$ at the point $C$, due south from him. The angle $C\,O\,S$ is $(90°-\theta)$ or the co-latitude. As we have seen, each 'fixed' star has its own plane of apparent rotation parallel to that of the celestial equator but the sun shows a declination which varies with the season, and is indeed responsible for them.

At the equinoxes the sun moves almost along the celestial equator, rising due east and setting due west and reaching at midday the point $C$, its maximum elevation above the horizon thus being $C\,O\,S$, the co-latitude. As the summer advances, however, the declination increases, the sun apparently spiralling higher and higher in the sky until its meridional angle reaches the limiting value of co-latitude plus the maximum positive declination $(+23°27')$ at the summer solstice. Lay off the angle $M_{sd}O\,C = 23\frac{1}{2}°$ and draw $M_{sd}M_{sn}$ parallel to $C\,L$; this line then forms the major axis of an ellipse $A\,M_{sd}\,D\,M_{sn}$ which represents the apparent path of the sun in midsummer. It will be noted that the sun then rises at $A$ and sets at $D$. Similarly, as winter approaches, the declination steadily decreases until it attains ($-23°27'$) and the path $B\,M_{wd}$ $F\,M_{wn}$ is drawn in analogous fashion to represent the sun's apparent path in midwinter. The sun then rises at $B$ and sets at $F$, attaining at midday the elevation $M_{wd}\,O\,S$ or the co-latitude plus maximum negative declination. From the point of view of the observer at $O$ at latitude $\theta$, at any given moment of time, the sun may be situated anywhere between the two planes $A\,M_{sd}$ $D\,M_{sn}$ and $B\,M_{wd}\,F\,M_{wn}$ according to season and time of day, but *it can never lie outside those limits*. Daytime is represented by positions above the observer's horizon $N$, $E$, $S$, $W$, and either twilight or darkness is experienced when the location is below this plane. If our figure were indeed a three-dimensional model the altitude of the sun (positive or negative) which corresponds with the angle of incidence of its rays with that plane, could be easily measured for any one of its possible positions. Failing a model we have recourse to the beautiful devices of spherical trigonometry. If we were able to keep the sun in view all the time, as we can indeed for a season near the poles, its path from solstice to solstice would appear as a spiral, finely threaded near its extremities but opening out to about 23 minutes of arc per day near the middle, i.e. at the equinoxes. We return to this point later in considering the implications of Fig. 4.

Figure $3(a)$ has been drawn for the latitude of Manchester $(\theta = 53\frac{1}{2}°\text{N})$. where the sun is more frequently visible than uninformed opinion may concede. In that latitude the sun thus reaches $90° - 53\frac{1}{2}° + 23\frac{1}{2}° = 60°$ above the southern horizon at midday in midsummer but only attains $90° - 53\frac{1}{2}° - 23\frac{1}{2}°$ $= 13°$ at the corresponding time in midwinter. This contrast is reinforced by the fact that the midsummer sun rises almost in the north-east (actually $48°$ east of north) and sets almost in the north-west (actually $312°$) thus swinging round $264°$ of the horizon; in midwinter, on the other hand, its daily

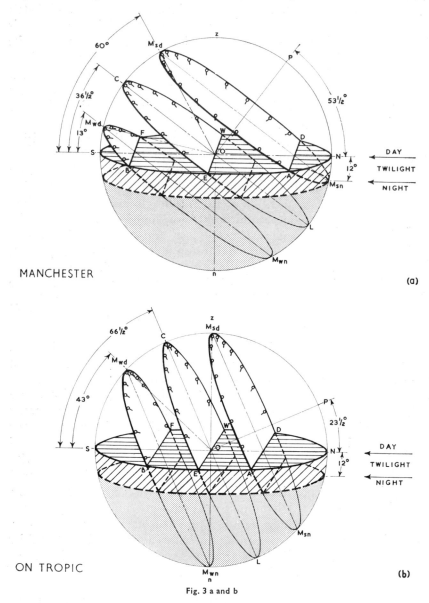

MANCHESTER        (a)

ON TROPIC        (b)

Fig. 3 a and b

FIG. 3. The apparent motion of the sun, (*a*) at Manchester, $53\frac{1}{2}°$N, (*b*) on the Tropic of Cancer, (*c*) on the Arctic Circle, (*d*) on the Equator. The diagrams must be interpreted as from the observer's position at 'O'

swing covers only 96° from *B* to *F*. This difference of swing clearly bears some relation to the duration of daylight which is 17 hours at Manchester on midsummer day but only $7\frac{1}{2}$ hours in midwinter. However, the solar clock is complex and cannot be read accurately from the horizon; it should be read with reference to the path planes $A\,M_{sd}\,D\,M_{sn}$ or $B\,M_{wd}\,F\,M_{wn}$ or whichever

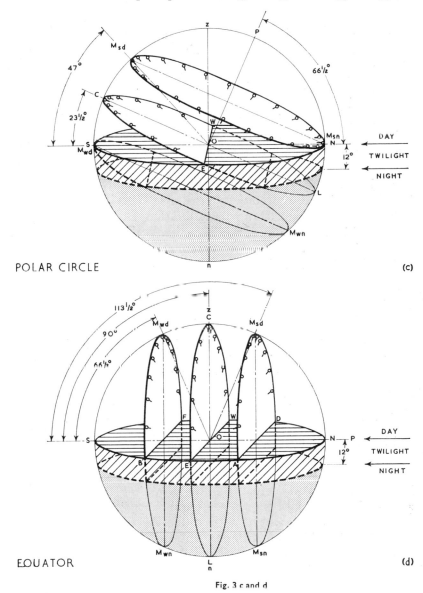

POLAR CIRCLE                                                                    (c)

EQUATOR                                                                          (d)

Fig. 3 c and d

of the 181 to 182 'planes'[1] between them is appropriate to the day in question. This would be possible on a three-dimensional model with a fair degree of

[1] Strictly speaking not quite planes—more analogous to threads of a screw or 'helix'.

9

accuracy. The rather unfamiliar fact that, although the sun is always due south (or north) at noon and due north (or south) at midnight, it is not necessarily due east or west at 6 a.m. and 6 p.m. respectively is illustrated by all the diagrams of Fig. 3 but it is, perhaps most readily recognised from the equatorial case (Fig. 3(d)). On the equator, although sunrise and sunset occur regularly at these hours throughout the year, the direction from the observer is clearly shown to swing from $23\frac{1}{2}°$ N to $23\frac{1}{2}°$ S of the cardinal points.

Before leaving this discussion of the geometry of insolation it may be worth while to say a word about the geometry of twilight, a phenomenon of little climatological significance but yet an interesting feature of our everyday experience. We are dealing here with the atmospheric scattering of light beyond the shadow line (the terminator), the degree of illumination depending broadly on the angular depression of the sun below the horizon. Figure 3 is therefore strictly relevant. It is true that we have so far ignored atmospheric refraction: when the upper limb of the sun appears above (sunrise) or disappears below (sunset) the true horizon, its *centre* is in fact about 50' below it. The day thus gains a bonus of 7 minutes or more from the night as compared with a strictly geometrical interpretation of the facts. This is a minor matter compared with the bonus which accrues from diffusion. It is by no means easy to specify when twilight ends and darkness begins so that three alternative definitions are available, all based upon the angular depression of the *centre* of the sun's disc below the true horizon.

| | |
|---|---|
| Civil twilight | depression not greater than 6° |
| Nautical twilight | depression not greater than 12° |
| Astronomical twilight | depression not greater than 18° |

In view of the small scale of the diagrams in Fig. 3 we have indicated the limits of nautical twilight but civil twilight has greater relevance to such practical problems as the determination of official lighting-up times.

It will be noted at once that the path planes traverse this zone at a more oblique angle in northern latitudes (Manchester, Polar Circle) than at the equator where it is crossed at right angles. The bonus of illumination thus added by twilight is particularly generous in polar climes. Civil twilight adds two periods of 22 minutes to the equatorial day but at Manchester it adds two periods of 35 minutes each at the equinoxes and two periods of 52 minutes duration in midsummer. The inequality of these two values may be a little puzzling at first since the path planes are parallel, but if the diagram for Lerwick (*c*. 60°N) be drawn it will be noticed that the angular depression of the sun at midnight in midsummer is only $90° - 60° - 23\frac{1}{2}° = 6\frac{1}{2}°$ so that it scarcely drops below the twilight zone the whole night through. Indeed the definition of civil twilight is rather conservative. Over most of the northern half of Scotland newsprint can be read at midnight, with some discomfort perhaps, on a few bright nights each summer. The bonus is, of course, exceptionally generous at the poles though few people remain there to enjoy it. At civil twilight level the Arctic night is cut down to barely 5 months and at nautical twilight level, adequate enough in the presence of snow and ice, it is reduced to only 4 months.[1]

This note on twilight may appear to be a digression from our main theme but the fact that it has been possible to refer to twilight as a 'bonus', a net gain by lambent day over the shadows of night for which no account is rendered, no countervailing losses are entailed, underlines a striking fact about the other features of the diagrams. In all other respects the requirements of geometry are inexorable. In each latitude a gain at one season of the year is balanced by a loss in the other. The compensation is complete.

## The seasons of the year

Since the march of the seasons is a direct consequence of changes in the declination of the sun we have plotted the course of these changes for a whole year in Fig. 4. The two solstices are recognisable as 'turning points' on this

THE DECLINATION OF THE SUN

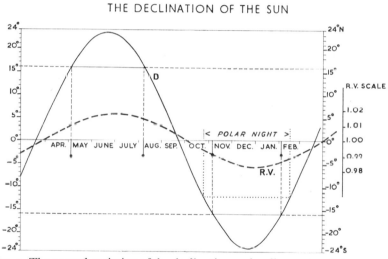

FIG. 4. The annual variation of the declination and radius vector of the sun. The declination of the sun is its angular distance north (+) or south (−) of the celestial equator. The radius vector is the distance from the centre of the earth to the centre of the sun expressed in terms of the semimajor axis of the earth's orbit

curve and the equinoxes are the points of maximum gradient between them. These features of a curve are known to delight the heart of a mathematician but it by no means follows that they provide the ideal method for dividing up the year for the purpose of climatological description. A division into a high-sun period and a low-sun period with intermediate periods of rapid change is not a whit less valid and if we wish to have four seasons of equal length (91 days) Fig. 4 shows that the times when solar declination is about ±16° are

[1] Declination below −12° from 25 October to 15 February.

the key dates to use. For lands north of the Tropic of Cancer we then have:

| Spring | waxing solar income | 5 February to 6 May |
| Summer | high solar income | 7 May to 6 August |
| Autumn | waning solar income | 7 August to 5 November |
| Winter | low solar income | 6 November to 4 February |

The same dates would apply south of the Tropic of Capricorn but the descriptive terms must be reversed. In his great *Manual of Meteorology* (I, 1926, 61) Sir Napier Shaw pointed out some of the advantages of such a division, calling it the 'May Year' in contrast to the Solsticial Year which is traditional.

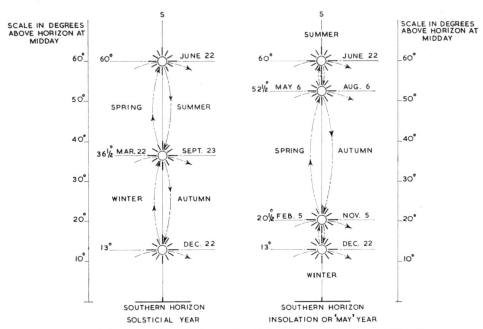

THE MID-DAY SUN AT MANCHESTER
53½° N. LAT.

FIG. 5. The Solsticial Year and Insolation or Daylight Year at Manchester. The diagrams indicate the angular height of the sun above the southern horizon at alternative interpretations of the four seasons

More strictly still it might be called the Insolation Year since this is indeed the criterion we have been using. The fact remains that, despite its obvious logic, this division of the year has never 'caught on'. Possibly it would have had better success had it been called the Daylight Year for we are all conscious that a long-day season and a short-day season are separated by periods when days shorten or lengthen with considerable rapidity.

In Fig. 5 we have expressed the differences between the two concepts in very simple terms—the figures quoted being the angular elevation of the midday sun above the southern horizon for an observer in Manchester, i.e. very near the centre of the British Isles ($53\frac{1}{2}°$N). For a station $10°$ of latitude south of Manchester the angular elevations would all be increased by that amount, for a station to the north they would be decreased correspondingly, but the essential features of the diagrams remain unchanged.

Within the tropics the terms 'summer' and 'winter' are best avoided and a fourfold division of the year has much less significance. Nevertheless, even on the equator, there is some point in distinguishing the 'northern sun' period (7 May to 6 August) from the 'southern sun' period (6 November to 4 February), the season between these having a near-vertical sun at midday.

## The distribution of insolation over the surface of the earth

Mankind dwells at the bottom of a vast ocean of air and the amount of solar energy received by the very narrow stratum which forms his immediate environment depends not only on latitude and the seasonal consequences of planetary motion but also on the varying degree of transparency or opacity of the mass of air overhead. It is a matter of common experience that this condition changes from day to day, indeed, from hour to hour. It is an essential element of what we know as 'weather'. A chance factor thus intervenes which, at first glance, appears to put an end to further geometrical analysis.

However, a considerable proportion of the loss of solar income by absorption, reflection and refraction takes place so far aloft that it may be regarded as incidental waste which can be written off from the earth's economy. Furthermore these losses in the upper air undoubtedly vary less through time than those sustained nearer the surface. Even on a bright sunny day with the sun high in the sky the *direct* solar radiation received at the surface rarely amounts to as much as three-quarters of the solar constant, and as the angle of incidence becomes more acute this proportion falls off rapidly. By making a systematic allowance for this loss, that is, by assuming a coefficient of transparency for the atmosphere of the order of 0.7, it is possible to carry the geometrical treatment yet one stage further. We shall be dealing thereafter with approximations but the step can be defended on the ground that further reduction in solar income at ground level, when it occurs, has less the nature of a total loss and more the character of a payment made into some other side of the final account.

If the atmosphere has a degree of opacity which can be broadly expressed in this fashion, the loss sustained by a ray will clearly vary with the length of its path through the atmospheric envelope. Figure 6 illustrates the nature of this relationship. *SA* represents a beam which meets the earth's surface at right angles (zenith angle $0°$) and *SB* a beam with an angle of incidence of $30°$ (zenith angle $60°$). Since *SB* equals 2 *SA* (for cosec $30° = 2$) mark the point *x* so that $Sx = xB = SA$. Represent the solar constant by *J*. Now, if, for initial simplicity, we assume the atmosphere to be completely homogeneous with a transmission coefficient of 0.7 throughout, then, 0.7 $J$ is received at $A$ and also, at right angles to the beam, 0.7 $J$ is received at $x$. This means that at $x$ the beam retains only 70 per cent of its initial intensity at $S$. On the path from $x$ to $B$, which equals $Sx$, it loses *an equal proportion of the intensity with which it started* at $x$, hence it arrives at $B$ with an intensity, normal to the beam, of $0.7 \ (0.7J) = (0.7)^2 J = (0.7)^{\text{cosec } SBA}J = (0.7)^{\text{sec } SBZ}J$. The formula is thus usually quoted as $a^{\text{sec } z}$ where $a$ is the transmission coefficient selected and $z$ the sun's zenith distance.

The reader who is aware that the atmosphere is never homogeneous but approximately stratified will notice that, by arguing the case stratum by stratum, the same result will be obtained.

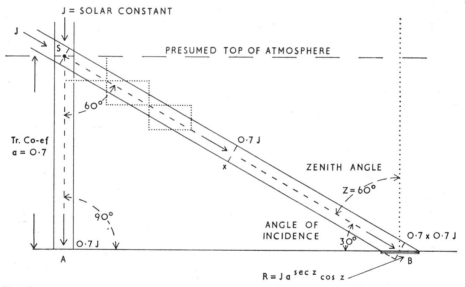

Fig. 6. The attenuation of the solar beam by the atmosphere at an angle of incidence of 30° assuming that the transmission coefficient is 0.7. The diagram illustrates the derivation of the formula given in the text for a simple case

Allowing for the presence of the atmosphere the equation for solar income on a *horizontal* surface thus takes the form

$$R = J \, a^{\text{sec } z} \cos z$$

where $z$ depends on the latitude, the season of the year (i.e. the sun's declination) and the time of day (expressed as the sun's hour angle).

Some of the results of this type of calculation are available in tabular form and all are capable of expression in beautiful diagrams which are easily interpreted even by those who find the mathematics unfamiliar. Figures 7, 8 and 10 are all drawn with the above basic assumption of an atmospheric transmission coefficient of 0.7.

Figure 7 illustrates the distribution of solar energy on a horizontal surface at the summer solstice (22 June) along the whole length of a meridian from pole to pole, the scale on the left being given in both gram-calories per square centimetre and in kilowatt-hours per square metre per day. The broken line shows values appropriate to the outer limit of the atmosphere where it will be noticed that the maximum solar income during the 24-hour period is actually received above the North Pole. This apparent anomaly is removed when a coefficient of transmission of 0.7 is applied but the curve retains several interesting features. Maximum insolation at the surface is received at about

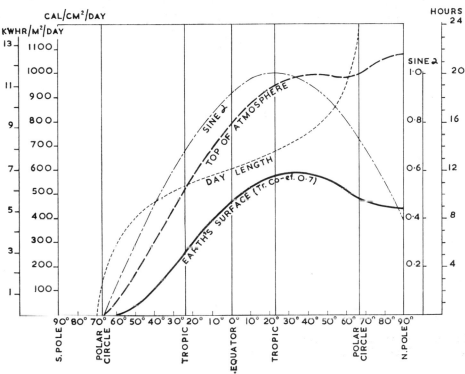

FIG. 7. The total amount of insolation received during 24 hours at the summer solstice at each point along any meridian. The diagram distinguishes between solar input (1) at the top of the atmosphere and (2) at the surface of the earth, assuming a transmission coefficient of 0.7. The form of the two curves is partially explained by the variation with latitude of (3) the length of daylight and (4) the sine of the angle of incidence of the sun's rays at midday, both shown on arbitrarily selected scales

32°N according to this estimate, but it is even more significant that the difference over the range from 22 to 42°N is so small as to be completely negligible. It has been shown (Fukui, 1954, 13, 22) that this statement remains substantially true even if we consider, not merely a single day, but the sum total of insolation available during the summer three months. As compared with an income of some 580 cal cm²/day in this peak region note that the equator receives 460 cal cm² or little more than the North Pole. Let us not overemphasise this contrast, the essential point is the slight range of variation

15

in the summer hemisphere as compared with the almost uniform rate of decrease in the winter hemisphere. From the equator to 50°S insolation falls off at this time at the rate of about 85 cal cm² (or about 18 per cent of the income at the equator) for every increase of 10° in latitude.

The reason for the ogive form of the insolation curves of Fig. 7 cannot be given in non-mathematical terms but it can be suggested graphically by the two dotted curves plotted upon arbitrarily selected scales. One of these shows the value of the sine of the angle of incidence of the *noonday* sun's rays at each point along the meridian—a fair measure of the peak of solar intensity—and the other shows the variation in length of day. Beyond 55°N the two curves are seen to trend in violently opposed directions. At the winter solstice

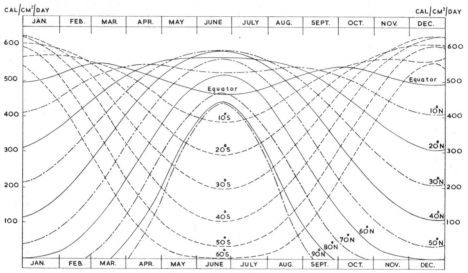

FIG. 8. The seasonal variation of insolation at various latitudes. The computation assumes a transmission coefficient of 0.7 throughout

of the northern hemisphere the insolation curve would be virtually the mirror image of that shown in Fig. 7.

Figure 8 is even more informative: in fact it illustrates the fundamental basis of practically all the seasonal contrasts in climate which we shall have to discuss hereafter. Assuming a transmission coefficient of 0.7 throughout, it shows the annual course of daily solar income on a horizontal surface for each ten-degree interval of latitude from 90°N to 60°S. The vertical scale is in gram-calories per square centimetre per day. Values for the middle of each month expressed as a percentage of mid March at the Equator are tabulated in Table 1.

Some of the features of the curves deserve special mention. Thus the *equatorial* curve has the double maxima and minima which are to be expected from the seasonal shift in the apparent path of the sun (Fig. 3*d*) but, although the two maxima are nearly identical, the minimum in June is rather lower than that in December. The reason for this can be seen in Fig. 4 where the variation in the declination of the sun is compared with the variation in the radius vector of the earth, i.e. the distance between earth and sun. We have tended to ignore this latter factor up to the present and it is indeed debatable whether it is really of much climatological significance. Indeed the scale of the seasonal variation of insolation at the equator (about ±8 per cent of the annual mean) is itself so low that the really significant point about the equatorial curve is its comparative uniformity throughout the year.

At 10°N the variation from March right through to September is quite negligible and the year falls into a period of 7 months during which insolation

TABLE I. *Computed direct insolation at the middle of each month ( Tr. Coef. 0.7) at intervals of ten degrees of latitude, expressed as a percentage of insolation in mid-March on the Equator*

| Latitude | JAN. | FEB. | MAR. | APRIL | MAY | JUNE | JULY | AUG. | SEPT. | OCT. | NOV. | DEC. |
|---|---|---|---|---|---|---|---|---|---|---|---|---|
| °N |  |  |  |  |  |  |  |  |  |  |  |  |
| 90 | — | — | — | 13 | 53 | 78 | 67 | 27 | — | — | — | — |
| 80 | — | — | — | 21 | 56 | 80 | 69 | 35 | 6 | — | — | — |
| 70 | — | — | 11 | 38 | 69 | 84 | 76 | 51 | 20 | 3 | — | — |
| 60 | 1 | 9 | 27 | 55 | 82 | 94 | 90 | 67 | 39 | 14 | 3 | — |
| 50 | 10 | 22 | 44 | 72 | 93 | 102 | 98 | 82 | 55 | 30 | 13 | 7 |
| 40 | 25 | 41 | 62 | 85 | 100 | 106 | 103 | 92 | 72 | 49 | 30 | 20 |
| 30 | 43 | 58 | 77 | 94 | 104 | 107 | 105 | 97 | 84 | 65 | 48 | 38 |
| 20 | 61 | 74 | 88 | 99 | 103 | 104 | 103 | 100 | 92 | 80 | 66 | 57 |
| 10 | 78 | 88 | 95 | 100 | 98 | 96 | 96 | 98 | 97 | 91 | 81 | 75 |
| Equator | 92 | 98 | 100 | 95 | 89 | 85 | 86 | 92 | 97 | 98 | 93 | 90 |
| °S |  |  |  |  |  |  |  |  |  |  |  |  |
| 10 | 104 | 103 | 98 | 89 | 76 | 71 | 72 | 82 | 93 | 101 | 102 | 102 |
| 20 | 110 | 104 | 94 | 78 | 61 | 54 | 56 | 69 | 86 | 99 | 108 | 110 |
| 30 | 111 | 102 | 84 | 62 | 45 | 36 | 39 | 53 | 73 | 94 | 108 | 114 |
| 40 | 109 | 96 | 72 | 47 | 28 | 19 | 22 | 36 | 58 | 85 | 104 | 113 |
| 50 | 104 | 82 | 55 | 29 | 12 | 7 | 8 | 19 | 42 | 70 | 97 | 109 |
| 60 | 91 | 67 | 38 | 14 | 2 | — | 1 | 6 | 26 | 55 | 84 | 101 |

is really strong and the remaining period of 5 months during which it drops to slightly lower intensities and then rises again. The same is true of 10°S though the range of extreme variation there is rather greater (±19 per cent) than in the north (±15 per cent). This is another minor result of changes in the radius vector of the earth.

By 30°N and S the seasonal rhythm is already strong enough to amount to some ±50 per cent of the annual mean. We have already seen that it is at about this latitude that maximum values are attained in each hemisphere— 588 cal cm² on 22 June in the north and 628 cal cm² on 22 December in the south. This difference also results from changes in solar distance and is partly

17

compensated by minimum values in midwinter of 210 and 196 cal cm²
respectively.

On middle latitudes from 30–60° in both hemispheres the peaks of the
successive curves fall away very gradually, emphasising the point made
above that midsummer insolation varies little with latitude. It is thus the
decreasing *duration* of the period of strong insolation which is the distinguishing
characteristic of the summers of middle lattitudes. The winters, on the other
hand, are distinguished by a steadily falling level of minimum insolation as
well as by a steadily increasing duration of the period of low solar income as
one moves towards the Arctic and Antarctic circles. This is a most important
contrast between summer and winter which is frequently overlooked.

By 60°N solar income remains almost negligible, i.e. less than 25 cal
cm²/day, during the 3-month period from about 6 November to 6 February and
at 60°S the period is at least a week longer (4 May to 10 August). A summer
with a near-tropical rate of insolation for 2 months is thus separated from a
winter with near-polar insolation for 3 months by two periods of rapid transition.

Polewards of 60°N and S the outstanding feature is the steady lengthening
of the period of negligible insolation until at the actual poles it occupies at
least 7 months out of the 12. In our discussion of twilight we have seen that
this does not mean that it is dark all that time, far from it. But the atmosphere,
which is responsible for lengthening the period of illumination, absorbs or
diffuses so much of the solar income at low angles of incidence that insolation
is a virtually negligible element in polar climates throughout this long period.
Summer insolation, on the contrary, is quite considerable, though short-
lived. At the poles as we know them it is rendered largely ineffective by the
presence of snow or ice on land and sea, the consequences, under present
conditions, of the very prolonged winter. There is an increasing body of
evidence, however, which suggests that this condition is by no means
inevitable and that polar climates have shown a wide range of variation in
recent geological times. Attempts to explain such changes simply in terms of
variation in solar income meet the obvious objection that, on the scale
required, such variations should have left devastating evidence within the
tropics. It would appear rather that what has changed is not solar income, as
such, but the uses to which it is put; a very real distinction recognisable to
every housekeeper. But more of this anon.

The student will have observed that, in the above discussion, we have very
carefully, it may appear obstinately, refrained from the use of terms implying
contrasts of temperature. It is likely that he has himself been tempted to
interpret what has been said in those terms because they are more familiar
to him. It is not denied, of course, that insolation is a potent factor affecting
the annual course of temperature. Indeed, no other factor influences temperature in such a systematic and all-pervading fashion. Nevertheless, as we
have hinted above, there are other factors and at times they may be of overriding importance. Variation of temperature is therefore rightly the subject
of a later chapter.

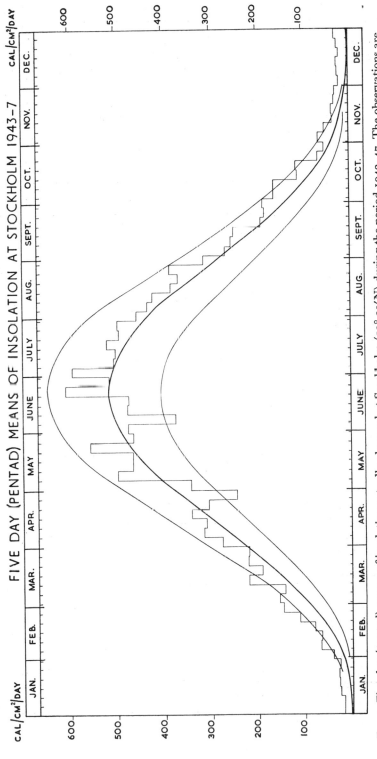

FIG. 9. Five-day (pentad) means of insolation actually observed at Stockholm (59° 21′N) during the period 1943–47. The observations are compared with computations based upon transmission coefficients (a) of 0.8, c.7 and 0.6

Figure 9 provides a welcome change from the somewhat theoretical speculations in which we have been indulging since it presents actually recorded measurements of insolation made at Stockholm Observatory during the 5 years 1943–47. The data have been summarised as a series of 5-day (pentad) means and the record is compared with computed curves of direct insolation at 60°N, using transmission coefficients ($a$) of 0.8, 0.7, and 0.6 respectively. We have before us, therefore, our first instance of the divergence between observed 'fact' and theoretical 'fiction'. Yet what we seek is truth and the young climatologist is wise to grasp at a very early stage that this may lie neither in the raw fact nor in the polished fiction; it may not even lie between them.

The first feature of the plot which catches the eye is the striking irregularity of the 5-day means, especially in the early summer months. Is this significant, does it represent some real distinction that can be anticipated with confidence, year after year, or is it merely the result of random variations which would disappear if the record were lengthened? This is a frequent question in climatology which we must investigate hereafter. If, until further evidence is forthcoming, we incline towards the latter view the next feature that we notice is the relatively close fit between the observations and the curve for 70 per cent transmission. Figure 9 can therefore be used to reinforce the assumption that has been implicit throughout the above discussion. Stockholm is in a land of clean air and moderate cloudiness. We cannot expect the same transmission coefficient to apply everywhere without exception; smoke, dust, haze and cloud vary widely from place to place but, in the absence of further observational evidence, it is shown to be a not unreasonable value for use in this general introduction. Again, the fit, though close, is by no means perfect. Particularly in the winter months the observations are consistently above the theoretical values. At this season the pyranometer must be recording something else besides direct solar radiation. This is almost certainly indirect or 'sky' radiation, a return to the earth by an indirect route of some proportion of the losses suffered in transit. Does no such return take place in the summer months? Of course it does but the proportion of loss to income is so much greater in the winter months that a return of but a small fraction of it is immediately recognisable in the daily budget. How large then is the fraction thus returned? This depends so much on the weather that no general statement is possible but data now becoming available from a few scattered stations as a result of improvements in instrumentation are quoted later in this chapter.

For a broad review of zonal contrasts in solar income computations of varying degrees of complexity are still necessary. We turn to consider its diurnal distribution. Thus Fig. 10 illustrates the variation of computed direct solar income, hour by hour, during selected days in different latitudes and at different seasons of the year. The mechanism underlying the contrasts shown has already been illustrated in Fig. 3 but the net results are rarely portrayed in this fashion although no aspect of climatology is more closely linked with everyday experience or can be more readily tested by personal observation. Indeed, it is even doubtful if any feature is of greater importance to mankind. The space-traveller on the moon, for instance, would soon appreciate the advantages accruing to the earth he had so rashly left behind of solar input in comparatively brief daily pulses. The fierce physical reactions which strong radiation can engender are kept under control by this rhythmic beat. All living creatures 'under the sun' respond in their innermost beings to this tempo and even the outcasts in deep caves or the oceanic abyss must feed on debris brought down from the world above which knows both day and night.

Each section of Fig. 10 illustrates the diurnal variation of solar income (1) at the top of the atmosphere (curves $A'$ and $B'$) and (2) at the surface assuming a transmission coefficient of 0.7, on the two days when it attains approximately maximum ($A$) and minimum ($B$) values at the latitude indicated. Sector (a) for the equator indicates both the uniformity of day length and the very limited nature of the contrast between 21 March, when the sun reaches the zenith (90°) at midday, and 22 June when its angular elevation above the northern horizon at that time is $66\frac{1}{2}°$. We have already seen (Fig. 8) that the March curve will be closely followed by that for a day in late September but that December values will not fall quite so far as those in June. Once again the figure emphasises the remarkably equable nature of equatorial conditions. Sector (b) illustrates conditions at 20°N on 21 May when the sun first attains the zenith at midday and on 22 December when it is over the Tropic of Capricorn. We recall from Fig. 8 that the May curve will suffer little modification until the sun passes south across this latitude again on 24 July, the daily solar income throughout this period being rather more than the highest figures attained at the equator owing to the slightly greater length of day. In actual fact the lower degree of cloudiness associated with the 20th parallel as compared with the equator reinforces this contrast considerably. At 20°N, a fairly marked seasonal rhythm is being imposed, the daily income of direct solar energy falling in December to little more than half of its value in May, June and July. We shall see later that seasonal variations in cloud amount may require some modification of this picture. Beyond the tropics the solstices become the key dates in the insolation rhythm and the two curves shown are limiting conditions which are not repeated in any one year. Already by 35°N the seasonal contrast is quite striking since summer days are lengthening and winter days are shortening as well as showing a marked reduction in peak intensities. In Fig. 10(c) the area under curve $B$, which represents the estimated solar income for 22 December, is little more than a quarter of the area under curve $A$ which represents the income in midsummer. By 55°N (Fig. 10(d)) contrast between midsummer and midwinter dwarfs all other features of the diagram; the income at the surface on a bright summer's day is estimated at more than fifty times that likely to be received on a clear day in midwinter.

We must recall that Fig. 10 is based upon estimate and not on measurement and that the estimate is for *direct* solar radiation on a clear day. Cloud will cut down the surface income considerably, especially when the sun is low and

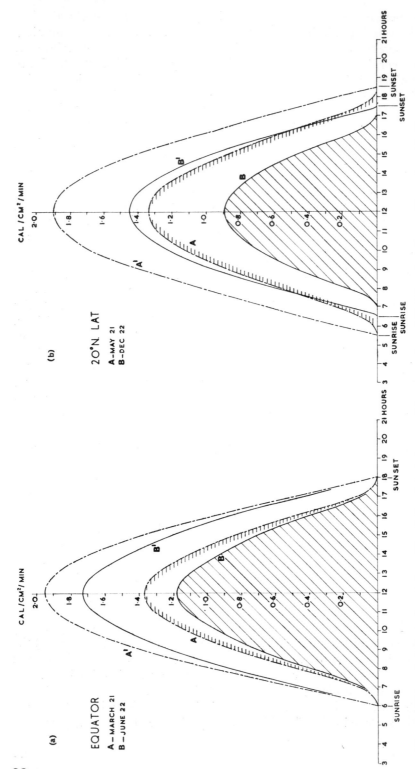

FIG. 10. Computed diurnal variation of insolation (I) at the top of the atmosphere (A′, B′) and at the earth's surface (A, B) at selected latitudes on selected days. *a.* At the equator on March 21 and June 22. *b.* At latitude 20°N on May 21 and December 22. *c.* At latitude 35°N on June 22 and December 22. *d.* At latitude 55°N on June 22 and December 22. A transmission coefficient of 0.7 is used throughout

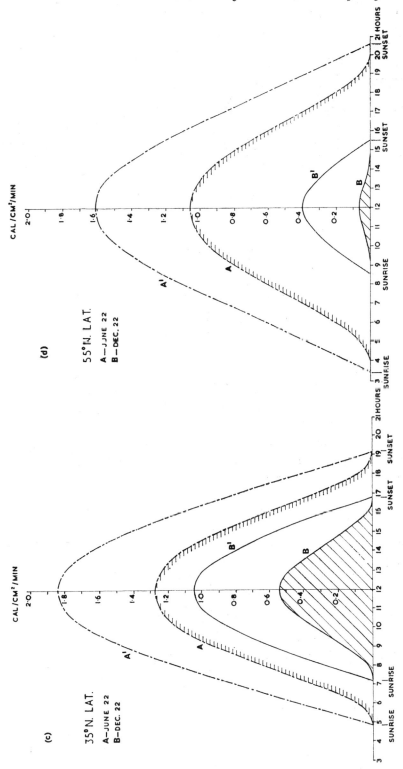

cloud shadows are long. On the other hand, some of the loss represented by the difference between the curves for the top of the atmosphere (transmission coefficient 1.0) and the surface (transmission 0.7) finds its way to earth as diffuse sky radiation, as we have already seen. This latter becomes particularly important in winter north of 45° N and thus plays some small part in mitigating the seasonal contrasts we have noted.

Figure 11 is drawn on a smaller scale for its base line must accommodate all the 24 hours of the day to illustrate the peculiar conditions prevailing north of the Arctic Circle. It also has two vertical scales, one in calories per square centimetre per minute like Fig. 10 and another, quite arbitrarily selected,

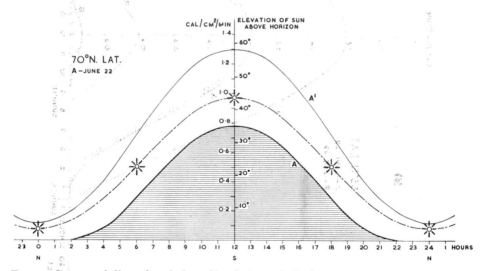

FIG. 11. Computed diurnal variation of insolation at latitude 70° N on June 22. Values are plotted for the top of the atmosphere (*A'*) and for the earth's surface (*A*). On a separate scale the angular elevation of the sun above the horizon is also indicated. This illustrates the midnight sun

representing angular elevation of the sun above the horizon. The day is the summer solstice. The diagram thus traces the path of the midnight sun at 70° N, showing how it dips towards the northern horizon but does not fall below it. Although daylight is thus continuous throughout the 24 hours the estimate of direct solar income at the surface from 2200 to 0200 hours is shown to yield values too small to plot. Even at the top of the atmosphere the income at this time is inconsiderable so that diffuse sky radiation can do little to remedy the defect. Despite the midnight sun therefore the diurnal pulse of solar income remains strong. It is not completely lost until the actual pole is reached. In midwinter, on the other hand, since the sun never rises at 70° N from late November to about the middle of January, there can be no direct solar income and hence no diurnal variation of the same. Midday during that period is represented by an hour or two of twilight, that is of indirect radiation of negligible power.

# Solar income and aspect

In the above discussion we have been dealing all the time with insolation upon an entirely horizontal surface. This is reasonable enough whilst we are thinking on a zonal scale but the curves will not represent the sensation of the sun on the face nor the solar income obtained by a peach tree trained along a sunny wall. Yet every suntrap casts a shadow and the gain in one spot is at the direct expense of some other section of the neighbouring environment. The same is true, on a much larger scale, of the southward-facing slope, but the areas then concerned are large enough to merit special consideration. What is the net value of a southerly exposure?

In cloudy Britain this term seems to be of little interest except to the house agent and his prospective customer but we have to travel no farther than the Rhineland to notice the compelling importance of this factor in the siting of orchards and particularly of vineyards. In the great interior valleys of Scandinavia and the Alps the contrast between sunny and shady slopes is one of the major features of the local geography. In the Alps, indeed, because of their scale and deep dissection, the problem is made still more complex by the shadows cast by one mountain upon the slopes of another (Garnett, 1937). Even in the Scottish Highlands, where aspect may often seem to play but little part in the distribution of natural vegetation, the ardent rock-climber will find himself talking of the 'north face' of Ben Nevis or of a hundred lesser peaks, thus expressing a contrast in physical form which has its roots in the effect of aspect upon the climate of past ages.

We shall see later that a hillside location may affect local climate indirectly through such factors as soil drainage, exposure to wind and local movements of the air, but its immediate and direct consequence is to change the angle of incidence of the sun's rays upon the surface of the earth. Since this effect varies with the latitude, the season, and the time of day, its investigation involves us in further spherical trigonometry but, given the basic astronomical data, the calculations are quite straightforward and the results are both significant and helpful. The curves in Fig. 12 have been drawn for latitude 45°N and refer to the three days when the sun is at the summer solstice (a), the spring equinox (b), and the winter solstice (c). Values for the autumn equinox are very similar to those in spring though a little (nearly 2 per cent) lower owing to the greater solar distance at that time. A transmission coefficient of 0.7 is assumed throughout. The outer envelope of the curves in this figure (P) is not solar input at the top of the atmosphere as in Figs. 10 and 11, that no longer concerns us. It represents solar income at ground level on a surface which moves with the sun so that the rays always impinge upon it at right angles whatever the time of day. A pyrheliometer mounted and moved by clockwork like an astronomical telescope so as to face the sun continuously gives a record of this nature. It is clear at a glance that such a record is of limited climatological value.

In each diagram of Fig. 12 the solar income received by a slope of 20°

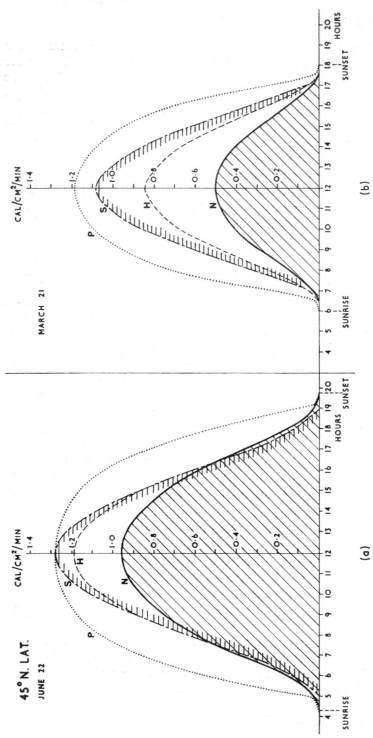

FIG. 12. Insolation and aspect in latitude 45°N. The diurnal variation of insolation on (a) June 22, (b) March 21, and (c) December 22 has been computed for a 20° slope facing due south (S) and a similar slope facing due north (N). In each case the values are compared with insolation on a horizontal surface (*H*) and the amount that would be received by a surface turning with the sun so as to meet the incoming rays always at right-angles (*P*)

(about 1 in 3) directed either due south (*S*) (azimuth 180°) or due north (*N*) (azimuth 0°) is compared with income at the same time and place upon a perfectly horizontal surface (*H*). In midsummer the northerly slope gains slightly in the early morning and later afternoon because the sun rises and sets north of the east–west line but this gain is more than offset by losses at midday. In midwinter the income on the northerly slope is almost negligible and indeed north of latitude $46\frac{1}{2}°$ direct insolation must be nil $(90° - 46\frac{1}{2}° - 23\frac{1}{2}° - 20° = 0)$. The relative gain or loss as a consequence of aspect is clearly seen to be least in midsummer and to increase progressively as midwinter is approached.

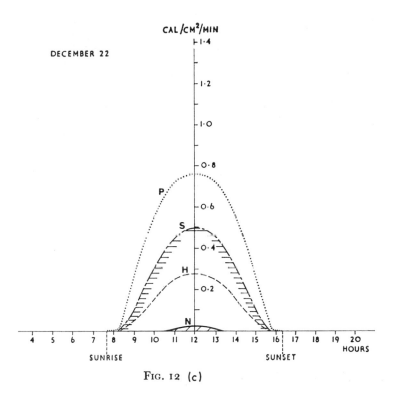

Fɪɢ. 12 (c)

This is expressed in numerical terms in Table 2 which gives for 45°N the total estimated daily income in calories on both slopes and the percentage relationship of these values to the income on a horizontal surface. In this latitude a southern exposure thus increases the *total* daily solar income (shown by the *areas* enclosed by the curves) by 2 per cent in midsummer, by 29 per cent at the equinoxes and by 93 per cent in midwinter. On the other hand, a northerly exposure decreases input by 14, 39 and 97 per cent at these respective periods. For slopes which are less steep than 20° the change is roughly, but not precisely, proportional. Thus for 10° slopes (about 1 in 7) which are more common in settled country, the above percentages must be reduced by about one half.

27

A more vivid way of expressing these differences is to regard a southerly exposure as yielding some of the advantages of a more southerly latitude and a northerly exposure as involving some of the risks of a more northerly latitude.

TABLE 2. *Comparison of daily solar income of north- and south-facing slopes in 45° N at three periods of the year (cal/cm²/day) and percentage of income on horizontal surface*

|  | HORIZONTAL SURFACE | S SLOPE OF 20° | N SLOPE OF 20° |
|---|---|---|---|
|  | (cal) | (cal) (%) | (cal) (%) |
| June 22 | 577 | 590 (102) | 495 (86) |
| March 21 | 315 | 408 (129) | 191 (61) |
| December 22 | 68 | 131 (193) | 2 (3) |

The effect of aspect can thus be expressed in terms of the 'equivalent latitude'. Since, as Figs. 7 and 8 have clearly shown, the change of solar income with latitude itself varies widely with the season of the year, this method of ex-

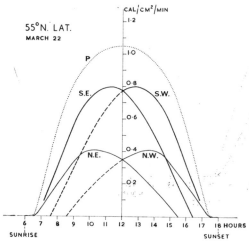

FIG. 13. Insolation and aspect in latitude 55° N. In this diagram for 22 March, slopes of 20° have been taken facing respectively in a north-easterly, south-easterly, south-westerly and north-westerly direction. Again the values are compared with those received on a surface turning with the sun so as to meet the rays always at right-angles (P)

pressing the effect of aspect gives different results at different seasons. Striking a broad average for the year it can be deduced that, in middle latitudes, a southerly slope of 20° is equivalent to a southerly shift of 8 to 9° of latitude and a northerly slope of the same steepness to a northerly shift of 12 to 15° of latitude. For 10° slopes the corresponding displacement is 4 to 5° southward and 6 to 7° northward. It would thus appear that, particularly in sunny regions, aspect is a decidedly important factor in local climate. In cloudy Britain, bathed in mild air from the western ocean, its effects are often masked by other climatic factors.

If the slope of the land is not either due north or due south the results of Fig. 12 have to be modified since the insolation curves will no longer be symmetrical. Figure 13, drawn for 55°N at the equinox, shows the nature of the curves for 20° slopes downward in a north-easterly, south-easterly, south-westerly and north-westerly direction (azimuths 45, 135, 225 and 315° respectively).

## Diffuse radiation

Of that fraction of short-wave radiation which is intercepted during its passage through the atmosphere not all is completely lost to the solar income at the surface of the earth. A proportion of the rays scattered in a clear sky or reflected by clouds passes on to reach the surface by devious, indirect routes as 'diffuse radiation'. Since it emanates from the whole dome of the sky diffuse radiation does not travel by parallel rays and hence its geometry is far more complex than that of direct radiation. Indeed it must be treated statistically rather than geometrically. Yet it is of undoubted importance since it brings illumination (and some heat) into rooms and all other shadowed areas and provides useful supplementary radiation in all situations where the angle of direct radiation is acute—in the morning and evening everywhere, in high latitudes particularly in winter, and on all slopes with an aspect directed away from the sun.

Although the sum total of direct and diffuse radiation is related to extra-terrestrial intensity and is thus subject to the geometrical rules outlined above, the relative *proportions* of the two elements of solar income vary widely in place and time. There are no simple rules though the height of the sun and the state of the sky are clearly significant factors.

In recent years improved instrumentation has supplied an increasing body of observational data on diffuse radiation. Table 3 summarises the mean daily values for total and diffuse radiation on a horizontal surface quoted in the *Monthly Weather Report* since 1956, the values for *direct* radiation then being obtained by subtraction. In Fig. 14 the vertical scale is in milliwatt-hours (1 milliwatt-hour = 0.86 calories) and the vertical bars show the mean (quartile) range of the values recorded during the decade 1956–65. For total radiation the extreme values are also indicated. It will come as no surprise that in Britain solar income is indeed variable and that most of the variation stems from the *direct* side of the account. The mean variation of *diffuse* radiation, indicated below the curves, is clearly much less significant. The asymmetry of the curves for direct radiation at both stations is a reflection of the relatively cloudy nature of British summers.

Table 3 also gives the percentage relationship between diffuse-total and diffuse-direct radiation at Lerwick and Kew. At Lerwick (60°09′N lat.) diffuse radiation accounts for some 64 per cent of the total during the summer, May to August, and very nearly 75 per cent of the total in winter, November to

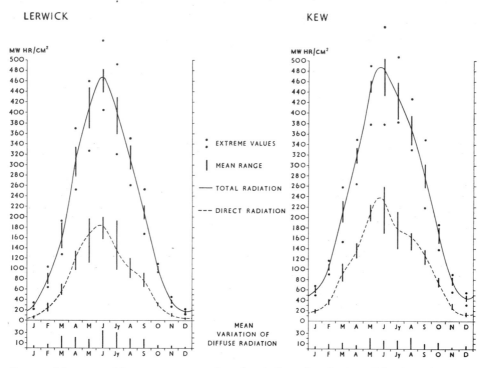

FIG. 14. Mean monthly measurements of total and direct (total minus diffuse) radiation on a horizontal surface at (*a*) Lerwick and (*b*) Kew over the period 1956–65. Mean values are shown by the two curves and the quartile range covered by the two sets of values is indicated by the vertical bars. For total radiation the extreme values are also indicated. The bars beneath the main diagram show the limited range of (quartile) variation of diffuse radiation

February. At Kew Gardens (53°29′N), affected to a limited extent by the proximity to London, the corresponding values are still as high as 55 and 65 per cent of the total radiation. Judging from the data for cloudless days during the first half of the decade quoted by Monteith (1962, 513), given in brackets, where corresponding values are seen to be 16.5 per cent and 28 per cent respectively, it is evident, as might well be anticipated, that diffused radiation in Britain is very largely a function of cloud cover.

This view is confirmed by data from sunnier climes. As an example we quote short period means for Pretoria (25°45′S. lat.) in similar units, transposing the months to facilitate comparison. Here the winter, May to August, is the season of clearest skies and diffuse radiation then amounts to only 20 per cent of the total whilst during the rainy summer season, November to February, though the mean proportion reaches 33 per cent it is still far below

TABLE 3. *Mean daily radiation per month (milliwatt-hours)*

| | JAN. | FEB. | MAR. | APRIL | MAY | JUNE | JULY | AUG. | SEPT. | OCT. | NOV. | DEC. |
|---|---|---|---|---|---|---|---|---|---|---|---|---|
| LERWICK (1956–65) | | | | | | | | | | | | |
| Total | 28 | 81 | 157 | 308 | 407 | 469 | 400 | 313 | 211 | 101 | 37 | 17 |
| Diffuse | 21 | 56 | 100 | 187 | 251 | 286 | 265 | 214 | 133 | 70 | 27 | 14 |
| *Direct* | 7 | 25 | 57 | 121 | 156 | 183 | 135 | 99 | 78 | 31 | 10 | 3 |
| Diffuse ÷ Total (%) | 74 | 69 | 64 | 61 | 62 | 61 | 66 | 68 | 63 | 69 | 73 | 82 |
| Diffuse ÷ *Direct* (%) | 290 | 220 | 176 | 155 | 161 | 157 | 196 | 214 | 169 | 226 | 272 | 450 |
| KEW (1956–1965) | | | | | | | | | | | | |
| Total | 59 | 104 | 209 | 315 | 445 | 483 | 430 | 371 | 277 | 164 | 75 | 45 |
| Diffuse | 40 | 68 | 118 | 181 | 238 | 250 | 251 | 209 | 150 | 91 | 47 | 31 |
| *Direct* | 19 | 36 | 91 | 134 | 207 | 233 | 179 | 162 | 127 | 73 | 28 | 14 |
| Diffuse ÷ Total (%) | 68 | 65 | 56 | 58 | 54 | 52 | 58 | 56 | 54 | 55 | 63 | 67 |
| Diffuse ÷ Total Cloudless days (%)† | (33) | (20) | (22) | (21) | (17) | (14) | (13) | (22) | (36)* | (22) | (27) | (31) |
| *Diffuse ÷ Direct (%)* | 212 | 189 | 129 | 135 | 115 | 108 | 140 | 129 | 118 | 123 | 171 | 231 |

| | JULY | AUG. | SEPT. | OCT. | NOV. | DEC. | JAN. | FEB. | MAR. | APRIL | MAY | JUNE |
|---|---|---|---|---|---|---|---|---|---|---|---|---|
| PRETORIA (1951–3) | | | | | | | | | | | | |
| Total | 406 | 504 | 600 | 626 | 666 | 682 | 740 | 612 | 575 | 474 | 414 | 389 |
| Diffuse | 81 | 101 | 125 | 187 | 220 | 236 | 215 | 217 | 181 | 130 | 87 | 79 |
| *Direct* | 325 | 403 | 475 | 439 | 446 | 446 | 525 | 395 | 394 | 344 | 327 | 310 |
| Diffuse ÷ Total (%) | 20 | 20 | 21 | 30 | 33 | 35 | 29 | 35 | 31 | 27 | 21 | 20 |
| *Diffuse ÷ Direct (%)* | 25 | 25 | 26 | 43 | 49 | 53 | 41 | 55 | 46 | 38 | 27 | 25 |

†After Monteith.
*Misprint?

the mean values at Kew. At stations with a 'Mediterranean' climate, such as Nice or Cape Town, the situation is reversed and comparable figures are of the order of 35 per cent in winter and 25 to 30 per cent in summer.

## World distribution of solar income

In view of the paucity of radiation records over much of the land surface of the earth and all of the oceans a world map of solar income, however desirable, must perforce be based to a very large degree upon inference and computation.

Soviet scientists have undertaken this formidable task and published their results in *The Atlas of the Heat Balance* (Budyko, 1955). Fig. 15 is a simplified version of the map of mean annual total radiation reproduced by Budyko (1958, 99). Despite the element of intelligent guesswork involved, this is unquestionably the most fundamental of all climatological maps. Note that it disposes once for all of the much-too-popular belief that solar income attains its maximum values in close proximity to the equator. This myth refuses to die although, over a century ago, Maury (1858, 209–19) coined the picturesque, even if not too accurate, phrase of the 'equatorial cloud ring' to describe conditions which must inevitably lead to heavy daytime shadows.

MEAN ANNUAL TOTAL RADIATION KG. CAL/CM²/YEAR

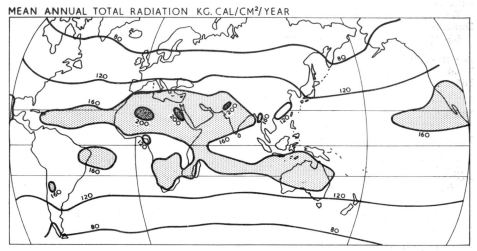

FIG. 15. A generalised version of Budyko's map of total radiation for the whole year. Units are in kg cal/cm²/year (after Budyko, 1958, p. 99)

Taking the year as a whole, the map shows that maximum total radiation is experienced over the tropical deserts and large sectors of the trade-wind zones. Locally in the northern hemisphere annual values exceed 200 kg cal/cm²/year as compared with values of about 180 kg cal in the Australian and Kalahari deserts. The 'humid tropics' on the other hand, including Amazonia, the Congo, and southeast Asia, are shown with values of between 120 and 140 kg. cal/cm²/year. This is of the order of 65 per cent of the maximum experienced on earth and the scale of this reduction is best appreciated when it is realised that a comparable reduction below the equatorial figure will carry us to the isoline of 85 kg cal/cm²/year characteristic of southern Scotland and central Patagonia. Clearly Maury's 'cloud ring' does play a major part in the distribution of solar income over the earth.

Budyko (1958, 101, 102) also presents maps of total radiation for the selected months of December and June. These have not been reproduced but their major characteristics are simply described. As might well be expected both show a major zone of high income (values in excess of 14 kg cal/cm²/month) in lower-middle latitudes in one hemisphere, south of the Equator in December and north of it in June, from which values decrease only slightly towards the summer pole but much more steadily towards the winter pole to reach zero a degree or two beyond the winter polar circle. This could have been anticipated from Fig. 7 but the maps add a number of interesting details. Both high-income zones show one area of comparatively low values, in December over the cloudy South Atlantic, and in June over the western Pacific and monsoon Asia where values fall to 12–13 kg cal. But still more significant are the outliers of high solar income found in the winter hemispheres. In December values in excess of 14 kg cal are found over the Sudan and in June similar values occur over south-central Africa and north-western Australia. These outliers are clearly tributes to the effectiveness of winter insolation in desert or semi-desert areas. Apart from these features it is not surprising, in view of the very different build of the two hemispheres, that the distribution is more clearly zonal in December than in June when values reach 20–22 kg cal over the Old World deserts and in Mexico as compared with the maxima of 18–20 kg cal attained in Australia and South Africa in December.

## Radiation and temperature

We have so far concerned ourselves with solar radiation as a whole and the distribution of this solar income in time and place over the surface of the earth. It is now necessary to explain that solar income is not uniform in *quality*, it is paid, so to speak, in a variety of currencies which, although closely related and readily convertible into energy, are sufficiently different from one another to play distinctive roles in the various accounts involved in the earth's economy.

In the first place, although we have emphasised that the sun is the prime mover dominating the living world, it is by no means the only radiator. Modern physics tells us that all matter is in a state of random molecular or atomic vibration which we measure as temperature (preferably in degrees Absolute or Kelvin; $0°C = 273°K$). As a result of this vibration all matter broadcasts radiation in the form of electromagnetic waves moving outwards in all directions in straight lines at the speed of light (300000 km or 186000 miles per second). Despite this common characteristic such radiation may vary over an immense range of wavelengths or, to express the same thing in another way, of frequencies. At one extreme we have the output of wireless stations capable of ringing the earth at wavelengths in excess of 30000 metres and at the other, wavelengths so short that even the micron—written $\mu$ (mu) and equal to one-thousandth part of a millimetre ($1 \mu = 10^{-4}$ cm)—ceases to be a convenient unit and has to be further subdivided into millimicrons ($1 m\mu = 10^{-7}$ cm) or Ångströms ($1 Å = 10^{-8}$ cm).[1]

Figure 16 represents this range on a logarithmic scale. It has been compared

[1] Under the new S.I. metric system, the micron becomes a 'micrometre' ($\mu m = 10^{-6}$m) and the millimicron is called a 'nanometre' (nm = $10^{-9}$m). What's in a name?

to an extraordinary pianoforte keyboard convering fifty-five octaves each of which occupies an equal subdivision of the scale. As on such a keyboard the frequency at the right hand of each subdivision is precisely double that at the left. It will be noted that the scale is open at both ends; further octaves may be added as research continues.

## ELECTRO—MAGNETIC WAVES

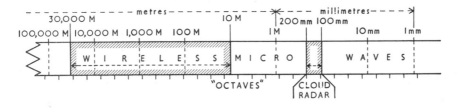

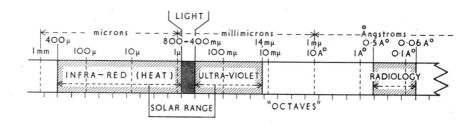

## THE VISIBLE SPECTRUM

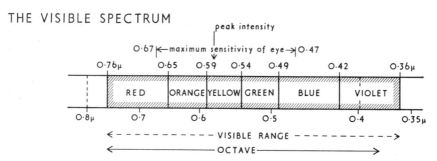

FIG. 16. The range of electro-magnetic waves and the visible spectrum

As climatologists our interest is mainly confined to some nine or ten octaves of this sequence, ranging from about 0.3 μ, the shortest ultraviolet waves which reach the earth from the sun, to about 120 μ, the longest waves emitted at the very lowest terrestrial temperatures. Even within this more limited range the human body is a relatively crude 'receiver'. The eyes are tuned to distinguish with some delicacy differences of wavelength from 0.36–0.76 μ, the range of visible light, but beyond that narrow band we have to employ a variety of photographic devices or else supplement our rather unreliable sensations of heat or cold by various methods of thermometry. Still farther afield we depend upon a host of ingenious electrical devices, each designed and tuned for some special purpose.

The lower scale of Fig. 16 shows the visible range in much greater detail (scale × 20). This is the spectrum of Noah's rainbow, a phenomenon which rarely fails to excite the interest of young and old. Rightly so, for it appeals to the mystic, the artist and the scientist in man and he who is not in some degree at least one of these is scarcely human. To the climatologist this is the original source of all the glory of colour in earth and sky and, at the same time, a valuable key to both plant and animal responses to solar radiation. Note that the human eye is not uniformly efficient throughout even this narrow range. We see most clearly within the more limited range from 0.47–0.67 $\mu$. The chlorophyll of living plants absorbs radiation in the process of photosynthesis most effectively in the orange-red range (0.59–0.76 $\mu$) whilst reflecting green light (0.49–0.54 $\mu$). Hence the predominant colour of vegetation.

## Full radiators

Gases, particularly at near-atmospheric pressures, are very selective radiators, each 'broadcasting' on a number of distinctive, very narrow bands. Most matter in liquid or solid form, however, radiates continuously over a considerable range of wavelengths, partly determined by its physical character but primarily related to its temperature. The sun itself behaves in this way, it radiates as a dense incandescent mass and not as a gas.

Physicists have shown that for a perfect radiator—a 'black body'—there is a precise relationship between intensity of radiation, wavelength and absolute temperature which can be portrayed by means of a curve of the form shown in Fig. 17. This is no idle theorising for, at wavelengths appropriate to their temperatures, even such unpromising subjects as the sun's surface and a snowfield have been shown to behave approximately as 'black bodies'. We shall avoid violence to the Queen's English if we call such bodies 'full radiators' for blackness is an optical property which has little or nothing to do with the present discussion.

We can handle Planck's Law with the minimum of mathematics if we accept the deduction from it that for a full radiator the peak of the curve or the point of maximum intensity of radiation occurs at wavelength $\lambda_m$ microns where

$$\lambda_m = \frac{2897}{T} \text{ or, more simply } \frac{2900}{T},$$

the temperature $(T)$ being quoted on the absolute or Kelvin scale. This provides a reference point from which the other characteristics of the curve appropriate to any given temperature can easily be found. Thus, although both the upper and lower limits of the wavelength scale are a little woolly, it appears that 99.8 per cent of the area enclosed by the curve lies between the range 0.3 and 8.5 $\lambda_m$. These two points thus cut off the lower and upper 'milles' of the frequency group and may be used to define its overall range. Its mean range, on the other hand, is given by the 'quartiles' which cut off the lower $(Q_1)$ and upper $(Q_3)$ quarters of the group. These are found to lie at 0.99 $\lambda_m$ and 2.09 $\lambda_m$; between these values lies the central half of the area enclosed by the curve. Although these results were obtained from a careful drawing there is some danger of introducing spurious accuracy at this stage and a more general statement that Planck's formula yields a skew curve with

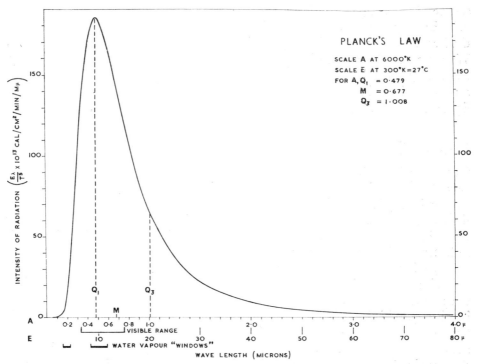

FIG. 17. Planck's Law

the lowest quarter of the frequencies below peak wavelength (the 'mode') and the upper quarter at wavelengths beyond twice the peak value will normally suffice.

Some idea of the use to which these measurements may be put is given by the data in Table 4 where our results are applied to alternative estimates of the temperature of the sun's surface ($A$ and $A'$) and to values representative of very hot ($B$) and very cold ($D$) conditions on the surface of the earth. To clinch the argument a temperature representative of mildly red-hot iron ($R$) is also given. It will be seen that this brings the lowest wavelengths within the visible range. In actual fact hot iron is visible in the dark in the neighbourhood of 1100° K, but the dull light thus emitted involves only the lowest mille of the total radiation, which our estimate has omitted.

It will be noticed that Fig. 17 has been given *two* horizontal scales in microns. Scale A is appropriate to a rounded figure for the temperature of the sun (6000°K) and, below it, the range of the visible spectrum is indicated. Scale E, appropriate to 300°K, is greater by a multiple of 60/3 = 20. It may be taken as representing terrestrial radiation on a warm summer's day (27°C or 80½°F). Below it is shown the range of the fully open and partially open water-vapour 'windows'. It is entirely accidental, though of considerable importance,

TABLE 4. *The wavelength of radiation at selected temperatures*

| | TEMPERATURE | | WAVELENGTH (*microns*) | | | | | |
|---|---|---|---|---|---|---|---|---|
| | °K | °C | Peak | Shortest | $Q_1$ | M | $Q_3$ | Longest |
| | *Factor* | | (1.00) | (0.3) | (0.99) | (1.4) | (2.09) | (8.5) |
| A | 6000 | 5727 | 0.48 | 0.14 | 0.48 | 0.68 | 1.01 | 4.1 |
| A' | 5793 | 5520 | 0.50 | 0.15 | 0.50 | 0.70 | 1.05 | 4.3 |
| B | 323 | 50 | 9.0 | 2.7 | 8.9 | 12.6 | 18.9 | 76.5 |
| C | 273 | 0 | 10.6 | 3.2 | 10.5 | 14.9 | 22.2 | 90.0 |
| D | 223 | −50 | 13.0 | 3.9 | 12.9 | 18.2 | 27.2 | 110.5 |
| R | 1200 | 927 | 2.4 | 0.72 | 2.39 | 3.38 | 5.05 | 20.5 |

*Note:* 'Shortest' and 'longest' values each exclude about one mille of total radiation.

that the widest of these coincides with radiation of relatively high intensity. Other scales appropriate to any given temperature may be similarly fitted, it being remembered that a corresponding adjustment ($= T^5$) must be made in the vertical scale.

The vertical scale denoting intensity of radiation is in terms of $T^5$ for Wien has shown that in full radiation the quantity of energy radiated at maximum intensity is proportional to the fifth power of the absolute temperature. Thus the peak intensity of radiation from the sun is $(60/12)^5 = 3125$ times greater than that from our red-hot iron and the peak from a hot desert surface will be about six and a half times $(323/223)^5$ that from the Siberian snows. The *total amount* of energy radiated per unit area in unit time is given, however, by the *area* below the curve. This varies as the *fourth power* of the absolute temperature, the sun emitting 625 times the energy of a mass of red-hot iron of equal area and the hot desert some four and a half times the output of the Siberian snow. The idea that winter snows are capable of radiating heat may strike the reader as a little bizarre at first encounter but a little thought will show that it is precisely for this reason that the temperature is able to fall so low.

This relationship is expressed quantitatively in Stefan's Law which states that total radiation $= \sigma T^4$ where $\sigma$ is Stefan's constant and is equivalent to $82 \times 10^{-12}$ gram cal/cm²/min or $5.71 \times 10^{-12}$ watts/cm²/min. From this it can be calculated that, with a vertical sun and 70 per cent effective transmission radiative equilibrium at the surface of the earth would not be attained until the temperature reached 359°K = 86°C = 187°F. This would certainly cook an egg. In-the-sun temperature records are notoriously unreliable but maxima of the order of 80–82°C (176–180°F) are claimed for Adelaide, Melbourne and Perth in January in Australia. (*Year Book of the Commonwealth*, 1963, 53–60).

## Partial or selective radiators

Whilst the above estimates are broadly valid for most terrestrial objects, particular mention must be made of gases, vapours, and smooth-surfaced or transparent solids or liquids. In so far as these either transmit or reflect radiation of any given wavelength they must fail to absorb it. Kirchoff has shown that failure to absorb radiation at a given temperature and wavelength also implies inability to emit radiation of the same character. If the body or gas is not tuned to act as a 'receiver' to a particular waveband it cannot act as a 'transmitter' within that band at the same temperature. A substantial change of temperature, however, may bring about considerable changes in the design of the 'set'. In meteorology this quality is met with mainly in the atmospheric gases. The selective radiation of water vapour is particularly important.

## The terrestrial redistribution of solar income

Solar income may thus be transmitted or reflected (both regularly and irregularly) without rise of temperature, or absorbed with consequent rise in temperature of the absorbing medium. This rise in temperature will stimulate the medium itself to more active emission and the stimulus is thus passed on to all other media within radiative range. The analogy with currency is not perfect since income is often received in one currency (or wavelength) and paid out in another but temperature is not dissimilar to spending power and what is expenditure to one body is income to a number of others. The exchanges are numerous and complex, they permit the work of the world to go on, but in this economy there appears to be no net accumulation (except for a very tiny fraction laid down as fossil fuel) and the fresh 'notes' issued in such profusion from the sun are precisely balanced by the withdrawal of used 'notes' into the infinitude of space. We are therefore concerned with a problem in accounting.

## The atmosphere

The first deduction[1] from solar income is made by a weak concentration of ozone high in the atmosphere, itself apparently a direct result of short-wave radiative action upon oxygen. This ozone belt, lying mainly between 40 and 10 km above the earth's surface, completely screens out all solar radiation of wavelength below 0.29 $\mu$ and since such radiation has been found to inflict

[1] Apart from some loss in the ionosphere of no climatic importance.

vital damage on living tissues it is a filter of supreme importance. Whether this zone has any direct climatological implications is still a matter for debate but it helps to produce remarkably high temperatures in the outer atmosphere with a diurnal range of the order of 6 to 8° C, a range not again approached until very near the surface of the earth. The deduction of solar energy involved is considered to amount to about 5 per cent. Variations in the depth and concentration of ozone are being actively studied as an indicator of large-scale atmospheric circulation at great heights.

Next follows a process of selective scattering of radiation by gaseous molecules or other very fine particles smaller than or of about the same diameter as the wavelengths of light. These are set oscillating at the same frequency as the light and send off radiation at that frequency, thus scattering the beam. Little absorption is involved in this process so there is scarcely any perceptible rise in temperature but at least half of the scattered light is tossed back into space and this represents a net loss of solar income for the earth as a whole. Furthermore, since scattering by small particles is inversely proportional to the fourth power of the wavelength of the incident radiation, blue (0.46 $\mu$) suffers twice the scatter of yellow-green (0.55 $\mu$) to which the eye is most sensitive, and four times the scatter of full red (0.65 $\mu$). This is the origin of the golden glow of sunset and the purple tint of distant hills. Beyond the visible range, ultraviolet (0.31 $\mu$) scatters five times more freely than the blue (we get our 'sun tan' from the sky) whilst infrared photography (0.84 $\mu$) secures pictures with less than a tenth of the atmospheric haze experienced when using normal film and a yellow filter (i.e. using actinic blue 0.46 $\mu$).

At still lower levels the incident radiation encounters suspended water droplets or ice crystals of diameters ranging from about 4–100 $\mu$ and in varying degrees of concentration. These are all large enough to produce irregular or diffuse reflection of white light and the blue of the sky is given a milky hue, particularly towards the horizon. From an aeroplane the upper margin of the haze zone is often clearly visible even on the best of days. Once again there is little absorption and hence little rise in temperature but a further fraction of radiation is being dissipated into the void of outer space. It has been estimated that the free atmosphere may experience a rise in temperature of the order of 0.5° C per day as a result of the absorption of direct solar radiation mainly by water vapour but, since this is no more than about a quarter of what it is likely to lose by night, the major factors responsible for the observed temperature of the upper air must be sought elsewhere.

The same process continues with still greater effectiveness when the water droplets are concentrated into active clouds. It is difficult to generalise since cloud is so variable but cloud from 300 metres to 1 000 metres in thickness is estimated to reflect some 70 per cent of the incident radiation regardless of angle and to absorb from 3 to 7 per cent of it. Deductions on this scale mean the virtual extinction of solar income at the surface on a really cloudy day, especially if the sun is never very high in the sky. Fog is no more than cloud at ground level and when once a thick blanket is well established it reflects so much radiation that it can defy the efforts of a winter sun to burn it away. It may be mentioned in passing that since fog reflects white light (i.e. all visible wavelengths), the advantages of an amber fog-lamp are largely illusory.

## The surface of the earth

The first reaction of terrestrial substances to radiation is to reject a proportion of it, to toss back part of solar income quite carelessly into space. This applies mainly to the visible part of the spectrum and it is by this process that we see the objects around us. We see a piece of coal or a glass of water mainly by the 'shine' at about equal intensities although the great bulk of the radiation is absorbed in the one case and transmitted in the other. The proportion thus rejected is known as the 'albedo' of the substance and it varies widely from about 8 to 15 per cent for wet earth and coniferous forest, through 15 to 25 per cent for dry soil or rock tundra, meadows, cropland and deciduous woods, 25 to 30 per cent for deserts, semi deserts and dry sand dunes, 30 to 40 per cent for sea ice, to a wide range of values depending on snow condition but reaching a maximum of 80 to 96 per cent for fresh, dry snow. On a snowy day we are in a stimulating world of light even when the sky remains heavily overcast. Smooth water reflects little when the angle of incidence is above 45° and from an aeroplane water bodies usually look darker than the surrounding ground. The glitter of water surfaces is produced by the inclined sides of waves or ripples which momentarily bring down the angle of incidence from sun or sky to the order of 20° or less which locally raises the albedo to between 30 and 50 per cent. Reflection involves no net gain of energy and therefore no change of temperature. However, substances which are good reflectors of light, like snow and, at low angles, glass or water, readily absorb the longer waves of heat when they are available.

The only really common transparent substance at the surface of the earth is clean water, itself a good deal less common than the city-dweller may suppose. Add to this a number of distilled liquids, the manmade glasses, a variety of crystals and a very considerable number of small marine creatures and we exhaust the list of substances which do not absorb light radiation within a very short distance of their external surfaces. Such substances transmit radiation to whatever lies beyond them, the bottom of a pool or the interior of a room, but there is usually some loss and if the water is deep enough or the glass thick enough it becomes virtually opaque. In clear ocean water only about 15 per cent of the incident short-wave radiation reaches below 10 metres (*c.* 35 ft) and by 100 metres (325 ft) the proportion is down to no more than 0.5 per cent. In water bodies exceeding 200 metres in depth there is complete extinction and the bottom lies in perpetual darkness—all the incident light has been absorbed and converted into heat. But transparency to light does not entail equal transparency to long waves which are fully absorbed near to the surface (and devoted mainly to evaporation). Every terrestrial object thus takes some share of the solar radiation incident upon it, with consequent rise in temperature.

The extent to which temperature is increased depends on a body's 'capacity for heat' as well as the share of the radiation which it is able to absorb. This physical property varies over a very considerable range for different substances. For unit mass (1 gramme) we define this property as 'specific heat' but for unit volume this must be multiplied by density. Thus water has specific heat 1 and density 1 whilst most dry rock or sand has specific heat 0.19 to 0.22

THE RELATIVE TRANSPARENCY OF WATER-VAPOUR (+ CO₂) TO VARIOUS WAVELENGTHS OF EARTH RADIATION

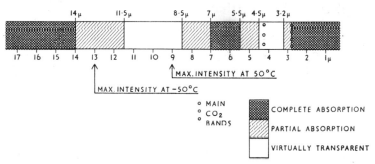

FIG. 18. Generalised version of the absorption spectra of water vapour and carbon dioxide. The diagram gives some indication of the extent to which the absorption bands of the two gases overlap and shows the position of the important water vapour 'windows'

and density 2.5 to 2.7. Rock thus responds to solar income absorbed about *twice* as much as an equal volume of water. Unequal response of the earth's surface to solar insolation, however, is mainly due to the presence or absence of water whether in compact masses such as seas, lakes or rivers or dispersed amongst the interstices of rock, sand, soil or plants. The high specific heat of water is partly effective here but a far more powerful coolant is the process of evaporation of which more anon. 'Warm' soils and 'cold' soils owe their character, therefore, almost exclusively to their water-retaining powers.

## Return radiation from the earth

The rise of temperature which follows the absorption of solar radiation at once stimulates more active return radiation on the part of the heated body. Solar income is promptly expended to all neighbouring bodies or to the open sky with complete impartiality, substances with a great thermal capacity responding more slowly but no less completely than substances whose capacity is slight. However, the repayment is made in a different waveband, in a type of currency appropriate to the temperature of the radiator, and not in solar coin. This fact is of the greatest importance in climatology since the atmospheric gases (water vapour and carbon dioxide in particular) which pass on solar currency with a very slight discount are much less careless about the returning payment. Earth radiation is heavily absorbed in the lowest layers of the atmosphere and its temperature is increased thereby. The absorption spectrum of water vapour is, in fact, highly complex but it may be simplified as in Fig. 18 which is sufficient to illustrate the basic facts. Comparing this figure with the data given in Table 4 it will be seen that the vapour heat-trap has a most significant escape hatch or 'window' in the 8.5–11.5 $\mu$ band. It is through this band that the earth cools off on a clear night but, even so, about half the earth radiation is retained in the shorter and longer wavelengths on either hand.[1]

Liquid water droplets, on the other hand, absorb at all terrestrial wavelengths so that haze and light cloud help to seal the leakage. Heavy cloud, particularly when it is low, checks the flow almost completely and there is a scarcely perceptible fall of temperature the whole night through at the surface of the earth though radiative loss continues from the cloud tops and the air above them.

## Return sky radiation

This brings us to a further exchange. In so far as the gases of the atmosphere, or any liquid or solid particles suspended in them, effectively absorb either direct solar radiation or the return radiation from the earth, they must rise in temperature. Increasing in temperature, they are stimulated to further radiation in all directions, both spacewards and earthwards. Spaceward radiation is lost from the terrestrial account but earthward radiation is again absorbed and re-emitted and thus the exchange goes on unceasingly.[2] The geometry of these exchanges is highly complex not only because all angles within the 180° arc of the sky are involved, but also because the thickness of the atmosphere, the absorbing and re-radiating medium, is effectively greater the more acute the angle of the ray path. Geiger (1965, 23) points out that at the earth's surface 'the least amount of counter-radiation is therefore received from the zenith direction and the radiation loss is greatest in that direction'. This is, of course, particularly true at night when intense scatter around the sun's disc does not further complicate the picture. The net result of all these exchanges is a steady loss by spaceward radiation from the earth, from clouds

[1] Infra-red radiation through this window is now used in remote sensing of the surface features of the earth by aircraft or satellite, even during the night. To penetrate cloud however various radar devices using longer wave bands must be employed (see Cooke and Harris, 1970).

[2] Geiger (1965, 20) has estimated that when the atmosphere contains some 14 millimetres of precipitable water (a fairly normal amount) as much as 75 per cent of the counter-radiation from air to earth occurs from the lowest 125 metres (*c.* 400 ft) of air whilst 90 per cent is derived from the lowest 580 metres (*c.* 1 900 ft).

and from the water-vapour in the air itself through the vapour 'window' open to the 8.5–11.5 $\mu$ band. This continues without interruption both by day and by night but during the day its effects are masked by the fresh surge of energy brought by the rising sun.

If radiative processes alone were responsible for the temperature stratification of the atmosphere it has been shown by computation that the 'lapse rate' or fall of temperature with height would be represented on the average by a broadly exponential curve; the rate of fall of temperature per unit height would be much greater in the lower layers (over twice the DALR, see p. 234) than farther aloft (Godske *et al.*, 1957, 100). In actual fact, over any considerable thickness of the atmosphere, such a pattern is never encountered. Clearly there must be other potent factors—turbulence, convection, condensation—which contribute to the redistribution of heat in the atmosphere.

Three final points conclude this section:

(*a*) Since the opacity of water vapour to long-wave radiation decreases with decreasing atmospheric pressure, the vapour 'window' is thrown more widely open as height increases. This facilitates the loss of heat to outer space.

(*b*) Clouds, though an effective screen to outgoing long-wave radiation, receive a net gain of heat towards their bases and show a net loss towards their tops. Turbulent overturning is thus stimulated and all types of cloud tend to disperse unless they are constantly regenerated.

(*c*) Low clouds are 'warm' clouds and high clouds are colder. Even when cloud-sheets are of the same density, high cloud is thus a less effective check upon nocturnal radiation from the earth since the intercepted radiation is paid into a colder 'sink' and return radiation is proportionately lower.

# The heat balance of the globe

It would be very convenient if it were possible, for any given geographical locality, to summarise all these complex exchanges in the form of a neat balance-sheet showing radiational gains and losses and indicating the manner in which a positive balance is disposed of or a negative balance recouped through other channels of heat transfer.

Numerous attempts have been made to do this for the earth as a whole. The problem is then simplified since the total income is fairly accurately known and, since there is no evidence that over the centuries the earth is getting appreciably warmer or cooler, it can be inferred that the income and expenditure *must* balance. Even so there is far from complete agreement as to the relative scale of the various factors involved.

Table 5 presents a summary of Budyko's (1958, 222-3) conclusions both in kg cal/cm$^2$/year and also in percentages of the mean solar income of the earth as a whole. Since there are 525 360 minutes in a year of $365\frac{1}{4}$ days, a solar constant of 1.94 or 2.0 cal/cm$^2$/min gives an annual income of 1 020 or 1 050 kg cal/cm$^2$/year to a surface normal to the incoming rays. But the earth is a revolving sphere with a total area four times greater than that of its shadow on a plane surface normal to the sun's rays. The mean solar income per unit area is thus only a quarter of the values just quoted—it is usually rounded off to 250 kg cal/cm²/year which thus becomes the starting-point of our analysis. The 'albedo' of the whole earth, its total reflecting power, is still a subject of

TABLE 5. *The heat balance of the earth (after Budyko)*

| PROCESS | EARTH Incoming | EARTH Outgoing | ATMOSPHERE Incoming | ATMOSPHERE Outgoing | SPACE To earth | SPACE Returned |
|---|---|---|---|---|---|---|
| *kg cal/cm²/year* | | | | | | |
| Total radiative income | | | | | 250 | |
| Reflected | | 35 ⟶ | | 65 | ⟶ | 100 |
| Effective total radiation | | | | | 150 | |
| Short-wave absorption | 111 | | 39 | | | |
| Net long-wave back-radiation from earth | | 43 ——————— | | | ⟶ | 43 |
| Evaporation and condensation | | 56 ⟶ | 56 | | | |
| Net turbulent transfer | | 12 ⟶ | 12 | | | |
| Net long-wave back-radiation from atmosphere | | | | 107 | ⟶ | 107 |
| TOTALS (effective radiation) | +111 | −111 | +107 | −107 | 150 | 150 |
| *Per cent of total radiative income* | | | | | | |
| Total radiative income | | | | | 100 | |
| Reflected | | 14 ⟶ | | 26 | ⟶ | 40 |
| Effective total radiation | | | | | 60 | |
| Short-wave absorption | 44.4 | | 15.6 | | | |
| Net long-wave back-radiation from earth | | 17.2 ——————— | | | ⟶ | 17.2 |
| Evaporation and condensation | | 22.4 ⟶ | 22.4 | | | |
| Net turbulent transfer | | 4.8 ⟶ | 4.8 | | | |
| Net long-wave back-radiation from atmosphere | | | | 42.8 | ⟶ | 42.8 |
| TOTALS (effective radiation) | 44.4 | −44.4 | 42.8 | −42.8 | 60 | 60 |

some debate but Budyko accepts a figure of 14 per cent for the surface of the earth plus 26 per cent for mean cloud cover. A total of 100 kg cal/cm²/year, or 40 per cent of the total income, is thus promptly dismissed from the account and the earth's *effective* solar income, available to run its domestic economy, is reduced to 150 kg cal/cm²/year.

This short-wave energy is absorbed by earth and atmosphere in the proportions shown and stimulates either a rise in temperature or an increase in evaporation where water is available. Alternatively it is shared between both of these activities in a proportion which varies with local conditions. At the earth's surface a rise in temperature stimulates increased long-wave back-

radiation but about half of this is intercepted by the atmosphere and partially
returned. It is simpler to sidetrack these interchanges by quoting only the *net*
loss from earth on long-wave account—about 17 per cent of the solar total.
Energy devoted to evaporation, from the oceans and other water-bodies and
from vegetation and surface soil, is temporarily withdrawn from the radiation
account and may be freely redistributed over the globe by the general circulation. Eventually, however, the water vapour is condensed and the latent
heat thus released is then free to join the complex atmosphere–earth and earth–
atmosphere exchanges mentioned above. The part played by turbulent
transfer, that is, of upward or downward convection of sensible heat as

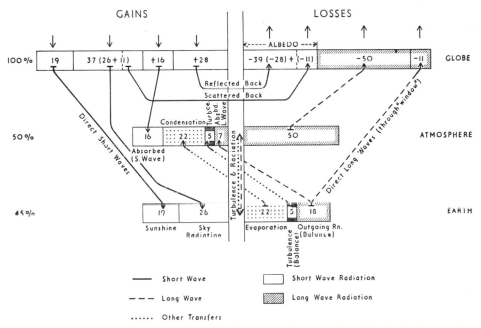

FIG. 19. The annual heat balance of the earth–atmosphere complex. Total solar income is
taken as 100 and the diagram shows the direction and relative proportions of the various
exchanges which take place (after Geiger, 1965, p. 225, slightly modified)

distinct from the powerful processes involving the presence of water-vapour,
has been a matter of prolonged debate. Budyko (1958, 215) is confident that,
on the average, 'the turbulent heat flux in all latitudinal zones (from 60° N to
60° S) is directed from the earth's surface to the atmosphere on land as well as
on the ocean' and regards the net figure of 12 kg cal/cm²/year or 4.8 per cent
of total solar income as a fair measure of its importance. The atmosphere,
thus warmed by the interception of short-wave radiation from above and of
long-wave radiation from below, as well as by the joint efforts of both moist
and dry convection, must lose its net balance by long-wave radiation into
space, particularly at night, to round off the account.

Figure 19 (after Geiger, 1965, 225) presents similar, but not identical data in diagrammatic form. The figure for turbulent transfer has been modified on Budyko's authority and the twin vertical arrows in the middle of the diagram serve to remind us of the turbulent and radiative exchanges that go on between earth and atmosphere, atmosphere and earth, which find no place in Table 5 or in the figures quoted within the panels.

To the geographer, interested above all in the *difference* between various parts of the world, such a global accounting may well appear indeed remote. His interest might very well parallel that of an inhabitant of Benbecula or Nantucket when faced with the National Budget—it concerns him but at so many removes that a detached approach is not unreasonable. Indeed the major object of this section is to underline the remarkable scale of the activities covered by the evaporation-condensation account. The geometry of radiation tends to throw undue emphasis upon latitude; it is the group of climatic factors associated with moisture-exchange that is largely responsible for the great wealth of azonal features—of 'climatic anomalies'—which do so much to give interest and variety to the surface of the earth (see Trewartha, 1961).

Budyko and his colleagues have carried the investigation of the heat balance very much further. We give one example of this in Fig. 20 which illustrates the latitudinal distribution of the four main elements of the balance for both land and water surfaces (Budyko 1958, Table 14, p. 214). Although still highly generalised the data are unquestionably of geographical relevance. It will be seen that over both surfaces total solar income ranges from about 150 kg cal/cm²/year near the equator to values of the order of 80–90 kg cal between the 50th and 60th parallels in both hemispheres. From the 10th to the 40th parallels, however, the radiational income received by the land (Fig. 20, b) is much in excess of that which reaches the sea (Fig. 20, a). Presumably this is in response to reduced cloud cover, especially over the deserts.

However, it is in the disposal of this total income that the most interesting contrasts between the two diagrams are to be found. From the *ocean surface* there is little variation in long-wave radiational loss with latitude, the overall mean of 51 kg cal thus being fairly representative of all of the twelve zones considered though values are rather less than this near the equator and rather greater near the tropics. Of the remaining balance over the ocean the great bulk is paid into the evaporation account. Between the 10th and 30th parallels this expenditure reaches mean values of 87 kg cal/cm²/year in the northern hemisphere and of 92 kg cal in the south as compared with a mean of 78.5 kg cal astride the equator and values of the order of 35 kg cal towards the Arctic and Antarctic circles. For all latitudes from 40° N to 50° S these figures are well in excess of a half of the total solar income received. On the other hand, values for net turbulent transfer are everywhere relatively small and where they are greatest, towards the poles, they can only be accommodated on the diagram by carrying it *below* the zero line. This part of the curve represents a net transfer of heat within the oceanic reserve, the supply being drawn from the

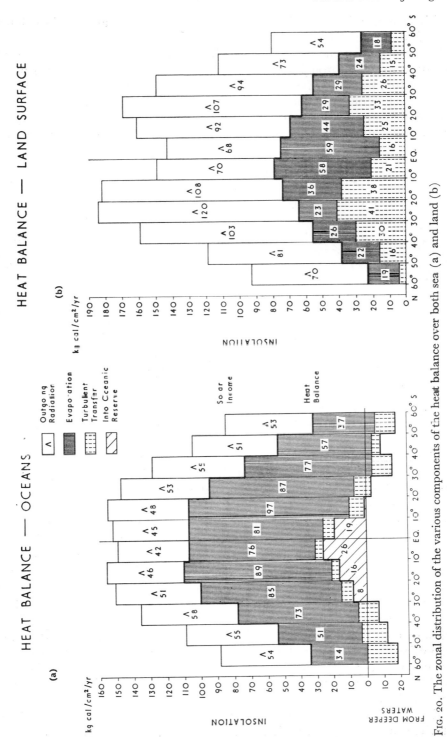

FIG. 20. The zonal distribution of the various components of the heat balance over both sea (a) and land (b)

positive balance shown, mainly from 30°N to 10°S, and transferred to higher latitudes by ocean currents.

On *land surfaces* the story is a very different one. Losses by long-wave radiation are high throughout the range of latitudes illustrated but the mean value of 86.5 kg cal/cm²/year is no longer representative of any one of them. Highest values occur in zones corresponding to the tropical deserts but even near the equator, values are half as large again as those for the sea. The 'radiation balance' for land surfaces thus emerges as a simple curve with a

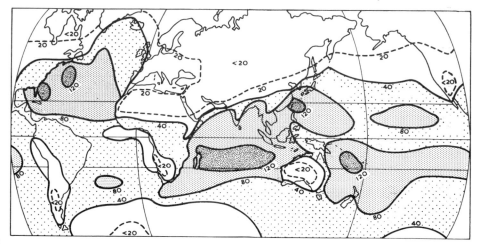

FIG. 21. A generalised version of Budyko's world map of the total amount of heat expended each year in evaporation. Units are kg cal/cm²/year (Budyko, 1958, p. 113)

maximum just north of the equator from which values fall steadily in a poleward direction. The disposal of this balance, however, has quite new characteristics. Over the humid equatorial zone, 10°N to 10°S, evaporation is still dominant though heat devoted to this account is only about three-quarters of that lost by the equatorial oceans and thus amounts to no more than 40 per cent of total solar income. But towards both poles, the heat devoted to evaporation from land surfaces diminishes rapidly, mainly for the want of a continuous surface supply, and values of the order of 20–30 kg cal/cm²/year are characteristic of middle and higher latitudes. In the zone of the Old World deserts, 20–30°N, the proportion of total solar income devoted to evaporation thus reaches its lowest zonal value of some 12.5 per cent. Even polewards of the 40th parallels in both hemispheres the corresponding proportion is rarely more than some 20 per cent. It naturally follows that the amount of heat available for turbulent transfer from land surfaces is considerable—a direct result of the correspondingly high temperatures stimulated by the daily heat-input from the sun. Since the heat reserve of the solid earth is negligible, at least on an annual account, and since horizontal transfer is out of the question, there are no negative values to Fig. 20(*b*).

Even this zonal analysis, it will be noticed, is on an *annual* basis so that much that is of seasonal interest, particularly in middle latitudes, is masked in Fig. 20. Budyko (1958, 118–35) handles this aspect by applying his methods to monthly data at a number of representative stations and to selected sections of the ocean surface.

Here we must content ourselves with a rather simplified version of his world map of the annual expenditure of heat for evaporation (Fig. 21). However tentative, this is a map of the greatest importance. Note that isolines show a discontinuity at the coast owing to the different evaporation regimes we have just emphasised. The most intense sources of atmospheric moisture emerge as the western Pacific, the Indian Ocean, especially south of the Equator, and the North Atlantic. Striking contrasts emerge between the northern Atlantic and Pacific oceans and again between the spheres of influence of the south-east trades in the Indian Ocean and the south Atlantic. Could one find stronger evidence of the importance of azonal factors in the distribution of climates?

# 2
# Temperature

A more traditional approach to the study of the effect of variations of solar income through time is via air temperature. The *Meteorological Glossary* defines temperature as 'the condition which determines the flow of heat from one substance to another'. Note therefore that, in this section, not only are we passing from causes to effects, but also that the concept of temperature is basically a *relative* one, to be measured on a relative scale. Our personal sensations of 'warmth' and 'cold' are relative; we receive them through the stimulation of nerve-endings in the skin. Our standard of reference thus ranges from blood-heat ($37°$ C, or $98.4°$ F on British clinical thermometers) to values as low as $10°$ C ($50°$ F) for skin temperature at the extremity of the limbs on a cold day. Our temperature sense is thus both crude and ephemeral—the initial sensation often fades very rapidly. Furthermore it covers only a limited range, for both severe heat and severe cold may destroy the skin tissue and produce a lesion requiring medical treatment.

Temperature is commonly measured by the relative expansion and contraction of a small volume of mercury (freezes at $-40°$ C) or of tinted alcohol (freezes at $-150°$ C). For a self-recording apparatus such as a thermograph or a radio-sonde, the differential expansion of a coiled bi-metallic strip is usually preferred. Today, for very delicate measurements, a variety of devices employing the variation of electrical resistance with temperature have been designed to give a reading on a convenient dial or even to transmit an impulse to a remote panel.

It should be understood that the scale of 'degrees' in which these instruments are calibrated is entirely arbitrary though use and wont have led to the almost universal acceptance of the scales devised by G. Fahrenheit (1686–1736) and A. Celsius (1701–44). Fahrenheit moved to the scale we know via a series of adjustments but his landmarks appear to have been the lowest temperature he was able to attain in a simple freezing mixture ($0°$ F) and blood-heat which he eventually took as $96°$ F—a number freely divisible without producing fractions. It seems by no means improbable that, working in Amsterdam and London, he also thought that air temperatures beyond this range were most unlikely. The rather odd values for the freezing and boiling points of water at sea level (32 and $212°$ F) which thus emerged did not distress him— why should they? In fact the interval of $180°$ F between them was reminiscent of degrees of arc—360 subdivisions would have produced units too small for the accuracy of his mercury thermometer. Celsius, working at Uppsala, must have been much more conscious of the occurrence of bitter winter cold and it may well have been with the object of avoiding negative values that he initially arranged his scale with 0° C at the boiling point of water and 100° C as the freezing point. Since this produced awkward negative values in metallurgy, the system was subsequently reversed to produce the centigrade scale which has carried all before it in the field of natural science.

Yet there were no real advantages to be obtained from the decimalisation of a *relative* scale. In contrast to measures of length or mass, no terms have emerged to express tens or hundreds of degrees. Furthermore when W. Thomson (Lord Kelvin) deduced from the gas laws the concept of an 'absolute zero' it was first stated as at −273° C,[1] a very odd figure indeed! Using centigrade intervals we thus have a third current scale, particularly appropriate to the free atmosphere, this is the Kelvin system where the freezing point of water is 273° K and its boiling point is 373° K. This finally disposes of the risk of negative values but at the cost of rather clumsy and unfamiliar figures.

Each of these systems has some advantages and the student should accustom himself to making ready conversions, remembering that each Celsius or Kelvin degree equals nine-fifths of a degree Fahrenheit—and not overlooking the appropriate additions or subtractions! For heavy conversions a table of equivalents is desirable though a sliderule can be useful. (See Appendix 3.)

The publications of the British Meteorological Office adopted the Celsius scale as recently as 1961 so as to fall into line with continental practice. During a long transition period it is proposed to quote both scales in the daily weather broadcasts. Indeed, at least for descriptive purposes, there was much to be said for the Fahrenheit scale. Unlike the units of the centimetre–gramme–second system which appear to be ridiculously small for geographical purposes, the Celsius degree is really rather large. One is thus almost invariably involved in decimals and the typographical difficulties and danger of misprints are by no means reduced by the frequent occurrence of negative values. Any mean value is subject to a range of 'statistical error' and it so happens that for temperature means over a fair span of years this would often justify quotation to the nearest degree Fahrenheit. (See Appendix 1.)

Temperature conditions may vary so widely over short distances that, in any general statement intended to be representative of a considerable area, very careful consideration must be given to the *exposure* of the thermometer. Normally we seek to express the temperature of the air stratum within which mankind lives and moves. For single readings this temperature is best obtained by whirling a thermometer in a sling but fixed instruments must be set in a screen, usually at breast height, so designed as to shield the instrument from solar and terrestrial radiation without seriously impeding free ventilation. An alternative method, popular in Scandinavia, is to suspend the thermometer on a northern wall. It cannot be too strongly stressed that the

[1] Now accepted as −273.15° C ± 0.02° C.

'shade' temperatures thus obtained have only a loose and limited relationship with the temperatures actually experienced, for instance, by a field crop growing within a few inches of the ground. Geiger (1955, 58, 161) quotes examples of ground temperatures from 20–30°C (36–54°F) higher than screen temperatures during brilliant sunshine in arid lands and ground frost is not uncommon on clear nights when the air temperature does not fall below 5 or 6°C (41–43°F).

Some extreme screen values near sea-level are expressed in the three scales as follows:

*For the world*

|  | °F | °C | °K |  |
|---|---|---|---|---|
| Max. temperature | 134 | 57 | 330 | Death Valley, California |
| Min. temperature | −96 | −71 | 202 | Oymiakon, Siberia |

*For Britain*

|  | °F | °C | °K |  |
|---|---|---|---|---|
| Maximum | 101 | 38 | 311 | Tonbridge |
| Minimum | −17 | −27 | 246 | Braemar. |

High on the Antarctic plateau a temperature of −126°F, −88°C or 185°K has been recorded at Vostok station, whilst in the free atmosphere the lowest temperature so far recorded was −133°F, or −92°C, or 181°K registered at 16.5 km above Agra in October.

Faced with this wide range of terrestrial temperature and accepting Hann's (1897: 1903, 3) view that climatology is concerned with 'meteorological conditions in so far as they affect animal or vegetable life' we may well ask, are there no critical points or thresholds in this long series near which important biological reactions can be expected to take place? Are there no milestones along the road which may help us to orient ourselves, no key values which may help us to assess the real significance of the statistical record? The following values are thought to have some significance:

| °F | °C | |
|---|---|---|
| −32 | −36 | Temperatures below this level kill many trees. |
| −22 | −30 | Stations with at least one month with mean temperature below this level are likely to suffer from a permanently frozen subsoil, known in Canada as 'permafrost'. This entails very serious problems of soil drainage in summer. |
| 23 | −5 | Air temperatures below this level are regarded as bringing a 'killing frost' in America. Such frosts involve great hazard to growing crops when they occur early in autumn or late in spring. |
| 30 | −1 | A common definition of frost in Britain. |
| 43 | 6 | The temperature suggested by De Candolle as giving a convenient indication of when a grain crop is likely to begin active growth in spring. Attempts have been made to assess the effectiveness of a growing season in 'day-degrees', the sum total of the |

excess of mean daily temperatures above this figure. This device has been both elaborated to take account of the diurnal range of temperature and generalised to give a result in 'month-degrees' with varying success.

50    10    It has been asserted with some justification that unless one month of the year has an average temperature above this figure, forest growth is impossible.

65    18    With outside air temperatures below this value it has been suggested that some kind of internal heating is normally required in buildings. This is a very approximate figure since standards vary widely in different countries.

86    30    When the air is humid, temperatures above this level are regarded as beyond the limits of comfort particularly for vigorous outdoor work or exercise.

98.4   37    Normal blood heat of the human body. Prolonged air temperatures above this level place a considerable strain upon the cooling mechanism of the body. Above this level a draught becomes a heating rather than a cooling mechanism and one may find it necessary to *close* the windows of a moving vehicle.

# The diurnal variation of temperature

It has been emphasised above that, except for two very restricted areas near the poles, the rotation of the earth involves the reception of thermal energy from the sun at the earth's surface in a regular, unremitting, rhythmic beat. This pulsation is expressed in a diurnal variation of temperature, often heavily damped by atmospheric influences, yet unquestionably the primary feature of surface temperatures. It is a feature so obvious, and apparently so elementary, that it has suffered from comparative neglect in the literature. It is nevertheless surprising to discover that the observational data available for its analysis are rarely adequate and often frankly misleading. Yet here is a factor of unquestionable importance in soil economy and of some considerable significance in regard to landing and take-off conditions at aerodromes. The demand for worldwide information for this latter purpose during the 1939-45 war could only be satisfied by a good deal of inspired guesswork.

There are reasons for this neglect. In the first place, some three-quarters of the earth's surface is covered by the ocean and both within and over the sea the solar impulse is damped down to such a degree that it is almost negligible—the diurnal range of temperature is of the order of 1–2°F or about 1°C. A low range of variation is therefore also characteristic of all coastlands which are under the sway of marine winds. This virtual elimination of diurnal variation over the sea follows mainly from (1) the comparative transparency of water bodies which spreads most of the daily income over a layer about 10 metres or some 35 feet deep (see p. 40) instead of retaining it within a few inches of the surface as on land, and (2) the mobility of liquids which permits convectional overturning especially when the surface waters are chilled by radiation at night. Three other factors, already mentioned above, are also brought into play, namely (*a*) the very heavy draft on income made by evaporation, a most effective method of 'banking' solar energy for release on some future occasion, (*b*) the large capacity for heat of water which is about twice that of the rock-forming minerals, volume for volume, and (*c*) the effectiveness of the moist air over the sea in intercepting back-radiation during the night. These are also brought into play with comparable effectiveness over forest and swamps in the hearts of continents.

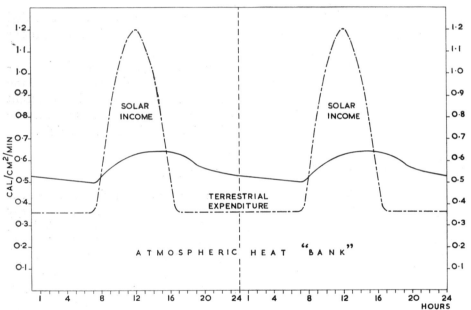

FIG. 22. The diurnal surge of solar income and its relation to the atmospheric heat bank

Secondly, heavy cloud, by reflecting solar radiation and absorbing earth radiation at night, effectively damps the diurnal variation of temperature—locally and for a limited time—as Brunt (1942, 33–5) has so clearly shown.

Finally in temperate and polar latitudes, particularly in winter when the sun is low, the *advection* or transport of heat (or 'cold'!) in moving air masses may so counteract the diurnal tendency that it is scarcely recognisable. Yet the tendency is always there so long as the sun rises above the horizon. Day is not identical with night, even over the sea, and a host of delicate sea creatures respond to the change. So do the larger fish which prey upon them, as every fisherman knows.

Figure 22, it is hoped, will help the student to appreciate the nature and origin of the diurnal variation of temperature. Strictly speaking it is only a diagram but actual values for a South African station (Maun 19°59′ S—Ht 945 m) with a diurnal range of 18° C or 33° F in June (a 'winter' but almost cloudless month) have played some part in its construction. The broken line represents radiative income received at the surface of the earth. It consists of two items, a uniform contribution of some 0.36 cal/cm²/min received day and night from an atmosphere assumed to remain unchanged at an average temperature of about $-15°$ C and the additional bursts of direct solar income received during the day. The full line represents outgoing radiation from the earth as estimated in accordance with Stefan's Law (see p. 37). A number of approximations are involved but the diagram is so drawn that the net solar income, received during 8 hours (of the 10-hour day), precisely balances net terrestrial expenditure during the remaining 16 hours. There is therefore no overall gain or loss. Such gains or losses are, of course, incurred from day to day but they are slight in comparison to the reserve in the atmospheric 'bank'; in fact they only become really effective when they occur progressively in spring and autumn. Although the diagram is not scaled for temperature the form of the diurnal curve is illustrated by the full line. Surface temperatures are seen to be highest at about 1500 hours as the solar bonus is drawing to its close. They remain high as the heated earth draws upon its own reserves by conduction but fall very rapidly (in fact at Maun by 7° C or nearly 13° F) between 1800 and 2000 hours. The fall then continues at a slower rate (8° C for the 11-hour period between 2000 and 0700 hours) as a better balance between the earth and the lower atmosphere is attained. The minimum is reached at sunrise. The rise is particularly rapid between 0800 and 1000 hours (actually at Maun as much as 9° C) and thereafter continues at a decelerating rate as convection currents are set up near the surface, i.e. as payments are made to replenish the atmospheric reserve.

The same essential features are to be observed in all diurnal curves though with infinite variety in detail. The practical difficulty is that to isolate the purely periodic element of temperature change we must be able to strike an average value for each of the 24 hours of the day, thus eliminating aperiodic or 'chance' variations. In many countries detailed hourly values are not published. They are often readily available only in the reports of the major astronomical observatories. Until recently only four stations in Britain supplied such data. We have used 10-year records from four west European observatories—Coimbra *c.* 40°N (1939–48), Potsdam *c.* 52½°N (1921–30), Uppsala *c.* 60°N (1941–50), and Abisko *c.* 68°N (1920–29)—to carry our analysis a stage further (see Figs. 23, 24, 25 and 26). Each graph shows the mean diurnal variation for the months of July and January,[1] calculated for the whole period, together with the means for the two months within this period which gave the highest and lowest daily mean values. The time interval between sunrise and sunset is also indicated.

The mean curves for *Coimbra* are typical for a comparatively sunny locality

---

[1] Except Abisko.

not far from the sea in middle latitudes. The diurnal range is about 12°C in July and 6°C in January. In July the temperature is rising for nine hours from 0500 to 1400 and falling during the remaining 15 hours. The morning rise is of the order of 1.5 to 2°C, 3 to 4°F per hour, the afternoon fall is only slightly less steep but from about 2 hours after sunset to the minimum at 0500 hours it amounts to no more than 0.3°C, 0.5°F, per hour. In January the temperature rises for seven hours at a rate which reaches 1.5°C per hour at midmorning, and falls for 17 hours, the rate of fall again slowing off appreciably about two hours

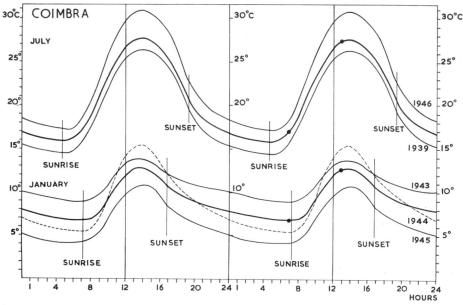

FIG. 23. The diurnal variation of temperature at Coimbra (40° 12′ N)

after sunset. All gradients are less steep than during the comparable times of day in July.

The diagram also shows the record for July 1946 and July 1939, respectively the warmest and coolest Julys during this comparatively short period. It is thus clear that the mean values obscure a variation from year to year of the order of ±2.5°C at midday and of ±1.5°C during the night. This is quite irrespective of any variation from day to day which is completely lost in this method of analysis and must be approached by an entirely different technique. It is well that the student of climatology should realise at a very early stage that even mean values are subject to a recognisable degree of variation which must be appreciated if they are to be interpreted aright. Beside the January curve we have also plotted the records for January 1943 and January 1945, again the months of highest and lowest mean daily values. Note that at this season there is apparently more variation during the night (±2.5°C) than during the shortened day ($\pm 1.5°$C). However this is not the whole story for in January 1944 (shown by the broken line), though the nights were cooler than the average, the days were appreciably warmer, so that the mean range during that month was 10°C, nearly as great as a normal July.

Figure 23 thus introduces us to a number of important concepts. (1) There is clearly the possibility of a variation of regional temperature from year to year, a variation which has no obvious connection with immediate solar activity and so is probably a result of the atmospheric circulation. (2) The

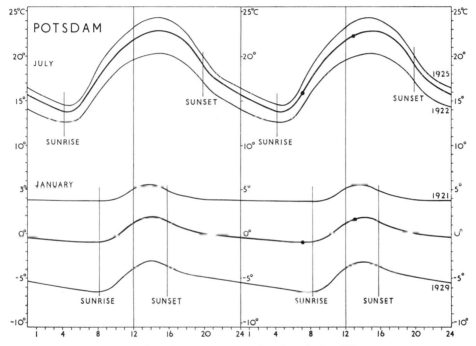

Fig. 24. The diurnal variation of temperature at Potsdam (52° 23′N)

solar cycle may be damped by maritime winds and cloud as in January 1943 (range scarcely 5°C) and yet the mean temperature for the month may be remarkably high. (3) In exceptionally clear seasons, such as January 1944, cool nights may be followed by unusually warm days, the range being twice that of the preceding year. (4) It has become clear that monthly temperatures, though variable in themselves over an appreciable range, may often conceal diurnal variations of considerable proportions and diverse types.

At *Potsdam* (Fig. 24) the effects of an increase in latitude of over 12° are at once apparent. Whereas at Coimbra midday temperatures in warm winter months occasionally exceed those recorded at night during cooler summers, here the two seasons fall clearly into quite different categories. There is no great contrast between the July curves though the mean diurnal range is

57

down to 9°C and, in response to longer days, the temperature rise extends from about 0500 to 1500 hours. In January, however, the mean daily range is down to less than 3°C and the morning rise occupies scarcely 6 hours. The subsequent fall is again most marked for two hours after sunset and then continues very slowly during the rest of the night. The curves for the July of 1925 and 1922 show a moderate range of variability, especially at night, but the range shown by the 1921 and 1929 curves in January is of the order of ±5°C.

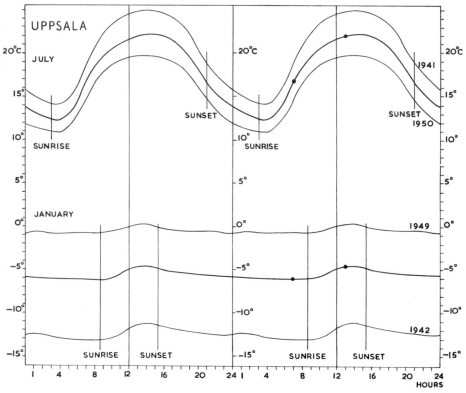

FIG. 25. The diurnal variation of temperature at Uppsala (59° 51′ N)

At *Uppsala* (Fig. 25) nearly twenty degrees nearer the pole than Coimbra and in a more continental location than Potsdam, the seasonal contrast is still more marked. Day now occupies 18 hours in July. The mean diurnal range is up to nearly 10°C (18°F) and the early morning rise in temperature is as rapid as at Coimbra but it is checked at about 0900 hours and then climbs at only some 0.5°C per hour until about 1500 hours. The break in the rate of fall two hours after sunset can be recognised, but it is almost lost in the comparative brevity of the night. The range of variation shown by the records of July 1941 and 1950 is closer to that of Coimbra than to that of Potsdam but the flat-topped curve of 1950 suggests that a cool July at Uppsala is cloudy,

particularly around midday. The mean diurnal range in January is down to
1.5° C, and temperatures below freezing point are characteristic of all years.
Yet temperatures in January 1949 were 11 to 12° C (22–22° F) higher than in
January 1942! It is even more remarkable that the short day had most effect
on the curve of the *colder* month. This carries the implication that cold winters
are clearer than mild ones but, even at best, it is evident that the winter
climate of Uppsala is almost divorced from solar control. The wide range of
winter temperatures at Potsdam and Uppsala points to advection of warm or
cold air from other latitudes as the dominant factor in the temperature

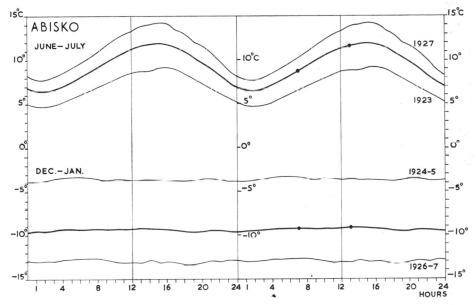

FIG. 26. The diurnal variation of temperature at Abisko (68° 20′ N)

régime. The sun may bring some light but its heating power is almost negligible.

Figure 26 rounds off the picture with the record for a station where the
sun scarcely sets in June to July and scarcely rises throughout December to
January. We have therefore plotted the hourly temperatures averaged over
2-month periods in this example. The mean diurnal range in June to July is
nearly 5.5° C, the maximum being as late as 1500 hours and the minimum as
early as 0200 hours. Temperatures thus rise for 13 hours and fall during 11, the
fall being slightly steeper than the rise, a rather unusual feature. Note that
combining two months should reduce the variability but the graphs show that
this remains at about the same scale (±2.5° C) as that recognised during the
summer at the other stations. In fact, during the period under review,

59

July was much more variable (±4°C) than either June or August. In December to January, on the other hand, diurnal variation has virtually disappeared and only a slight random variation of the hourly means is recognisable. The dominant fact again is the variation from year to year, the mean value for the period December 1924 to January 1925 being some 9°C (16°F) above that for December 1926 to January 1927. Such prolonged contrasts must result entirely from the pattern of the general circulation. A strange tendency for temperatures during the polar night to be slightly *lower* between 1000 and 2000 hours than at other times is faintly revealed by the records of the coldest months at Abisko. The range involved is so slight (less than 1°C) that one would be tempted to dismiss it as insignificant but for the fact that similar observations have been made on the fringe of the Antarctic continent by G. Simpson (Conrad, 1936, 170-2). We mention it here merely as a climatological curiosity not yet completely understood though it is probably related to atmospheric pressure.

Table 6 illustrates how the diurnal variation of temperature can be most conveniently expressed in numerical form. The values quoted at 2-hourly intervals are the 'deviations' of mean temperature at that time from the overall mean for 24 hours, quoted above each section. To facilitate comparisons with England, Potsdam has been replaced by Kew (London). For Abisko, January being featureless, we have quoted February values instead. On the righthand side of the table the *true diurnal range* of temperature is compared with the difference between mean daily maxima and minima which is often mistakenly regarded as a useful substitute. The evident anomalies are discussed later (p. 65-66).

It will be wise to supplement this analysis of the diurnal variation of temperature at four different latitudes by a brief inspection of Fig. 27 which shows the daily variation in the warmest and coolest months of the year at five South African stations. All five stations lie between 28½ and 30°S but whereas Alexanderbaai and Durban are near sea level on the Atlantic and Indian Ocean coasts respectively, the other three stations, O'okiep, Kimberley and Ladysmith, are in the interior at elevations of 930, 1197 and 1069 metres respectively. The diagram contains a number of interesting features which anticipate to some degree later parts of our discussion but it is useful to enumerate them at this point.

1. Although Durban lies about 85 miles farther from the Equator than Alexanderbaai mean temperatures there are a few degrees higher than at the latter station throughout the year. This reflects a contrast between the eastern and western coasts of continents which is almost universal in these latitudes. We shall find that it is related to the oceanic circulation.

2. At both Alexanderbaai and Durban the mean diurnal range is rather larger during the cooler season than during the warmer season. This is also by no means uncommon in tropical and near-tropical latitudes where the relatively slight seasonal changes in solar income are easily overcome by seasonal contrasts in rainfall and its associated cloud cover. It will be observed

that the same is true of two of the three interior stations. O'okiep is the exception which gives point to the rule since it is the only station with maximum rainfall in the cooler season (precipitation at Alexanderbaai being negligible) and the only station with less midday cloud in February than in July.

3. At the three interior stations in the warmer season, despite their very considerable elevations above sea-level, afternoon temperatures rise well above those at Durban whilst night temperatures are no lower than at Alexanderbaai. The mean diurnal range is thus of the order of 12°C or 22°F.

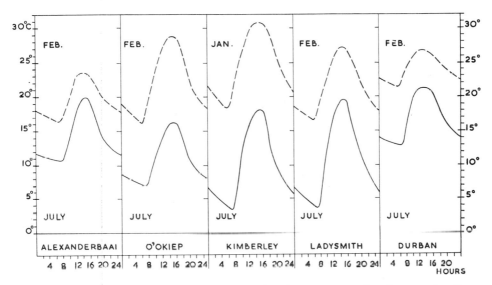

FIG. 27. The diurnal variation of temperature during the warmest and coolest months of the year at five stations along a cross-section of southern Africa

4. In the cooler season, on the other hand, although afternoon temperatures in the interior are only 3 or 4°C below those at coastal stations, night temperatures especially at Ladysmith and Kimberley fall to within striking distance of the freezing point. The mean diurnal range at Ladysmith is then 16°C (29°F). Note that we have been dealing throughout this discussion with *mean* temperatures. In actual fact Kimberley records about eight or nine frosty nights each July whilst Ladysmith records six or seven; the average number of frosty nights per year at the two stations being twenty and fourteen respectively. In view of the great daily range at these stations it is not to be anticipated that low temperatures will persist for more than a few hours nor that the frost should ever become really severe. The lowest temperatures ever recorded in July at these places were −6.8 and −4.3°C respectively.

5. In conclusion, note that all five stations record midday temperatures in 'winter' that are either approximately equal to or in excess of the night

TABLE 6.   *The true diurnal variation of temperature (deviations in °C)*

| HOURS | 0 | 2 | 4 | 6 | 8 | 10 | 12 | 14 | 16 | 18 | 20 | 22 | RANGE (°C) | MAX– MIN (°C) |
|---|---|---|---|---|---|---|---|---|---|---|---|---|---|---|
| ABISKO 68°20′N (Means: *February* −10.5°C; *July* 11.5°C) | | | | | | | | | | | | | | |
| February | −0.3 | −0.3 | −0.3 | −0.4 | −0.3 | 0.2 | 0.8 | *0.9* | 0.2 | 0.0 | −0.1 | −0.2 | 1.3 | 8.2 |
| July | −2.6 | −3.0 | −2.3 | −1.3 | −0.1 | 1.2 | 2.1 | *2.5* | 2.7 | 1.8 | 0.3 | −0.3 | 5.7 | 9.0 |
| UPPSALA 59°51′N (Means: *January* −5.7°C; *July* +17.6°C) | | | | | | | | | | | | | | |
| January | −0.1 | −0.3 | −0.4 | −0.5 | −0.6 | −0.3 | 0.6 | *1.0* | 0.5 | 0.2 | 0.1 | 0.0 | 1.6 | 5.7 |
| July | −3.9 | −4.9 | −5.3 | −3.2 | −0.7 | 2.9 | 3.8 | *4.5* | 4.3 | 3.1 | 0.4 | −2.3 | 9.8 | 11.9 |
| KEW 51°29′N (Means: *January* 5°C; *July* 17.2°C) | | | | | | | | | | | | | | |
| January | −0.45 | −0.6 | −0.65 | −0.8 | −0.8 | −0.1 | 0.95 | *1.6* | 1.1 | 0.4 | 0.05 | −0.2 | 2.4 | 5 |
| July | −2.4 | −3.2 | −3.8 | −2.9 | −0.9 | 1.0 | 2.4 | *3.4* | 3.5 | 2.8 | 0.65 | −1.2 | 7.3 | 9 |
| COIMBRA 40°12′N (Means: *January* 9.0°C; *July* 20.7°C) | | | | | | | | | | | | | | |
| January | −1.1 | −1.7 | −2.0 | −2.2 | −2.1 | 0.0 | 2.9 | *3.7* | 2.5 | 0.8 | −0.1 | −0.7 | 5.9 | 8.3 |
| July | −3.9 | −4.6 | −4.8 | −4.6 | −1.8 | 2.7 | 6.0 | *7.0* | 5.8 | 3.1 | −1.2 | −3.0 | 11.8 | 14 |

temperatures recorded in 'summer'. When this occurs the distinction between the seasons, at least on temperature grounds, is clearly becoming weak—hence the inverted commas and the avoidance of these two terms in the above paragraphs. This suggests a useful criterion for distinguishing between near-tropical lands on the one hand from warm-temperate lands on the other.

# Estimates of the diurnal range of temperature

In the source-books of climatological data fully tabulated hourly temperatures are comparatively rare, partly because the material is bulky and partly because the information is not often available. We are supplied instead with two alternative sets of data: (*a*) temperatures at fixed hours, often at 0700 and 1300 hours local time though with increasing frequency at 0800 and 1400 hours, and (*b*) mean daily maxima and minima derived from standard maximum and minimum thermometers. How close do such records take us to the true periodic range? To what extent do they offer a complete measure of the thermal effects of the diurnal pulsation of solar income?

It may clarify the presentation if we state the answer to these questions in advance. Method (*a*) depends on the relation of the two hours selected to the current length of day; it may thus be adequate at one season but not at another. It can never give a result in excess of the true range but may often err on the small side. Method (*b*), on the other hand, can never give too small a range but almost everywhere it yields values in excess of the variation which we are endeavouring to assess. Where both types of information are available we may thus take comfort in the thought that it is at least within our powers to straddle the target! What may surprise the inexperienced investigator is the extent to which the two answers may differ, especially in middle and high latitudes.

(*a*) *Fixed-hour observations.* A glance at the righthand sides of Figs. 23–26, where the 0700- and 1300-hour values have been clearly marked, reveals clearly enough the weakness of this approach. Observation hours are selected to satisfy certain standard formulae for the derivation of daily means, they are not intended to mark the turning points of the diurnal curve. At *Coimbra* in January it happens that they do this well enough but in July the thermometer has been rising since sunrise, i.e. about 0500 hours, and the difference between the values at 1300 and 0700 hours omits this part of the curve. At both *Potsdam* and *Uppsala* in July when the sun rises early the morning error is considerably greater and since maximum temperatures are not recorded until 1500 hours there is another slight omission in the early afternoon. At Uppsala it is apparent at a glance from Fig. 25 that the true diurnal variation is nearly twice the difference between the 1300- and 0700-hour readings. It is equally clear that in June–July at *Abisko* the time interval between 0700 and 1300 hours arbitrarily cuts away a sector of the diurnal curve entirely lacking any logical relationship either to its turning points or to its zone of maximum slope.

63

Similar criticisms apply, indeed with greater force, to the use of observations at 0800 and 1400 hours. It is equally evident that observations at the evening hour, usually either at 1800 or 1900 local time, add nothing of value to our present purpose.

How far is the difference between observations at the morning and afternoon hours a *fixed proportion* of the true diurnal range at any one time and in any one latitude? To the extent that this is true it should be possible to apply empirically determined factors to fixed-hour observations and thus arrive at a close approximation to the range we desire to express. Since the dominant

TABLE 7. *Relation of fixed-hour observations to true periodic range of temperature*

| | JAN. | FEB. | MAR. | APRIL | MAY | JUNE | JULY | AUG. | SEPT. | OCT. | NOV. | DEC. |
|---|---|---|---|---|---|---|---|---|---|---|---|---|
| *a.* Multiply 1300–0700 range by (°N) | | | | | | | | | | | | |
| 70 | 1.0 | 1.0 | 1.1 | 1.6 | 1.8 | 1.8 | 1.8 | 1.8 | 1.3 | 1.1 | 1.0 | 1.0 |
| 60 | 1.0 | 1.1 | 1.1 | 1.4 | 1.7 | 1.8 | 1.7 | 1.5 | 1.2 | 1.1 | 1.0 | 1.0 |
| 50 | 1.0 | 1.1 | 1.1 | 1.2 | 1.4 | 1.5 | 1.5 | 1.3 | 1.1 | 1.1 | 1.0 | 1.0 |
| 40 | 1.0 | 1.1 | 1.1 | 1.2 | 1.3 | 1.4 | 1.3 | 1.2 | 1.1 | 1.1 | 1.0 | 1.0 |
| 30 | 1.1 | 1.1 | 1.1 | 1.2 | 1.3 | 1.3 | 1.2 | 1.2 | 1.1 | 1.1 | 1.1 | 1.1 |
| 20 | 1.1 | 1.1 | 1.1 | 1.1 | 1.2 | 1.2 | 1.2 | 1.1 | 1.1 | 1.1 | 1.1 | 1.1 |
| *b.* Multiply 1400–0800 range by (°N) | | | | | | | | | | | | |
| 70 | 1.0 | 1.1 | 1.4 | 1.9 | 2.2 | 2.2 | 2.2 | 2.0 | 1.6 | 1.3 | 1.2 | 1.0 |
| 60 | 1.0 | 1.0 | 1.2 | 1.6 | 2.2 | 2.2 | 2.2 | 1.8 | 1.4 | 1.1 | 1.0 | 1.0 |
| 50 | 1.0 | 1.0 | 1.2 | 1.4 | 1.8 | 1.9 | 1.8 | 1.5 | 1.3 | 1.1 | 1.0 | 1.0 |
| 40 | 1.1 | 1.1 | 1.2 | 1.4 | 1.7 | 1.7 | 1.6 | 1.5 | 1.3 | 1.2 | 1.1 | 1.1 |
| 30 | 1.2 | 1.2 | 1.4 | 1.5 | 1.6 | 1.6 | 1.5 | 1.5 | 1.3 | 1.2 | 1.2 | 1.2 |
| 20 | 1.2 | 1.2 | 1.4 | 1.5 | 1.5 | 1.5 | 1.4 | 1.4 | 1.3 | 1.2 | 1.2 | 1.2 |

*Read for southern hemisphere:*

| JULY | AUG. | SEPT. | OCT. | NOV. | DEC. | JAN. | FEB. | MAR. | APRIL | MAY | JUNE |
|---|---|---|---|---|---|---|---|---|---|---|---|

control on the form of the diurnal curve is exercised by the times of sunrise and sunset this possibility has been explored by the author and the resulting factors are given in Table 7. These factors will yield comparatively satisfactory estimates for most geographical locations.

(*b*) *Mean daily maxima and minima.* Since the maximum and minimum thermometers are read once a day and thus give the highest and lowest temperatures experienced during a 24-hour period, they would seem, at first glance, to give just the information we are seeking. There is indeed nothing inherently at fault in any such single reading, it states two precise facts about the last twenty-four hours which may well be of considerable climatological importance. The 'error' involved is a purely statistical one; it is introduced when these values are *averaged* out over a month for convenient quotation in the yearbooks. An average has validity only when it is composed of like quantities, that is, of measurements of similar things, of things with some attribute in common. This is true of temperature records at any fixed hour but it is *not* strictly true of 'minimum' temperatures which may occur on one day at 0200 hours, on the next at 0700 hours and on the day following perhaps at 2000 hours. As long ago as 1897 J. Hann distinguished clearly between the 'periodic' variation we have been discussing and the 'aperiodic' or inter-diurnal variation of temperature which is of a much more random nature and associated with an entirely different set of causal factors. The real difficulty with the maximum–minimum range is that it is sensitive to *both* of these variations at one and the same time and thus gives no clear picture of either.

TABLE 8. *Relation of maximum–minimum range to true periodic range of temperature*

| | JAN. | FEB. | MAR. | APRIL | MAY | JUNE | JULY | AUG. | SEPT. | OCT. | NOV. | DEC. |
|---|---|---|---|---|---|---|---|---|---|---|---|---|
| Apply the following percentages to Maximum–Minimum range: | | | | | | | | | | | | |
| (°N) | | | | | | | | | | | | |
| 75 | 0 | 0 | 20 | 45 | 55 | 60 | 55 | 50 | 45 | 10 | 0 | 0 |
| 70 | 0 | 10 | 30 | 55 | 60 | 60 | 60 | 55 | 50 | 25 | 5 | 0 |
| 65 | 10 | 30 | 55 | 70 | 70 | 70 | 70 | 70 | 65 | 50 | 15 | 5 |
| 60 | 30 | 50 | 70 | 75 | 75 | 75 | 75 | 75 | 75 | 70 | 35 | 20 |
| 55 | 40 | 55 | 70 | 75 | 80 | 80 | 80 | 80 | 75 | 70 | 50 | 30 |
| 50 | 55 | 75 | 75 | 80 | 85 | 85 | 85 | 85 | 80 | 75 | 65 | 50 |
| 45 | 65 | 75 | 75 | 80 | 80 | 85 | 85 | 85 | 80 | 75 | 70 | 65 |
| 40 | 70 | 75 | 75 | 80 | 80 | 85 | 85 | 85 | 80 | 80 | 75 | 70 |
| 30 | 80 | 85 | 85 | 85 | 85 | 85 | 85 | 85 | 85 | 85 | 80 | 80 |
| 20 | 90 | 90 | 90 | 85 | 80 | 80 | 80 | 80 | 85 | 90 | 90 | 90 |

If the reader turns to any curve in Figs. 23–26 and tries to visualise how it would run if the second day had a mean temperature a few degrees above or below the first, he will not find it difficult to appreciate that the distortion involved in both cases would produce a greater contrast between the maximum and minimum values than is justified by the periodic variation which the diagrams are actually designed to show. Whatever the trend, therefore (and interdiurnal variations are usually upward for a few days and then downward again in a rather irregular fashion) the maximum–minimum range will have captured a fraction of this variation *in addition* to the periodic variation we are seeking to examine. This is why the mean periodic range is always less than the maximum–minimum range. Under no circumstances can it be greater.

Yet this type of information is readily available even for third-order climatological stations. Can we not devise some convenient reduction factors which will make it possible to effect an approximate conversion? The task is less promising than the conversion of fixed-hour data since random interdiurnal variations of temperature lack the necessary close relationship with latitude that we had the right to expect of the times of sunrise and sunset. Nevertheless an examination of a considerable number of records in the western hemisphere suggests that the factors of Table 8 have a broad validity.

It will be noted that the mean maximum–minimum range is particularly deceptive during the winter months in middle and high latitudes (see Table 6). A particularly striking instance of this is provided by the record at Grön-fjord in Spitzbergen (78°02′N 14°15′E) where fixed-hour observations at 0200 hours provide a useful check on both methods. For want of better information we have applied the factors quoted for 75°N. Results are given to the nearest degree Fahrenheit in Table 9. Note the wide maximum–minimum range during the months of November, December and January when the sun never appears above the horizon at Grönfjord. These figures must thus be due exclusively to interdiurnal variation. Appearing where it is

TABLE 9. *The diurnal variation of temperature at Grönfjord, Spitzbergen (78°02′N, 14°15′E) (°F)*

| | FIXED HOUR OBS. | | | MEAN DAILY | | RANGE | | ESTIMATED RANGE | | RANGE |
|---|---|---|---|---|---|---|---|---|---|---|
| | 1400 | 0800 | 0200 | max. | min. | 1400–0800 | Max.–Min. | (a) | (b) | 1400–0200 |
| Jan. | 3 | 3 | 4 | 10 | − 4 | 0 | 14 | 0 | 0 | 1 |
| Feb. | 1 | 1 | 0 | 8 | − 7 | 0 | 15 | 0 | 0 | 1 |
| Mar. | 0 | − 3 | − 3 | 5 | −12 | 3 | 17 | 4 | 3 | 3 |
| April | 12 | 6 | 3 | 15 | − 3 | 6 | 18 | 11 | 8 | 9 |
| May | 28 | 23 | 18 | 30 | 15 | 5 | 15 | 10 | 8 | 10 |
| June | 38 | 36 | 33 | 40 | 31 | 2 | 9 | 5 | 5 | 5 |
| July | 44 | 42 | 39 | 46 | 38 | 2 | 8 | 4 | 4 | 5 |
| Aug. | 43 | 40 | 38 | 44 | 37 | 3 | 7 | 5 | 4 | 5 |
| Sept. | 34 | 32 | 31 | 35 | 29 | 1 | 6 | 3 | 3 | 3 |
| Oct. | 22 | 21 | 21 | 25 | 17 | 1 | 8 | 1 | 1 | 1 |
| Nov. | 11 | 11 | 11 | 17 | 6 | 0 | 11 | 0 | 0 | 0 |
| Dec. | 7 | 7 | 7 | 14 | 1 | 0 | 13 | 0 | 0 | 0 |

(a) Results from application of factors in Table 7b (70°N)
(b) Results from use of Table 8 (75°N)

not wanted this factor entirely obscures the most interesting characteristics of the true diurnal variation at this polar station shown by the last three columns. April and May are comparatively clear but fog and mist damp down the effects of insolation as soon as the sea ice begins to disperse.

# The interdiurnal variation of temperature

We now leave the regular daily pulsation of temperature related to insolation and hence ultimately to the facts of astronomy and turn to the irregular, day-to-day changes in temperature which must evidently stem from an entirely different set of causes. We have seen that an increase in the density of the cloud-screen has some effect upon surface temperatures, lowering the after-noon maximum and raising the early morning minimum, but substantial changes in mean daily temperature from one day to another must be the result of a much more potent factor. This is the wholesale importation (advection) of large volumes of warmer or colder air from distant parts of the earth's surface. It is thus a function of the wind and of all that wind movement involves. For the present we are concerned only to recognise its temperature effects.

Figure 28 illustrates the nature of this variation at two Swedish stations, Gällivare (67°20′N) and Jönköping (57°47′N), during the 365 days from 1 November 1945 to 30 October 1946. Since the values plotted are mean temperatures for each day the diurnal factor has been completely eliminated.

At both stations the curve is extremely irregular though the range of variation is evidently greater in winter than in summer and greater at the northern than at the more southerly locality. This at once suggests that we are dealing with a factor which increases in significance at precisely those times and places when insolation is least effective. Occasionally the variation swings from day to day—high, low, high, low; but more frequently the upward or downward tendency persists for two or more successive days—high, lower, low, higher, and so on in a variety of patterns. Often the swings are not symmetrical and although each fall is eventually followed by a recovery there is no evidence at all of a regular rhythm except that shown by the year as a whole. At Gällivare three very striking features are the fall of 23°C from 25 November to 1 December, followed by a recovery in two days, the fall, rise and fall between 3 and 20 January with a range of 25°C, and the fall of 19°C and rise of 25°C between 5 and 14 February. Note that not one of these features is clearly reflected in the curve for Jönköping—in fact, as the temperature rose 14.5°C between 11 and 15 January at Gällivare, it fell 13°C at Jönköping some 700 miles farther south. On the other hand, the cold waves of late February and mid-March and the first onset of winter at the end of September were registered at both stations with about equal intensity. Clearly the operative factors behind such variations must vary widely in duration, direction and scope. They are also obviously of the greatest importance.

Statistical analysis of such irregular changes must clearly follow a pattern entirely different from that employed in the preceding section. J. Hann (1903, 20) suggested that we 'determine the differences of temperature from one day to the next for a whole month and obtain the mean of these differences', the mean of such values for the same month over a series of years being called the 'normal variability of temperature' for the given station and the given month. This has been done for the 5-year period 1943-47 at Gällivare, Jönköping, and Zurich and the results are given in Table 10 along with similar values for Vienna and Naples (for a much earlier period) quoted by Conrad (1936, 100-1). Maps of interdiurnal variability for Siberia in this latter volume (p. 101) and in Visher (1954, map 350), suggest that the most variable temperatures in the whole world are experienced in the Ob River basin and on the north-central prairies of Canada and the United States where mean values reach 5.5°C (10°F) during the winter months.

The reader is entitled to feel that even these values are not very impressive. It must be confessed indeed that Table 10 gives a disappointing representation of the very striking features portrayed in Fig. 28. Without consulting the

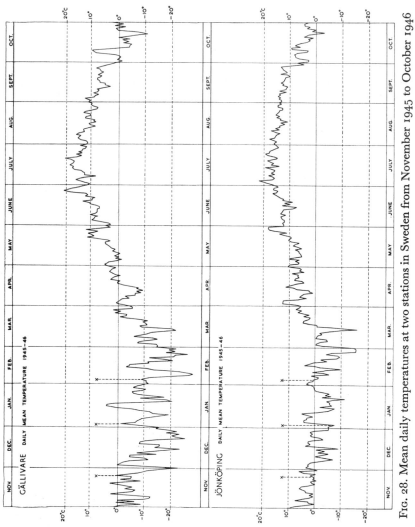

FIG. 28. Mean daily temperatures at two stations in Sweden from November 1945 to October 1946

graph it would be virtually impossible to visualise what the tabulated values entail; they have taken on much of the nature of abstract indices and their significance is almost entirely comparative. This may be partly because the comparison of successive days involves too short a time-span to do justice to the type of variation involved. Section (*b*) of Table 10 shows the results of a similar comparison of *alternate* days at our three stations. This certainly increases the scale of the contrasts and underlines the significance of 'weather', especially in winter, but without similar, and very laborious, computations for other stations the values remain of limited interest. In fact the values in the new series approximate very closely to those in Section (*a*) multiplied throughout by the square-root of 2. (See Appendix 1, p. 524.)

We are compelled to conclude that the force of this type of analysis is largely lost if the results are subjected to the process of averaging. Indeed Hann (1903, 21) continues: 'There are certain other data which are of even greater importance than the mean of the differences between successive daily means, especially when we wish to make comparisons of two or more climates, and

TABLE 10. *Mean interdiurnal variation of temperature* (°C)

| | JAN. | FEB. | MAR. | APRIL | MAY | JUNE | JULY | AUG. | SEPT. | OCT. | NOV. | DEC. | YEAR |
|---|---|---|---|---|---|---|---|---|---|---|---|---|---|
| (*a*) *Successive days* | | | | | | | | | | | | | |
| Gällivare | 3.5 | 3.6 | 3.1 | 1.9 | 1.8 | 2.1 | 1.9 | 1.5 | 1.4 | 2.5 | 3.6 | 3.9 | 2.6 |
| Jönköping | 2.5 | 2.2 | 2.3 | 1.9 | 1.9 | 1.6 | 1.5 | 1.4 | 1.6 | 2.2 | 2.0 | 2.1 | 1.9 |
| Zürich | 2.2 | 1.5 | 1.9 | 2.1 | 2.1 | 2.2 | 2.1 | 1.6 | 1.5 | 1.4 | 1.9 | 1.7 | 1.8 |
| Vienna | 2.9 | 2.7 | 2.6 | 2.1 | 1.9 | 1.9 | 1.7 | 1.8 | 2.2 | 2.7 | 2.3 | 2.6 | 2.3 |
| Naples | 1.6 | 1.9 | 1.6 | 1.2 | 1.4 | 1.3 | 1.2 | 1.2 | 1.3 | 1.6 | 1.6 | 1.8 | 1.5 |
| (*b*) *Alternate days* | | | | | | | | | | | | | |
| Gällivare | 4.9 | 5.1 | 4.4 | 2.7 | 2.5 | 3.0 | 2.9 | 2.1 | 1.8 | 3.2 | 4.7 | 5.3 | 3.5 |
| Jönköping | 3.3 | 3.1 | 3.2 | 2.6 | 2.0 | 2.0 | 1.8 | 2.0 | 2.0 | 2.9 | 2.9 | 2.6 | 2.5 |
| Zürich | 3.4 | 2.4 | 2.8 | 3.2 | 3.3 | 3.1 | 3.1 | 2.2 | 2.3 | 2.1 | 2.5 | 2.6 | 2.7 |

to give a vivid description of the variability of the temperature. These additional data concern the number of times in each month that these differences . . . reach a given value.' In modern statistical terms he is stressing the significance of 'frequency'. Such an approach not only extracts much more information from the data, it is also much more amenable to diagrammatic presentation.

Figures 29 and 30 show the percentage frequency of various temperature differences between both successive and alternate days at Gällivare and Jönköping over the two periods May, June, July, August and November, December, January, February. The heavy line represents the 5-year period 1943-47 and the broken line represents the single year November 1945 to October 1946 already plotted in detail in Fig. 28. The results for the summer period have been grouped by intervals of one Celsius degree but the more widely scattered winter values have been grouped into two-degree intervals so as to improve the smoothness of the gradation. Since the vertical scale of the winter diagrams is precisely half that of the summer diagrams the whole set remain directly comparable. It will be noted at once that during the specimen year plotted in Fig. 28 the variability during the winter months, particularly at Gällivare, was rather greater than usual, the curve spreading more widely to the right than that for the period 1943-47. In the summer, however, the variability at both stations during this year was less than usual and the broken curve is piled more steeply to the left.

The diagrams show at once how inadequate are the average values quoted in Table 10, particularly in winter. At Gällivare for instance, where the mean interdiurnal variation over the four winter months was 3.6°C (6.5°F), it is surely much more significant that on 10 per cent of successive days the contrast was in excess of 8.3°C (15°F) or that on 25 per cent of occasions it exceeded 5.5°C (10°F). The extreme quarter of each curve has been cross-hatched to emphasise this point. The corresponding values for alternate days are 11°C (20°F) and 7.4°C (13.3°F). The other graphs should be read

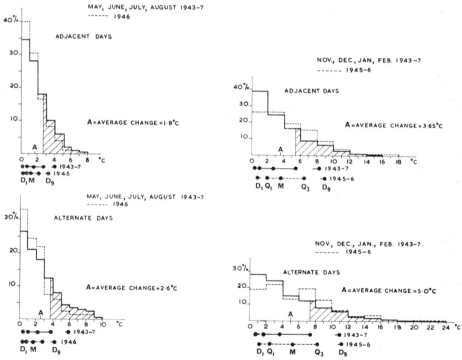

Fig. 29. Percentage frequency groups of the interdiurnal variability of temperature at Gällivare (1943–47) in summer and winter

in similar fashion. Is it not evident that this method of presentation does much greater justice to the variability actually experienced though in a manner a good deal more compact than the historical diagram presented in Fig. 28?

An even more compact method is illustrated by the bars placed beneath each curve where the points marked $D_1$, $Q_1$, $Q_3$, and $D_9$ indicate the points on the temperature scale above which cut off the lowest and highest ten per cent of the instances ($D_1$ and $D_9$) and the lowest and highest quarters of the total number of instances ($Q_1$ and $Q_3$). These points are known to statisticians as the 'deciles' and 'quartiles' of the group. The range between the two quartiles thus includes the most centrally placed half of the total number of observations. It forms a very simple way of expressing the mean scatter of the values re-

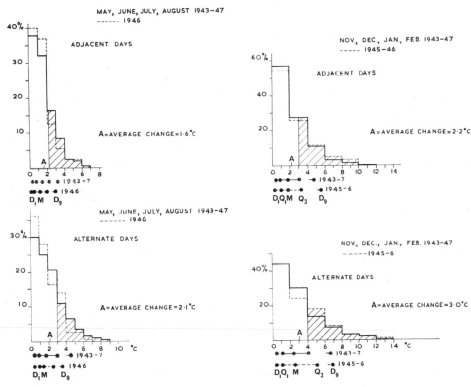

Fig. 30. Percentage frequency groups of the interdiurnal variability of temperature at Jönköping (1943–47) in summer and winter

presented by the various curves. Similarly the point marked $M$, so situated that half of the instances lie above it and half below, is known as the 'median' value. It forms a very useful substitute for the arithmetic average when the analysis is being conducted by graphical rather than by arithmetical processes. Note that it does not necessarily coincide with the arithmetic average. It may be none the worse for that (see Appendix 1).

If we are interested in the most extreme examples of aperiodic variation of temperature regardless of the time interval over which they occur, yet another method of approach is necessary. We must return to an inspection of the daily data, usually after plotting them in an historical diagram like Fig. 28, and count the instances in excess of some selected range. Thus the great variability of winter temperatures at Gällivare is well summarised in the statement that during the 5 years 1943–47 there were eighteen cases of an unbroken rise of mean daily temperature (unbroken, that is, save possibly by a slight diurnal interruption which we are ignoring) in excess of 15°C (27°F) and twenty-six cases of falls of the same magnitude. Alternatively we can express this as an average of about 3.5 such rises and 5 such falls per annum though the pattern varies from year to year. It may not be without meteorological significance that the rise usually covered this 15°C range within 2 to 3 days (mean 2.5 days) whilst the fall occupied 3 to 4 days (mean 3.3 days). At Jönköping changes on such a striking scale were rare indeed but there were 20 rises and 20 falls (i.e. an average of four per year) in excess of 10°C (18°F) during the winter and even two or three such cases in the summer months. Here too the rises were rather more rapid than the falls.

Although these methods of analysis bring us close to reality it is evident that they are time-consuming and rather cumbersome. A survey on a wider canvas, whether in time or space or both, demands a more generalised approach. This is possible via the analysis of mean monthly temperatures to which we now turn. It must now be clear to the reader that the expression of the temperature record in this compact form must conceal much but, properly handled, mean monthly temperatures can be made to give some indication of the features outlined above.

## The seasonal variation of temperature

Turning now to the broad sweep of temperature change through the various seasons of the year we reach the second type of periodic variation related more or less closely to insolation. Figure 8 has demonstrated the seasonal course of insolation for various latitudes and temperature would follow a parallel series of patterns if it were controlled by insolation alone. In actual fact the temperature patterns are damped and distorted in a variety of ways—by the influence of heat storage in large water-bodies, by the presence of land or sea ice, by the advection of warm and cold air through the mechanism of the general circulation, and also by the complex though more local effects of cloud, fog, and precipitation. Nevertheless the overriding influence of the annual course of solar income is usually apparent.

The analysis of the diurnal cycle of temperature was facilitated by the division of the day into twenty-four uniform intervals (hours) but astronomical considerations make a uniform subdivision of the year much more difficult. The year contains approximately 365¼ days and, even if we shelve the problem of leap years, the factors of 365 (5 × 73), are two prime numbers. Some observatories summarise their data as 5-day (pentad) means (see Fig. 9) and a period of this length has proved convenient in making generalisations about pressure distribution but 73 is a cumbersome figure for columns in a table or for class intervals in a graph. A 73-day period, on the other hand, has never proved attractive enough to be given a name, partly because of its length and partly because the odd number of these in the year makes it impossible to compare opposite seasons, e.g. high summer with midwinter. We are thus compelled to fall back upon the twelve traditional months despite their unequal length, their lack of any obvious relationship to the solstices, and the anomalous character of February. The annual cycle of temperature is thus normally presented by tabulated values of the twelve monthly means and the difference between the highest and lowest of these is known as the mean annual range.

It is clear that an average covering 30 or 31 days will go a long way towards smoothing out the aperiodic variations described above and illustrated in Fig. 28. However, such an average is not completely insensitive to this type of variation. A month with two cold spells, e.g. February 1946 at Gällivare, will have a lower mean temperature than a month when only one or no such spell occurred. Warm spells in the summer months, e.g. in May at Jönköping and in June 1946 at Gällivare, will register a parallel effect. Indeed, when we speak of a 'cold winter' or a 'warm summer' we usually mean that one or more such spells stand out clearly in our recollection. Since the occurrence of such spells is largely of a random nature the mean temperatures of each successive January or July, or indeed of any other selected month, may be expected to show a considerable range of variation. To ignore this variation and pass at once to the consideration of long-period means is to throw away a most important part of the climatic record.

We select for analysis in the first place data for two stations in America where parallel and comparatively homogeneous records for the 84-year period 1873-1956 are available for Key West (24°33′N) and Winnipeg (49°53′N); the first in an extremely maritime situation amidst the warm waters of the Gulf of Mexico and the second in the heart of the continent. The records for January, February, July and August at these stations are briefly summarised in Table 11 and the January and July records are analysed graphically in Fig. 31. It will be noticed that the overall range of each of the groups of 84 values for July and August at Key West is no more than 5.0°F (2.8°C). This may be considered of little significance and a mean value provides a satisfactory summary of the group. Extreme ranges of monthly values of the order of between 4–8°F (2.2–4.4°C) appear to be characteristic of most intertropical stations throughout the whole year—clear evidence of the paramount importance of insolation over a wide zone on either side of the equator. However at Key West in 'winter' (temperatures remain so high that the use of this term seems scarcely justified) the range of variation is very much greater owing to the very vigorous air mass movements which occur over the American continent at that season. Nevertheless, the variation at Key West in winter is only of the same order as that experienced in temperate latitudes in summer (compare with July or August at Winnipeg), the season when insolation is particularly effective in those latitudes. The most striking fact

TABLE 11. *Some characteristics of monthly temperatures at Key West and Winnipeg 1873–1956 (to nearest 0.5°F)*

| | KEY WEST | | | | WINNIPEG | | | |
|---|---|---|---|---|---|---|---|---|
| | Jan. | Feb. | July | Aug. | Jan. | Feb. | July | Aug. |
| Highest | 77°('37) | 76°('49) | 86°('81) | 86.5°('56) | 13°('43) | 23°('78) | 75.5°('36) | 70.5°('49) |
| $Q_3$ | 71.5° | 73.0° | 84.0° | 84.5° | 4.0° | 6.0° | 68.5° | 66.5° |
| $Q_1$ | 68.0° | 68.5° | 83.0° | 83.5° | − 8.0° | − 2.0° | 65.5° | 63.0° |
| Lowest | 63°('05) | 63.5°('95) | 81.5°('01) | 81.5°('01)('89) | −16°('83) | −15°('75) | 60.5°('84) | 59.5°('85) |
| Median | 69.8° | 71.0° | 83.5° | 84.0° | − 2.5° | 2.0° | 67.1° | 64.4° |
| Average | 69.3° | 70.7° | 83.6° | 83.9° | − 2.1° | 2.2° | 67.2° | 64.6° |
| Extreme range | 14° | 12.5° | 4.5° | 5.0° | 29° | 38° | 15° | 11° |
| Quartile range | 4° | 4.5° | 1.0° | 1.0° | 12° | 8° | 3° | 4° |
| σ of group | ± 2.8° | ± 2.7° | ± 1.0° | ± 1.0° | ± 7.4° | ± 7.6° | ± 2.6° | ± 2.5° |
| σ of mean (84 yrs) | ± 0.3° | ± 0.3° | ± 0.1° | ± 0.1° | ± 0.8° | ± 0.8° | ± 0.3° | ± 0.3° |
| σ of mean (35 yrs) | ± 0.5° | ± 0.45° | ± 0.15° | ± 0.15° | ± 1.25° | ± 1.3° | ± 0.45° | ± 0.45° |

σ = Standard deviation (see Appendix 1, p. 522)

is the remarkable range of monthly values in January and February at Winnipeg.

Clearly to express a group of values covering a range of 38°F or 21°C (February) or even of 29°F or 16°C (January) by means of a simple average is to indulge in a gross generalisation which hides more information than it yields. If it is felt that this emphasis on the extreme range of the groups leaves us too much at the mercy of the highly exceptional or freak record (to say nothing of possible misprints which are never a remote contingency in material of this character), we may refer to the interquartile range which covers the most central half of the values. The wide scatter of January and February values at Winnipeg is clearly much too significant to be ignored. Fig. 31 makes the same point with a clarity that cannot be gainsaid. The July values have been grouped in intervals of one degree (F) though it is clear that this interval is not very appropriate at Winnipeg. The January values are grouped in two-degree intervals but the vertical scale is halved so that all four curves are directly comparable. Note the close clustering about the mean at Key West in July; this value is truly representative of the group. At Winnipeg in July and at Key West in January, although the spread is wider, there still remains a marked tendency for the values to be heaped about the central mean, which thus retains some significance. What shall we say of the January record at Winnipeg? The curve is not a hill, it is a plateau. No particular significance attaches to the mean value of approximately −2°F (−18.9°C). Is it not clear that any value lying between 9°F (−12.8°C) and −13°F (−25.0°C) is just as representative of January temperatures at Winnipeg? This is the *real* fact which the exclusive use of mean values only serves to obscure.

The scatter of monthly temperature data has implications beyond those simply of the accuracy of climatic description. These are of a statistical nature and are discussed in greater detail in Appendix 1. It is a widely held belief that means for a period as long as 35 years, where they are available, have a peculiar degree of reliability and this period is often selected by official authorities. As a comment on this widespread misconception we quote two of our statistical conclusions in advance.

1. Outside the tropics the accuracy of long-period means of monthly temperature rarely justifies their quotation beyond the nearest degree Fahrenheit or half-degree Celsius. Within the tropics quotation to the nearest half-degree Fahrenheit or quarter-degree Celsius may be in order if the record is a long one.

2. In comparing the 35-year means of temperature for any two months with approximately the same degree of scatter, a difference of less than 1°C (2°F) can scarcely be regarded as significant evidence that one month is really warmer than the other if the months are comparable to July at Winnipeg or January at Key West in their range of variation. In continental interiors in the winter season, the Winnipeg record suggests that differences between such means as great as 2–3°C (4–6°F) may sometimes be no more than the product of pure chance. The interpretation of statistical theory on this matter is not straightforward since it involves the question whether there is any

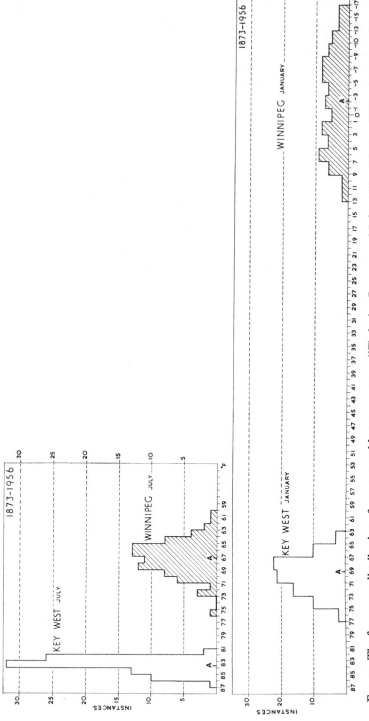

FIG. 31. The frequency distribution of mean monthly temperatures (°F) during January and July at Key West and Winnipeg over the period 1873 to 1956

correlation between the two months (e.g. between two different months at the same station or between the same months at two different stations). The closer the relationship the smaller the range of the chance factor. Nevertheless, Figs. 28–31 have made it quite evident that there is an important element of random variation in temperature data and it is wise for the student to be warned against what is known as 'false accuracy' at the earliest possible stage of his enquiry.

We now turn to Fig. 32 showing the range of variation of monthly temperatures during July and January over a standard period (1891–1940) at twenty-one stations circuiting the earth in moderately high latitudes in the northern hemisphere. Note that the seven Canadian stations lie several degrees south of the others for the want of lengthy records in the northern territories. Temperature is given on the horizontal scale in degrees Celsius and the vertical columns show the number of years out of fifty when the mean monthly temperatures for July and January fell within the indicated two-degree intervals. (For Thorshavn where March is cooler than January and for Petropavlovsk where August is warmer than July, these months are superimposed on the diagram.) The broken horizontal lines indicate the levels corresponding to 50 and 25 per cent of the instances.

A number of remarkable facts emerges. Clearly in midsummer when insolation is paramount the range of values is rather limited. In fact only at Surgut is the overall range in excess of 8° C though the employment of two-degree intervals causes an apparently wider spread in two or three other diagrams. The average range for the twenty-one stations is 6° C (11° F) and though this seems slight on the scale of the diagram it represents a not inconsiderable variety of midsummer temperatures since we are dealing with monthly means. Nevertheless every July graph contains one column representing 40 per cent of the instances and except across continental Eurasia well over half the record falls within one two-degree interval. The most closely packed group of July values is at Vestmannaeyjar, on the southern coast of Iceland, where the overall range is only 2.5° C.

In winter, on the other hand, when insolation in the zone 50–60° N latitude is far from potent, there is a much greater degree of regional variety. Frequencies in excess of 40 per cent still occur at stations on the shores of the Atlantic and Pacific Oceans but elsewhere the range of variation is much wider. At Yenisseysk and Kirensk, in Siberia, the mildest Januarys, though still very cold by British standards, have mean temperatures for the month that are over 24° C (43° F) higher than the most severe Januarys and in central Canada (some hundreds of miles farther south) the overall range is of the order of 15–20° C. At these continental stations it is very rare for as many as ten Januarys to fall within any one of our two-degree categories. In these circumstances there is clearly no single typical winter value and conditions at that season can only be adequately expressed in terms of a range. Much of the detail on these January diagrams is of a random nature and the presentation would be improved by doubling the class interval. Yet there are some

4

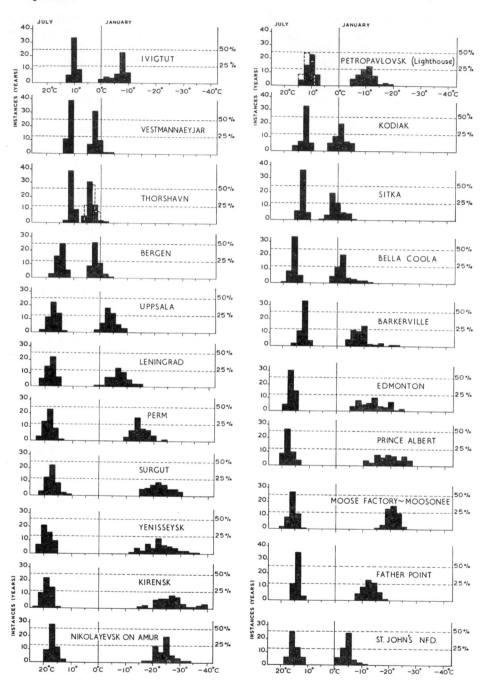

FIG. 32. Frequency groups of mean January and July temperatures at a series of stations which rings the northern hemisphere in high-middle latitudes. The data employed are for the common 50-year period 1891 to 1940. The diagram also illustrates the wide variation in seasonal range between continental and oceanic stations

interesting features which would clearly survive this process. Notable amongst these is the difference between the west coast climates of the Atlantic (Bergen) and the Pacific (Sitka and Bella Coola). The asymmetrical curve for the latter station, reflecting the occasional occurrence of exceptionally cold

TABLE 12. *Temperatures in July and January at a ring of stations in high latitudes in the northern hemisphere (Standard Period 1891-1940)* $(°C)$

| JULY | HIGHEST | LOWEST | EXTR. RANGE | $Q_3$ | $Q_1$ | MEAN RANGE | MEDIAN |
|---|---|---|---|---|---|---|---|
| Ivigtut | | | | | | | |
| 61°12′N 48°10′W | 12.8 | 7.6 | 5.2 | 10.5 | 9.1 | 1.4 | 9.9 |
| Vestmannaeyjar | | | | | | | |
| 63°24′N 20°17′W | 12.7 | 10.2 | 2.5 | 11.9 | 10.9 | 1.0 | 11.5 |
| Thorshavn | | | | | | | |
| 62°03′N 6°45′W | 12.1 | 8.8 | 3.3 | 11.1 | 10.1 | 1.0 | 10.6 |
| Bergen | | | | | | | |
| 60°24′N 5°19′E | 17.7 | 11.0 | 6.7 | 15.4 | 13.7 | 1.7 | 14.3 |
| Uppsala | | | | | | | |
| 59°51′N 17°38′E | 21.4 | 14.2 | 7.2 | 18.0 | 15.8 | 2.2 | 16.9 |
| Leningrad | | | | | | | |
| 59°56′N 30°16′E | 21.9 | 14.3 | 7.6 | 19.1 | 16.5 | 2.6 | 17.7 |
| Perm | | | | | | | |
| 58°01′N 56°16′E | 21.5 | 15.1 | 6.4 | 19.9 | 17.1 | 2.8 | 18.1 |
| Surgut | | | | | | | |
| 61°15′N 73°24′E | 20.4 | 11.7 | 8.7 | 18.8 | 16.0 | 2.8 | 16.8 |
| Yenisseysk | | | | | | | |
| 58°27′N 92°11′E | 22.9 | 15.0 | 7.9 | 20.1 | 17.3 | 2.8 | 19.1 |
| Kirensk | | | | | | | |
| 57°47′N 108°07′E | 22.5 | 15.3 | 7.2 | 20.0 | 17.5 | 2.5 | 18.5 |
| Nikolayevsk | | | | | | | |
| 53°08′N 140°45′E | 19.6 | 13.5 | 6.1 | 17.4 | 15.8 | 1.6 | 16.7 |
| Petropavlovsk | | | | | | | |
| 52°53′N 158°42′E | 13.7 | 7.6 | 6.1 | 11.8 | 9.3 | 2.5 | 10.5 |
| Kodiak | | | | | | | |
| 57°47′N 152°22′W | 15.8 | 9.3 | 6.5 | 12.9 | 11.3 | 1.6 | 12.3 |
| Sitka | | | | | | | |
| 57°04′N 135°19′W | 15.4 | 10.3 | 5.1 | 13.2 | 12.3 | 0.9 | 12.7 |
| Bella Coola | | | | | | | |
| 52°40′N 126°54′W | 19.3 | 14.2 | 5.1 | 16.7 | 15.3 | 1.4 | 16.0 |
| Barkerville | | | | | | | |
| 53°02′N 121°35′W | 15.4 | 9.8 | 5.6 | 13.2 | 11.4 | 1.8 | 12.3 |
| Edmonton | | | | | | | |
| 53°33′N 113°30′W | 19.1 | 14.6 | 4.5 | 17.4 | 15.8 | 1.6 | 16.6 |
| Prince Albert | | | | | | | |
| 53°10′N 105°38′W | 20.7 | 14.7 | 6.0 | 18.7 | 16.8 | 1.9 | 17.9 |
| Moose Factory | | | | | | | |
| 51°16′N 80°56′W | 19.8 | 12.6 | 7.2 | 16.9 | 15.0 | 1.9 | 15.9 |
| Father Point | | | | | | | |
| 48°31′N 68°10′W | 16.1 | 11.9 | 4.2 | 15.2 | 13.7 | 1.5 | 14.5 |
| St John's | | | | | | | |
| 47°34′N 52°42′W | 19.0 | 11.8 | 7.2 | 16.4 | 13.7 | 2.7 | 15.3 |

TABLE 12—*continued*

| | HIGHEST | LOWEST | EXTR. RANGE | $Q_3$ | $Q_1$ | MEAN RANGE | MEDIAN |
|---|---|---|---|---|---|---|---|
| **JANUARY** | | | | | | | |
| Ivigtut | 1.0 | −10.2 | 11.2 | −6.0 | −8.8 | 2.8 | − 7.6 |
| Vestmannaeyjar | 4.1 | − 3.5 | 7.6 | 2.7 | 1.1 | 1.6 | 1.8 |
| Thorshavn | 5.4 | − 1.1 | 6.5 | 4.3 | 2.4 | 1.9 | 3.5 |
| Bergen | 4.8 | − 4.1 | 8.9 | 2.8 | 0.6 | 2.2 | 1.9 |
| Uppsala | 1.5 | −10.0 | 11.5 | − 2.4 | −5.2 | 2.8 | − 3.7 |
| Leningrad | 0.8 | −15.3 | 16.1 | − 4.9 | −9.4 | 4.5 | − 7.5 |
| Perm | − 9.3 | −20.7 | 11.4 | −13.3 | −17.1 | 3.8 | −14.5 |
| Surgut | −15.6 | −30.2 | 14.6 | −19.6 | −25.5 | 5.9 | −22.3 |
| Yenisseysk | −11.0 | −35.1 | 24.1 | −18.8 | −26.5 | 7.7 | −22.2 |
| Kirensk | −14.6 | −39.7 | 25.1 | −23.3 | −29.5 | 6.2 | −26.5 |
| Nikolayevsk | −17.0 | −32.8 | 15.8 | −21.8 | −25.6 | 3.8 | −24.5 |
| Petropavlovsk | − 4.9 | −18.7 | 13.8 | − 8.1 | −11.6 | 3.5 | −10.1 |
| Kodiak | 3.5 | − 5.2 | 8.7 | 0.8 | − 2.1 | 2.9 | − 0.9 |
| Sitka | 6.2 | − 5.1 | 11.3 | 1.9 | − 1.0 | 2.9 | 1.0 |
| Bella Coola | 2.9 | −12.2 | 15.1 | − 0.1 | − 3.2 | 3.1 | − 2.2 |
| Barkerville | − 3.1 | −22.8 | 19.7 | − 6.7 | −10.1 | 3.4 | − 8.3 |
| Edmonton | − 5.4 | −24.7 | 19.3 | − 9.5 | −17.3 | 7.8 | −13.9 |
| Prince Albert | −10.7 | −27.8 | 17.1 | −15.4 | −23.4 | 8.0 | −19.1 |
| Moose Factory | −15.1 | −25.4 | 10.3 | −18.8 | −22.2 | 3.4 | −20.7 |
| Father Point | − 7.0 | −17.2 | 10.2 | −10.8 | −14.4 | 3.6 | −12.4 |
| St John's | − 0.4 | −11.1 | 10.7 | − 2.7 | − 5.7 | 3.0 | − 4.6 |

winters, is characteristic of stations as far south as Seattle though it is undoubtedly emphasised by Bella Coola's fjord-head location.

Broadly speaking, it will be observed that summer values vary more widely the higher the mean summer temperature whereas winter values vary most where the mean temperature is lowest.

This brings us to another aspect of the diagram, the distance between the July and January curves, an index of the contrast between the seasons. The numerical data given in Table 12 give greater precision to such seasonal comparisons. The contrast between maritime and continental locations is at once apparent. Here is a major factor in the heat economy of northern climes compared with which the differences in latitude or altitude of the various stations are but minor details. The pattern in Eurasia is of particular interest. In July there is an increase of over $6°$ C as we pass eastwards from Thorshavn to Uppsala and little further change till we move from Kirensk to Petropavlovsk; the variation is thus disposed symmetrically about the continent. In January, on the other hand, the variation with longitude is both greater and more asymmetrical, winter temperatures falling steadily from Thorshavn $(3.5°C)$ to Kirensk $(−26.5°C)$ and then recovering more rapidly, though Petropavlovsk $(−10.1°C)$ is still much colder than Thorshavn despite an advantage of some $9°$ of latitude. A similar pattern is indicated in North America though it is partially disrupted by the exceptional elevation of Barkerville $(1274 m)$.

The difference between the long-period mean temperature of the warmest and coldest months at any one station is known as the mean annual range of temperature, It is a very coarse statistic but serves to describe a most important climatic characteristic. Using the median values quoted in Table 12 and subtracting the figures for January from those for July we could thus reduce the basic features of Fig. 32 to four simple figures: mean annual range of temperature at Thorshavn 7.1°C, at Kirensk 45.0°C, at Sitka 11.7°C, and at Prince Albert 37.0°C. Nowhere else on earth are such wide contrasts found within such a limited range of latitudes, for the corresponding zone in the southern hemisphere is an almost unbroken expanse of ocean where seasonal contrasts are both slight and much the same in all longitudes. At inter-tropical stations, of course, July and January may not be the months with the most extreme values. To a lesser degree this is true of two of our selected stations, thus at Thorshavn March (3.3°C) is rather cooler than January whilst at Petropavlovsk August (11.7°C) is normally more than 1°C warmer than July. These facts throw interesting sidelights on the climate of these two localities but the major function of Fig. 32 is the commentary it offers on the validity of mean values in general.

## The period of observations

In view of the innate variability of temperature the important question arises as to what extent the evidence presented in Fig. 32 is limited in validity

TABLE 13. *Distribution of mean temperatures in July and January over periods of different length at four selected stations* (°C)

|  | | YEARS | HIGHEST | $D_9$ | $Q_3$ | $M$ | $Q_1$ | $D_1$ | LOWEST |
|---|---|---|---|---|---|---|---|---|---|
| **JULY** | | | | | | | | | |
| Edinburgh | (a) | 187 | 18.4 | 16.1 | 15.5 | 14.7 | 14.1 | 13.4 | 12.4 |
| | (b) | 50 | 16.6 | 15.9 | 15.3 | 14.6 | 14.2 | 13.5 | 12.9 |
| Berlin | (a) | 187 | 23.6 | 20.6 | 19.9 | 18.9 | 17.7 | 17.0 | 15.4 |
| | (b) | 50 | 21.1 | 20.5 | 20.0 | 18.9 | 18.0 | 17.3 | 16.2 |
| New York | (a) | 134 | 27.4 | 25.1 | 24.2 | 23.4 | 22.8 | 22.3 | 20.8 |
| | (b) | 50 | 25.9 | 24.4 | 23.9 | 23.4 | 22.7 | 22.1 | 21.3 |
| Charleston | (a) | 134 | 30.4 | 28.7 | 28.1 | 27.4 | 26.6 | 26.0 | 25.3 |
| S.C. | (b) | 50 | 28.9 | 28.0 | 27.6 | 26.8 | 26.3 | 26.0 | 25.6 |
| **JANUARY** | | | | | | | | | |
| Edinburgh | (a) | 187 | 7.0 | 5.0 | 4.1 | 3.2 | 1.9 | 0.3 | − 3.1 |
| | (b) | 50 | 6.8 | 5.3 | 4.5 | 3.8 | 3.1 | 2.2 | − 0.7 |
| Berlin | (a) | 187 | 6.5 | 3.0 | 1.3 | − 0.5 | − 3.0 | − 5.8 | −11.9 |
| | (b) | 50 | 5.0 | 3.6 | 2.4 | 0.4 | − 2.1 | − 3.6 | − 9.6 |
| New York | (a) | 134 | 5.8 | 2.7 | 1.0 | − 0.5 | − 2.1 | − 3.8 | − 6.9 |
| | (b) | 50 | 5.8 | 4.1 | 1.2 | − 0.1 | − 1.3 | − 3.8 | − 5.9 |
| Charleston | (a) | 134 | 16.4 | 13.2 | 11.4 | 10.1 | 8.9 | 7.4 | 3.4 |
| S.C. | (b) | 50 | 16.3 | 13.9 | 11.4 | 10.1 | 9.2 | 7.4 | 4.2 |

because it is based upon a standard period of 50 years. Has this period been long enough and has its climatic history been typical enough to give a truly representative picture, particularly of the probable occurrence of extreme values? This is an interesting but rather difficult question to answer.

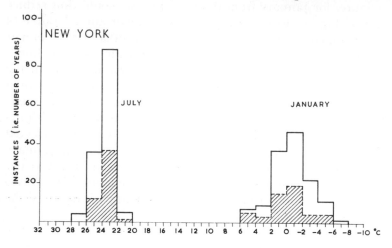

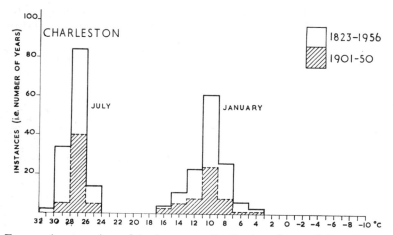

FIG. 33. A comparison of the frequency distribution of mean July and January temperatures over a long period with the distribution during 1901–50 at Edinburgh, Berlin, New York and Charleston S.C.

When a really long record is available it may be examined for 'trend' though trends are usually difficult to establish in view of the innate variability of temperature and of possible changes in the exposure of thermometers at any one place over a long period of time. A simple and compact method is to compare the frequency-groups for a long and a short period as in Fig. 33 and Table 13. There the 50-year record for 1901–50 has been compared with the 187-year records (1769–1955) available at Edinburgh and Berlin and the 134-year records (1823–1956) of New York and Charleston. In Fig. 33 the data are presented in two-degree (C) grades and it must be remembered that the unit under analysis is the mean temperature for a whole month. Visual

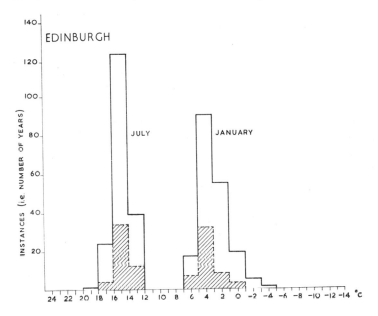

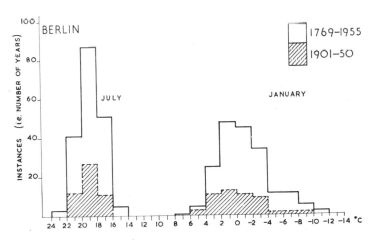

comparison would be easier if the vertical scale were expressed in percentage frequencies but it was thought wiser, at this stage, to retain the actual number of years in each grade. In Table 13 the values of the upper decile ($D_9$), the upper quartile ($Q_3$), the median, the lower quartile ($Q_1$), and the lower decile ($D_1$) are given in degrees Celsius along with the highest and lowest monthly mean in each group.

Naturally enough, the longer the period under review the greater is the chance of capturing really extreme values. Yet at the European stations in July both the interquartile and interdecile ranges show a remarkable degree of consistency between the long and shorter records. Along the eastern seaboard of America however, though there is a similar consistency in the frequencies of cooler Julys, it does appear that recent years have shown some falling-off in the proportion of really warm summers. Rather remarkably in January the situation is reversed, at New York and Charleston the long and shorter records are reasonably consistent whilst in Europe the lower quartiles and deciles indicate a marked reduction in the chance of a really severe winter in recent years. At Berlin indeed the mean value for 1901–50 is nearly a degree above that for the longer period. How real are these apparent changes? Cooler summers in eastern America and warmer winters in western Europe— are they related in some way? Here there is some danger of being drawn into realms of pure speculation. Suffice it to add that statistical tests have been devised to determine how far such differences can be the product of mere chance. Intuitively it is difficult to dismiss the 1901–50 records for January in this fashion though the winters of 1940 and 1947 were sharp reminders that an amelioration is only a relative concept.

The question of the probable frequency of extreme values may also be approached via statistical theory. At its simplest this involves the basic assumption that values are 'normally' distributed, i.e. that the frequency-curve is symmetrical about its mean value. Temperature data approach this form of distribution much more closely than rainfall data but a glance at Figs. 31–33 will reveal some notable exceptions.

## The annual temperature cycle in the northern hemisphere

Most modern atlases contain maps showing the global distribution of temperature, at least in January and July, by means of isotherms often 'reduced to sea-level'. The patterns which emerge are sometimes interesting but it is doubtful whether presenting temperature data in this way is really superior to a tabulation of monthly values for a carefully-selected group of stations. We conclude this section with a new set of maps derived from the computations of Tucker (1965).

Tucker's aim was 'to represent the geographical distribution of the annual rate of change of total heat content of the troposphere', i.e. of the lowest four-fifths of the atmosphere where most weather processes occur. To do this for the whole of the northern hemisphere he used upper-air data for seventy-eight stations over the 5-year period 1955–9. His maps thus describe the atmosphere *in depth*, a distinct advantage for many purposes over the earth-bound data derived from screen observations. To get a broad view, to iron out some inconsistencies, and to present the maps in a more compact form we have reduced his twelve maps to six by a process of graphical addition.

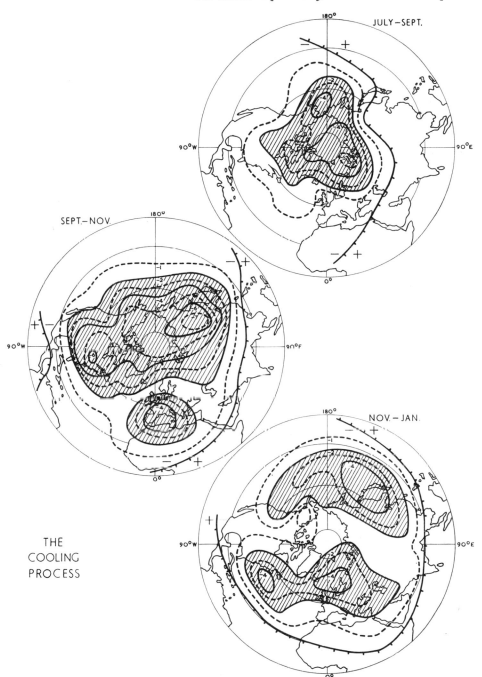

THE
COOLING
PROCESS

FIG. 34. Stages in the cooling of the whole troposphere in the northern hemisphere from July to January. A comparison of alternate months has been made by combining pairs of Tucker's (1965) original maps. His units were $10^{-4}$ g cal/cm/sec; as a result of the addition, the isopleths plotted show rate of change over *two-second* intervals

4*

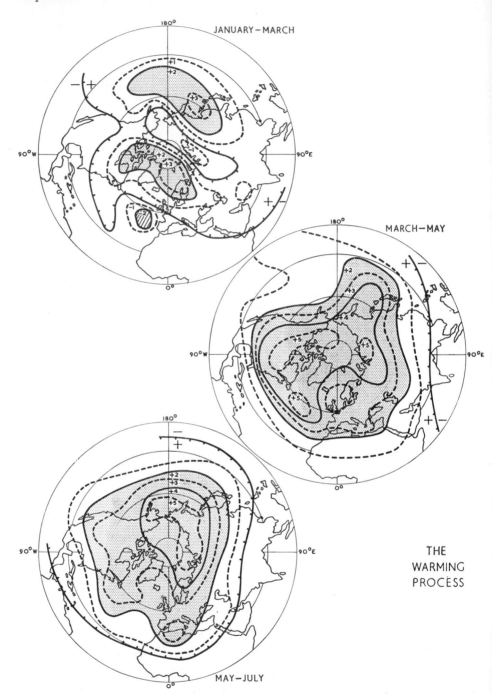

FIG. 35. Stages in the warming of the northern troposphere from January to July. Again the comparison is between alternate months to illustrate the major features of the changes which occur

The results are presented in Figs. 34 and 35, the former showing the stages by which the atmosphere is cooled from late summer to midwinter and the latter how it is warmed again during the early half of the year. It should be borne in mind that radiation (incoming and outgoing) is only one of several processes that effect this transformation; advection, convection and the release of latent heat at condensation all play their part. The maps should thus become more fully intelligible as this text proceeds but we welcome them here as presenting a general picture of what climatology is all about.

The maps should be viewed in broad terms since much of the detail is probably valid only for the comparatively short period employed. Tucker himself questions their validity south of the 20°N parallel. To facilitate comparison with Tucker's original diagrams we have not divided by two after the addition. The values plotted may thus be interpreted as follows:

$$2 \text{ units equal } 8.6 \text{ cal/cm}^2/\text{day}$$
$$4 \text{ units equal } 17.3 \text{ cal/cm}^2/\text{day}$$
$$6 \text{ units equal } 25.9 \text{ cal/cm}^2/\text{day}$$

However, as with most data involving temperature, it is the *relative* scale of changes that is of greatest importance.

### The Cooling Stages (Fig. 34)

In his month-by-month analysis Tucker (1965) shows that the winter cooling in the atmosphere begins within the Arctic Circle as early as July–August. The July–September map shows the area enlarged and intensified. Two features are particularly noteworthy: (1) the asymmetry of the distribution  Asia south of its mountain backbone is so far not affected, and (2) the lobate form of the area of maximum chill. The maritime coastlands of British Columbia and western Europe are but slightly affected and a deep re-entrant marks the province of the Manchurian 'monsoon'. Between these re-entrants, continental chill extends at least to the fiftieth parallel over Canada, European Russia and the lands around the Bering Sea. The relation of this striking pattern to the general circulation brooks no question, but how far it is a cause or how far an effect it would be premature to discuss.

By September to November, whilst the Arctic continues to lose heat at some 17–20 cal/cm²/day, it is the land masses of Siberia and North America which dominate the picture. Thus are produced the twin poles of cold that play such an important part in the winter climate of the northern hemisphere. The Manchurian re-entrant has disappeared and that over Pacific Canada has sadly weakened but an anomalous pattern over northwestern Europe is still much in evidence. Tucker wonders whether this 'may be a feature merely of the five years chosen' but it clearly survives our process of generalisation and is not unreasonable. The relatively isolated area of chill over central Europe and the Mediterranean is rather surprising but it is the accumulated result of two of Tucker's diagrams.

By November to January winter comes finally to the atmosphere over the mid-latitude oceans and the lands lying under their sway. The map thus shows two broad zones of marked heat loss whilst between them, along the Canadian–Siberian axis of greatest cold, temperatures through a great depth of the atmosphere fall at a greatly reduced rate. This may well be due to general subsidence (see p. 233-4).

## The Warming Stages (Fig. 35)

Tucker's maps covering the period from December to March are the most complex of his series since each of them shows heating and cooling occurring side-by-side in the northern hemisphere, sometimes with no great consistency of location. Even when expressed in the simpler form of our January to March diagram the pattern remains complicated and it seems clear that direct insolation has very little to do with the first stages in the reheating of the atmosphere. Recovery begins over the northern Atlantic and northwest Pacific, in the former well above the sixtieth parallel, whilst across the inner Arctic heat loss continues uninterrupted. Tucker associates these contrasts with strong zonal winds in the North Atlantic.

By the period March to May vigorous heating has become general north of the thirtieth parallel though it is most marked over Canada and the north-west Atlantic. The main re-entrant at this stage is in the eastern Pacific. The process continues between May and July, more or less symmetrically about the pole, though the major increase is now found in the Siberian sector.

Whatever the validity of their details, these maps clearly pose a number of interesting and difficult questions. We have evidently moved a long way from the comparatively simple pattern of solar income. Nor is there, as Tucker himself points out, any very obvious relation between many of the longitudinal contrasts displayed and the broad distribution of land and sea. Screen temperature data have their place when we are concerned with events near (but not at) the earth/atmosphere interface where much of life's activity goes on, but atmospheric behaviour depends upon its characteristics in depth. Temperature, density and air motion all have to be considered *in bulk*. In generating the bulk contrasts which stimulate atmospheric motion, water vapour plays a major and most intricate part. To this we now turn.

# 3
# The role of water vapour

If the ultimate source of all effective thermal energy on earth is the sun (radioactive sources are relatively so negligible that their presence was long entirely unsuspected), it is no less true that, in the capture, utilisation and redistribution of that energy, water in its various states plays a most fundamental part. We have already noted the purely physical responses of liquid water surfaces to incoming short-wave radiation and the extent to which terrestrial long-wave back-radiation is intercepted both by water vapour and by clouds (tiny liquid water droplets or 'water smoke'). It is also commonplace that without water there can be no life since all living organisms can only absorb nutriment in aqueous solutions.

The relationship between humidity and heat in the atmosphere is indeed so intimate that it extends to their regional, and even local, distribution. This is the result of two closely related groups of physical facts.

In the first place, within the range of temperatures and pressures common on earth it is familiar to all that water may exist concurrently in its three phases, solid, liquid and gaseous. Transformation of phase, forward or backward between these alternative states, locks up or releases massive quantities of thermal energy.[1] Water thus provides one of the most potent heat-exchange mechanisms of the atmosphere and at times water vapour would appear to act almost as if it were the 'fuel of the atmospheric engine'. The scale of the exchanges involved is given by the latent heats of fusion, vaporisation and sublimation quoted, page 90.

Hence the familiar generalisations that it takes 80 times as much heat to transform a gram of ice at $0°C$ into water at $0°C$ as it does to raise that amount of water $1°C$ whilst it takes 539 times as much heat to transform a quantity

---

[1] These transformations may be regarded as following two directions and it has proved helpful to illustrate them as follows:

(1) SOLID .....*melting* ———→LIQUID....*evaporation* ——→GAS
(2) GAS .......*condensation* ——→LIQUID....*freezing*———→SOLID

Direction (1) is associated with increasing temperature but, since heat is locked away in latent form, the result of the transformation is to check the rise in temperature. Direction (2) is associated with decreasing temperature but, since latent heat is unlocked by the transformation, the result is to check the fall in temperature.

of water at 100°C into steam at 100°C as it does to raise that quantity of water from 99–100°C. Such transformations into 'latent' heat provide facilities for heat storage not only of great capacity but also free, for the most part, from the ceaseless losses entailed by radiation processes described above. Hence the importance of evaporation and condensation in the process of heat transference from warmer to cooler latitudes.

Secondly, while the proportions of the dominant constituents of dry air (by mass, nitrogen 75.5 per cent; oxygen 23.2 per cent) not only remain remarkably constant at all levels up to 23 kilometres (*c.* 75000 ft) but also would provide a stable mixture whatever the ratio, this is emphatically not true of a mixture of dry air and water vapour. The vapour content of the

| °C | *Fusion* (g cal) | LATENT HEAT OF *Vaporisation* (g cal) | *Sublimation* (g cal) |
|---|---|---|---|
| At  −30 | 63 | 615 | 678 |
| 0 | 80 | 597 | 677 |
| 30 | | 580 | |
| 60 | | 563 | |
| 100 | | 539 | |

atmosphere is rarely as high as 3 per cent by mass, often very much lower. It is greatest within the tropics, no matter how arid the region may appear, and usually falls off steadily with height to reach very low values above 6 kilometres (*c.* 20000 ft). Even under moist intertropical conditions the amount of 'precipitable moisture' in the whole air column is only sufficient to produce 2 to 3 inches of rain. (Only a small fraction of this—about 10 per cent—is in fact 'available' at any one time and rainfall in excess of this figure must be due to a supply brought in from neighbouring columns.) This variability of water vapour content of the atmosphere arises primarily because of the part played by temperature in the vapour–liquid equilibrium. For any given temperature there is a limiting value of water vapour concentration which can be exceeded only temporarily, to a small degree, and then under rather special conditions. Normally the anomaly is promptly relieved by the condensation of liquid water—clear evidence that the air is more than saturated. This concentration is entirely a function of temperature and therefore although we are sometimes tempted to speak in terms of 'the quantity of water vapour which air will hold' it is much more scientific (and therefore, in the long run, much more satisfying) to express it in terms of 'vapour pressure'. Dalton's Law states that in a mixture of gases each gas exerts its own partial pressure as if the other elements of the mixture were not present. The vapour pressure of water is thus independent of the pressure of the dry gases. *Saturation vapour pressure* over water is thus defined as the pressure of aqueous vapour when in a state of neutral equilibrium with a plane surface of pure water at its own temperature. It is a function of temperature and of temperature only and may be expressed as such in tabulated or graphical form (see Appendix 2, Table A.2.2). Note that 'neutral equilibrium' in the above definition means that for each molecule that happens to be knocked out of the liquid into the vapour (evaporation) a molecule is knocked back from the vapour to the liquid (condensation). The intensity of the bombardment is measured directly by the vapour pressure and indirectly by the temperature (since heat is molecular motion). Note also that tabulated values of saturation vapour pressure are strictly valid only for a plane surface of water. Many natural water surfaces, particularly those of droplets, do not satisfy this condition. Equilibrium vapour pressure is increased by curvature so that tiny cloud droplets show a strong tendency to evaporate and may thus produce a low degree of 'supersaturation'.

Although the vapour pressure ($e$) is independent of the dry air pressure ($p - e$), it is the sum of these two pressures ($p$) which is recorded by the barometer. Since saturation vapour pressure increases rapidly with temperature it follows that, as temperature rises, saturated vapour takes over an increasing proportion of the atmospheric burden until boiling point is attained when $e = p$. We use this fact in the laboratory when we evacuate the air from a flask by boiling a little water in it whilst, on a mountain top, a careful measurement of the boiling point of water was formerly considered the most accurate method of obtaining $p$ and hence of estimating altitude.

The reader may be excused if he finds the concept of 'partial pressure' a little unfamiliar and rather abstruse at first encounter. It must be confessed that we have no method of measuring this pressure in any direct manner. Let us turn therefore to a concrete example and express ourselves in the familiar terms of volume and mass. We assume throughout a mean atmospheric pressure of 1013.2 millibars (760 mm or 29.92 in of mercury) and quote for a typical range of terrestrial temperatures the gross weight of air contained in a room such as might be found in any ordinary house (dimensions $15\frac{1}{2} \times 10 \times 8$ ft, volume 1240 ft³ or very nearly 35 m³). (See Table 14.)

If this room is filled with completely dry air at 0° C (weighing 1.293 kg/m³) the total weight of air is shown to be 100 pounds (*c.* 45 kg). Merely changing the temperature of this sample of air in a hermetically sealed chamber cannot affect its gross weight but it will certainly affect its pressure. To adjust for our assumed constant pressure we shall thus have to allow some of the sample to escape as temperature rises or add to it as temperature falls, thus decreasing and increasing the weight of the sample. Charles's Law states that if the temperature of a gas is increased from 273 to 303° K its volume at constant pressure will be increased by 30/273 or, shall we say, from 273 to 303 units. At the new temperature therefore of the 303 units (which still weigh 100 lb) we must allow 30 units to escape from the room to maintain our assumed pressure and the weight of the air remaining will be very nearly 90 pounds. Similarly with a fall of temperature to 243° K the volume of the 100 pound sample at our standard pressure will be only 243 units and we must admit to the room 30 units at 243°K, i.e. 30/243 × 100 or another 12 pounds of air in order to avoid a fall of pressure. The values tabulated thus give a broad indication of the variation in density of dry air over a normal range of terrestrial temperatures.

Returning to our original sample of air at 0°C, let us now permit water to evaporate into it until it is fully saturated. Dalton's Law states that in so doing we shall be adding to the original dry air pressure of 1013.2 millibars the independent partial pressure of water vapour at 0°C, which line (*d*)

TABLE 14. *Weight of air and water vapour (at saturation) in a room $15\frac{1}{2} \times 10 \times 8$ feet (i.e. volume 1240 ft$^3$) at standard pressure (1013.2 mb) and at various temperatures*[1]

| Temperature | °K | 243 | 258 | 273 | 288 | 303 | °K |
| --- | --- | --- | --- | --- | --- | --- | --- |
| | °C | −30 | − 15 | 0 | 15 | 30 | °C |
| | °F | −22 | 5 | 32 | 59 | 86 | °F |
| *Weight* | | | | | | | |
| (*a*) Dry air | | 112 | 105.5 | 100 | 94.5 | 90 | lb |
| (*b*) Saturated air | | 112 | 105.4 | 99.8 | 94 | 88 | lb |
| of which water vapour | | 0.03 | 0.12 | 0.37 | 0.99 | 2.34 | lb |
| and hence air alone | | 111.97 | 105.28 | 99.43 | 93.01 | 85.66 | lb |
| (*c*) *Mixing Ratio* (parts per 1000) | | 0.3 | 1.2 | 3.8 | 10.7 | 27.3 | g/kg |
| *Partial pressures* | | | | | | | |
| (*d*) Water vapour | | 0.5 | 1.9 | 6.1 | 17.0 | 42.4 | mb |
| (*e*) Dry air | | 1012.7 | 1011.3 | 1007.1 | 996.2 | 970.8 | mb |

1 m$^3$ = 35.3147 ft$^3$; 1 kg = 2.2046 lb.

[1]If the volume of the room is increased to 77 cubic metres (35 × 2.2) values given in pounds can be read as kilograms. Ratios and pressures remain unchanged.

of the table shows may attain 6.1 millibars. To avoid a consequent rise of pressure in the room to 1019.3 millibars we shall thus have to remove from it a small proportion of the dry air. The table shows that a balanced situation will be attained at saturation when the room contains, in round figures, some 99.4 pounds of dry air plus 0.4 pounds of water vapour, the contents of the room thus now weighing 99.8 pounds. The table gives comparable values for saturation at other temperatures. It will be thus seen that moist air is rather lighter than dry air at the same temperature and pressure but that the contrast becomes considerable only when temperature permits a large 'water content' at saturation, i.e. at temperatures well above 0°C. Since water vapour is only 18/28.9, i.e. 5/8 as dense as dry air, this phenomenon has sometimes been explained in terms of the dry air having to be removed to 'make room' for the water vapour. Strictly speaking this is not accurate and the above example makes it quite clear that the dry air is extracted only to satisfy the original assumption of constant pressure. There is a real distinction here though it may require some moments of contemplation.

It thus becomes apparent that instead of expressing vapour content in terms of a partial pressure which we are unable to measure directly, we can turn to the relatively concrete comparison of weight of vapour compared with weight of dry air with which it is associated. This is known as the *Mixing Ratio*. It is normally quoted in parts per thousand, or more specifically in grams per kilogram. The values given in line (*c*) of the table are thus the saturation mixing ratios at the appropriate temperatures and, let us not forget, at our standard pressure of 1013.2 millibars. Since it would appear that fully saturated air under natural conditions rarely if ever reaches a temperature above 30°C (86°F) it may be concluded that the vapour content of the atmosphere rarely exceeds 2.7 per cent by weight. This is equivalent to rather more than 4 per cent by volume. Returning for a moment to our room, the total vapour content at saturation in terms of liquid water thus ranges from about a tablespoonful at −30°C to approximately a quart at 30°C.

In the above discussion we have confined our attention to the two extreme cases of the air being 'dry' in the very strictest sense of that term (i.e. containing no water vapour whatsoever), or of being saturated (i.e. containing the maximum proportion of water vapour normally possible at the given temperature). In fact, of course, a sample of air may contain any amount of water vapour between these two limits. If the partial vapour pressure in a sample at 0°C is actually 3.05 millibars, or that of a sample at 30°C is 21.2 millibars, it will be noted that these pressures are precisely half the vapour pressure quoted for saturation. The 'relative humidity' of both samples is then said to be 50 per cent although the air contains only 1.9 g/kg of water vapour in the first case as compared with 13.65 g/kg in the second. The relative humidity is thus the percentage relationship of the actual vapour pressure with the saturation vapour pressure. Clearly a comparison of the actual mixing ratio with the saturation mixing ratio (at the same pressure) will yield an identical result. In the past the relative humidity has been a very popular method of expressing the humidity condition of the air, it has been regarded as giving an indication of the 'feel' of the air and some measure of its 'drying power'. Within any limited range of temperatures this is broadly true but the concept is in fact a *percentage relationship* and all such relationships must be handled with considerable care. We could have expressed the drying power of our two samples in more absolute terms if we had said there was a deficit of 3.05 millibars at 0°C and a deficit of 21.2 millibars at 30°C. The difference between the actual vapour pressure and the saturation vapour pressure is thus known as the *saturation deficit*. It will be seen at once that the picture it gives of the relative drying power of these two samples of air is very different from that conveyed by their identical relative humidities. Both methods of expression are equally

93

valid; it is in drawing conclusions from them that care must be exercised.

Let us note, since the point has been raised, that the 'feel' of the air can be most deceptive. We have no way of distinguishing how far the chill at the skin is due to conduction rather than evaporation. Furthermore evaporation from the body takes place at skin and mucous temperatures often differing widely from that of the surrounding air. Finally, and above all, the body is sensitive to ventilation and our subjective sensations at all temperatures depend largely upon air movement.

## The measurement of humidity

This apparent digression brings us quite logically to the problem of humidity measurement. We have noted that there is no way of measuring partial vapour pressure. If we turn instead to the mixing ratio we are faced with the chemical absorption of vapour and involved in slow and delicate measurements of weights and volumes. Until recently the nearest approach to a continuous direct measurement, and that only of relative humidity, was obtained from the behaviour of dead organic tissues—a bundle of human hair in the hair hygrometer which stretches as humidity increases and contracts as it falls (about $2\frac{1}{2}$ per cent between 0 and 100 R.H.). Most geographers in the west country will certainly have had the opportunity to notice that maps pinned up for display behave in precisely the same fashion. The tightening of ropes after rain is an exception that proves the rule for it is the *swelling* of the hemp fibres which increases their twist and thus shortens the rope. The hair hygrograph is not very reliable and has now fallen into some disfavour. For continuous readings, for instance, from a radio-sonde we now employ the varying electrical resistance of a hygroscopic film with appropriate adjustments for changes of temperature.

For measurements at observation hours at ground stations the best approach to humidity is still provided by simultaneous readings of the wet and dry bulb thermometers. These are two carefully matched thermometers mounted side by side, the bulb of one, the 'wet bulb', being enclosed in moistened muslin. Evaporation into unsaturated air leads to a 'depression of the wet bulb' and the differences between the two readings is interpreted in terms of relative humidity by means of tables. For example, with the relative humidity at 50 per cent, the 'depression of the wet bulb' amounts to 5.7°C (10.3°F) at 20°C (68°F); 4.0°C (7.2°F) at 10°C (50°F); and 3.3°C (5.9°F) at 5°C (41°F). The method thus becomes less sensitive at low temperatures. A major problem is the standardisation of ventilation and hence a variety of devices for swinging the two thermometers through the air or for drawing across them a known volume of air by a clockwork fan.

The tables used in the above method had to be arrived at empirically by determinations of the 'dew point', the temperature at which the quantity of water vapour in the air would produce saturation. The indication that this point has been reached is usually the clouding of a highly polished surface,

The dew point is always a little lower than the wet bulb temperature for reasons which will be explained later (see p. 252).

It is evident that the water vapour content of the air is analogous to the working reserves of a great bank. It is being drawn upon continually by condensation and replenished continually by evaporation. In so far as it tends to remain substantially the same over any considerable period of time this is no proof of inactivity, only of a temporary balance between opposing forces. Atmospheric vapour is thus one stage in what has been called the hydrological cycle; the vapour is condensed into snow or rain which refreshes the vegetation, recharges the springs and fills the rivers but eventually returns to vapour by a variety of circuits, some short, as in the transpiration of plants, and some very long, as when the evaporation loss is from a distant sea. There is no logical point at which to break into such a circuit. The oceanographer may prefer to begin with evaporation since the seas suffer a net loss which is made good by the annual tribute of the rivers. The agronomist, on the contrary, may well start with precipitation, partly because it is easier to measure and partly because he may wish to set evaporation loss against precipitation income. However *all* evaporation is not loss, any more than is *all* expenditure unprofitable. Indeed the circulation through the hydrological cycle makes possible all the processes of life and growth.

## Evaporation

Of the two sides of the evaporation–condensation exchange it can be argued that the former takes pride of place. It proceeds continuously and universally wherever humid surfaces (water, vegetation, the soil) are in contact with air not fully saturated at the temperature of that surface. At ground level the heavy heat demand it entails is supplied for the most part by insolation so that its diurnal and seasonal cycles follow closely the regular insolation curves with which we have now become familiar. Aloft it disperses clouds only a little less rapidly than they are generated and takes a considerable toll from falling raindrops; here it draws heat from the air itself and the consequent fall of air temperature occasionally produces momentous results. It is indeed via evaporation that the atmosphere recoups its precipitation losses and maintains the supply of moisture upon which many of the active weather processes ultimately depend. The other side of the exchange, condensation and particularly precipitation, though of immense importance, is much more sporadic in distribution and intermittent in occurrence. What nature thus draws from humid surfaces all over the earth she subsequently returns with a most uneven hand both in space and time.

Yet, surprising as it may seem, evaporation has proved remarkably difficult to measure with any degree of precision. In fact, Penman (1950, 372) who has long specialised in this field of enquiry, has been compelled to admit that 'direct measurement of evaporation from natural surfaces is almost impossible'. Devices abound, from crude open pans of water, through soil-filled containers with a plant cover, to comparatively delicate instruments (atmometers) exposing a continuously-moistened bulb or plate, but the records they yield are inconsistent and their relationship to evaporative loss on a regional scale is often indeterminate and obscure. Open pans, for instance, often yield values two or three times the observed loss from large reservoirs despite the fact that large water-bodies must lose a proportion of their volume by percolation through the underlying natural floor. Atmometer records too are often unrealistic. An even more exasperating fact is that apparently identical devices exposed within a few yards of one another often fail to agree by surprising margins.

Reasons for this difficulty are not far to seek. Though almost universal wherever moisture is present, evaporation varies widely in intensity over very short distances, with contrasts in albedo, relative shade and ventilation. The natural world is full of such contrasts. This is because, in undisturbed conditions, evaporation is virtually self-defeating. Beginning in response to a disequilibrium across the interface between the air and a moist surface, it tends to restore the balance in two ways: (1) it produces a saturated 'vapour blanket' in the ambient air, and (2) it cools the evaporating surface. On both counts, in theory at least, the process should be soon brought to a close. In nature a disturbance normally occurs, in the form of a fresh supply either of heat (from sun or sky radiation or from earth or water heat-storage) or of a movement of the air which tears the 'blanket' to shreds. Yet each of these events is apt to be local in both incidence and intensity. A close crop can retain its vapour blanket against a powerful sun or a fresh breeze and we become aware of this fact on entering a dense woodland on a hot or breezy day. In other words evaporation is highly sensitive to exposure, as we all know. A further difficulty, when we are dealing with *actual* evaporation, is that after prolonged drying the available moisture supply may be cut off by the drying-out of the surface soil or the wilting of its vegetative cover; this introduces a new factor independent of subsequent meteorological conditions—until the next rain.

It follows that a shallow pan of water on land gives excessive readings owing to insolation effects on the pan base and sides, to conduction through its base, and to the easy removal of its meagre vapour blanket by the drier air on all sides. Floating the pan on a lake has been tried but it brings problems of raft motion and splash. The soil evaporimeter is misleading unless it is surrounded by a wide buffer zone with a similar vegetation and, of necessity, its installation must lead to disturbance of soil structure. An atmometer, set up in isolation, has an all-round ventilation comparable only to a single free-standing plant which is never allowed to wilt; yet, if placed within a growing crop, it measures a microclimate which changes as the shade increases. None of the readings thus obtained have any great relevance to what is happening over a square mile of the surrounding countryside.

At one time great hopes were pinned on yet another solution. If we cannot measure evaporation at the surface from which it is occurring, can we not sample it en route to the free air aloft? This should be equally applicable to *all* types of surface, all that is required are careful measurements of temperature and humidity at two or more fixed levels. The method is particularly associated with the names of Thornthwaite and Holzman (1939), but others have contributed to its theoretical framework. Unhappily there are great practical and even theoretical difficulties. Turbulent transport is a condition of incessant change and overall means have no validity outside the very narrow range of conditions specified in their computation. Formulae abound but coefficients are so flexible that this approach falls more in the field of estimate than of measurement.

## The estimation of evaporation

In view of these vagaries of measurement a great amount of ingenuity has been expended in devising methods by which evaporation may be estimated from other climatic parameters, notably from temperature, humidity and radiation data with or without some consideration of mean air motion. Sooner or later however such estimates have to be compared with measurements and the decision as to the reasonableness of the answer must retain an element of subjective judgment.

It is first necessary to be quite clear as to what is being estimated. Is it *actual* evaporation from terrestrial surfaces as we know them under the wide variety of climatic conditions which prevail on earth, or is it *potential* evaporation, the loss which would occur if those surfaces were never short of water? The latter is a 'model' that differs widely from the facts in deserts and during long dry spells elsewhere but it is usually preferred for first approximations since it corresponds with conditions over the oceans and other natural bodies of water and since estimates can then be compared (with reservations) with the data from open pans. Experience has also shown that soil evaporimeters yield less inconsistent records when the growing crop is supplied at frequent intervals with small measured quantities of water. The relationship which then exists between potential loss from a water surface and potential loss from a cultivated crop has been a matter of warm debate. Furthermore is the type of crop important? H. Penman (1963, 34) has argued these points at length and concludes that, despite differences of opinion, there is 'a common concept, namely that when a full crop cover is kept plentifully supplied with water, the rate at which the water is transpired is dictated primarily by the weather, with plant and soil factors playing only secondary roles'. Still more explicit were his 'important broad generalisations' stated seven years earlier, namely,

'(i)  For complete crop covers of different plants having about the same colour . . . the potential transpiration rate is the same, irrespective of plant or soil type.

'(ii)  This potential transpiration rate is determined by prevailing weather.

'(iii)  The transpiration of a short green cover *cannot exceed* the evaporation from an open water surface exposed to the same weather' (Penman, 1956, 20, 23).

In applying his own method for estimating evaporation, discussed below, Penman has thus used a reduction factor for land as compared with water surfaces.

It is surely self-evident that the method to be used for estimating evaporation partly depends on the use to which the estimate is to be put. If we seek only a broad annual total to serve, for instance, in a definition of the arid lands of the earth, a comparatively crude estimate will suffice, particularly since annual precipitation, the other side of the balance, is known to vary over a magnificent range—from nearly zero in parts of Chile to 11000 millimetres (435 in) or more at Cherrapunji. Even the most elaborate methods outlined below, those of Penman and Budyko, set their sights no higher than this in the first instance. For geographical purposes, however, a breakdown of this total into monthly means is highly desirable and here difficulties begin to multiply—though, again, monthly precipitation covers so wide a range from place to place, from month to month, and even at any one place in any one month, from year to year, that a broad agreement as to the balance is possible from methods varying widely in degree of sophistication. To estimate water loss from a specific plot of land over a limited period of time given the weather conditions then prevailing is a very different question which we must leave to the agronomist. Approximations, at least to the potential loss, can certainly be arrived at but it is much too often assumed that a method designed to give answers to the first two questions can be applied to the third.

(*a*) *Estimates based upon mean air temperature.* The idea that the *effectiveness* of precipitation is to some degree dependent upon the temperature prevailing in the region in which it falls goes back at least a century in European literature. The object, from the first, was to establish what would now be known as a working $P/E$ ratio rather than to obtain an actual estimate of evaporation itself. There is a real distinction here though it may require a moment's thought. Indeed, it is a distinction that has been partially obscured by Thornthwaite's two papers (discussed below), for in 1931 his statement was exclusively in terms of $P/E$ whilst in 1948 he plots evaporation in absolute terms though stressing its relationship with precipitation throughout the work. For a simple working ratio to delimit 'arid' lands or the length of the 'dry' season, a comparatively crude approach will suffice. Thus Lang's (1920) *annual* 'rain factor' of $P/T$ was amended to $P/(T + 10)$ by de Martonne (1925; 1947, Vol. I) when he introduced it as an 'index of aridity'. Transformed for use with *monthly* data this index becomes $12P/(T + 10)$ but Köppen (1922; 1931, 122–36), after a series of attempts to fit the index to observed distributions, eventually came out in favour of a flexible index ranging from $10P/T$ to $10P/(T + 14)$ according to whether the region in question experienced predominantly winter or summer rain. In each of these statements $P$ is in millimetres and $T$ in degrees Celsius and it is clear that their authors were thinking in round numbers. It is worth noting that when Bagnouls and Gaussen (1953) were seeking a definition of a 'dry month' they found that the still simpler $P/2T <1$ gave quite useful results though they then analysed the period thus defined much more searchingly by a consideration of the number of completely dry days.

It cannot be denied that these attempts to weight rainfall according to the prevailing temperature represented a step in the right direction since any combination of what Miller (1947, 6) has called the 'elements' of climate is at least one move towards the complete synthesis which is climate itself. Yet it is evident that the weights proposed by the above methods were both subjective and arbitrary. It is greatly to the credit of Thornthwaite that he endeavoured to distil a system of weighting that was both objective and flexible from the observations that were to hand. In his first paper to achieve wide currency (1931), based upon evaporation data from twenty-one stations in the more arid parts of the United States for the period April to September, he sought to establish a relationship between temperature $(T)$ and the precipitation/evaporation ratio $(P/E)$ on a monthly basis for use in a rational classification of American climates. We have already noted that results expressed in this way give no immediate clue to the absolute values of evaporation either used or inferred. Yet the question is neither irrelevant nor unfair. In a critical analysis of this method the present author (Crowe, 1954, 47) has shown that, bearing in mind the inherent 'errors' of monthly temperature and rainfall data, Thornthwaite's rather formidable formula

$$\frac{P}{E} = 11.5 \left( \frac{P}{T-10} \right)^{10/9},$$

can be effectively simplified to

$$\frac{P}{E} = \frac{9P}{T-10},$$

where, in both cases, $P$ and $E$ are expressed in inches and $T$ is in degrees Fahrenheit. Within the quite tolerable limits of this generalisation, it is evident that the inferred evaporation-temperature 'gradient' is of the order of $E = (T-10)/9$. This implies zero evaporation when the mean monthly temperature is as low as $10°F$ $(-12°C)$, a monthly total of 2.4 inches (61 mm) when mean temperature is $32°F$ $(0°C)$, and a virtual ceiling to evaporation of some 10 inches (254 mm) per month when the mean temperature attains $100°F$ $(38°C)$. We may note in passing that the highest monthly temperature quoted in Kendrew (1953, 134) is $99°F$ at In Salah in July. Whilst this ceiling value appears to be by no means unreasonable, it is perhaps unfair to lay too much stress on the results for low temperatures since Thornthwaite's (1931) basic data were for the summer months. Some of the subjective formulae quoted above are clearly not intended to accommodate negative Celsius values though it has been argued against both Lang's and de Martonne's *annual* formulae that cold winters must have an unhappy effect upon mean annual temperature (*T*). Dzerdzeevskii (1958, 317) asserts that 'actual evaporation under negative temperatures is insignificant'. We shall see below that Thornthwaite later accepted such a view but by implication only and perhaps too readily for, in a continental climate, a *mean monthly* temperature of 0°C may very well involve many daytime temperatures well above this figure. For the moment, however, it is wise to return from this digression and to note that, over a significant range of temperatures, Thornthwaite's first method involves a change of evaporation with temperature (conveniently described as an 'evaporation gradient') of some 1.1 inches per 10°F or of 2 inches (51 mm) per 10°C.

At a later date, Thornthwaite (1948, 56) began to experiment with soil-filled and grass-covered evaporimeters, a device that had apparently been invented earlier in France. Supplementing the new observations with data on water-use in irrigation projects, he now endeavoured to estimate 'potential evapotranspiration' (*PE*) or 'the amount of water . . . which would transpire and evaporate if it were available'. Penman (1963, 33) has argued that neither the vegetative cover nor this rather ugly term was really necessary since 'the evaporation rate from the bare soil is at least as great as the transpiration rate from a "complete" crop cover of a short crop'. This was not known for certain at the time and, as we have already noted, the use of the word 'potential' certainly made for clarity of thought. The author has subjected this system also to searching analysis (Crowe, 1954, 50–61) but he would be the last to deny either Thornthwaite's ingenuity or his flair for graphical presentation. His criticisms have been confirmed by a more directly mathematical approach by J. Sibbons (1962).

The method by which a relationship between potential water loss and temperature was established by Thornthwaite (1948) defies brief description. In retrospect one is inclined to wonder how far it was illuminated by a mischievous sense of humour! Factors involving ten places of decimals and power indices to four significant figures are hardly likely to be taken seriously by anyone who is aware that if we know either precipitation or evaporation to the nearest half-inch per month we are doing pretty well. But we must forget about inches for now Thornthwaite had 'gone metric', a regrettable step since, plotting his data on log-log paper, he could find no place either for 0°C or for zero evaporation. The general implication of the text, however, is that these points will coincide, a very different picture from that given by him in his earlier method. The only other fixed point on his scale is the 'point of convergence' at 26.5°C; 13.5 cm (80°F; 5.3 in) and it is of interest to note that the mean gradient between these two points thus works out at 51 mm/10°C, precisely the same as was encountered above. Carried on as a straight line this gradient will yield a ceiling value of 19.4 centimetres (7.6 in) at 38°C (100°F). It must be emphasised that this simple conclusion was certainly not that drawn by Thornthwaite himself.

The complexity of his second method arises from two new features.

(1) His log-log graphs suggested the existence of a multiplicity of evaporation

gradients and he concluded that the appropriate scale for any station could be selected by a 'heat index' derived by the summation of twelve factors, each a function of mean monthly temperature. This elaborate device is the very essence of the new system. The present author tried to master its implications by transforming Thornthwaite's 'nomogram' to simple linear scales (1948, Fig. 2, p. 55). It then became visually evident that, as Sibbons puts it, 'the mean annual temperature effectively controls the form of the moisture-need curve' which the method employs at any selected station. J. Gentilli (1953, 83) had already stressed the same point more piquantly: 'On pourrait arriver à la conclusion absurde que l'évapotranspiration potentielle d'un mois puisse diminuer par le réchauffement des autres mois, et qu'elle puisse augmenter par leur refroidissement.' The general result is that estimates are low during the cooler season in hot climates and excessive during the warmer season in cold climates. Whatever the relationship between evaporation and temperature, it is likely to be at once both simple and imprecise; a rough-and-ready parallelism defying refined techniques.

(2) The method then provides for the adjustment of the crude values by a factor 'for day and month length'. Day length is certainly important for evaporation is mainly a daytime phenomenon. This innovation is welcomed since it supplements the evidence of the thermometer by a series of values which move in closer harmony with insolation. Its general effect is to add a bonus at higher temperatures, thus steepening the evaporation gradient to about 65 mm/10°C (1.4 in/10°F) and thus raising the potential ceiling at 38°C (100°F) to very nearly the same level as that implied by his first method.

Stripped of their manifold complexities the main contrast between Thornthwaite's two methods is thus revealed as the very real difference between the evaporation implied at a mean monthly temperature of 0°C, namely about 60 millimetres (2.4 in) in the first method and zero in the second. The maximum possible evaporation under terrestrial conditions, i.e. 240–255 millimetres (9.5–10 in), is the same in both.

Offered the perfectly fair challenge by Thornthwaite to 'produce a better' the author endeavoured to comply (Crowe, 1957). Since this journal is now unhappily defunct the essentials are given in Appendix 2 but the original paper should be consulted for the argument and some healthy self-criticism. The effort never became a labour of love and, in his more depressed moments the author felt he was 'producing another', which was not quite the same thing! Yet, even in retrospect, the stress on *maximum*, rather than on mean, temperatures seems to be on the right lines if temperature data are to be used at all and no operations are involved which require more than a slide-rule. Comfort has also come from a most unexpected quarter. Penman has declared 'the annual cycle of evaporation is very nearly a pure sine wave' (in Peake-Jones, ed., 1955, 59). Expressed in these terms, the method aims at locating the absolute value of the crest of this wave and derives its amplitude from a combination of true maximum–minimum range and day-length, both of them insolation factors. The method is least satisfactory in India, that is, just where Penman's generalisation is invalidated by the occurrence of the monsoon.

Figure 36 offers a concluding comment on the relationship between observed pan evaporation and mean monthly temperature. At each station the months are numbered onwards from July (1) to June (12). The 'closed loop' form of the curves was familiar to hydrologists long before Thornthwiate began his work and it is remarkable that he never mentions it. Indeed the feature is so nearly universal that evaporation data that do not exhibit this

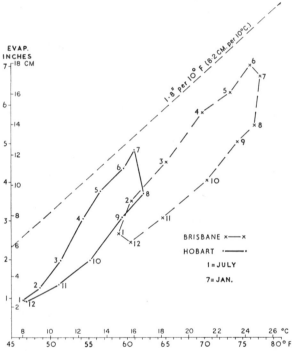

FIG. 36. Open-pan evaporation measurements plotted against mean monthly air temperatures at Hobart and Brisbane. Note the looped form of both diagrams which suggests that any postulated relationship between evaporation from an open-water surface and air temperature must be different in spring and autumn

form are highly suspect. It is thus evident that, for the same temperature, evaporation is greater in spring than in autumn; this is despite the fact that sensible heat is accumulating during the former season and being dispersed during the latter. It has been argued that for deep water-bodies the curve should loop in the opposite direction but I know of no observations which confirm this. A potent factor in producing this form is undoubtedly the difference in phase between the annual insolation and temperature curves—

of which more anon. A contributory cause may be the broad contrast in atmospheric stratification between the two seasons; the clarity of spring skies as compared with the 'season of mists'. At first glance the figure would appear to vitiate all that has gone before but it is one of the advantages of a 'length of day' factor that it produces such a loop, albeit in a somewhat attenuated form. It may be noted in passing that the general gradient of the two curves is (a) approximately rectilinear, and (b) of the order of 1.8 in/10°F or 41 millimetres per 5°C. This is 25 per cent greater than that deduced above from Thornthwaite's 1948 method, but we are now dealing with *water* surfaces (see below, p. 105). Both curves lie below the straight line which will pass through the point 32°F; zero *E*, but both stations have considerable rain at all seasons, an argument in favour of incorporating a rainfall factor into our method. If the reader cares to plot the values for Perth, W.A., from the *Australian Year Book* he will find that values well above the line are recorded from November to February, that is when monthly rainfall is considerably less than an inch. It is under such conditions that open pans are most subject to exaggeration.

It should now be abundantly clear that the derivation of evaporation estimates from temperature data is a rather tricky matter. The best we can hope for is an approximation, though it does not follow that our estimate is much more off the mark than attempts at direct measurement. If better methods were available the attempt might long ago have been abandoned. We now turn to show that alternative methods are either (i) no better, or (ii) if better, particularly in the sense of being more delicate or more logical, they require parameters not universally available and are inherently so much more cumbersome that they are scarcely appropriate for general geographical description.

(*b*) *Estimates based upon the saturation deficit.* At first glance it would appear that the most logical approach to an estimate of evaporation would be via some index of the 'drying power' of the air. The saturation deficit is a function of both the temperature and the moisture content of the air and thus would seem to offer just the index we are seeking. It has been shown by experiment that 'the evaporation value from a water-surface usually increases considerably slower than the saturation deficit' (Budyko, 1958, 154) but this is not difficult to take care of. Thornthwaite and others have raised difficulties regarding the use of a moisture-index; 'atmospheric moisture is not a conservative property of the air', it is as much a result as a cause of evaporation—but these arguments lose much of their force when applied to a general estimate over considerable periods of time. The real difficulties lie elsewhere.

This method too was first applied to *annual* data. As long ago as 1911 a Russian, E. Oldekop, suggested that the annual loss from a water surface could be estimated by $E = 232\ sd$, where $E$ was in millimetres of water and *sd* was the saturation deficit expressed in millimetres of mercury, i.e. in terms of vapour pressure. Clearly the formula remains unchanged when *E* is in inches and *sd* is in inches of mercury. In this form it was introduced into English literature in 1931 where it is known as the 'Waite formula' after the Waite Agricultural Research Institute, Adelaide. Subsequently the co-efficient has been changed, to 258 in 1949 and 263 in 1951. Using the data in the *Australian Year Book* it is easy to show that the correct coefficient for the annual record at Hobart should be 202 whilst at Canberra it should be no less than 50 per cent greater, i.e. 303. Clearly this delightfully simple method has its drawbacks even within the confines of the more humid parts of Australia. The next-lowest and next-highest coefficients amongst the Australian capitals yield little comfort for they are 207 at Adelaide and 274 at Perth, both stations with a pronounced dry season.

TABLE 15. *Factors to be applied to saturation deficit to give pan evaporation*

|  | JAN. | FEB. | MAR. | APRIL | MAY | JUNE | JULY | AUG. | SEPT. | OCT. | NOV. | DEC. | SUM | ANNUAL DATA |
|---|---|---|---|---|---|---|---|---|---|---|---|---|---|---|
| Hobart | 21 | 17 | 16 | 13 | 12 | 10 | 10 | 11 | 14 | 18 | 20 | 21 | (183) | 202 |
| Adelaide | 19 | 16 | 15 | 14 | 12 | 11 | 11 | 14 | 16 | 18 | 19 | 20 | (185) | 207 |
| Perth | 26 | 21 | 22 | 17 | 16 | 14 | 15 | 17 | 22 | 27 | 26 | 27 | (250) | 274 |
| Canberra | 24 | 21 | 23 | 24 | 20 | 17 | 16 | 20 | 23 | 26 | 25 | 24 | (263) | 303 |
| Mean of seven Australian capital cities | 23 | 19 | 19 | 17 | 15 | 13 | 13 | 15 | 18 | 21 | 22 | 23 | (218) | — |

It seems that Oldekop regarded the method as equally applicable to monthly data, suggesting as appropriate a coefficient of about 23 for the warmer half of the year and of 16 for the colder half (Budyko, 1958, 153). He was not far off the mark for the seven Australian capitals yield mean values of 21 from October to March and 15 from April to September. Yet Prescott *et al.* (1952) have used a flat rate of 21 for each month of the year. The author has already indicated the scale of the error involved (Crowe, 1957, 60). Table 15 gives the monthly coefficients for the four Australian stations mentioned above. The other three capitals yield values within these limits. Note that the monthly coefficients cannot be added to give the appropriate annual figure as a 'weighted' mean is involved. All the observations cover a considerable span of years at well-maintained stations. The range of variation revealed may well arise from (*a*) the vagaries of pan data or (*b*) the difficulty of estimating mean saturation deficit but if it is due, as some critics assert, to (*c*) the omission by this method of significant meteorological factors, it must be confessed that the geographical distribution of the extreme stations offers no clue as to their nature.

Prescott has elaborated this approach in his treatment of water loss from different types of vegetation but the basic difficulty remains. Of all the elements of climate, with the exception perhaps of cloudiness, relative humidity is the least amenable to the derivation of meaningful average values.

*(c) Estimates based upon the available energy supply.* We have already seen that the great bulk of the solar energy received at the surface of the earth is ultimately paid into the evaporation account. Furthermore, as Penman (1963, 34) puts it, 'one inescapable overriding condition for maintained evaporation is the provision of a source of energy'. Hence Budyko's (1958, 76) assertion that solar energy balance is 'the most significant factor in the evaporation process'. The logical force of this argument is inescapable but difficulties arise in the attempt to assess such a balance since evaporation is itself the largest of the unknowns that one seeks to identify.

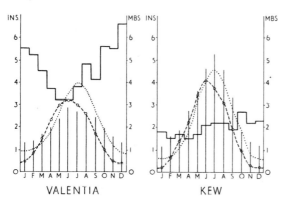

VALENTIA                KEW

Fig. 37. A comparison of the estimates of mean monthly evaporation at Valentia and Kew by Thornthwaite (dotted line) and Penman (broken line). The stepped graph shows mean monthly rainfall and the vertical columns give mean monthly saturation deficit in millibars

In the absence, over most of the earth, of adequate direct observations of incoming and outgoing radiation, Penman (1950, 374) has designed an elaborate formula to assess the available energy supply and the ability of the air to take up more vapour from observations of the 'mean duration of bright sunshine, mean air temperature, mean vapour pressure and mean wind speed'. The method neatly avoids the physical problem of unknown *surface* temperatures but, since this approach must relate to *potential*, rather than to actual, evaporation, the computation is regarded as giving an estimate of loss from an open-water surface. His results for British stations suggest that the method works better in summer than in winter and there is some suspicion that his values may be rather excessive in spring. Plotted against temperature in a diagram similar to Fig. 36, his monthly results thus yield a generous 'loop'. Throughout subsequent refinements and applications Penman has adhered resolutely to the view that potential losses from vegetated land surfaces are *less* than those yielded by his formula. Evapotranspiration is thus less than simple evaporation by a factor of at least 0.8

The method has found wide acceptance and it might well have made all other methods obsolete but for two difficulties: (1) in the absence of computer facilities, the necessary calculations are long and arduous, and (2) over much of the world, the basic data required are just not available.

Figure 37 illustrates the difference in phase implied by the methods of Penman and Thornthwaite and compares both with the annual cycle of the saturation deficit at two contrasting British stations.

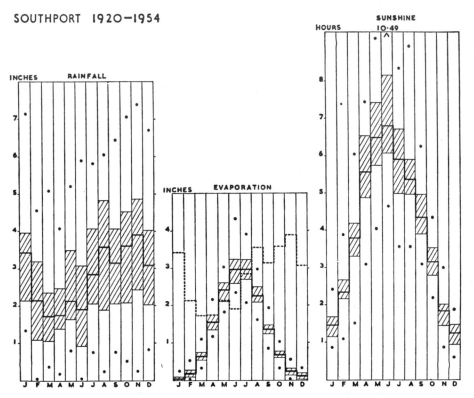

SOUTHPORT 1920–1954

FIG. 38. The variability of mean monthly rainfall, evaporation and sunshine at Southport during the period 1920 to 1954. Heavy lines indicate median values and the rainfall medians are superimposed on the evaporation diagram for comparison

Budyko (1958) has approached the evaporation problem along parallel but, it would appear, more general lines. He seeks to evaluate the 'radiation balance', namely the total radiant energy absorbed at the earth's surface minus the energy lost by outgoing radiation. These are computed by methods which take into account the effect of cloud upon both processes but difficulties arise because 'in the same region values of radiation balance for various underlying surfaces can differ by scores of percent' (p. 91). He is mainly concerned with the mean annual evaporation from moist surfaces and admits that monthly estimates involve an intractable problem in the soil storage factor. His results are reached rather circuitously via the tabulated relationship between two ratios—that between radiation balance and the heat required to vaporise mean annual precipitation, and that between evaporation and precipitation. The derivation of this table (Budyko, 1958, 72) is apparently described elsewhere. We have already reproduced his world map of the heat used in evaporation (Fig. 21). Let us note in passing that the maximum values of 140 kg cal/cm²/year shown over the south Indian Ocean corresponds to an evaporation-loss of 2300 millimetres (90 in) which is very close to the observed fall in the level of Lake Eyre in 1951–2. Over land he does not consider that the *actual* loss (involving further computations) is ever much in excess of 1000 millimetres (40 in).

We conclude this long and rather difficult survey with a quotation from Sibbons (1962, 292): 'The problems raised by the study of potential evapotranspiration are little nearer their final solution than at the time this concept was first formulated by Thornthwaite.' Returning for comfort to hard facts let us contemplate Fig. 38 which illustrates the annual cycle and different degrees of variability of rainfall, evaporation and sunshine at Southport in north-west England.

## Condensation and precipitation

Although these terms express two aspects of the processes brought into play when air is cooled to its dew point and thus attains supersaturation, it is very necessary to distinguish between them. Cloud formation and falling precipitation are certainly associated in the public mind but careful scientific enquiry suggests that they give rise to two quasi-independent groups of problems. One set of physical conditions is responsible for the formation of water-smoke, the fine emulsion of air and water or air and ice which we know as cloud (or fog). Quite another complex set of conditions must be satisfied if this emulsion is to be resolved into falling raindrops or snowflakes. This is the field of 'cloud physics', a subject of most active research and experiment at the present time. Any attempt to summarise its conclusions can be little more than an interim report.

(*a*) *Condensation*. It was in 1875 that Coulier, experimenting with what would now be known as a 'cloud chamber', discovered that the fog produced by a sudden expansion of air would not reappear after repeated experiments with the same sample. He also established that filtered air would not fog but that it could be reactivated by contamination with unfiltered air. He thus decided that condensation in fog form required the presence of 'dust'. Five years later Aitken (1881) rediscovered the same fact but carried the enquiry much further—he counted the droplets, showed that the fog nuclei were not the visible motes but 'a much finer form of dust', distinguished a hygroscopic group of nuclei which would produce results *before* the air reached saturation, and presumed that they originated in dried sea-spray. G. Simpson (1941, 103) explained the need for condensation nuclei in the free air by showing that, even if as many as 5000 water molecules could be imagined as brought together by pure chance, the resulting droplet would have a diameter of only a millionth of a centimetre (0.01 microns, well below the limit of microscopic vision) and that such a droplet must immediately evaporate unless the relative humidity of the air around it had reached the highly improbable value of 125 per cent.

We now know that nuclei are present in the atmosphere in vast numbers and that their size distribution covers a very wide range indeed. Generalisation is difficult but the following classification gives some idea of the magnitudes involved (Mason, 1962, 23–7):

|  | *Diameter* $(\mu)$ | *Mean concentration* (per metre$^3$) |
|---|---|---|
| 'Aitken' nuclei | 0.01–0.4 | $4 \times 10^{10}$ |
| Large nuclei | 0.4–2 | $10^8$ |
| Giant nuclei | 2–60 | $10^6$–$10^3$ |

Aitken nuclei appear to be mainly products of combustion.[1] They are usually in gaseous form and only a small proportion of them play an active part in cloud formation. Large nuclei are usually of sulphuric acid, ammonium sulphate or ammonium chloride. 'Here solid particles may play an important role by absorbing the gases and water-vapour and thereby concentrating the substances' (Mason, 1962, 28). Giant nuclei are mostly of sea salt, derived for the most part from the bursting of foam bubbles. The really important conclusion from these investigations is that there is no part of the lower atmosphere where the gross number of potential nuclei is so small as to suppress, or even seriously to check, condensation in the water phase when other conditions are favourable.

In recent years there has been much interest in a special group of nuclei which appear to play an important part in the transition from the water to the ice phase. They are presumably solid and their concentration varies widely from day to day. On a poor day cloud-chamber tests may reveal only 1/m³ at −10°C; 100/m³ at −20°C; and 1000/m³ at −30°C; but on another day the concentrations may be a hundred times greater. The most abundant material is kaolinite from terrestrial clays. A marked feature of these ice nuclei is that their number increases very rapidly at temperatures below −33°C, bringing the concentration up to $10^6$ per metre³, i.e. much closer to the values quoted above. There is growing evidence that ice nuclei then arise spontaneously, especially when the fall of temperature is sharp. Schaefer

[1] Went (1966) has made the interesting suggestion that natural nuclei are also produced by the action of light with the aid of a catalyst such as nitrogen dioxide from terpenes given off by vegetation at the death of a cell. These are recognisable in the air as the smell of pines, meadows, seaweed, etc. If proved correct, this will relieve us of the logical difficulties arising from the implied connexion between cloud and fire. Marine algae are unquestionably of very great age.

(1946, 1951) considers that the ice crystals form directly from the vapour phase but Mason thinks that the freezing of super-cooled water droplets is their more probable source. We shall see below that their appearance in quantity can have most important results.

These untold myriads of invisible nuclei begin to acquire meteorological significance when they acquire a water-jacket which increases their diameter. The most strongly hygroscopic are able to begin this operation when the relative humidity exceeds 75 per cent and if enough of them grow in diameter to 0.5 microns or more they begin to intercept light and produce 'haze'. The transition from haze to water smoke, whether it be called mist, fog or cloud, is quite abrupt; all these require that the air should reach saturation, either generally or locally. A vast number of nuclei are then set into action and the growth in droplet size is comparatively rapid. We see the analogy with smoke most convincingly when there is a marked contrast in temperature; in the 'steam' from a kettle, in our breath on a cold day and, even more remarkably, from a cold-water jet in the open during the sub-zero weather of an American or Russian winter. In each case the water-smoke *rises* but is rapidly dispersed by diffusion and evaporation. When Nature produces precisely the same effect it does so on a scale that gives quite a different subjective impression, clouds appear to 'float' majestically across the sky whilst fogs may 'hang about' for days together. In fact, in all these instances, we are the victims of an illusion. Water droplets are subject to universal gravitation, hence as soon as they form they fall, but tiny droplets fall very slowly so that the motion is easily neutralised by any upward movement of the invisible air about them. Furthermore such droplets are very easily evaporated; a cloud or a fog thus acquires persistence only by continuous regeneration.

In his delightful book on the *History of the Theories of Rain*, W. Knowles Middleton (1965) describes the difficulties found by early scientists in accepting such a view. He quotes (p. 58) Dalton (1793) as one of the first of the moderns—'a multitude of exceedingly small drops form a cloud, mist or fog; these drops, though 800 times denser than the air, at first descend very slowly, owing to the resistance of the air'. So far so good, but when Dalton continued 'if the drops in falling enter into a stratum of air capable of imbibing vapour, they may be redissolved, and the cloud not descend at all', he was not quite so near the mark. Clouds dissolve on all sides and the flat base so common in cumulus clouds shows rather the level at which they are being generated in a column of rising air.

Water-smoke is thus produced by cooling moist air to its dew-point and the type and density of the cloud-form depends primarily on the manner in which the cooling is effected. If the cooling is slow—as in most fog and stratus—the more efficient nuclei get a good start and then dominate the process. Droplets with diameters of the order of 20 to 50 microns in concentrations of $5 \times 10^6$ to $30 \times 10^6/m^3$ are then characteristic. In more active clouds rapid cooling sets off a much wider range of the spectrum so that droplet-size falls to 5 to 20 microns whilst the concentration rises to between $100 \times 10^6$

5

and $300 \times 10^6/m^3$. Such statements can only be made in the most general terms but even the most extreme concentrations observed suggest that no more than a small fraction of the Aitken nuclei are effective in nature. The opposite trends of droplet size and droplet number suggests that the liquid-water content of clouds is less variable. Estimates range from less than 0.1 g/m³ to a maximum (in the absence of rain) of about 2.0 g/m³. The general mean appears to be less than 0.5 g/m³ even within the tropics. Note that this implies that the liquid water in a cloud occupies no more than $\frac{1}{2\,000\,000}$th of its volume. Furthermore, if *all* the liquid water in a 2000 metres-thick column of such a cloud were to fall to the earth's surface (which never occurs) we should find 1000 cubic centimetres of water spread over an area of 10 000 cm² or a 'rainfall' of only 1 millimetre. Falls greatly in excess of this amount are so common that we are compelled to accept the view, despite all appearances to the contrary, that rain does not 'come from clouds'. It comes from the water vapour in the air via a cloud-forming process, which is quite a different thing.

The reader is referred to meteorological texts for a description of cloud types and for the outlandish names that it has been seen fit to give them (see Scorer, 1963, and Kington, 1969). Seen from the *inside* they are all mists or fogs of very variable density—note the flickering view of the wings obtained from inside an aircraft. From the ground their appearance depends a great deal on illumination and shadow but two basic forms emerge. Clouds can be lumpy or layered, towering or stratified, local enough to be regarded as entities or so general as to screen the whole sky. When seen from a satellite they are found to occur either as great irregular, tattered masses or in long lines or 'streets' which may extend for hundreds of miles. All are produced, and maintained, mainly by the uplift and consequent chilling of moist air and their broadly twofold nature depends on whether that uplift is local or general, rapid or slow. Cumulus clouds have been described as local 'air fountains' but their frequent occurrence in 'streets' suggests the play of forces at least on a subregional scale. Stratus decks are regional phenomena but they may often obscure a good deal of internal cumulus activity.

As a towering cloud builds up to well beyond the freezing level its water droplets become 'supercooled' i.e. they first remain in the liquid phase. This was first inferred by H. de Saussure, a pioneer Alpinist, from the nature of rime.[1] We now know indeed that much of the condensation which occurs at temperatures below freezing point actually takes place in the water phase. We normally have to ascend to levels where the temperature is about $-40°\text{C}$ to reach genuine ice-cloud which has a blurred or fuzzy outline because the crystals are comparatively small in number and yet are more resistant to evaporation when mixed with the drier environmental air. Hence the flat-topped 'anvil cirrus' above an active thunderstorm and the long fallstreaks of 'marestail' cirrus which are slight snowstorms well within the ice-cloud zone.

[1] Heavy rime deposition was apparently the cause of the collapse of a 1260 foot radio mast in northern England on 19 March 1969. A similar process is responsible for icing on aircraft.

(*b*) *Precipitation.* All this detail leads to the important question, why does it rain or snow? Yet it clearly fails to provide an adequate answer. H. Byers has written that precipitation is not 'merely intense condensation, rather it is another phase of the water cycle' and we have stressed this point in our opening paragraph. The mass of a raindrop only 1 millimetre (1000 $\mu$) in diameter is one million times that of a cloud droplet of diameter 10 microns and the problem of precipitation is to grasp how this concentration of mass can occur within a strictly limited period of time. Nature proves again and again that it is capable of performing this miracle within a period of no longer than 30 to 60 minutes.

Best (1957, 32) has shown convincingly that continuous condensation is not enough; even under very favourable conditions, although 'a drop life of one hour will permit the drops to grow to a radius between 10 and 20 microns . . . it may take 6 to 12 hours for drops to reach a radius of 50 microns (diameter 100 $\mu$) by condensation alone'. This is partly because hygroscopic activity weakens as the droplets become more dilute and partly because we are dealing with the growth in *volume* of tiny spheres. Such a process may thus produce the finest drizzle ('Scotch mist') under humid conditions but it cannot produce genuine rain, capable of falling any considerable distance through clear air (see Table 16). This must involve some kind of trigger mechanism, preferably of the 'chain reaction' type, so that momentous consequences follow from relatively minute beginnings. We seek a process, or a series of related processes, which will disturb the relatively stable nature of the emulsion of water and air which we know as cloud to produce the 'fall-out' that we experience as rain or snow.

A deep cloud system must include elements of the following series of strata— and may indeed include them all.

(1) With temperature below −40°C—ice cloud.

(2) With temperature between −40°C and −10°C—a mixture of ice crystals and supercooled water droplets.

(3) With temperature between −10°C and 0°C—supercooled droplets only.

(4) With temperature above 0°C but below the appropriate dewpoint— normal water droplets.

(5) Below condensation level—rising clear air approaching saturation in which the hygroscopic nuclei have not attained visible dimensions.

Although the term 'strata' is justified we must not view this as a static picture for the cloud is maintained by uplift, of the order of 1 m/sec. or some 200 ft/min. The air is thus rising through the series of layers from (5) towards (1) and the cloud it generates is being transformed in the process. The uplift may indeed be highly irregular, involving large eddies and a vigorous churning action, but the temperature strata remain relatively undisturbed since they are the result of the fall of pressure with height (see pp. 228–32). If the uplift is weak or not sustained only the layers (5) to (3) may be involved and the clouds then remain entirely in the water phase. If the impulse is vigorous layer (2) is invaded and ice nuclei become active.

In temperate latitudes the −10° C level is normally attained at 4000 metres (*c.* 13000 ft), though it naturally varies with season and weather situation,

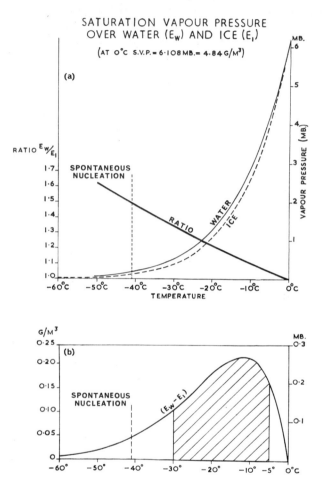

FIG. 39 (a). The actual values of saturation vapour pressure over water and ice at temperatures below 0° C. The straight line shows the variation of the ratio $E_w : E_i$. (b). The difference between saturation vapour pressure over water and ice shown on an enlarged vertical scale. The vertical scale is expressed both in terms of pressure (millibars) and in terms of saturation mixing ratio (grams per cubic metre). It will be observed that the contrast is greatest between −5° C and −30° C

and it was the observation that cloud tops above this level were often associated with rainfall that gave the first clue to the precipitation process. As early as 1911 Wegener pointed out that a mixture of ice crystals and supercooled water droplets is not a stable partnership, and Bergeron (1933) elaborated this into the rainfall theory which now bears his name.

TABLE 16. *Terminal velocities—distilled water (Pressure 760 mm. $T = 20°C.$, R.H. 50%)*

| | DIAMETER | M/SEC | MASS *micrograms* $= g \times 10^{-6}$ | FT/MIN | |
|---|---|---|---|---|---|
| | ($\mu$) | | | | |
| Cloud | 4 | 0.0005 | 0.000032 | 0.1 | 1 ft in 10 min |
| | 20 | 0.012 | 0.004 | 2.4 | 24 ft in 10 min |
| | 100 | 0.27 | 0.524 | 53.0 | 0.6 mph |
| Rain | 200 | 0.72 | 4.19 | 143 | 1.6 mph |
| | 500 | 2.06 | 65.5 | 406 | 4.6 mph |
| | (mm) | | | | |
| | 1 — 1000 | 4.03 | 524 | 795 | 9.0 mph |
| | 3 = 3000 | 8.06 | 14 140 | 1590 | 18.0 mph |
| | 5.8 = 5800 | 9.17 | 102 200 | 1810 | 20.5 mph |
| Hail | 25 | 20.0 | | 3940 | 45 mph |
| | 50 | 30.0 | | 5900 | 67 mph |
| Snow-flakes—dry | | 0.6–1.5 | | 120–300 | |
| —wet | | 1.0–2.2 | | 200–440 | |

*Note*: Limit of unaided vision: 50 $\mu$. Limit of microscopic vision: 0.2 $\mu$.
*Source*: *Smithsonian Meteorological Tables*, 1951, Table 114, p. 396.

(*a*) *The Bergeron theory. Rainfall and icing.* Layer (2) has a mean depth of no less than 4600 metres (*c.* 15 000 ft). Throughout this deep layer clouds contain both ice crystals and supercooled water droplets though in very unequal proportions—often one ice crystal to several hundreds of thousands of water droplets. It was already known that vapour pressure over ice ($e_i$) was *lower* than the vapour pressure over a supercooled water surface ($e_w$) at the same temperature. Fig. 39(a) shows the absolute values of the two sets of pressures at various sub-zero temperatures whilst the difference between them in millibars is illustrated, on a much larger vertical scale, in Fig. 39(b). The progressive increase in the ratio $e_w/e_i$ as temperature decreases may also be read from the lefthand scale of Fig. 39(a). Bergeron's thesis was that in an intimate mixture of ice crystals and water droplets the latter, by evaporation, will tend to maintain the vapour content of the air at the water-saturation level. This will represent considerable supersaturation for ice so that crystals will grow rapidly by sublimation, removing vapour from the air and thus stimulating further evaporation of the droplets. Fig. 39(b) shows that the vapour pressure behind this process is at a maximum at about −12°C and that it remains considerable from about −5°C to −30°C. There is indeed a balance in hand at all sub-zero temperatures but above −5°C few ice crystals are likely to be present to take advantage of the fact whilst below −30°C crystals tend to predominate and the water reserve represented by the droplets will be readily exhausted. At temperatures below −40°C droplets are rapidly frozen and the process of exchange comes to an end. This process provides a most efficient mechanism for producing differentiation of drop size and it may operate continuously as long as fresh droplets (and ice particles) are fed into the system by rising air currents.

The gorged ice crystals thus produced fall at increasing speed as their mass is increased (Table 16) and this sets in train the second process required by the Bergeron theory, further growth by collision and coalescence[1] as they pass through the lower layers of the cloud en route for the ground. Their history during this fall is a complex function of temperature, cloud density and fall path. Low temperatures and comparatively low droplet densities will favour snowflake production whilst a higher water content in the cloud may produce graupel or soft hail. The fact that hilltops are frequently whitened when the lowlands around them have experienced only chilly rain is clear proof that melting occurs during a long descent. Raindrop size depends partly upon the water content of the cloud but mainly on the fall path in relation to the rising air, that is, the time during which the coalescence process is able to proceed.

A great attraction of this theory is its flexibility. Bergeron does not say that, given the stipulated conditions, rain *must* fall. If ice crystals are in short supply only a few active drops will be initiated and these may never reach the soil. If there are too many crystals their growth will be inhibited and the cloud will become glaciated and therefore stable. If the cloud layer is thin coalescence may be ineffective and the elements may evaporate in stratum (5) to produce no more than a lowering of cloud base. Furthermore, in very active cloud it is possible to imagine a number of *supplementary mechanisms* which may augment raindrop production when once the process has been set in motion. At contact with a supercooled droplet an incipient snowflake is thought to throw off a number of minute ice spicules, thus renewing the supply of ice nuclei aloft. Splash from breaking raindrops and flooded hailstones may multiply drop production at lower levels. Falling rain may drag down cool downdraughts which trigger off updraughts elsewhere and thus the process may go on. Little wonder that this theory held the field for nearly twenty years. Indeed it is still thought to be a substantially valid explanation of precipitation throughout the temperate zone except possibly in high summer.

(*b*) '*Warm cloud*' *precipitation.* In recent years, however, increasing evidence has come to hand that quite vigorous precipitation does in fact occur from clouds with summit temperatures as high as 5–10°C. In such clouds therefore the first three of our five strata are completely missing and the Bergeron process simply cannot apply. America, Australia, India and Hawaii have

---

[1] This term applies strictly to water droplets only. The crystals may 'aggregate' and ice and supercooled water leads to 'accretion'.

yielded such evidence of 'warm cloud' precipitation and it may be inferred that the phenomenon is by no means unusual in tropical regions. An important factor would thus appear to be the temperature of cloud base. Where this is high we are usually dealing with an air mass of high water content (see Table 14) and therefore with cloud of high droplet density. Nevertheless the problem of initiating droplet amalgamation still remains.

A method depending on vapour pressure difference, and therefore rather analogous to the above, is available if it can be accepted that cloud droplets of different temperatures can remain intermixed in the cloud emulsion long enough for cool droplets to grow at the expense of warmer ones. The pressure difference in the Bergeron mechanism has been shown to be about 0.2 milli-bars; this would be exceeded at a droplet difference of only 0.3°C at about 5°C, and of only 0.2°C at about 15°C. Such differences are small enough to be not unreasonable but we can do little more than infer that they may occur. What is quite clear is that they cannot be maintained since the process of evaporation and condensation is a most powerful agent for restoring equilibrium. This process would therefore require a considerable degree of small-scale turbulence to maintain continuity and, once again, the mixture would have to contain just the right proportions or the reaction would quietly peter out. Note also that this process is likely to be most effective towards cloud base. For effective growth to raindrop size we must therefore assume a net upward movement in the first stage of the droplet's history.

Ludlam (1951) has used this double sweep process, first upwards and then downwards through the cloud with droplet growth continuing all the time, in a theory of warm rain formation which has received wide acceptance. He notes that the life of a warm cloud is rarely more than half an hour so that in any theory time is a most vital factor. With vertical ascents of from 60 to 240 metres per minute such a cloud is likely to be composed mainly of droplets less than 40 microns in diameter in a concentration of some hundreds of millions per cubic metre. We now know that such droplets are too small to collide, how then can raindrops be generated in the very limited time available? Ludlam notes that sea spray may generate nuclei of the order of 60 microns though in concentrations as low as only 1000/m³. Postulating that these may be swept upwards into the cloud base, he calculates their subsequent growth. Given a cloud depth of some 2000 metres he estimates that they may attain 300 microns near the cloud tops. He regards this size as critical for 'a droplet of about this size can survive a fall of several hundred metres from the evaporating summits with little change of radius and can therefore settle back into the cloud bulk and resume rapid growth' (Ludlam, 1951, p. 411). Many of these growing droplets may survive to emerge from the cloud as rain and, in favourable circumstances, some may break up during their downward path and induce a chain reaction so that the ultimate number of raindrops may be far in excess of the triggering nuclei.

This theory implies that warm cloud rainfall is exclusively the privilege of maritime air masses. What about the summer showers so important in continental interiors? Are these only of the thunderstorm type where cloud reaches far into the freezing layers? East (1957) has maintained that this is by no means necessary. He argues that, in an updraught, 'condensation and co-alescence occur simultaneously' so that the liquid-water content increases as the cloud matures till it reaches values of the order of 6 g/kg towards active cloud tops. Collisions which lead to droplet growth are thus as much a result of random accelerations produced by turbulence as of free fall under gravity. This turbulence is most effective towards the upper limit of the cloud and he thus explains the occurrence there of the first radar echo which requires that droplets should have attained a diameter of not less than 100 microns. Although Ludlam's sea salt mechanism is not ruled out as a contributory factor, he regards condensation-coalescence as 'a basic process which must always be active in sufficiently deep clouds' (p. 73) East's paper may be regarded as reopening the discussion since most earlier workers had tacitly rejected turbulence as a factor except to the degree that it might lengthen the fall path.

(*c*) *The 'balance level' in convective storms.* Recent investigation of the actual fall-speed of precipitation elements by the use of Doppler radar has suggested the presence in many growing storms of an important vertical discontinuity. This has been called the 'balance level', i.e. the level where the terminal velocity of a significant proportion of the falling droplets is approximately balanced by the speed of the updraught. Below this level the net motion of (larger) droplets is downwards whilst above it the net motion (of smaller droplets) is upwards. If the drop-size spectrum and the upward acceleration were both continuous no such discontinuity would be anticipated but contributory factors appear to be wind shear with height and the drag of the larger droplets on the lower sector of the rising column of air. The balance level is observed to rise as the storm develops but its normal position is roughly 'a third of the way down from the storm top' (Atlas, 1966, 649). Its importance in cloud theory is that whilst the droplets are thus almost suspended (rather like a ping-pong ball in a fountain) they may grow with remarkable rapidity whilst covering (upwards and downwards) only a very limited range of elevation.

# The evaporation of falling rain

We have noted that J. Dalton invoked the evaporation of falling cloud droplets to explain cloud base. Even if he was a trifle mistaken, what of the same process in relation to falling rain? Since measured precipitation is the net amount caught by the gauge at surface level there has been a tendency to dismiss this matter as irrelevant; what is not captured by the gauge might never have been precipitated, so why worry about it? This is certainly not the attitude of the cloud physicist and it ought not to be that of the geographer, particularly if he is interested in the rainfall of lands with strong relief. Why are the mean annual totals recorded in the deep Scandinavian valleys (Laerdal 760 mm, Lom 280 mm) or in the great Alpine troughs (Sion 580 mm) so much lower than those recorded on the mountains *on all sides*? What also of the sequence, Jerusalem 400 mm, Jericho 125 mm, Amman 280 mm? Such features are far too readily dismissed as 'rain shadow' phenomena but what do we mean by this term? *Local* showers may be stimulated by relief features but in so far as general rain is a response to large-scale disturbances and is generated high in the troposphere, it is hard to see that minor crevices in the earth's crust can have much effect. The evaporation of falling rain would offer

TABLE 17. *The evaporation of raindrops*

| ICAN (SFC = 15° C = 59° F) RH 70 % | | | | SUMMER TROPICAL (SFC = 41 °C = 106° F) RH 70% | | RH 50% | |
|---|---|---|---|---|---|---|---|
| Initial size mm | | | | | | | |
| D | R | r | Depletion % | r | Depletion % | r | Depletion % |
| *Fall of 1000 metres (3280 ft)* | | | | | | | |
| 0.7 | 0.35 | — | 100 | — | 100 | — | 100 |
| 0.8 | 0.4 | 0.20 | 88 | — | 100 | — | 100 |
| 1.0 | 0.5 | 0.35 | 66 | 0.30 | 78 | — | 100 |
| 1.2 | 0.6 | 0.48 | 49 | 0.45 | 58 | 0.35 | 80 |
| 1.5 | 0.75 | 0.65 | 35 | 0.63 | 41 | 0.50 | 70 |
| 2.0 | 1.0 | 0.93 | 19 | 0.90 | 27 | 0.80 | 49 |
| 3.0 | 1.5 | 1.44 | 13 | 1.41 | 17 | 1.40 | 19 |
| 4.0 | 2.0 | 1.95 | 7 | 1.92 | 12 | 1.90 | 15 |
| *Fall of 2000 metres (6560 ft)* | | | | | | | |
| 0.8 | 0.4 | — | 100 | — | 100 | — | 100 |
| 1.0 | 0.5 | 0.20 | 93 | — | 100 | — | 100 |
| 1.2 | 0.6 | 0.35 | 80 | 0.29 | 96 | — | 100 |
| 1.5 | 0.75 | 0.60 | 49 | 0.40 | 85 | | 100 |
| 2.0 | 1.00 | 0.90 | 27 | 0.83 | 43 | 0.65 | 73 |
| 3.0 | 1.5 | 1.41 | 17 | 1.36 | 25 | 1.30 | 35 |
| 4.0 | 2.0 | 1.93 | 10 | 1.88 | 17 | 1.84 | 22 |

*Source:* After Best (1952, 210) and N. H. Fletcher (1962, 192–3).
$R$ = initial drop radius.     $r$ = reduced drop radius.

at least a partial explanation of such dry 'islands' but the author is unaware of any detailed studies at an ideal site. That evaporation losses occur has long been known and the name 'virga' has been given to falling trails of precipitation that fail to reach the ground. The rim of the Grand Canyon of the Colorado often offers a grandstand view.

Estimation of such losses is far from easy since the factors at work include length of fall path, relative humidity of the sub-cloud layers, drop size, and hence fall speed, the latter decreasing as the drops dry out. Best (1952) has undertaken an elaborate mathematical analysis and we have derived the data in Table 17 from two of his diagrams to give some idea of the magnitudes involved. Two model atmospheres are selected, the ICAN 'standard atmosphere' (see p. 227) with a surface temperature of 15°C, and a 'summer tropical atmosphere' with similar characteristics except that surface temperature is taken at 41°C. Best gives curves for relative humidities of 90, 70 and 50 per cent but we have used only a selection of these and concentrated attention on the lowest two kilometres. Best's results are given in terms of droplet radius; we have added an indication of what this implies in terms of percentage *volume* of water depleted.

It is evident that the tabulated data will not supply a direct answer to the apparently simple question, if a given amount of general rainfall is recorded on high ground what proportion of this amount can be expected to reach a valley floor 1000 or 2000 metres below? The proportion would clearly depend upon the drop size spectrum involved. Nevertheless by employing Best's (1957, 38) useful generalisation, 'the drop size which contributes the greatest amount of water varies from slightly under 1 mm diameter at 0.5 mm/hr to slightly over 2 millimetres at 25 mm/hr', the reader may attain at least some feeling of the losses entailed. They are far from negligible, particularly in a dry tropical environment. Note the extent to which the smaller droplets are completely eliminated as fall-path increases or relative humidity decreases. This throws further emphasis on the importance of large-droplet production within the cloud when cloud base is at a considerable altitude.

## The artificial control of precipitation

The above details become strictly relevant when the question is raised, to what extent, if any, can direct human intervention be expected either to stimulate or inhibit the rain-making process? As a mystical art rain-making appears to be almost as old as man himself, but the methods employed for many centuries fell more in the realm of public relations than in the field of science. The real gift required was the power of convincing the customer that the showers provided by nature were procured entirely through the good offices of the exponent. No mean achievement this!—and before we smile patronisingly it is well to recall that medicine men were simply the advertisers of pre-industrial societies.

The barometer tells us that the total mass of the column of air over any square mile of the earth's surface is of the order of $26\frac{1}{2}$ million tons and that a cubic mile of its lowest and most humid layers weighs about 5 million tons. Any frontal assault on the problem by mass cooling, mass uplift or mass churning would thus involve the expenditure of energy on an astronomical scale. Even if practicable it would certainly be uneconomic.

Scientific interest has thus centred on the trigger mechanisms involved.

Given that nature has prepared a situation in which precipitation is not highly improbable, the question then becomes to what extent can human intervention either release or restrain the natural triggers and thus achieve some measure of control over weather processes? Let us therefore run briefly over the above account to look for points where human ingenuity might be best applied.

(*a*) *Hygroscopic nuclei.* Since these are normally present in a more-than-adequate number for cloud formation the only point in attempting to add to them would be to increase the proportion of large nuclei to stimulate raindrop production. Finely ground salt has been blown into the air in the Punjab in the hope of increasing the rainfall from the monsoon showers which drift across the plains from the east (Fournier d'Albe *et al.*, 1955). This involves general seeding of the atmosphere and it is hard to prove success owing to the inherent variability of such shower conditions.

(*b*) *Freezing nuclei.* These offer a much more delicate trigger in cold clouds for between the $0°C$ and $-15°C$ levels they are composed for the most part of supercooled water droplets ready to respond to the Bergeron process if ice nuclei are introduced or multiplied. The first attempts to do this were made in 1946 when pellets of solid $CO_2$ (which sublimes at $-72°C$) were dropped into selected clouds from the air. In its fall each pellet freezes myriads of cloud droplets before it disappears but this does not necessarily produce rain. The Bergeron process depends upon a delicate balance of numbers; if ice nuclei are in short supply we can expect no more than a few drops of rain but if their number is excessive the cloud 'glaciates' and becomes inert. This was soon discovered by experience. Success in producing showers yielding up to 5 millimetres (0.2 in) of rain has been claimed in Australia but, even with careful radar scanning, it is difficult to be sure that 'seeding' was alone responsible for the result. The use of a plane is expensive and, since the showers are local, it is difficult to serve any particular customer unless estates are on an Australian or Texan scale.

Shortly afterwards Vonnegut (1949) discovered that minute crystals of silver iodide, introduced into supercooled cloud in the form of a smoke, would produce a parallel effect. This appears to be a result of crystal shape though the process is still not completely understood. The smoke can be injected into individual clouds by plane or by explosive rockets or balloons but the great advantage of this method is that a continuous screen can be sent up from a number of burners on the ground. Ludlam (1955) has seriously suggested that a series of such burners set up at 10–20 km intervals might make so considerable an addition to the winter snowfall of the Scandinavian highlands that the cost would be profitably covered by the increase in hydro-electric potential. There is risk in such enterprises. Already one American operator has had to seek a court ruling that he could not possibly have been responsible for a downpour of seven inches which did much more harm than good. Subsequently the British Meteorological Office broke off a series of carefully controlled tests on Salisbury Plain in response to protests from the farming community; unluckily what turned out to be an unusually wet season had been selected for the trial! On the other hand, when the situation is ripe for shower development, premature seeding may well destroy the updraughts before fallout has reached its maximum potential. Thus, in Project Whitetop in Missouri, the release of silver iodide from aircraft on a random selection of days over a period of five years apparently led to a *reduction* of rainfall by 20 to 50 per cent over a very extensive area. (Mason, 1969, 483.)

Yet the method remains the most promising yet known and experiments continue. It costs about £1.25-1.50 per hour per burner but no cheaper material has yet been discovered that gives as satisfactory results. Unfortunately daylight reduces the efficacy of the smoke to one-fiftieth within an hour of emission and if the burners are used at night the natural updraughts necessary to carry the smoke to cloud base are much weaker. Obviously the method is useless for 'warm clouds' and, even when supercooled water is present, the range of activity is limited to the $-4°C$ to $-15°C$ layer. At temperatures below the latter figure ice nuclei are produced naturally in adequate numbers and to multiply them will only stabilise the cloud.

A proposed massive attempt to devitalize an incipient hurricane by this method was recently called off at the last minute by the American hurricane service. It is likely to be attempted again when conditions are more propitious.

(*c*) *Cloud droplets.* This process relies upon Ludlam's (1951) double-sweep theory and is thus suitable for warm clouds. The notion is to spray water-droplets of about 50 microns diameter into the lower layers of deep clouds so that they may grow by sweeping up cloud droplets during their subsequent upward and downward paths. The hope is that for every 1000 pounds of spray emitted some 100000 tons of rain may be released; a 200000-fold return not to be despised by the most sanguine investor. Bowen (1952), who has experimented with this method in Australia, claims good results from clouds exceeding 1500 metres in depth and the tests are being continued. There appears to be little danger of overseeding but the method is of no use for cloud suppression.

To conclude, whatever the ultimate success of these devices, it is most important not to lose sight of two quite basic considerations: (1) They all require the prior existence of cloud, usually of considerable depth and, in the case of ice nuclei, of cloud already in a rather delicately critical condition. The stage has thus to be set by purely natural processes. These alone can produce condensation upon which subsequent developments depend. (2) As N. Fletcher puts it, 'the liquid water content of convective clouds tends to be self-limiting' at a figure which rarely exceeds 1 $g/m^3$ (Fletcher, 1962, 30). This is a measure of the total supply available whatever the technique employed and hence the stress upon cloud depth. Cloud seeding cannot unlock the reserves of water vapour in the atmosphere since 'none of the seeding techniques contain any mechanism for renewing the cloud' (Best, 1957, 154; see also Mason, 1957, 1962).

# 4

# The amount of precipitation

To understand the mechanics of precipitation we have had to carry our thoughts a long way above the surface of the earth but to the practically minded man, be he farmer, engineer or geographer, the vital questions are how much rainfall reaches the ground, how is it distributed normally and under exceptional conditions both in space and time, and at what intensities is it to be anticipated? On the analogy with a bank, we are concerned with reserves, deposits and withdrawals. What limits are imposed by nature on the volume of business transacted? To what extent must we consider processes of transfer from one regional bank to another in order to balance the account?

In the first place let us attempt to get some idea of the magnitudes involved by an elementary deductive approach. The atmospheric reserve of humidity is mainly in the form of invisible water vapour. Clouds are only temporary phenomena and, as we have seen (p. 110), except when actively precipitating, the liquid water they contain rarely represents more than a tenth of an inch of rain even when the cloud bank is many thousands of feet thick. If the air were completely saturated at all levels the humidity reserve expressed in terms of inches of precipitable water would be given by the following table. Temperatures are quoted at 1000 millibars, not far above mean sea level, and the range of values given covers likely night temperatures over most of the habitable earth. The close relationship of vapour content with temperature thus means that the possible reserve falls off rapidly from the tropical zone towards the poles, from summer to winter, and from the surface to the upper atmosphere. The great bulk of the vapour reserve is thus located in the lower half of the atmospheric envelope, i.e. below the 500 millibars level or below 5000 to 5800 metres (16500–19000 ft) according to surface temperature.

An atmosphere fully saturated at all levels is a meteorological rarity but, even if it occurred, the mechanism of precipitation is such that *only a fraction of this reserve is readily available*. This can be illustrated by a simple though rather artificial model. Thus, imagine that a sector of an air mass with properties like those described by any of the lines of Table 18 is lifted up progressively en masse by the intrusion beneath it of a wedge of unsaturated air (itself yielding no precipitation in the process) and consider the situation when layer $A$ has attained position $B$. It is not too unreasonable to argue that layer $B$ would be found in position $C$, whilst layer $C$ would then find itself forming the outermost quarter of the atmosphere since the air formerly above it would have been dispersed (hydrostatically) to other parts of the world thus maintaining the standard pressure. The total yield in fallout from this transformation is then given by $(A - B) + (B - C) + C = A$ mm, i.e. it is unlikely to exceed the values given in column $A$ which represent some 50 per cent of the total reserve in the warmest air mass and reach over 70 per cent of the reserve in the cooler air masses whose yield however is comparatively slight. Of course, a still greater yield could be obtained by increasing the range of uplift but, for a variety of reasons, it is unwise to push this very elementary model too far. Note for instance that, as the intrusive unsaturated layer deepens, rain falling through it will suffer increasing losses from evaporation.

In the above example the rainfall has been derived exclusively from the air already immediately overhead, i.e. the motion of the moist air has been assumed to be entirely vertical. The introduction of a wind, or, to put it more technically, of the horizontal advection of moisture, can produce much heavier rainfalls but then only over *preferred localities*. The highly stylised diagram of Fig. 40 may help to make this point. A fully saturated air mass with any of the properties given in Table 18 is shown moving uniformly at 20 m.p.h. up the sloping edge of a 2100 to 2400-metre (7000–8000 ft) plateau.

TABLE 18. *Precipitable water in a saturated pseudo-adiabatic atmosphere (millimetres)*

| Temperature* at 1000 mb (°C) | A In layers 1000–750 mb | B In layers 750–500 mb | C In layers 500–250 mb | TOTAL (mm) |
|---|---|---|---|---|
| 25 | 43.5 | 27.5 | 10.0 | 81.0 |
| 20 | 30.5 | 17.5 | 4.5 | 52.5 |
| 15 | 21.5 | 10.0 | 2.0 | 33.5 |
| 10 | 14.5 | 6.0 | 1.0 | 21.5 |
| 5 | 10.0 | 3.5 | 0.5 | 14.0 |
| 0 | 7.0 | 1.5 | 0.2 | 8.7 |
| − 5 | 4.5 | 1.0 | 0.1 | 5.6 |
| −10 | 3.0 | 0.5 | 0.0 | 3.5 |

*Temperature quoted is thus the wet bulb potential temperature (see p. 251)
A represents lowest quarter of atmosphere
B represents the second quarter of atmosphere
C represents the third quarter of atmosphere
The water vapour content of the top quarter of the atmosphere (D) is negligible.
*Source*: Generalised from *Smithsonian Meteorological Tables*, 1951, Table 82, p. 327.

Rainfall will occur mainly in the region of the windward slope and, if that zone is about twenty miles wide, sustained precipitation at mean *intensities per hour* of the order of the values under column *A* is probable for as long as the motion persists. Withdrawal from reserves accumulated far upwind is indeed characteristic of the rainfall of all the wetter parts of the world. Note, once again, that this model will not work unless at least all the air in stratum *D* is continuously disposed of to some other part of the world. In fact, to maintain the presumed forward motion of the humid air mass rather more air than layer *D* contains must be continuously removed from above the plateau surface by what Sir Napier Shaw (*Manual IV*, 1931, 307) picturesquely termed the 'scavenging currents' of the upper atmosphere. This introduces us to a most important principle. Air will no more readily flow uphill than water unless it is impelled to do so by considerations of hydrostatic balance. In the absence of an adequate scavenging process stratum *A*, faced by a relief obstacle, may well lie stagnant and inert whilst the wind moves forward aloft. There will then be no uplift and no orographic precipitation though the thermal influence of the plateau surface may well produce local effects of a type to be discussed later.

Yet another potent source of precipitation is the situation where convergent airflow in the lower layers of the atmosphere is stimulated by divergent airflow aloft. This involves a complex, 'live' system with no obvious landmarks and no predetermined dimensions; an elementary model is thus difficult to construct and even for mathematical analysis it is necessary to throw an arbitrary boundary round such evidence of the system as may be revealed by the weather map. Some idea of the consequences can be obtained, however, if we make the not unreasonable assumption that convergence is general below

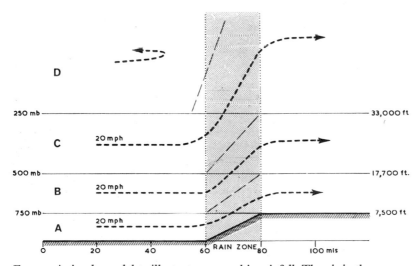

FIG. 40. A simple model to illustrate orographic rainfall. The air in the uppermost layer over the lowland is assumed to be swept away laterally, either into or out from the paper so as to be no longer encountered by the cross-section over the plateau

500 millibars and that divergence is general above that level. Two approaches are then possible. A converging air stream must occupy a progressively smaller area of the earth's surface; the various layers of which it is composed must therefore get thicker and thicker. This is equivalent to a relative uplift of the upper sectors of each layer *plus* a bulk uplift of all layers except the very lowest. Thus, returning to our broad stratification of Table 18, at the moment when the converging layers $A$ and $B$ occupy only one half of their former area, it is evident that the air formerly in layer $A$ must now occupy the levels $A + B$ whilst the air from $B$ will now be found in levels $C + D$. To make this possible the air formerly in layers $C$ and $D$ must have been completely removed by horizontal divergence. This gives a fairly vivid qualitative impression of the close relationship between low-level convergence, uplift and high-level divergence but the apparent fallout by this stage, viz. $(A - B) + (B - C) + B$ $= A + B - C$, gives no measure of the intensity of precipitation for three very good reasons. Note (1) that the area of the system has been progressively decreasing, (2) that we have specified no time for the reduction in area, and (3) that there is no obvious reason why the process should cease at this stage. To achieve quantitative results the alternative approach must be adopted. This consists in throwing an imaginary wall round the system and holding the area thus enclosed constant, at least for a time. The mean rate of updraught over the whole area will then be a fraction of the observed mean rate of entry of air through the 'wall', the fraction being given by

$$\frac{\text{Area of the side wall}}{\text{Area of the enclosed space}}$$

(if the rate is measured in mass per unit time we are freed from the complexities arising from the compressibility of the air). In our own elementary model the wall must extend up to the 500-millibar level, that is it must be about $5\frac{1}{2}$ kilometres high. In a circular area of radius 110 kilometres the mean updraught would thus be only one-tenth of the component of motion of the inflowing air towards the centre. The larger the system the smaller this fraction becomes. If, by way of illustration, we take a mean ascent of 2 kilometres per hour (0.6 m/sec) as characteristic of such systems, we are faced with the upward evacuation at this speed of columns of air containing $(A + B)$ inches of precipitable water in a depth of $5\frac{1}{2}$ to 6 kilometres. A rainfall intensity of the order of nearly $1/3$ $(A + B)$ per hour would thus seem to be indicated. Such systems may thus produce local rainfall of considerable intensity if the air is very warm and humid, but if the system moves, as is usually the case, the total yield may be spread over a much larger area.

All these models are readily adjusted to accommodate layers which are not fully saturated. Uplift will bring such layers nearer to saturation and the net yield from the process will be diminished by the amount of humidity thus added to reserves. The reader is warned that atmospheric motion is usually much more complex than any of the comparatively simple patterns suggested above but the exercises will have served their purpose if they have accustomed

him to think in three dimensions and alerted him the to intimate relationship which exists between the circulations of the upper and lower
atmosphere.

In the three cases considered so far we have made it a rule not to disturb the
relative positions of the various strata of the atmosphere. New features arise
when we consider the phenomenon of 'overturning', the exchange of one
layer with another as a consequence of thermally induced circulations. This
is a common source of precipitation usually described as 'convectional' but it
involves a pattern of motion which is by no means as simple as is often assumed.
We start with a model atmosphere which is completely homogeneous horizontally and is lying virtually inert over an extensive uniform plain. Incoming solar radiation warms the soil and, through it, the lower layer of the
atmosphere until, due allowance having been made for the pressure difference,
it is less dense than a layer at a higher level. What happens? There is plenty of
evidence that the first answer is 'absolutely nothing'! A thin unstable layer
finds the greatest difficulty in leaving the ground—hence mirage phenomena
—and its first function is merely to increase progressively in depth. Given
a depth of at least two or three hundreds of metres, the stage is set for
overturning or 'penetrative convection' but how is this to be organised
between two homogeneous layers of limited depth but of great horizontal
extent? A basic principle here, far too often overlooked in elementary accounts,
is that *updraught must be related to downdraught*. Warm air does not rise of its own
volition; it must be 'buoyed up', an expression which implies that cooler,
denser air comes in to give it an impetus. In the model before us the only
source of cool air is aloft and hence our emphasis on the *two-way* nature of the
process. Clearly the reciprocal currents must displace the same volume of air
in unit time though they may operate at different speeds if their cross-sections
are inversely proportionate to speed. Secondly, they must achieve this
exchange at an *effective distance* from each other. It is not possible to specify *a
priori* what is an effective distance; it must vary with the scale and energy
of the circulation. Yet it should be evident that if the opposing motions are too
close they may lose efficiency by mutual interference whilst, if they are too
far apart, a time factor, a frictional factor, and the distracting effects of
possibly competing systems may all cooperate to prevent the achievement of
the 'continuity' upon which any circulation depends.

Nature appears to solve this problem most effectively by establishing a
cellular pattern, best seen during the break-up of a uniform cloud sheet where
the uniformity of scale is often most remarkable. In our predicated model
there is no obvious reason why updraught or downdraught should begin at
one point rather than another. What appears to occur is that when one cell
is set off, perhaps by pure chance, it rapidly stakes a claim to an effective
territory and, at the same time, triggers off comparable cells at an appropriate
distance. Under uniform conditions, competition between cells may be
expected to weed out the least efficient and thus make for regularity of pattern
but if widely scattered areas have some initial advantage (a warm slope or a

patch of sandy soil) a composite pattern is likely to emerge and the giants may ultimately damp out their smaller neighbours.

In a theoretical treatment of a system of regular cells it is convenient to regard them as a pattern of regular hexagons since this is the geometric figure obtained by drawing straight boundary lines halfway between a number of equidistant centres. In such a system it is a matter of indifference whether updraught is near the centre of each hexagon and downdraught round its margins or vice versa. Both types of circulation are observed in nature but where the turnover is deep enough to involve the release of latent heat at condensation the former would appear to be the more probable. Under such conditions updraught speed is likely to be considerably in excess of the rate of subsidence so that the *area* beneath updraught elements must involve only a fraction of the area of the whole system. Fig. 41 illustrates how a pattern where the updraughts (shaded) occupy (*a*) one-third and (*b*) one-quarter of the total area can be obtained graphically and the process may be extended to evolve still smaller fractions. This approach makes the relative areas self-evident but it is clear that, starting with a pattern like Fig. 41 (*b*) for instance, we could easily expand or shrink the updraught cells to represent any desired fraction of the total area.

To estimate the probable yield in rainfall of such a process of cellular overturning we need to know the characteristics of the air mass involved and particularly (1) the vapour content of the layer likely to be displaced, and (2) the height to which the overturning will extend. We shall return to the complex problems involved in atmospheric 'instability' at a later stage (see pp. 235–51) but, as a rough numerical illustration, let us take an extreme case and assume that a fully-saturated layer between the 1000 and 875 millibar level (i.e. roughly the bottom 1000 metres of the atmosphere) is in the process of being lifted to the 500 to 375 millibar level (i.e. to above 5.5 km). The precipitable-water content of these two layers of the atmosphere at saturation is given below, and, if the rest of the atmosphere plays an entirely passive role in the process, the maximum available rainfall will be the difference between them.

| Temperature (°C) at 1000 mb | 25 | 20 | 15 | 10 | 5 | 0 | −5 | −10 |
|---|---|---|---|---|---|---|---|---|
| Precipitable water | | | | | | | | |
| 1000–875 mb (mm) | 23.0 | 17.0 | 12.0 | 8.5 | 6.0 | 3.5 | 2.5 | 2.0 |
| 500–375 mb (mm) | 7.5 | 3.5 | 2.0 | 0.8 | 0.5 | 0.2 | 0.1 | 0 |
| Difference (mm) | 15.5 | 13.5 | 10.0 | 7.8 | 5.5 | 3.3 | 2.4 | 2.0 |

The interpretation to be placed on the last line of this table must clearly depend upon a number of circumstances. If the updraught cells move relatively to the air mass to tap successively each sector of it, then the figures give some notion of the maximum total fall, spread regionally or more or less evenly over the whole area. If, on the other hand, the updraughts remain stationary but draw upon the reserves beneath the downdraught areas, local convergence will have occurred and *local* falls three, four or five times as great as the figures quoted may be recorded though the overall total, including areas which experience no rain, must remain the same. Similarly, if the whole air mass is drifting slowly over the landscape carrying the active cells with it the local falls will be spread over a wider area and 'millimetres of rainfall' must be reinterpreted accordingly. Finally, if two or more active cells merge and, thus reinforced, set out to poach upon the territory of their neighbours, they may generate extremely heavy local falls but this cannot affect the regional total; it is purely a matter of its distribution.

In fact it is most unlikely that the whole of the potential supply of moisture indicated by our table will indeed become available. Reasons for this will

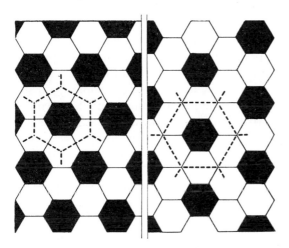

(a) ONE IN THREE      (b) ONE IN FOUR

FIG. 41. Two examples of hexagonal cell patterns

be discussed later (see p. 234; 248) but the basic difference between this type of circulation and the three types discussed above is that the 'spent' air *remains overhead;* it is not removed from the region. Uplift of any portion of the surface layer thus involves subsidence i.e. increasing pressure, in the layers over which the lifted air must move. An increase in the temperature of these layers is thus involved which, at one and the same time, increases their 'capacity for moisture' and decreases the degree of latent instability with which the process has been presumed to begin. A damper or brake is thus an inherent feature of such a system, increasing in power as the process develops until the conditions favouring precipitation are finally eliminated. A time element is clearly involved here though the diurnal surge and decline of solar insolation may well impose its own time-scale on this type of circulation.

We may conclude from this discussion (*a*) that from a regional point of view convectional rains are unlikely to be heavy unless temperatures are high— even then the humid surface layer, once exhausted, must be recharged either by local evaporation or by advection of moisture from elsewhere; and (*b*) though local downpours are possible, and indeed common, this effect is achieved by a process which robs Peter to pay Paul so that purely convectional rainfall is notoriously unreliable. With prevailing low temperatures such a process must be relatively unproductive. Indeed we shall see that really cold air masses have such a vertical structure that the process usually is impossible.

# The processing of rainfall data

These general speculations are a long way from the routine tasks of making the daily readings from rain gauges upon which most of our knowledge of the actual distribution of precipitation is based. Such readings (usually made from gauges with a funnel diameter of 5 or 8 inches, set 12 inches above the ground) are customarily expressed as a depth of water (given in inches or millimetres) such as would accumulate upon a level surface from each day's precipitation if there were no loss from either evaporation or percolation. In British practice the 'day' so employed is a period of 24 hours beginning at 9 a.m. when the gauge is normally emptied. Values are now available from some 5500 sites in Great Britain and Northern Ireland, the great majority of them run by volunteer enthusiasts to whom both scientists and practical administrators owe a debt it is far too easy to forget. In other lands the general pattern is very similar for a rainfall network must be much closer than that required for routine forecasting.

Apart from its sheer bulk, the material which thus comes to hand for analysis would thus appear to be essentially simple in character. In fact, its treatment presents a variety of peculiar difficulties, mainly of a statistical nature.

1. In the first place, the basically local character of individual rainstorms, even when they are quite extensive, means that a map of *daily* rainfall has an interest which is quite ephemeral—except perhaps in the case of some particularly notable storm.

2. Equally important is the fact that rainfall is *temporary* in character, it is not a continuous function like temperature or pressure, and both its onset and its cessation are usually comparatively abrupt. A time factor is thus always involved and for some purposes the day is much too long. On the other hand, because of its local character, no systematic variation of precipitation with time, no seasonal regime, is likely to emerge unless the material is assembled for time periods of the order of from 5 to 30 days (pentads or months).

3. Thirdly, precipitation is measured as an *amount*. This has two important consequences, (*a*) Amounts of precipitation must often be considered cumulatively, as is so obviously the case with snowfall, but the effective period of accumulation varies widely according to the matter under review. In considering a single river flood we shall approach the data with quite a different emphasis from a student of artesian water reserves. (*b*) Furthermore the amount may be zero in the sense that either absolutely no rainfall was recorded or that the fall was too small to measure (sometimes recorded as 'trace'). Zero readings of this type do not occur in connection with either temperature or pressure so that the methods of statistical generalisation appropriate to these elements of climate are not necessarily suitable for rainfall analysis. There are

some nice points here. Thus, if the diminishing series, 3, 2, 1, 0, represents the number of cars owned respectively by four of our acquaintances A, B, C, and D, do we not regard D as in quite a special category? He is a 'have-not' and to include him in any statistical assessment of the 'average motorist' would be patently absurd. Is a day without recorded rainfall to be interpreted as one on which an infinitesimally small amount must have fallen or as a quite separate and distinct 'have-not'? The question has its meteorological as well as its statistical implications. A parallel issue arises in regard to calms and very light winds but since winds are usually analysed in terms of frequency from various directions, a separate column is allotted to 'calms' and there is less danger of statistical misrepresentation.

Yet a single rainless day does not constitute a drought; what then is the logical period over which daily rainfall amounts can be most usefully accumulated for statistical analysis? A *five-day* period has an attraction to the tidy mind since the normal year can be divided into seventy-three such periods. Ramamurthy (1965) has analysed long records for 168 stations in India on such a basis and reached some interesting conclusions. However, the method has its disadvantages which are worthy of enumeration: (i) the great number of the periods makes both the tabulation and diagrammatic representation of the data extremely cumbersome. (ii) There is a high incidence of zero values and hence a high degree of 'scatter' of the pentad values. This makes it difficult to distinguish the significant features of the data from the element of random 'noise'. (iii) It is doubtful whether, even in India, the rainfall seasons are sufficiently clear-cut to be amenable to such fine analysis. Certainly this would be true of most other parts of the world. (iv) Is a period of five days not too short for effective compensation by soil storage to be ruled out? This is really a question for the agronomist who may wish to apply the data thus presented to some practical end.

To illustrate the effect of a shortened period of reference at a station with very striking characteristics we have combined pairs of pentads to give a 10-day interval[1] in Fig. 42 (a) and placed beside it a diagram (b) for the standard months on a vertical scale one-third of that for (a). The periods covered by the two diagrams are not precisely the same but it is clear that a time interval of less than one month is highly desirable in the analysis of Indian rainfall. K. Ramamurthy (1948) has already employed *fortnightly* data in a valuable paper on the rainfall of Tamilnad.

[1] To combine pentads properly one should return to the original data and add pairs before plotting. In the absence of the detailed record I have added the appropriate medians and quartiles as given by Ramamurthy and then reduced the mean scatter by $1/\sqrt{2}$. As a consequence the mid-values may be a trifle low but experience has shown that the scatter will be fully representative.

The fact that the year does not divide conveniently into ten-day periods offers no real difficulty. An overlap into the following year during the 37th period between 27 December and 5 January can be handled as in the diagram, whether rain is actually recorded during that period or not. The boundary dates of the ten-day periods must, of course, remain the same for all years.

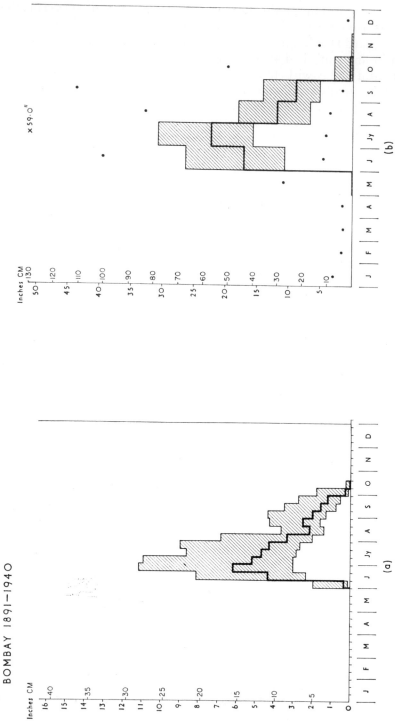

Fig. 42. The annual rainfall regime at Bombay as revealed by an analysis (a) in terms of ten-day periods and (b) by the usual monthly grouping. The heavy line shows median values and the shaded area covers the interquartile range, i.e. it covers the central half of the records in each column. Note that the vertical scale of the monthly diagram is one-third of that employed for the ten-day periods though the base-lines are identical. Unfortunately the periods covered by the two analyses are not quite the same. It is clear, nevertheless, that at a station with such extreme rainfall contrasts interesting details may be lost by the use of the traditional monthly totals. In the monthly diagram the highest and lowest monthly totals are indicated by dots

The most common method of presenting a summary of rainfall observations today is to tabulate them as a series of *monthly totals*. The inequalities in the number of days in the month are accepted as an awkward fact for which due allowance can be made, if necessary, when the data are being employed. Such a summary should give the monthly values for each separate year of the

CAMDEN SQUARE, LONDON 1920-1960     RAINFALL DATA

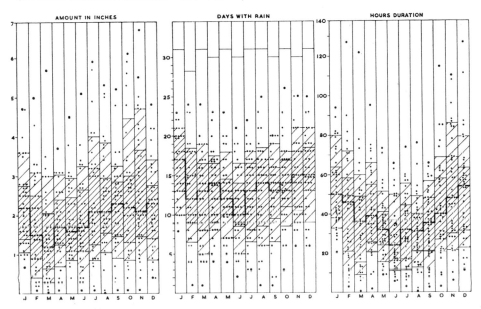

FIG. 43(a). The annual regimes of rainfall amount, days with rain and duration of rainfall (in hours) at Queen's Park, Bolton (1920–57) and Camden Square, London (1920–60). It will be noted that the graphs of rain duration show a more orderly seasonal transition than either of the other two methods of assessment

record. This will provide ample evidence to put us on our guard against over-facile generalisation. Fig. 43(a) has been plotted from the 1920–60 data for Camden Square, London (for long the home of the British Rainfall Organisation), and Fig. 43(b) for Bolton, in a wetter part of England. The degree of variability shown by both diagrams is such that many minor differences between average values for each of the monthly columns are the products of pure chance and completely without significance (see Appendix 1). No statement of monthly means for the whole period is of much value unless it is accompanied by some indication of the degree of scattering of the values from which each mean has been derived. As a broad generalisation for the British Isles it can be said that differences of less than about 10 millimetres (or about 0.5 inch) in long-period monthly means are unlikely to be reliable. In tropical lands where rains are heavier and records shorter the range of statistical error may be very much larger.

Rainfall may also be expressed in terms of the number of *rain days* per month. Figure 43 shows that this number has also varied widely from year to year at both Camden Square and Bolton. But before rain days can be

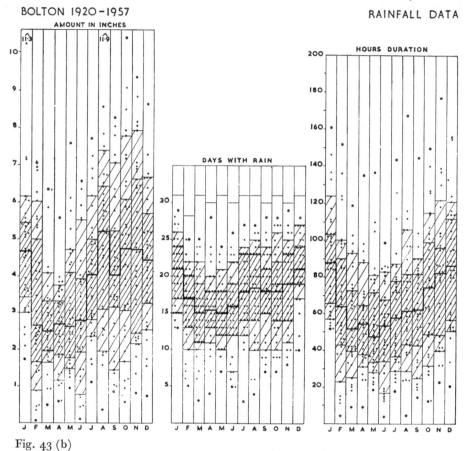

Fig. 43 (b)

counted they must be defined. In Britain when the measurement was made in inches a 'rain day' was defined as the period of 24 hours from 9 a.m. during which at least 0.01 inches of rain was recorded. Since the record was made to two decimal places this in fact included all measurements of 0.005 inches or more. It is argued that amounts smaller than this figure may well be the results of direct condensation on the gauge. Experience indicates that the use of so low a limit produces marked inconsistencies between neighbouring stations. The British Rainfall Organisation also distinguishes days with falls of at least 0.04 inches (1.0 mm) as 'wet days'. This produces results which are consistent enough to be profitably mapped; it is also closely in accord with continental practice.[1] The Indian Meteorological Service takes yet another step and distinguishes 'rainy days' with falls of at least 0.1 inches (2.5 mm).

[1] Data for days with 0.20 inches (5 mm) or more are now also collected.

It is interesting, at any given station, to carry the analysis still further (see Fig. 45). Note that a single shower occurring when the gauge is due to be read may be spread over two days even if its actual duration is only an hour or two.

A much more refined statement of the frequency of rainfall is possible if it can be expressed in *hours*. This also requires an agreed minimal rate of fall per hour[1] and before self-recording apparatus became available such data could be little more than personal estimates. Fig. 43 indicates that this approach also encounters a wide range of variability though the course of the monthly means is rather more consistent than in the case of either amount or rain days at both stations.

All three of these methods can thus be used to give an indication of the mean distribution of rainfall in time, i.e. to show the seasonal regime of precipitation, but the results must be interpreted with care. In Table 19 the diagrams of Fig. 43 are summarised by a statement of a mean value for each column (the median) together with an index of the mean scatter (the quartiles). The maximum value in each line is indicated by heavy type and the minimum value by italics. Notice how misleading undue emphasis upon the month of maximum or minimum values could be. Thus, at Camden Square, in one or other of the nine lines of the table, maximum values occur in August (1), September ($\frac{1}{2}$), November (2), December (3), and January ($3\frac{1}{2}$),[2] whilst minimum values occur in March ($4\frac{1}{2}$), June (4) and July ($\frac{1}{2}$). From these facts we can develop no more than a rather confused picture of the general run of events from 1920 to 1960 with November, December and January as a fairly compact period of heavier and more prolonged rains whilst March and June, on the whole, appear to be comparatively fine. At Bolton, on the other hand, maxima occur in August (1), November (1), December (3), and January (4) whilst minima are found in February (1), March (3), April (2), May ($2\frac{1}{2}$) and June ($\frac{1}{2}$). Here then the most consistently wet season would appear to be December–January,[3] the heavy falls in August being of comparatively short duration, whilst the drier season occurs in March, April and May.

This is an unusual but perfectly valid approach to the problem of seasonal differentiation. It could be further elaborated by tabulating the 'octile' values or by taking account also of the second-highest and second-lowest value in each row of Table 19. How far this is a true picture of the climate of either station over a much *longer* period of time it is very hard to say. Analysis of earlier records has shown that the rainfall for a single month can be persistently anomalous in Britain for as much as half a century (see Crowe, 1940, 307).

*Rain spells and dry spells.* It can be argued with some cogency that the search for a seasonal regime in this fashion is unreal on the grounds that 'the normal

---

[1] In Britain a rainfall hour is defined by precipitation at the rate of at least 0.001 inches per hour.

[2] Yet August had only one rain day in 1940 and September only one in 1941, though 24 rain days were counted in March 1947, and 22 in July 1922.

[3] With November a close runner-up.

never occurs'. The features of climate which really affect man, the features which he recalls when he thinks climatically, are the exceptional occurrences—not the normal or average rainfall but the floods and droughts. There is much more than a grain of truth in this assertion but it is a view that must not be

TABLE 19. *Rainfall at Camden Square, London, and Bolton, Lancashire*

| | JAN. | FEB. | MAR. | APR. | MAY | JUNE | JULY | AUG. | SEPT. | OCT. | NOV. | DEC. |
|---|---|---|---|---|---|---|---|---|---|---|---|---|
| **CAMDEN SQUARE, LONDON (1920–60)** | | | | | | | | | | | | |
| *Amount (in)* | | | | | | | | | | | | |
| $Q_3$ | 2.75 | 2.5 | 2.05 | 2.4 | 2.45 | 2.65 | 3.2 | 2.95 | 2.85 | 3.0 | **3.65** | 2.75 |
| $M$ | 2.2 | 1.5 | 1.2 | 1.7 | 1.6 | 1.7 | 2.1 | 2.1 | **2.3** | 2.2 | 2.1 | **2.3** |
| $Q_1$ | 1.4 | 0.9 | 0.65 | 2.25 | 1.1 | 1.1 | 1.25 | **1.55** | 1.45 | 1.3 | 1.45 | 1.4 |
| *Rain Days* | | | | | | | | | | | | |
| $Q_3$ | **20** | 18 | 16.5 | 16.5 | 14.5 | 14 | 16.5 | 17 | 16 | 17 | 18 | 18.5 |
| $M$ | **17** | 12 | 13 | 14 | 12 | 10 | 13 | 14 | 13 | 14 | 15 | 15 |
| $Q_1$ | **13** | 9 | 8 | 9 | 10 | 8.5 | 8 | 8.5 | 9.5 | 11 | 10.5 | **13** |
| *Duration (hours)* | | | | | | | | | | | | |
| $Q_3$ | 64 | 58 | 49 | 55 | 41 | 37 | 44 | 40 | 48 | 58 | **69** | 63 |
| $M$ | 50 | 46 | 36 | 39 | 32 | 24 | 32 | 31 | 35 | 40 | 48 | **54** |
| $Q_1$ | **38** | 21 | 15 | 26 | 24 | 16 | 17 | 21 | 21 | 27 | 30 | 31 |
| **BOLTON, LANCASHIRE (1920–57)** | | | | | | | | | | | | |
| *Amount (in)* | | | | | | | | | | | | |
| $Q_3$ | 5.45 | 5.0 | 3.3 | 3.3 | 4.1 | 3.9 | 5.0 | 6.45 | 5.2 | 6.1 | **6.6** | 5.7 |
| $M$ | 4.65 | 2.65 | 2.5 | 2.7 | 2.65 | 2.8 | 4.05 | **5.2** | 4.0 | 4.75 | 4.7 | 4.45 |
| $Q_1$ | **3.4** | 1.65 | 1.95 | 1.85 | 1.75 | 1.9 | 2.8 | 3.1 | 3.15 | 3.05 | 2.45 | 3.25 |
| *Rain Days* | | | | | | | | | | | | |
| $Q_3$ | **24** | 20 | 19 | 18 | 18 | 19 | 23 | 23 | 21 | 22 | 23 | **24** |
| $M$ | **21** | 17 | 15 | 15.5 | 15 | 16 | 18 | 18.5 | 18 | 18 | 19 | **21** |
| $Q_1$ | 17 | 13 | 11 | 13 | 12 | 12 | 14 | 14 | 15 | 14 | 14 | **19** |
| *Duration (hours)* | | | | | | | | | | | | |
| $Q_3$ | 103 | 90 | 73 | 65 | 72 | 68 | 78 | 82 | 82 | 99 | 96 | **112** |
| $M$ | **88** | 64 | 52 | 54 | 47 | 53 | 58 | 62 | 62 | 75 | 82 | 87 |
| $Q_1$ | **66** | 43 | 40 | 38 | 34 | 34 | 37 | 43 | 43 | 49 | 48 | 57 |

carried too far. To do so would be analogous to assessing a man's wealth from his occasional gains and losses on the stock-market or the turf without asking whether he has a regular income. Nevertheless there is an element of gambling, of taking calculated chances, in both agriculture and engineering which it would be very foolish for the geographer to ignore. Of any place on earth it is an entirely valid question to ask what is the longest period which has elapsed without appreciable rainfall—or, broadening the interest, when are such periods likely to occur and what is their frequency? Equally reasonable are such questions as, How long can it go on raining, hour after hour, or day after day? or, What is the maximum amount of rainfall likely to occur within any given period of time? The difficulty, of course, is that special questions require special answers, and special answers usually involve special investigations. We can deal here with some of the issues involved only in rather general terms.

The basic information for our own country is contained in the annual volumes of *British Rainfall*. Here are listed the times of occurrence at various stations of 'absolute droughts', 'dry spells', 'rain spells' and 'wet spells'.[1] But one cannot digest a long list, it is still necessary to *operate* even on this class of material.

(*a*) A simple first step is to extract some of the most exceptional occurrences. Thus for three stations with good records from 1921 to 1960 representing respectively the drier, wetter and very wet sections of Great Britain we have the results tabulated below.

| | Longest dry spells | | Duration | Total amount (in) |
|---|---|---|---|---|
| Camden Square | 28 July – 11 Sept. | (1940) | 46 days | 0.06 |
| | 21 Aug. – 28 Sept. | (1929) | 39 days | 0.02 |
| | 5 Aug. – 11 Sept | (1947) | 38 days | 0.04 |
| | 17 Nov. – 24 Dec. | (1933) | 38 days | 0.09 |
| Greenock W. W. | 29 July – 3 Sept. | (1947) | 37 days | 0.01 |
| | 4 July – 7 Aug. | (1955) | 35 days | 0.03 |
| | 1 Feb. – 3 Mar. | (1932) | 32 days | 0.00 |
| Fort William | 19 Jan. – 25 Feb. | (1947) | 38 days | 0.01 |
| | 29 July – 3 Sept. | (1947) | 37 days | 0.02 |
| | 5 Oct. – 2 Nov. | (1946) | 29 days | 0.07 |

| | Longest wet spells | | Duration | Total amount (in) |
|---|---|---|---|---|
| Camden Square | 19 Oct. – 4 Nov. | (1960) | 17 days | 4.83 (only instance) |
| Greenock W. W. | 17 Jan. – 18 Feb. | (1928) | 33 days | 12.99 |
| | 8 Jan. – 6 Feb. | (1938) | 30 days | 10.78 |
| | 13 Feb. – 13 Mar. | (1926) | 29 days | 10.69 |
| Fort William | 8 Jan. – 13 Feb. | (1938) | 37 days | 20.37 |
| | 16 July – 18 Aug. | (1946) | 34 days | 10.41 |
| | 23 Dec. – 21 Jan. | (1929–30) | 30 days | 10.78 |

Note that there is little variation in the *intensity* of dry spells as between the wetter and drier sections of the country—this is partly induced by the definition

[1] An *absolute drought* is a period of at least 15 consecutive days, to none of which is credited 0.01 inches (0.2 mm) or more of rain.
A *dry spell* is a period of at least 15 consecutive days, to none of which 0.04 inches (1.0 mm) or more of rain is credited.
A *rain spell* is a period of at least 15 consecutive days to each of which is credited 0.01 inches (0.2 mm) or more of rain.
A *wet spell* is a period of at least 15 consecutive days to each of which 0.04 inches (1.0 mm) or more of rain is credited.

(the spell must not be broken by a day with 0.04 inches (1 mm) or more). There is some contrast in their *duration* though the reader may be surprised to learn that a full month may pass in the western Highlands without appreciable rain. The effectiveness of such a drought, of course, is influenced by the season of its occurrence. Here the contrast is more marked—three of the extreme cases at Camden Square were in summer when evaporation is high whilst two of the three longest dry periods at Fort William were in the winter half of the year.

There is, on the other hand, a very marked contrast in the regional distribution of wet spells. Camden Square experienced 15 consecutive days with 0.04 (1 mm) inches or more of precipitation only once during this period. At Greenock and Fort William, however, a full month of unbroken wet days is occasionally to be expected and at the latter station one of the more extreme cases occurred in midsummer.

(*b*) A useful second step is to count the frequency of such occasions and to tabulate the results. Thus,

*Number of dry or wet spells in forty years* (1921–60)

| | Dry spells | | | Wet spells | | |
|---|---|---|---|---|---|---|
| | Summer[1] | Winter[1] | Total | Summer | Winter | Total |
| 1. Camden Square | 47 | 53 | 100 | 0(2)[2] | 1(4)[2] | 1(6)[2] |
| 2. Greenock | 28 | 20 | 48 | 1 | 22 | 23 |
| 3. Fort William | 12 | 16 | 28 | 13 | 43 | 56 |

| | Max. no. in any one year | No. years with no such spell | Max. no. in any one year | No. years with no such spell |
|---|---|---|---|---|
| 1. Camden Square | 5 (1921, 1942, 1955) | 1 | 1 (1960) | 39 |
| 2. Greenock | 4 (1933, 1941) | 12 | 3 (1928) | 23 |
| 3. Fort William | 3 (1947) | 17 | 4 (1928, 1950) | 9 |

[1] Half-years, April–September and October–March.
[2] In brackets, figures for 'rain spells'.

It will be observed that at Camden Square five *dry spells* have occurred in each of 3 years (1921, 1942 and 1955) though 1923 recorded none. The mean frequency is about five dry spells over a 2-year interval, there being a slight preponderance in the winter half-year (October to March) when spells are rather shorter. At Greenock 12 years recorded no such spell so that the average for the remaining years was less than two per year, though four were recorded in 1933 and 1941. At Fort William nearly half of the years had no dry spell and there was an average of rather more than one per year during the remainder of the period. The winter half-year was again slightly favoured and under the very exceptional conditions of 1947, two spells fell in winter and one in summer.

As might well be expected, *wet spells* are predominantly a feature of the winter half-year yet even at Fort William where three can be expected every 2 years, as many as 9 years during the period made no such record. It is also

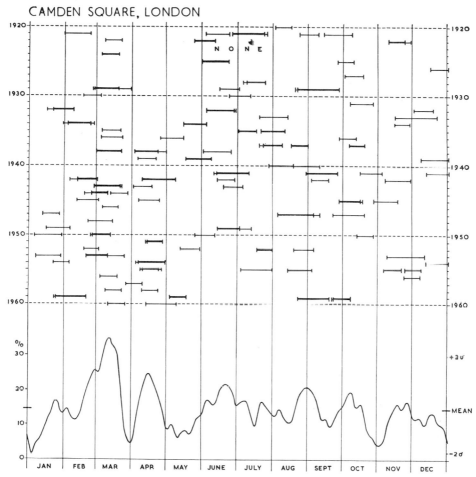

FIG. 44. 'Droughts' and 'dry spells' at Camden Square, London during the period 1920 to 1960. See text for definitions. The duration of each spell is shown by the horizontal bars, heavy for droughts and light for dry spells which often include drought periods. The curve in the lower part of the diagram is the result of a day-by-day count of dry spells vertically across the upper section but for reproduction on this scale the figures have been slightly smoothed by the use of a five-day running mean. Some remarkable singularities in the occurrence of dry weather in London are thus revealed

noteworthy that at Fort William the mean fall during wet spells in winter (11.3 in) was nearly twice the mean fall in such spells in summer (6.5 in).
(c) A third method of analysis is to plot the data as we have plotted the record for droughts and dry spells at Camden Square (1920–60) in Fig. 44. From the upper part of the figure it is evident that, in London over this particular

period, 'droughts' were much less frequent during the 3 mid-winter months (November–January) than at any other season of the year. Indeed the drought of mid-November 1922 appears to have been a highly exceptional event.[1]

On the other hand, *dry spells* are of much more widespread occurrence and only the 3 days 3–5 January were never involved in such a spell. The number of occasions in which any one day was included in a dry spell can be obtained by counting the number of bars crossed by any given vertical cross-section of this diagram. Results, expressed in terms of percentage frequency, are shown in the lower part of the figure.[2] The curve must clearly contain a random element but it shows some remarkable features which point to 'singularities' in the climate of London. Taking the year as a whole, the mean chance that any single day may be involved in a dry spell works out as 14.5 per cent, i.e. about 1 year in 7 but from 7–20 March the observed frequency is over twice that figure i.e. over 29 per cent whilst from 1–13 January; 28 March–3 April; and 28 October–8 November it is less than half the normal, i.e. 7.2 per cent or about 1 year in 14.[3] The causes of these singularities are not understood (see Brooks, 1954; Lamb, 1964, 147–56 and 175–89) but the curve has a practical relevance to organisers of out-of-door functions, particularly those which require firm ground underfoot, such as agricultural shows and similar field exercises.

# The intensity of precipitation

This important subject can be handled only in very general terms where the record is derived from *daily* readings of a rain gauge. Nevertheless, since the effects of more than one downpour in fairly close succession are cumulative from the point of view, for instance, of the drainage engineer or bridge builder, an examination of daily records is worth while.

At the world's wettest station, Cherrapunji with an annual average of 425 inches of rain on 176 rain days, the mean rate of fall is 2.4 inches per day (2.7 in per 'rainy' day). Here 36.4 inches was recorded on 21 June 1934 and over 10 inches per day has been recorded at one time or another in all months except December, January and February. The world's record fall in a single day is 73 inches,[4] a catastrophic figure which might well be questioned.

In the British Isles, justly famous for its moderation in all things climatological, daily falls of over 2.5 inches are designated 'noteworthy', falls over

---

[1] There were no droughts at all between 26 November and 21 January and again, for shorter intervals, from 24 March to 2 April and from 22 October to 10 November.

[2] The curve has been smoothed by a 5-day running mean—this introduces some slight modification of sharp peaks and troughs, notably the trough in early January.

[3] Periods with frequencies half as large again as the normal (over 21.7 per cent) and two-thirds of the normal (less than 9.7 per cent) are respectively 22 February–22 March; 12–20 April, and 30 December–15 January; 26 March–4 April; 1–25 May; 22 October–10 November.

[4] At Cilaus, Réunion, 15 March 1952 (Lane, 1967).

3.75 inches are 'remarkable', and falls of over 5.6 inches per day are described as 'very rare'. Nevertheless a review, over the period 1920–55, of the number of stations (of a mean total of about 5000) which recorded, in any given year, values in excess of 4 and 6 inches in a day gave the following results.

| Rain per day (in) | Number of stations per year | | Minimum |
| | Maximum | Mean range | |
| --- | --- | --- | --- |
| Over 4 | 114 (1931) | 24–12 | 2 (1925) |
| Over 6 | 13 (1955) | 2– 0 | 0 (16 years) |

Four occasions, outstanding both as records and in their devastating results, were as follows:

  11.0 inches at Martinstown, Dorset—1955, 18 July
   9.6 inches at Bruton, Somerset—1917, 28 June
   9.4 inches at Cannington, Somerset—1924, 18 August
   9.0 inches at Longstone Barrow, Somerset—1952, 15 August

The fact that all of these occurred during the warm season in the humid and hilly southwestern part of Britain is doubtless significant. (See also Rodda, 1970.)

A more general treatment of the frequency of daily rainfall records is given in Fig. 45. Points on the various curves express the mean percentage of rain days experiencing more than a certain specified fall (note that the vertical scale is logarithmic). The curve for Bombay is based on the records for the monsoon months (June to September) over the period 1931–40. Of the 1220 days thus involved, 963 were rain days with more than 0.01 inches. The curves for Keswick and Mildenhall are derived from the full 5-year period from March 1953 to February 1958, that is a period of 1826 days of which 1051 were rain days at Keswick as compared with 752 at Mildenhall. It should be emphasised that the percentages are based upon the values 963, 1051 and 752 respectively, *all rainless days having been omitted.* All three curves clearly belong to the same family, the intensity per day at Keswick being approximately twice that at Mildenhall over the whole range of frequencies. At Bombay, however, during the 8 per cent of occasions experiencing the most extreme falls, intensities ranging from three to three and a half times the Keswick values were recorded. The heaviest daily fall at Bombay during the period in question was 10.1 inches.

When self-recording gauges are available we can extend the analysis by considering *hourly* rainfall. Table 20 presents mean values for the whole year from the 1933–55 records of five British stations. Snowdon, Manchester and Cranwell represent a cross-section of the country near the 53° parallel whilst Kew and Loch Sloy–Glenleven stand for the more continental and more oceanic sectors of the country respectively. The table is worth close inspection as it contains a number of interesting features. Thus it is evident that Manchester's worldwide reputation as a rainy city is scarcely deserved. Its mean annual total is by no means excessive and only 5 days out of 10 qualify as

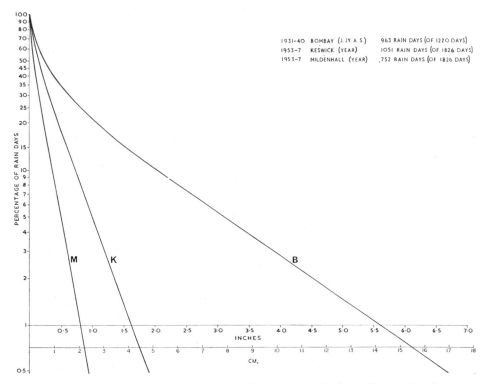

| 1931–40 | BOMBAY (J. JY. A. S.) | 963 RAIN DAYS (OF 1220 DAYS) |
| 1953–7 | KESWICK (YEAR) | 1051 RAIN DAYS (OF 1826 DAYS) |
| 1953–7 | MILDENHALL (YEAR) | 752 RAIN DAYS (OF 1826 DAYS) |

FIG. 45. The probability (per cent) that rainfall during a single day will exceed a given amount at Bombay, Keswick and Mildenhall. The diagram is designed mainly to illustrate the type of curve which emerges. Data for Bombay cover only the monsoon months, June to September 1931–40. Data for Keswick and Mildenhall are derived from the whole of the five-year period 1953–57

'rain days' despite the meagre requirement of only 0.01 inches per day; in the 'drier' east over 4 days in 10 also fall into this category. If Manchester receives about a third more rain than the eastern stations, it is due primarily to the greater frequency of 'wet days' (with over 0.04 inches per day) and these occur mainly because the rain lasts longer, the difference in intensity is slight. Even so, less than 4 days in 10 qualify as wet days in Manchester as compared with 3 days in the east and nearly 6 days in 10 in the notorious west country. Mountainous Snowdonia records four times the annual total of Manchester because both the duration and the intensity of the rainfall are doubled;[1] in the West Highlands the mean intensity is not so great but the duration of the rains

[1] See Pedgley (1970) for a discussion of heavy falls over Snowdonia.

is even longer. Nevertheless it may come as a surprise to those familiar with western Scotland that, on the average, rain falls for no more than 4 hours of the 24; even during the worst years of the period (1938, 1948 and 1954) the mean duration of rain did not exceed $5\frac{1}{2}$ hours. If these values seem to be out of

TABLE 20. *Rainfall at five British stations 1933-55*

| | LOCH SLOY-GLENLEVEN[1] | SNOWDON CWM DYLI[2] | MANCHESTER AIRPORT | CRANWELL AERODROME | KEW OBSERVATORY |
|---|---|---|---|---|---|
| Mean annual rainfall (in) | 99 | 123 | 31 | 23 | 23 |
| Extreme totals (in) | 141–71 | 177–83 | 41–23 | 29–18 | 31–18 |
| Mean duration of Rainfall (hours) | 1470 | 1189 | 632 | 532 | 428 |
| Extreme duration (hours) | 1998–1015 | 1594–901 | 801–433 | 722–401 | 574–298 |
| Mean fall per hour (in) | 0.07 | 0.10 | 0.05 | 0.04 | 0.05 |
| Mean number of rain days | 256 | 233 | 188 | 163 | 153 |
| Extremes (days) | 316–201 | 268–203 | 222–152 | 189–132 | 185–124 |
| Mean number of wet days | 212 | 212 | 138 | 109 | 107 |
| Extremes (days) | 261–181 | 250–174 | 172–112 | 134–85 | 135–79 |
| Mean fall per rain day (in) | 0.39 | 0.53 | 0.17 | 0.14 | 0.15 |
| Mean hours of rain (all days) | 4.0 | 3.3 | 1.7 | 1.5 | 1.2 |
| Mean hours of rain (rain days) | 5.7 | 5.1 | 3.4 | 3.3 | 2.0 |
| Mean percentage | | | | | |
| $\frac{Wet\ days}{Days\ in\ year}$ | 58 | 58 | 38 | 30 | 29 |
| $\frac{Rain\ days}{Days\ in\ year}$ | 70 | 64 | 51 | 45 | 42 |
| $\frac{Wet\ days}{Rain\ days}$ | 83 | 91 | 73 | 67 | 70 |

[1] Glenleven (Blackwater Dam), 1933–45; Loch Sloy, 1948–55.
[2] At only 310 feet but below the mountain.

accord with subjective impressions it is partly because of the natural tendency for wet spells to impress themselves on the memory and partly because heavy cloud slows down evaporation and thus prolongs the impression of wetness for hours after the rain has ceased to fall.[1]

A really fruitful discussion of rainfall intensity must go beyond these annual means. In Fig. 46 daily rainfall in millimetres has been plotted against its duration in hours at Mildenhall for the two periods May to August and November to February inclusive. In *summer* the median duration was about 2 hours and only a quarter of the rain days had more than 4 hours rain; in

[1] The Manchester region is cloudy rather than wet. Yet the Airport sunshine record of some 1320 hours per year is not unduly low by British standards. It would appear that, in fine weather during the summer half of the year, much of this is recorded early in the morning before most people are up and about. Cloud often builds up between 11.00 and noon. Manley has suggested that the impression of dampness in Lancashire towns is enhanced because cloud plus smoke slow down evaporation.

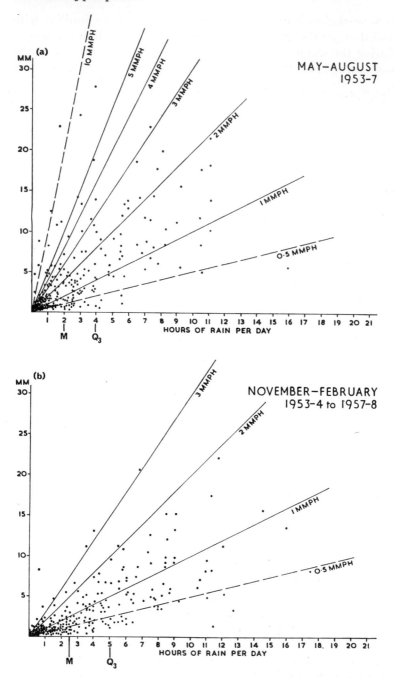

Fig. 46. Daily rainfall amount (in millimetres) plotted against rainfall duration (in hours) at Mildenhall, Suffolk, over the two periods, May to August and November to February. From an analysis of the daily records over a five-year period. It is clear that there is a real difference in type between summer and winter rainfall in eastern England

*winter* the corresponding values were 2.5 and 5.0 hours. However, the radiating lines indicate selected intensities per hour and it is quite clear that some summer rains are much heavier than those experienced in winter. Over the period of 5 years the mean intensity in summer was 1.65 mm (0.065 in) per hour and in winter it was 0.90 mm (0.035 in) per hour. At Keswick, on the other hand, for exactly the same period, there was no contrast in the intensity of summer and winter rains, both periods yielding an average of 1.40 mm (0.055 in) per hour.

Exceptionally heavy rainfalls in short periods are classed by the British Rainfall Organisation as 'noteworthy', 'remarkable' and 'very rare' according to whether they exceed the following limits:

|  | *Noteworthy* | | *Remarkable* | | *Very Rare* | |
|---|---|---|---|---|---|---|
|  | (in) | (mm) | (in) | (mm) | (in) | (mm) |
| 10 minutes or less | 0.54 | 13.8 | 0.85 | 21.6 | 1.31 | 33.2 |
| 30 minutes or less | 0.77 | 19.7 | 1.19 | 30.3 | 1.82 | 46.2 |
| 1 hour or less | 0.97 | 24.5 | 1.48 | 37.6 | 2.24 | 56.8 |

Falls at this level of intensity are by no means common anywhere in the world though they are doubtless more frequent in the humid tropics than in Britain. Speaking for this country only, E. G. Bilham (1935) concluded that the incidence of such rains was 'very fortuitous' and that 'there is no very clear evidence of geographical influence in respect to the frequency of intense falls'. J. Grindley (1953) has listed such falls over the period 1860 to 1953. He notes falls in excess of 2 inches in 30 minutes (i.e. very rare falls) in eleven counties ranging from Dumfriesshire to Hampshire and from Glamorgan to Lincolnshire.

## Two examples of rainfall analysis

We conclude this chapter with examples of two very different types of graphical analysis of rainfall data.

Figure 47 presents a generalised cross-section of the Scandinavian peninsula near the parallel of 62°N upon which is superimposed, at appropriate longitudes, a series of vertical columns representing the mean annual rainfall of twenty-seven stations in that vicinity. The proportion of the annual total which falls during the summer and winter half-years (April–September and October–March) is also indicated. The elevation of each station is shown by the horizontal strokes across or above the columns. The observations do not cover a uniform period but this is a matter of minor importance in a generalised study of this type.

The effect of the general build of the peninsula is self-evident. On the Atlantic slope annual totals are in excess of 1000 mm (40 in) and maximum values of over 2250 mm are recorded some 25 to 30 miles from the open coast.[1]

[1] Totals of up to 3300 mm are recorded in this zone at stations not far south of the line of section.

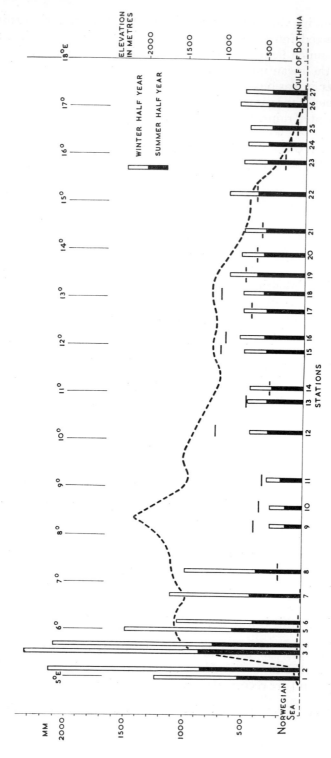

FIG. 47. A rainfall cross-section of the Scandinavian peninsula at about 62°N. A key to the twenty-seven stations employed is given in the text. The broken line is a generalised relief section of the peninsula along the 62nd parallel. Many of the stations are in narrow valleys but the elevation of each is given in the text and indicated on the diagram by a horizontal stroke. The vertical columns express mean annual totals and show the relative proportions contributed by the winter and summer half-years

In the interior, with notable exceptions, an annual fall of 500 to 600 milli-metres (20–24 in) is characteristic as far as the coast of the Gulf of Bothnia. The three exceptional stations, Skjåk, Lom and Sandbu, with annual totals of barely 270 to 300 millimetres (11–12 in), all lie within a deep valley which drains eastward through the highest part of the mountain system. Many stations on the western flank get more rain than this in a single month, for example in January. Here then is a very striking example of a 'rain shadow' effect intensified doubtless by persistent foehn winds (see pp. 429–31).

The seasonal distribution of precipitation also shows some notable features. Thus at the fjord stations in the west, from Ulversund to Opstryn, 60 per cent or more of the annual fall occurs during the *winter* half-year, the proportion reaching 65 per cent at Ålfoten. In the interior of the peninsula, on the other hand, from Sandbu to Bjuråker, all stations record 60 per cent or more of their annual total during the *summer* half-year, the proportion reaching 65 per cent at several localities. Values rather less than 60 per cent are found at the two most easterly stations in the neighbourhood of the Bothnian coast.

The diagram thus suggests that rainfall on the two flanks of this massive peninsula is produced by different mechanisms and it is also possible that moisture reaches the two regions by fairly distinct routes.

The location and elevation of the stations plotted are given below:

|  |  | *North* | *East* | *metres* |
|---|---|---|---|---|
| 1. | Kråkenes Fyr | 62°2′ | 4°59′ | 39 |
| 2. | Ulversund | 61°58′ | 5°9′ | 1 |
| 3. | Davik | 61°53′ | 5°32′ | 32 |
| 4. | Ålfoten | 61°50′ | 5°41′ | 20 |
| 5. | England | 61°55′ | 5°59′ | 15 |
| 6. | Gloppestad | 61°48′ | 6°10′ | 20 |
| 7. | Stryn | 61°55′ | 6°44′ | 6 |
| 8. | Opstryn | 61°56′ | 7°14′ | 205 |
| 9. | Skjåk | 61°54′ | 8°12′ | 424 |
| 10. | Lom | 61°50′ | 8°34′ | 380 |
| 11. | Sandbu | 61°53′ | 9°9′ | 354 |
| 12. | Atnasjø | 61°53′ | 10°9′ | 744 |
| 13. | Alvdal | 62°1′ | 10°48′ | 485 |
| 14. | Øvre Rendal | 61°53′ | 11°5′ | 293 |
| 15. | Gløtvola | 61°51′ | 11°51′ | 706 |
| 16. | Flötningen | 61°52′ | 12°11′ | 670 |
| 17. | Idre | 61°52′ | 12°43′ | 450 |
| 18. | Storfjäten | 61°59′ | 13°7′ | 700 |
| 19. | Storhärjeåvallen | 61°52′ | 13°29′ | 500 |
| 20. | Linsäll | 62°10′ | 13°55′ | 405 |
| 21. | Sveg | 62°2′ | 14°25′ | 363 |
| 22. | Los | 61°43′ | 15°11′ | 405 |
| 23. | Föne | 61°50′ | 15°50′ | 175 |

|      |            | *North* | *East*  | *metres* |
|------|------------|---------|---------|----------|
| 24.  | Stenegård  | 61°43′  | 16°13′  | 135      |
| 25.  | Bjuråker   | 61°52′  | 16°34′  | 73       |
| 26.  | Bergsjö    | 61°59′  | 17°3′   | 50       |
| 27.  | Strömsbruk | 61°52′  | 17°19′  | 10       |

Figure 48 is obviously a much more elaborate affair. It presents an analysis of *daily* records at Bombay from late April to early November, a period which includes both the first preludes and last echoes of the south-west monsoon. Daily rainfall is shown by the black columns and the totals in excess of 5 inches for 7-day periods with some rain every day are given in the square brackets above. It will be noted that the difference between the monsoon of one year and another depends a great deal upon the *number* of such wet spells which occur. It is useful to tabulate these results, quoting the percentage of (*a*) monsoon rains (total June to October inclusive) and (*b*) the annual total rainfall, which fell during the limited wet periods thus defined.

|                    | *a* (%) | *b* (%) |
|--------------------|---------|---------|
| 2 periods in 1936  | 49      | 47      |
| 3 periods in 1939  | 68.5    | 68      |
| 4 periods in 1932  | 63      | 62      |
| in 1933            | 70      | 65      |
| and 1934           | 70      | 67      |
| 5 periods in 1935  | 67.5    | 67      |
| and 1937           | 70.5    | 70      |
| 6 periods in 1931  | 73      | 71      |
| 7 periods in 1938  | 70      | 68.5    |
| 8 periods in 1940  | 84      | 82      |

It will be noted from Fig. 48 that only two of these wet periods occurred in October and there is some question whether these should be included in 'monsoon rains'. Yet, even during the generally accepted monsoon months, the spells are usually separated by drier periods, often 2 or 3 weeks in length. Only in 1931 and 1940 are more than two wet periods actually contiguous. Furthermore, the above table shows that, except in the most extreme instances (1936 and 1940), the proportion of total rainfall falling during such spells is remarkably constant, irrespective of their actual number in any one year. The mean annual total rainfall recorded at Bombay during this decade ranged from 54 and 58 inches in 1939 and 1936 respectively to 98 inches in both 1931 and 1940 but, in view of likely losses from both run-off and evaporation, it is surely evident that the number and distribution of wet spells must be of much greater significance to the farmer than the crude seasonal totals.

The contrast between the two halves of the true monsoon period (June–July and August–September), already evident from Fig. 42, is also borne out by wet-spell data despite the relative brevity of the decade analysed. Thus,

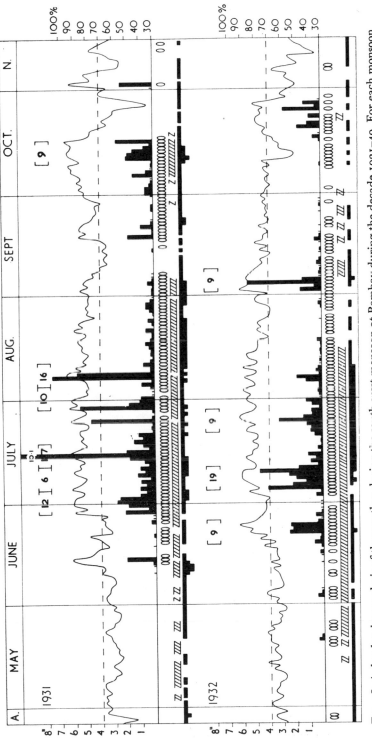

FIG. 48. A day-by-day analysis of the weather during the south-west monsoon at Bombay during the decade 1931–40. For each monsoon season the following information has been plotted:

(a) The relative humidity at 1400 hours—a light line.

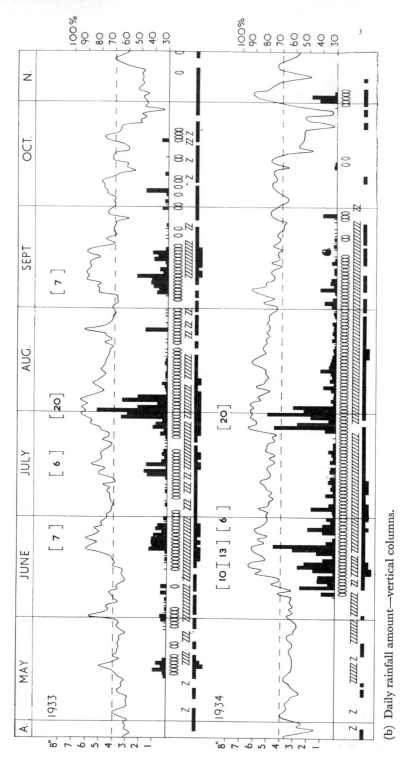

(b) Daily rainfall amount—vertical columns.

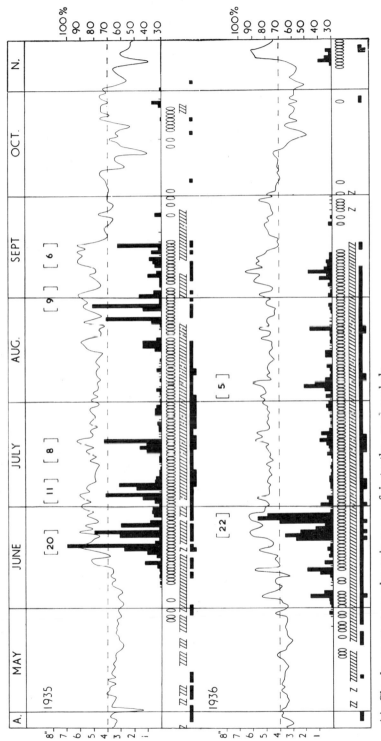

(c) Cloud amount at 1400 hours in excess of six-tenths—oval symbols.

*The amount of precipitation*

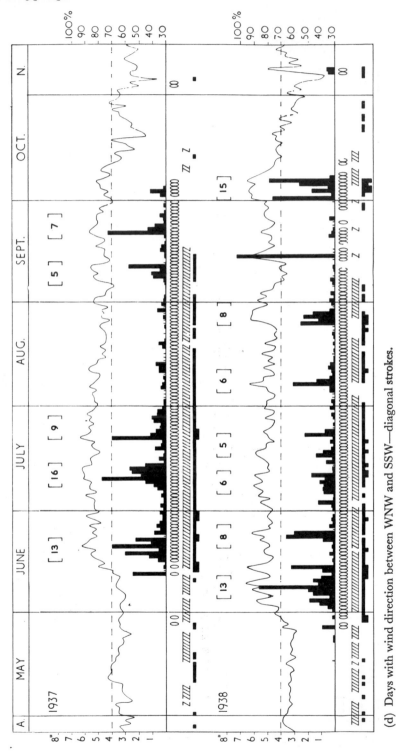

(d) Days with wind direction between WNW and SSW—diagonal strokes.

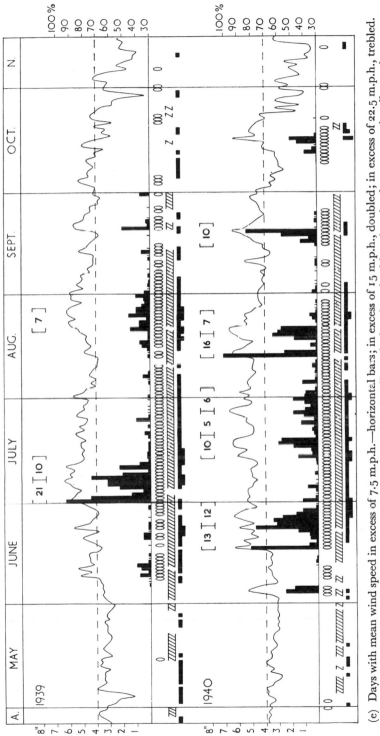

(e) Days with mean wind speed in excess of 7.5 m.p.h.—horizontal bars; in excess of 15 m.p.h., doubled; in excess of 22.5 m.p.h., trebled. Though the diagram is bulky it is presented as evidence that the climatologist who takes his task seriously must occasionally get down to the daily record

within this decade, June–July experienced a total of 29½ such spells as compared with only 16½ in August–September. The mean fall per spell was also greater (11.7 in) in the former than in the latter 2-month period (8.9 in).

It should now be clear that the famous 'burst' of the monsoon is no more than the arrival of the first wet spell. This occurred as early as 5 June in 1938 and as late as 1 July in the following year. However, Fig. 48 presents a great deal of supplementary information which is relevant to a definition of the 'southwest monsoon' and which may be employed in a determination of the dates of its onset or cessation.

Thus (1) the irregular line shows relative humidity at 1400 hours, a time when at most places values are comparatively low because temperature is then near its daily maximum. Clearly the monsoon brings exceptional humidity even when rainfall is light. The period when values remain consistently above some stipulated figure, such as 70 per cent, for weeks together, would thus offer a possible definition.

(2) Symbols beneath the rainfall graphs indicate each day when the cloud amount at 1400 hours was in excess of six-tenths. Cloudy spells may both precede and follow the real monsoon but they are then broken by quite long sequences of clearer weather. We could thus define the monsoon in terms of 'continuous cloud' though this must obviously be a relative term; it would seem appropriate to ignore breaks in the cloud which did not last longer than 3 days.

(3) The remaining two sets of symbols indicate wind direction—days with wind blowing from between WNW and SSW—and mean wind speed. It is evident that at Bombay winds from the southwest quadrant often precede both the sharp rise in humidity and the actual outbreak of rain. When fully established, however, the southwesterlies are also 'continuous' in the above sense (with breaks no longer than 3 days) at least until early September. Later rains may or may not be associated with their return and humidities usually remain high long after the wind has shifted to another quarter. October occasionally experiences brief storms with unusually violent winds though not invariably from the southwest quadrant.

Over the decade 1931–40 the mean duration of the 'monsoon' according to these various criteria was as follows:

|  | Onset |  | Cessation |
|---|---|---|---|
| SW wind without break of more than 3 days | 23 May | to | 7 Sept. |
| Cloud cover 6/10 without break of more than 3 days | 7 June | to | 19 Sept. |
| RH over 70 per cent for at least a week | 14 June | to | 8 Oct. |
| Heavy rain spells | 19 June | to | 1 Sept. |

It is evident from the diagram, however, that there is a wide variation in these dates from year to year.

For a general review of the southwest monsoon, see Durand-Dastès (1961).

# 5
# Pressure and wind

Wind is air in motion. Laminar flow in air is difficult to produce and even more difficult to maintain; the air normally moves in a complex of eddies ranging in scale from the tiny curls revealed by cigarette smoke to swirls of regional, if not continental, dimensions. Each eddy, however ephemeral, is an entity conforming to known physical laws and no part of it is fully intelligible away from its context, yet it is a long-established tradition that the term 'wind' implies only the horizontal component of the motion—or perhaps, a little more broadly, the component essentially parallel to the surface of the earth. We are all made early aware of the importance of this component in walking, cycling and sailing and we are all attracted by windmills and balloons but when we consider the raising of dust, the dispersal of seeds, or the diffusion of smoke it is quite clear that motions in other than a horizontal plane are of no mean significance. The fundamental difficulty in generating power from the wind by modern techniques is to be found in its inherent gustiness—the driving force is not delivered consistently either in direction or strength so that 'sweet running' is extremely hard to achieve.

To understand the wind we therefore require a three-dimensional model. The difficulty of making three-dimensional measurements, of presenting them when made, and of comprehending them when presented, makes this subject one of the most difficult in meteorology. There are also formidable problems of scale. The atmospheric ocean is so vast as to defy comprehension but it is also one and indivisible so that a major disturbance in one section has repercussions throughout the whole. Furthermore, it is a spherical envelope wrapped round a rotating earth and broad-scale motions within such an envelope are both involved and peculiar. Little wonder then that the problem of the 'general circulation' still remains something of an enigma.

Yet to mankind in general the wind is much more than a problem in applied physics. At one time it is a disembodied spirit whispering in the ear, at another a blustering bully roaring under the eaves. Clearly this visitant has a 'personality' and men have tried to express their sense of its quality in terms which range from the picturesque figures of Greek mythology to the abstruse categories of modern air mass climatology. To the Greeks the qualities were inherent in any given wind direction but modern science looks for origins, for causes, for evolutionary trends—it seeks indeed to trace 'the life-history of surface air-currents' (the title of a classic paper by Sir Napier Shaw (1906; rep. 1955). This poses a serious problem of *identification*. At any one point it is easy to specify a wind in terms of direction and force and to give its momentary characteristics of temperature and humidity, but this is very much a worm's eye view of the facts. We have already seen that the surface air responds readily to insolation, radiation, evaporation and precipitation so that temperature and humidity are far from conservative properties; they change over both time and space. At all land stations even direction and force are often a reflection of the presence of obstructions which produce local channelling, deflection and eddying. Yet even when we observe the wind aloft, its momen-mentary direction is no clear key as to its ultimate origin. In Britain, for instance, few northerly winds bring air which has actually crossed the pole and few southerly winds hail from the Sahara. We now know that to identify an airstream we must comprehend the vast swirl of which it is a part—but the classical concepts die hard.

These introductory paragraphs are not designed to show that wind analysis is impossible, or even unprofitable—far from it. It is self-evident that the wind is a most significant climatic element and some attempt must be made to resolve the intricate problems it involves. Rather do we seek to explain why the meteorological approach to this apparently so simple and tangible phenomenon has usually been indirect, pseudomathematical, and sometimes esoteric in the extreme. Few branches of modern science invite the student to transfer the terms of the argument so readily from a concrete to an abstract plane.

In temperate latitudes, where all the fundamental research has been conducted, it has been found convenient to approach the problem via atmospheric *pressure*, imagining that the wind field is given (above 600 m or about 2000 ft) when once the pressure pattern has been established. As a basis for forecasting this device has achieved much success; from changes in pressure pattern we foresee changes in wind. Yet the common man is not unjustified in feeling some resentment when he finds the discussion thus transformed. The wind he can feel, but to normal changes in the barometric pressure at ground level he is physically supremely indifferent. Furthermore, a pressure *pattern*, emerging as it does from the miracles of modern telegraphy, is a mystery of the new meteorological priesthood in which even the confirmed barometer-tapper can have only a most attenuated share. As is so very often the case, the common man is right! At best the pressure pattern is only a tool and it is a tool that has proved inadequate throughout the tropical girdle of the earth where so much that is of importance to climatology has its origin. Furthermore the answer to the question—'why the pressure pattern?' is rarely any simpler than that to the more obvious question—'why the wind?' As we shall see, steep pressure gradients can persist only if they are *maintained* by the wind movement, it is thus only in a most limited sense that the air motion is *caused* by the pressure difference. We have here a balance of mutually dependent forces and in attempting to break into the circuit pedagogically we must do some little violence to the truth, even if only in emphasis. This is not to deny that air will move only if it suffers an impulse, nor that the root impulse must be traced to the force of gravity acting upon air masses of different density. At any given level of the atmosphere density contrasts depend, above all, on temperature. Thus the atmospheric heat engine drives the winds but it does so through such a complex series of systems within systems, of wheels within wheels, that this basic relationship is as much concealed as revealed in any single local situation. It is thus just as valid to consider the convergence of air streams as the cause of an eddy as it is to regard an eddy as producing a convergence of air streams.

## Wind and temperature

Let us now attempt to show how temperature contrasts throw both pressure and wind into this intimate relationship. In Fig. 49 we give diagrammatic cross-sections of some portion of the lower atmosphere, isolated, for the purposes of this discussion, from all outside effects. In (*a*) the situation is in a state of equilibrium. Surface temperatures and the vertical lapse rates are uniform throughout and the fine horizontal lines thus represent 'surfaces' of uniform pressure. Pressure decreases with height on a logarithmic scale indicated by the numbers on the left hand side. It will be useful to consider the atmosphere as composed of a series of strata and to carry the argument forward by stages, separating temperature effects (on the right) from pressure readjustments (on the left).

In (*b*) we imagine that the area at *C* is chilled whilst that at *W* is warmed. (On the curved earth such a change can be easily produced by insolation-radiation balance.) At first only the lowest layer of the atmosphere is affected and we have illustrated this in diagram (*b*) by contracting the 9–10 layer at *C* and expanding it at *W*.[1] For the moment pressure distribution at the *surface* remains unchanged for there has been no shifting of the atmospheric load but all the surfaces of uniform pressure aloft have been given a tilt and a pressure gradient from *W* to *C* is established throughout the free atmosphere. Air will thus tend to move hydrostatically, that is, under gravity, from right to left at all levels except at the actual surface of the ground.

This wholesale shift of the atmosphere at once raises all pressure surfaces over *C* and correspondingly lowers all pressure surfaces over *W*, the change in pressure being most marked at the lower levels owing to the cumulative effect of the shift in each of the overlying strata. We have tried to catch a moment of the process in (*c*). At level 5 equilibrium has been re-established but, below that level, pressure has been raised above *C* and lowered above *W*. A gradient in the *lower* layers of the atmosphere has thus been induced which will tend to promote air movement from left to right to restore a situation analogous to (*a*).

[1] To an exaggerated degree to make the result visible.

In diagram (*d*) however it is assumed that the thermal contrast is maintained. Once again the air near the ground is expanded at *W* and contracted at *C*, and once again a gradient towards *C* is produced aloft though near the ground the gradient towards *W* is not completely neutralised. The two-way circulation thus established will tend to be maintained as long as the differential in heat income between *W* and *C* persists and, since we have assumed a closed circulation, there must be a net upward movement of air over *W* and a net subsidence

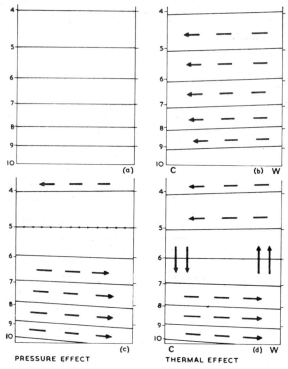

PRESSURE EFFECT                    THERMAL EFFECT

FIG. 49. An attempt to illustrate in a very elementary fashion the interrelationship between temperature, pressure, wind and vertical motion in the atmosphere

over *C*, as shown. If the differential remains constant a steady state will ultimately be achieved in which the surplus heat energy at *W* is fully absorbed in overcoming the frictional losses of the system and warming up the cooler air from *C* which constantly threatens to 'damp its fires'. A delicate balance of forces will then be achieved in which insolation and advection, temperature and pressure, surface wind and upper wind, are all in close relationship. A disturbance of one of these factors must then involve a change in the others.

It should thus be clear that *both* the pressure pattern and the wind system are induced phenomena. Although the surface wind arises as a response to a pressure difference between *C* and *W*, the high pressure at *C* is itself a direct result of the movement of the upper air. The whole system is one and in-divisible and root causes must be sought in the thermal contrast between *W* and *C*. Furthermore the inherent balance of the system acts as a universal governor; checks are immediately imposed on excessive thermal contrasts, excessive pressure differences, and excessive wind speeds. The forces at work in the atmosphere are so vast that beside them a hydrogen bomb is a mere bagatelle, yet violent weather phenomena are so rare that they immediately catch the headlines.

With all its simplifications Fig. 49 has thus served a useful purpose but we shall make a great mistake if we think we can substitute the words 'pole' and 'equator' for *C* and *W*, as it is so easy to do in a thoughtless moment. Whilst the pole is a point (or spot) on the earth's surface, the equator is a line (or zone) approximately 40076 kilometres or about 25000 miles long. The true cross-section of the earth's atmosphere from pole to equator is thus the volume lying between neighbouring meridians. It is what Ferrel (1889, p. 97) called a 'half-lune', a curved wedge-shaped sector halved at latitude 30° and with one-quarter of its volume poleward of latitude 48°36′ and another equatorward of 14°29′ (see Fig. 50). This form must have a marked effect upon any broad system of circulation set up within it. Thus in the movement from a tropical 'heat source' to a polar 'heat sink' the updraught will take place from a *zone* and the descent over a *cap*. We compare various zones and caps in the following table:

| Area of zone | | Areas of cap | Ratio of length of bounding parallels |
|---|---|---|---|
| Equator to | 5° | = Pole to 65°54′ | 100 : 41 |
| | 10° | = 55°44′ | 100 : 57 |
| | 15° | = 47°50′ | 100 : 69½ |
| | 20° | = 41°9′ | 100 : 80 |
| | 25° | = 35°16′ | 100 : 90 |
| | 30° | = 30° | 100 : 100 |

Hence if air rises consistently over the zone between the equator and 5°N and descends *at the same rate* over the pole it will require for its accommodation not less than the whole area north of 66°N. If the zone of ascent extends from the equator to 10°N the margin of the corresponding downflow must reach out to at least 56°N. The minimal extent of the subsidence cap is thus brought well into middle latitudes by quite modest estimates of the areas of updraught. Furthermore the ratio of the parallels across which this hypothetical flow is taking place will make it necessary for the air to enter (aloft) and leave (below) the margin of the polar cap at horizontal speeds, in the first case 2½ times and in the second case 1¾ times, greater than that at which it leaves (aloft) or re-enters (below) the equatorial girdle. Without involving ourselves in the dynamics of a rotating earth we thus find ourselves with a circulation which would be more active, both aloft and near the ground, in middle latitudes, than towards the equator. It would also be reasonable to expect that, deep within the polar cap, where the direction of an escape route is comparatively indeterminate, the air would lie relatively stagnant until it became so deeply

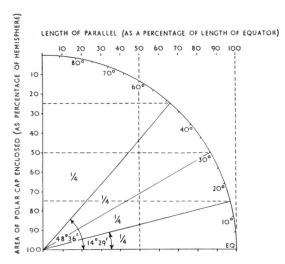

FIG. 50. The relationship between the relative lengths of parallels and the proportions of the total area of a hemisphere which they enclose

chilled that it spread outwards under gravity. In such circumstances its release might well be intermittent and 'polar outbreaks' might well occur at rather irregular intervals, particularly along some favourable channel or channels.

In the above estimates we have made no allowance for the reduced volume of chilled air[1] but we have also ignored both surface and internal friction. Higher wind speeds involve greater frictional losses and this factor would tend to expand the polar caps. Without involving ourselves in complex computations it may be taken that the figures quoted above are broadly valid and they certainly make a point which is far too often overlooked when the atmosphere is viewed in meridional section.

## Wind direction on a rotating earth

We now enter upon what is unquestionably the most difficult aspect of climatology. The shortest distance between a heat source and heat sink on a globe is along a 'great circle' but, of the infinite number of such circles which envelop the earth, there is only one, the equator, which offers a path for the

[1] Air at −30°C has a volume which is 243/303, or 80 per cent, of that of air at +30°C at the same pressure.

free motion of air which implies no distortion from the spin of the earth about its polar axis. Since both major sources and sinks are most unlikely to occur simultaneously at the same elevation along the equator, except perhaps at the vast distance which separates day from night, it may be taken that the effects of this rotation are universal.

The real difficulty arises from the natural attempt to *visualise* this effect for, though it can be demonstrated mathematically, this process involves assumptions, isolating one factor at a time, which present images or 'models' which often seem not a little unreal. In the history of the subject at least three distinct lines of approach can be recognised and all are encountered with varying degrees of emphasis in current literature. Yet the rotation of the earth is one single and indivisible fact!

(*a*) *Momentum and inertia.* This approach starts with Newton's well-known rule that a force continues to operate unless and until it meets with some other force. It is usual to picture an imaginary 'ring' of air round some selected parallel and convenient to postulate that initially it has no east-west component of motion with respect to the earth's surface. At the equator therefore such a ring would be actually turning *in space* with the earth beneath it at a speed of 1037 mph (900 knots). If such a ring of air, it is argued, were now to be shifted bodily due north so that it passed successively the parallels of 10, 20 and 30°N, it would find itself above points on the solid earth that were turning eastward at only cos 10° (0.985), cos 20° (0.940) and cos 30° (0.866) times this speed, i.e. at 886, 846 and 779 knots respectively. Hence, it was first argued that, *in so far as the initial momentum of the ring was conserved*, it would appear to overtake the motion of the solid earth and produce at these latitudes a *westerly* wind of 14, 54 and 121 knots respectively. The argument can be reversed and if the ring of air starts at 30°N and moves equatorwards it will tend to lag behind the solid earth and reveal itself as an *easterly* wind of 67 knots at 20°N, 107 knots at 10°N and 121 knots at the equator. Even at this stage the wind speeds predicted by the theory seemed highly fanciful but note the effect of the changing westerly (or easterly) component on the apparent track of any single element of the imaginary ring. We had postulated that it should move due northward (or southward) but, from the earth's surface, it would appear to be deflected progressively towards the right in the northern hemisphere and towards the left in the southern hemisphere.

Ferrel (1889), who popularised this idea of 'deflection' in his famous Law (see below), at first took pains to correct the above computations. He showed that Hadley was at fault in thinking in terms of simple momentum, the problem is one of *angular momentum*, of what he called the Principle of the Preservation of Areas, already familiar to Newton. Angular momentum varies inversely with the radius of curvature and the *radii* of the various parallels as well as their length vary as cos $\theta$. Our imaginary ring will thus actually accelerate as it moves polewards and decelerate as it moves equatorwards. Ferrel understood, though he does not say explicitly, that the computed wind speeds must therefore be derived from a comparison of $\cos^2\theta$.[1] The corrected values are thus 28, 112 and 260 knots for westerlies derived from northward motion and 128, 201 and 255 knots for easterlies derived from southward motion over the range of latitudes quoted above. One cannot resist the thought that this only makes the exercise seem still more unrealistic.

In favour of this approach it can be argued that what is true of a highly improbable 'ring' is also true of any sector of it, however small. Furthermore it can be argued that it is crude to think in terms of bulk transfer, we should interpret the theory as a tendency to be viewed infinitesimally in the delightful way that mathematicians have. Frictional losses may also be invoked to bring theoretical wind speeds into closer line with what is actually observed.

Against the approach it can be validly argued that there is a tacit assumption that the air motion is exclusively horizontal. A 'ring of air' is only a valid model if it retains some standard width and depth during the transformation. During a shift in latitude on the earth this is clearly impossible. If the width is held to a standard, its depth must increase with poleward motion and decrease as it moves towards the equator. We shall see below that such changes are ignored at peril in a fluid medium.

As a student the author was much puzzled by the fact that to explain the deflection of a current moving either due east or due west in middle latitudes it was necessary to change the model completely and to argue in terms of centrifugal force. Clearly this too was involved on a rotating earth but why invoke it only in this unique instance? He now sees quite clearly that since such motion is *not* along a great circle, it is not a valid presumption. All great circle routes except that around the equator involve some northing or southing and, to that degree, the arguments of this section apply.

(*b*) *Deflection.* Ferrel summarised his conclusions in his Law, viz. 'If a body moves in any direction upon the earth's surface, there is a deflecting force arising from the earth's rotation, which deflects it to the right in the northern

---

[1] The formula for computing changes in zonal velocity in knots for air moving without frictional loss from latitude $\theta_1$ to latitude $\theta_2$ is (Reiter, 1963, 16):

$$\text{Change in velocity} = \frac{900(\cos^2\theta_1 - \cos^2\theta_2)}{\cos\theta_2}.$$

Values for $\cos^2\theta$ at five-degree intervals are given below:

| (°) | | (°) | |
|---|---|---|---|
| 0 | 1.000 | 45 | 0.500 |
| 5 | 0.992 | 50 | 0.413 |
| 10 | 0.970 | 55 | 0.329 |
| 15 | 0.933 | 60 | 0.250 |
| 20 | 0.883 | 65 | 0.179 |
| 25 | 0.821 | 70 | 0.117 |
| 30 | 0.750 | 75 | 0.067 |
| 35 | 0.671 | 80 | 0.030 |
| 40 | 0.587 | 85 | 0.0075 |
| | | 90 | 0.000 |

hemisphere, but to the left in the southern hemisphere' (Ferrel, 1889, p. 78). Furthermore he estimated the magnitude of this force and showed it to vary as the sine of the latitude and the speed of the wind. Here he had been anticipated by the French physicist, G. Coriolis, who had represented the force as an acceleration directed at right angles to the path of a moving body of magnitude $2\Omega v \sin\theta$, where $\Omega$ is the angular rotation of the earth ($15°$ per hour but for mathematical purposes better expressed in radians, $2\pi$ per day), $v$ is the wind speed and $\theta$ the latitude.

Ferrel does not make clear that this is quite a fresh formulation of the problem as compared with section (*a*), it is a new model. In (*a*) stress has been laid upon the apparent acceleration of the air with respect to the solid earth, in this new approach the forward motion is presumed constant and the stress is upon the apparent curvature of its path. In this form the theory is undoubtedly much more flexible. Indeed it has been 'proved' by long-range projectiles and other ballistic missiles. The 'Coriolis acceleration', often described as an 'imaginary force', has also found an honoured place in meteorological thought though the plain man may harbour some intuitive doubts as to whether a continuous medium like the atmosphere should behave precisely like a 'body'. Yet it appears to work!

Hare (1953, 50) has commented that 'to try to visualise the physical explanation for this curious force . . . is to run the risk of hopeless confusion'. Byers (1959, 203) has tried to clarify the situation by distinguishing between inertial and non-inertial forces. Ferrel (1889, 79) used the analogy of a man walking across a narrow swing-bridge. It may be useful to elaborate on this a little. Imagine two men, *A* and *B*, standing facing the same way at the opposite ends of a stationary swing-bridge. If the bridge now moves in an anticlockwise direction, *A* will be impelled against the left handrail and *B* against the right. This is an inertial force which will cease to operate when the whole system of bridge-plus-men reaches a steady state. But if *A* now begins to walk towards *B* he will find himself mysteriously impelled against the right handrail *all the way*[1] and, if he tries to run, the force may involve quite serious inconvenience. This would be Byers' non-inertial force. To *A* it would be a very real experience, he would be conscious of being 'deflected' and would regard the adjectives 'imaginary' or 'apparent' as scarcely appropriate, yet *B*, who is presumed to remain stationary within the same system, would be quite unconscious of *A*'s difficulty.

The relevance of such an analogy to motion on the rotating earth is only obvious across the poles. Elsewhere what has been aptly described as the turntable component of the earth's rotation is much more difficult to visualise

---

[1] On the near side of the central pivot each step will take him to a part of the bridge which is swinging to the right on a smaller radius (and circumference) than his body was before. On the far side, each step will bring him to a point on the bridge moving to his left more rapidly than his body was before. He will also carry his relative righthanded motion to the central pivot itself. The result will therefore be uniform throughout as long as his forward speed is constant.

though it is beautifully illustrated by Foucault's pendulum experiment. In 1851 he set up a heavy pendulum on a long wire under the dome of the Pantheon in Paris (lat. 48°50′N) and showed that its plane of vibration, as indicated by a ring of sand, apparently turned in a clockwise direction at the rate of nearly 11.3 degrees per hour. The laws of physics demand that a free pendulum should maintain its plane of vibration *in space* and so he concluded that the floor must be turning in space in an anticlockwise direction to produce the effect. Set up over the North Pole and given an initial swing in the plane of the 0 to 180° meridians, it is evident that one hour later the plane of the pendulum's vibration would be observed to lie over 15°W to 165°E and two hours later over 30°W to 150°E, and so on. Over the equator, on the other hand, if swung in an east–west direction, i.e. in the plane of the equator, the pendulum would show no deviation, for the plane of the equator is not disturbed by the earth's rotation. It is not quite so easy to grasp why the pendulum also maintains its direction over the equator if swung in a north–south (or any other) direction. The plane of vibration then clearly moves in space (with the earth) but it moves laterally and there is no torque, no turntable component of motion of the earth's surface. Riehl (1965, 121) has rather happily expressed the rotation of the earth as viewed from the pole and the equator in the sentence: 'A person standing near the pole turns around *on his heels* once in 24 hours, while a person on the equator makes one *somersault* in space.' It is the former type of motion, motion in the plane of the horizon, that has the major effect on the direction of a missile and hence presumably on that of the wind. It is a safe presumption that, on a sphere, any variable which has the relative value of unity at the pole—the angular velocity of the earth as a whole, usually denoted by $\Omega$ or $\omega$ (omega)—and the value of zero at the equator, will vary over the sphere as the sine of the latitude. Foucault's experiment demonstrated this to be so (sin 48°50′ = 0.7528). The turntable component of motion on the earth thus varies as $\Omega \sin \theta$ where $\theta$ is the latitude.

Even when these mysteries are grasped, it must be confessed that it is by no means self-evident why the Coriolis acceleration should amount to $2\,\Omega\,v \sin \theta$. The '2' can be shown to be derived from the '1/2' inherent in all acceleration formulae which involve an average between two speeds, initial and terminal, yet if one were plotting deflection stage by stage, one would surely use $\Omega \sin \theta$? It is one of the great merits of our third model, discussed under (*c*) below, that it makes this point immediately comprehensible.

A final difficulty in appreciating the effect of the earth's rotation in terms of a deflection or deviation is that, as applied to the wind, it leaves unanswered the basic question, deviation from what? In the case of a projectile an initial path is uniquely given by the training of the gun; can any such path be postulated for motion within a fluid medium such as the air? . . . it is a matter of common experience that, in the northern hemisphere, winds often appear to follow a curved path but if the curvature is righthanded out of a high-pressure system it is no less lefthanded into a low. Such impressions are deceptive when the systems are in motion but Shaw's attempt to trace actual air 'trajectories' (see Figs. 54, 105) reveals a marked preponderance of *lefthanded* curvature in air streams irrespective of whether their motion is poleward or equatorward. Clearly then, the concept of *righthanded* deflection or deviation is emphatically not to be interpreted as *from a previous path*. The analogy with the motion of bodies, such as projectiles, would thus appear to be a very doubtful one.

The solution to this paradox lies in the intrinsic nature of the model being applied. The meteorologist is accustomed to think in terms of a field of forces and his basic data have long been supplied by the pressure map. The initial direction of air motion is thus taken as down the 'pressure gradient', that is, at right-angles to the isobars from high to low. The Coriolis deviation is thus from a purely imaginary path which air in the mass never follows. Furthermore, this model leads one to the concept of a balance between the pressure gradient and Coriolis forces such that the air in the main moves *along* the isobars as a 'geostrophic wind'. This concept has proved of great value as a first stage in forecasting though it carries with it some very remarkable logical corollaries: (1) Such a wind *maintains* the pressure pattern just as much as it may be considered to arise from it. This is a useful property since it is observed that pressure patterns often persist for days even though they may move over the surface of the earth. (2) Yet, if the wind were never ageostrophic, the pressure pattern would remain frozen for all time. There could be no weather change and no 'development'. Clearly this is far from the facts. Hence (3) causes for ageostrophic motion have therefore to be sought. Today these are found mainly in dynamical effects generated in the upper air but the author can remember the days when major emphasis was placed upon surface friction. That all the weather of the world arose from surface friction it seemed indeed a very tall order.

The essence of this model of the atmosphere was effectively stated in Buys Ballot's Law: 'Stand with your back to the wind; in the northern hemisphere the pressure is then higher on your right hand than on your left; in the southern hemisphere it is higher on your left hand than on your right.' This is a purely empirical and qualitative statement of the facts, immediately intelligible and with no nonsense as to implied causation about it. It is the plain man's refuge if this section should have proved elusive but it scarcely meets the demands of physical science. Furthermore it preserves a distinction between the hemispheres which a new model described in the following section will show to be unnecessary.

(*c*) *Vorticity.* This is a feature of fluid media which is familiar to all in everyday life but which requires a certain amount of discipline if it is to be thought about logically. Indeed it has generated a type of mathematics with a symbolism all its own. Nevertheless its basic ideas are capable of expression in non-mathematical terms and we hope to give the reader something of the feel of

163

this model so that he can decide for himself whether to explore the subject more deeply.

The concept of the vortex entered meteorology at first in connection with comparatively small features of the circulation such as tornadoes and water-spouts where the analogy with eddies in water seemed pretty obvious. In a typical phrase Sir Napier Shaw refers to it as 'the vitality of spin' (*Manual*, IV, 1931, 53) and, though he mentions that it is a 'characteristic of turbulence', he was already thinking in a much wider context. Scorer (1958, 49–82) describes vorticity as 'the very stuff of atmospheric motion' and we owe it to him that its basic principles are now intelligible to the general reader.

In its widest application this model starts with a grand generalisation; the earth rotates *cyclonically*. This is true of both hemispheres and it puts the traditional distinction between anticlockwise and clockwise rotation in proper perspective as depending entirely on the point of view.

This carries an immediate corollary; the whole atmosphere, when stationary with respect to the earth beneath it, is charged with cyclonic rotation *in space*. This is a property which will tend, like momentum, to be conserved and if, by any means, the motion can be enhanced it will reveal itself as cyclonic motion with reference to the earth beneath. Observed cyclonic systems are therefore not merely 'eddies', which conceivably could spin either way, they are the visible evidence of an innate property of the whole atmosphere. In mythical terms they are the daughters of Air, sired by Global Spin, the brother of Time! This imagery has a comprehensive sweep which is notably lacking in references to pressure gradients and deviating forces.

To understand this lineage, however, it is first necessary to master the technique by which rotary motion in fluids is analysed. We are all familiar with flat turntables and are accustomed to regard them as 'wheels' in solid rotation. The *rotary* element of the motion however can equally be regarded as embodied in an infinite number of small discs and the turntable surface is then viewed as a field of such discs all turning at the same rate together. The truth of this surprising image is easily confirmed by scattering a number of small compasses over a slowly-moving turntable and observing the result. All the compasses, whatever their position, turn at the same rate—the rate of rotation of the turntable as a whole. Now if the turntable is in fact the base of a cylindrical tank of water, we have an additional dimension in depth and the imaginary discs become vertical 'tubes'. In uniform rotation all these tubes are turning at the same rate and the water is behaving precisely like a similar block of ice. But it is the essential property of fluids that particles are capable of *relative* motion. Imagine a small bundle of the tubes to be 'stretched' to twice their former length; the rate of rotation of the bundle would then be doubled. Equally if the length of the tubes is halved, the rate of rotation of the bundle is similarly diminished. The easiest way to 'stretch' the imaginary tubes in a tank is to produce convergent flow by drawing off some water from above or below; at once a vortex is produced which rapidly extends from top to bottom, that is, along the tubes thus stretched.[1]

On a spinning globe the field of rotation is far from uniform. The turntable component of each disc must be referred to its projection upon the plane of the equator (Fig. 51(d)) and thus it varies, as we have seen, as the sine of the latitude, reaching zero at the equator. The vortex tubes of a stationary atmosphere thus vary in their degree of cyclonic spin from zone to zone. Polar air has a greater degree of inherent cyclonic vorticity than tropical air. When the air is at rest this difference is perfectly matched by the turntable motion of the earth beneath and so we are supremely indifferent to it but if the air is set in meridional motion or if the vorticity is enhanced by convergent flow (or reduced by divergence) we may expect the result to become evident in the wind.

The results are indeed complex but it is much easier to visualise the operation of the latter contingency than the former, it is generally more local in its incidence and corresponds more closely to what is possible in experimental models. If vortex tubes are stretched the increase in cyclonic spin will generate a recognisable 'cyclone' above the earth's surface and, since there is virtually no limit to the degree of stretch, the resulting system may be extremely violent. Note, however, that there is a real difference between the atmosphere and the bath-plug model. There the water is evacuated *downwards* and completely removed from the system. In the atmosphere the only road of escape is *upwards* and the system will only work with full efficiency if this load is rapidly removed from the area of the disturbance by strong currents aloft. Near the equator, of course, there being little inherent cyclonic spin, the effect will be much less notable. This corresponds with experience. A reduction of cyclonic spin below the inherent turntable motion of the earth's surface will, on this view, generate a relative anticyclonic motion. This will be associated with divergent flow which will shorten and widen the vortex tubes but there is a clear limit to this effect for, though shortening can reduce the intensity of cyclonic rotation, it cannot actually *reverse* it. Apparent anticyclonic rotation on earth is thus a lag-effect and most anticyclonic air still retains an element of cyclonic vorticity in space. This residuum can therefore always be rejuvenated when circumstances permit. Since Galton invented the term 'anticyclone' in 1863 we have thus been tempted to think in terms of what has proved to be a false dichotomy. Strong cyclones, weak cyclones, and anticyclones are better viewed as members of a single diminishing series varying only in their degree of cyclonic

[1] A good deal of ingenuity and patience has been spent in trying to determine whether the bath-plug vortex spins preferentially anticlockwise, i.e. cyclonically, in the northern hemisphere. Inherently improbable on account of scale, such a preference could only be revealed (*a*) if the water were absolutely still and (*b*) if a plug like a camera iris were designed so that it could be opened with the minimum disturbance. In this homely instance the tube-stretching is immense and an almost imperceptible rotation is thus exaggerated until it becomes visible. The spin can be in either direction and if the water at two ends of a bath is deliberately set in motion in opposite directions, the vortex can be seen to flood and reverse.

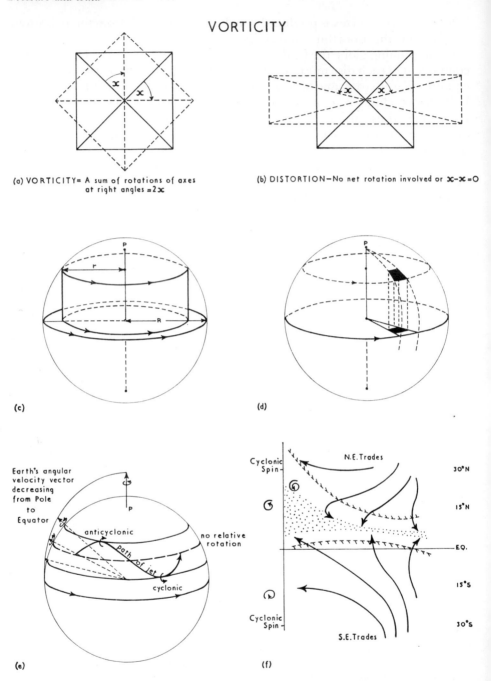

## VORTICITY

(a) VORTICITY= A sum of rotations of axes at right angles =2$x$

(b) DISTORTION—No net rotation involved or $x$-$x$=0

(c)

(d)

Earth's angular velocity vector decreasing from Pole to Equator

anticyclonic

path of jet

no relative rotation

cyclonic

(e)

Cyclonic Spin

N.E.Trades

30°N

15°N

EQ.

15°S

Cyclonic Spin

S.E.Trades

30°S

(f)

FIG. 51. Various aspects of vorticity. (a) and (b). Rotation as compared with distortion, showing that rotation in a fluid medium must be expressed as the *sum* of the rates of rotation of any two lines of particles at right-angles. (c) The basis of the apparent 'deflection' of the wind as Ferrel and others have seen it. It was argued that a 'body' in

vorticity in space. The limit of the series is zero cyclonic vorticity in space, and there is much evidence that, actually, this limit is rarely attained except near the equator where it is endemic. A little difficult to grasp at first encounter, since it involves the *relative* rotation of earth and air, this interpretation has the signal merit that it makes abundantly clear why anticyclonic circulations are much weaker than most of their cyclonic counterparts.

The effects of meridional motion (or of motion along any great circle except the equator, see p. 160) according to this model are most revealing. If we assume that the motion is entirely horizontal, a stream of air moving equatorwards 'will possess an excess cyclonic rotation around the vertical over the earth itself when it reaches its destination' (Rossby 1941, 616). Rossby points out that in the northern hemisphere this may reveal itself either in a cyclonic (left-handed) curvature or in a right-handed 'shear' (the right flank of the air-stream accelerating as compared with the left). It is clear that he considers the former the more probable.[1] Note that this is tantamount to saying that for a narrow current of air, viewed in relation to its known previous path, Ferrel's rule for bodies does not necessarily apply! In actual fact equatorward motion is usually much more complicated than this. As Rossby says subsequently (*ibid*, 619): 'It is a well-established fact that cold currents from the north gradually sink and spread out next to the surface of the earth.' This fundamentally modifies their vorticity distribution and the pattern which emerges is that of a rather asymmetrical fan (Fig. 54). On a much larger scale much the same pattern is revealed in each of the trade-wind cells (Fig. 51 (f)). It would be indeed difficult for a 'body' to achieve this feat.

Similarly, a stream of air moving poleward will carry with it a deficit of cyclonic vorticity as compared with the invaded environment. On the above

[1] Nevertheless we must not forget the possibility of shear. Thus it has been suggested that the wind pattern in a tropical hurricane where the speed near the core is much greater than that towards the edge of the system (in the northern hemisphere, a lefthanded or anticyclonic shear) is a device to make the maximum use of the limited amount of cyclonic vorticity available in low latitudes.

---

steady eastward motion in, say, latitude 45°N would appear to lag relatively to the surface of the earth if it were impelled equatorward and to accelerate if it were impelled poleward. At first the change was expressed merely in terms of the varying lengths of the parallels but Ferrel showed that it must be computed from the area swept by '*r*', the radius of rotation. The same principle holds for initial motion in any other direction. (d) The 'turntable motion' of any section of the earth's crust is proportionate to its projection upon the plane of the equator. For equal areas on the surface it is therefore greatest at the poles and becomes zero at the equator, in fact it varies as the sine of the latitude. A mass of air of constant area (and depth) tending to conserve its absolute rotation in space, will thus appear to gain in cyclonic vorticity as it moves equatorward and to lose cyclonic vorticity (develop an apparently increasing anticyclonic rotation) as it moves poleward. (e) Scorer's representation of the Rossby theory of long waves in the westerlies. (f) A diagrammatic representation of what actually happens to the Atlantic trades in September–October. Vorticity is profoundly affected by divergence (subsidence) and by convergence (uplift). The pattern of the winds is thus much more complicated than Ferrel's Law would lead one to suppose

assumptions therefore, it should tend to swing anticyclonically, that is right-handed in the northern hemisphere as Ferrel's Law suggests. But, in middle latitudes where such a direction is most common, tropical air is almost inevitably involved in convergence and uplift. In open 'warm sectors' its flow is thus characteristically 'straight' but its ultimate fate is to make its contribution to the cyclonic vorticity of mid-latitude depressions.

Rossby also employed the conservation of absolute vorticity to explain the laterally-waving course of the high-level jet-stream often found in association with such depressions (Fig. 51(e)). Such a current, turning alternately righthanded and lefthanded, is certainly difficult to fit into any other of our models. However, since similar 'waves' arise in a flat rotating pan where the vorticity field is uniform (see pp. 164; 189) it would seem that other factors may be at work here.

A final word on the expression of vorticity in mathematical terms. Fig. 51 (a) and (b), after Scorer, shows that it must be expressed as the sum of the rates of rotation of any two lines of particles at right angles. Thus it is indeed *twice* the angular velocity.

It will be noted that each of these models is more flexible, more comprehensive and less dogmatic in its conclusions that its predecessor. Such is the road of scientific progress. The remarkable fact, however, is that the reader will find all of them in current use. (See Lorenz, 1970.)

# 6

# The general circulation: historical

We are now in a position to review some of the ideas held about the general circulation of the atmosphere. Today this group of problems has become very much the happy hunting-ground of applied mathematicians but, though a variety of models has been evolved, the actual situation is so complex that a complete solution still evades us. When a question remains open in this manner there is much to be said for an historical approach since we advance upon the shoulders of our predecessors.[1]

However, there is general agreement that the energy of the winds is ultimately drawn from the differential distribution of solar radiation. Let us therefore remind ourselves of one or two basic facts which are beyond dispute. In Fig. 2 we showed that *at any one moment of time* one half of the intercepted beam is being received by the earth at an angle of incidence in excess of 45°. The area involved is circular and it has a diameter of 90 great-circle degrees or approximately 10 000 kilometres or 6210 miles. Its area is thus 78 million square kilometres or about 30 million square miles but this is barely 15 per cent of the earth's total surface. The other half of the intercepted beam is spread over the remainder of the sunlit hemisphere, some 35 per cent of the earth's surface, but it is received there at comparatively ineffective angles, both towards the poles and at morning and evening in middle and intertropical latitudes. At the equinoxes this 'hot spot' is bisected by the equator where it travels from east to west at a speed of 1037 miles per hour or 900 knots, taking 6 hours to pass overhead. At the northern solstice, however, its meridional cross-section extends from 68½°N to 21½°S and only about 17 per cent of its total area then lies south of the equator. At the southern solstice, of course, the situation is reversed and between these extremes its centre moves with the declination of the sun (Fig. 4). The zone from 21½°N to 21½°S thus experiences the passage of this circle of comparatively intense insolation *on every day of the year*, the duration of its passage increasing very rapidly a few degrees inwards from its polar limits. Towards the poles the expression of this duration in *hours* is complicated by the converging meridians (the summer days are long) but near the equator it is broadly given by the chord

---

[1] For early history see Hildebrandsson and Teisserenc de Bort (1898–1905).

intersected. Clearly from 10°N to 10°S we shall not expect any very significant variation of solar income throughout the year (see Fig. 8).

If the effects of insolation on pressure and hence on air-movement were instantaneous, this 'hot spot' would be recognisable as an area of low pressure and inflowing winds on a synoptic chart. In fact in 1686 Halley propounded the very fanciful idea that its westward progress was responsible for the generally westward motion of air in the trade winds. This would indeed be a hare-and-tortoise race with the tortoise (the wind) finding after every few hours that it was being overhauled by the hare *from the rear*! Yet there are some purely local pulsatory effects, such as land-and-sea breezes and mountain-and-valley winds, which can be directly related to its passage. It would appear however that diurnal variation of pressure in the tropics, with a mean range of 2.5 to 3 millibars and minimum values at 0400 and 1600 hours, local time, is much more a tidal phenomenon.

In fact, as we have seen, many of the effects of insolation are paid into the terrestrial thermal bank so that the periodicity of solar income is damped and the moving area of maximum insolation builds up quasizonal contrasts, more going into reserve over the sea than over the larger continents. The first interpretations of the general circulation (Hadley 1735, Dalton 1793 and Maury 1855) all laid heavy stress on zonal contrasts in temperature since the distinction between insolation and temperature was not then clearly recognised. All were comparatively happy with the trades but the variable mid-latitude westerlies presented an enigma. They appeared to encroach upon the poleward margin of the trades 'for some reason', as Maury (1858, 76) put it, 'which does not appear to have been very satisfactorily explained by philosophers'.

In the following account it will be observed that three rather distinct approaches have been adopted in the attempt to relate the pattern of air circulation to the observed thermal contrasts. For a long time major stress was laid on the 'heat source', often erroneously assumed to lie on or very near to the equator. Sir Napier Shaw made a clean break with this view and emphasised the operation of the 'heat sinks' provided by the polar ice caps. More recently attention has been increasingly directed towards the two mid-latitude zones of most rapid thermal gradient. Each of these approaches has a logical basis but it is interesting to note how the difference in emphasis affected the ultimate conclusions. We shall illustrate these views from a selection of authors.

## William Ferrel

To Ferrel must go the credit for the first really scientific exploration of this problem for he supplemented his qualitative concepts with some quantitative estimates. He was intrigued by the existence of zones of comparatively high pressure in the tropics and as early as 1856 he suggested that the circulation in each hemisphere was basically tricellular (Fig. 52). His full treatment, however, must be sought in *A Popular Treatise on the Winds* published in 1889.

His approach is deductive. Starting with the observed 'distribution of temperature over the earth's surface . . . for the different seasons of the year' (p. 98), he infers a circulation essentially similar to that of Fig. 49, that is with a poleward movement aloft and an equatorward movement at the surface. In the presence of 'the deflecting force arising from the earth's rotation' these

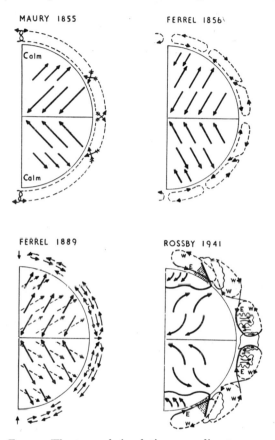

Fig. 52. The general circulation according to Ferrel, Maury and Rossby (1941). In Ferrel, 1889, broken arrows indicate inferred upper winds

currents are converted into apparently eastward and westward movements and 'the general motions of the atmosphere . . . are similar to those of machinery run by a steam engine and controlled by a governor'[1] (pp. 109–110). 'There is a certain limit beyond which the speed cannot go . . . for when this is reached all the force is cut off.' In its day this was a most valuable image though it carries less force in an age when most machinery is controlled by throttles or rheostats.

[1] James Watt's 'universal governor'.

His explanation of the mid-latitude westerlies is much less happy. It involves the introduction of the frictional factor and progresses by three stages, not clearly distinguished by Ferrel himself and not necessarily consistent.

(*a*) Apart from losses due to internal turbulence, the upper westerlies suffer frictional loss of energy only from their *lower* surface, their contact with the lower airstream. The lower easterlies, on the other hand, suffer frictional loss on two fronts, *above* to the airstream aloft and *below* to the earth's surface which is turning in a west-to-east direction against them. Hence, in all latitudes except the immediate neighbourhood of the equator, the west-to-east component of motion aloft is greater than the east-to-west component near the surface. Although we are now aware of the existence of several layers of air above Ferrel's 'upper' airstream, they are certainly tenuous and it would be carping not to go with him thus far.

Now, introducing the effects of 'vertical motions' (i.e. ascent in low latitudes and subsidence in high latitudes), Ferrel argues that these will involve an algebraical sum of both components in each locality, thus yielding a residual west-to-east motion *aloft* in the tropics and *at the surface* in high and middle latitudes. His phraseology has to be interpreted with care but we quote for the latter instance—'as the air in the upper strata of the higher latitudes, with its large east component of velocity above, settles down toward the earth's surface where this component is less, it gradually loses its momentum and the effect of this lost momentum, transferred by means of friction from one stratum to another' maintains 'a *small* east component of velocity of the air at and near the surface' (p. 112). The italics are the author's but the idea is reiterated on p. 149 where Ferrel refers to 'the gentle southwest winds in the northern hemisphere and northwest winds in the southern hemisphere in these latitudes'. The argument is ingenious and was substantially repeated by Rossby as late as 1941, yet though it justifies surface westerlies in middle latitudes, they emerge only as a residual, a notion that can scarcely be held to do justice to the 'roaring forties'.

(*b*) Ferrel then changes his ground, going back to a concept of Hadley's: 'All motions in any direction must have their counter-motions, else the effect upon the earth's surface would be to change the earth's rotation upon its axis.' He clearly realised that the leverage available to such implied frictional forces depended on the radius of the parallel at which they were applied: 'The quantity of air is equally divided by the parallel of 30°, and therefore the east components of the higher latitudes at the earth's surface *must be* much greater than the west components in lower latitudes, or else the dividing parallel . . . *must be* nearer the equator, on the average, than the parallel of 30°'[1] (p. 118). The relative absence of frictional bite over the South Seas is thus enlisted to explain both the strength and comparatively low latitude of the 'forties' above mentioned. But we must not allow this to obscure the

---

[1] Again, the author's italics. The term 'east component' denotes a *westerly* wind.

fact that the cogs of this argument will just not mesh with those of section (*a*). It was surely rather naughty to derive the westerlies from the comparatively friction-free upper airstream by a kind of subtraction sum in which the friction factor had a negative sign and then to demand that they should be strong enough to keep the old world turning against the drag of the laggard easterlies. Yet Ferrel was not the first scientist to demand the best of both worlds, nor is he likely to be the last to disguise the fact, to himself and to others, by the introduction of an interesting and plausible aside.

(*c*) The third stage in Ferrel's argument is 'to examine the effects of these . . . motions . . . in causing gradients of atmospheric pressure . . . between different parallels of latitude' (p. 133). Note the direction of this reasoning, from motion to pressure and not vice versa. His conclusion that 'there is then . . . a zone of high pressure in each hemisphere around the globe with its maximum near the parallel of 30°, an equatorial zone of low pressure with its minimum at the equator, and an area of low pressure around each pole with its minimum at the pole' (p. 136) is thus derived directly from his inferred wind directions and therefore adds nothing to the picture. The subtropical high is found in the lower layers of the atmosphere because the air motion demands (or perhaps we should say, produces) it and disappears in the upper layers because the upper westerlies require a continuous gradient from equator to pole. This does not carry us very far but, once again, Ferrel's picture bears a close resemblance to that which emerged from actual observation except for the omission of the shallow thermal arctic and antarctic highs which his 1856 diagram had suggested. He realised however that it was also legitimate to consider the effects of thermal contrast on pressure and recognised that 'there is . . . a little more air, and consequently a little greater pressure at the earth's surface, in each hemisphere in winter than in summer' (p. 142). To be fair we must admit that the sentences—'it is probable that the polar motion of the lower part of the atmosphere in the middle latitudes does not extend . . . beyond about the parallel of 60° in either hemisphere' and 'it seems, therefore, that in the polar regions of this hemisphere (the southern) the winds, upon the whole, have a component of motion from the pole' (p. 151) are far from dogmatic. He inclined to the view that around the poles there was 'very nearly a calm, unless there is some abnormal disturbance' (p. 152), but quite frankly he did not know.

The deductive approach based upon observed zonal contrasts in temperature is a field now cultivated for the most part by exponents of the higher mathematic but it can be claimed for Ferrel that he defined many of their basic attitudes.

## Sir Napier Shaw

Shaw was a British meteorologist of remarkable insight who associated himself with the inductive school of Buys Ballot, Teisserenc de Bort and Hann. He rests his notion of the general circulation upon observations, regarding observations of pressure as among the most important. His difficulty with the work of the physical school, of Maury, Ferrel and others, is that 'they divide the earth into zones of latitude in which parallel transverse winds are shown all round the several zones, regardless of the difficulty or impossibility of adjusting such an arrangement to any admissible scheme of pressure' (*Manual* I, 291).

He is confident that 'the determination of forces from observations of motion is a more trustworthy process than the computation of the motion from assumed values of the forces' (IV, 324). Yet to discover his own views it is necessary to roam freely through all four volumes of his massive yet stimulating *Manual of Meteorology* first published between 1919 and 1929. In some respects this work must naturally be 'dated' but it is a pleasure to quote. We do so generously though, it is hoped, with discrimination.

First, as to general principles: 'Entropy (with water-vapour as an accomplice) and air-motion are the joint rulers of the atmosphere; gravity, pressure, the centrifugal force of rotation . . . are their obedient servants.' 'In the free air, motion is the controlling dynamic feature of the circulation. Pressure gradient . . . appears as a static index of the motion . . . that is to say the pressure distribution is regarded as the "banking" required for the maintenance of the air-currents.' 'Motion is derived ultimately from convection with the aid of the conservation of momentum but the derivation is intricate' (IV, vii–viii). Then as to the manner in which these forces should be analysed; 'There can be no justification for the tacit assumption that the natural motion of air is horizontal. . . . Atmospheric motion must be considered with due regard for isentropic surfaces which will guide the air upwards or downwards.' The weather-map should be supplemented by 'a vertical section in order that the dynamical conditions may be exposed'. (IV, 323). Indeed, since 'the motion of the air at the surface is a composite picture of the motion of the air in every layer above the surface . . . the only hope of a true picture lies in well-directed co-operative effort in the investigation of the upper air' (IV, 325). During and since the Second World War the techniques of such an investigation have been vastly improved but, as far as the nature of the general circulation is concerned, it seems a fair comment that more new questions have been raised than old ones answered.

Next let us try to build up a brief picture of the surface circulation as Shaw saw it. He begins with the *polar* end of the system since 'downward convection of cold air combined with displacement of the upper layers, which would be persistent even "if the earth went dry", is responsible for the primary circulation of the atmosphere' (IV, 333). In high latitudes he considers there are two great pools or 'underworlds' of cold air situated permanently over Greenland and the arctic ice and over Antarctica respectively. They vary in extent and intensity with the season, the northern cap in particular extending widely over northern Canada and Siberia in the winter. Their upper surface is represented by a temperature discontinuity which, he suggests, may lie at 6 kilometres above sea-level in the central regions but slopes at a low angle towards the margins where it is recognised as the Polar Front. As this cold air tends to spread southward it 'cannot leave the surface . . . on account of the defect of the necessary entropy' and 'it will turn westward unless otherwise guided by pressure' (IV, 332). The strong deflective force experienced in high latitudes 'amounts to an imprisonment of the air' but 'the occasions when the guard fails to restrain the cold air, in consequence of lack of velocity or from some other cause, mean a good deal in the matter of the weather of the eastern Atlantic, the Mediterranean and the Baltic' (IV, 330) and likewise, of course by implication, in the middle latitudes of the southern hemisphere. 'For the drainage in the polar regions necessary to maintain the circulation' he brings in 'the effect of a slope losing heat to the sky by radiation. . . . The slopes of Greenland, Alaska and other high northern lands, or the slopes of the Antarctic continent, working day and night, would make no difficulty about introducing a few billion tons at the right time' (IV, 332).

The idea of an inherent conflict between polar and equatorial air goes back many years but, whilst Shaw was writing, V. Bjerknes (1920) and his son J. Bjerknes (1921, 1930) revived it, gave precision to the concept and, above all, provided it with a vocabulary. In his later volumes Shaw commends the new terms and adds that if the conflict is an expression of the reaction between the 'underworld' (of polar air) and the 'overworld' (of equatorial air) then the travel of depressions which is so characteristic of the outer margin of the polar cap 'may be that of some deformation of the isentropic surface which separates the two worlds'. 'Personifying the acknowledged difference . . . the aim of polar air is to create or enhance anticyclones, and so far as we can tell it succeeds, unless it loses speed . . . the aim of equatorial air is by its motion to fill up and destroy depressions' but 'water-vapour intervenes and not merely frustrates that aim but actually secures the opposite—it is water-vapour that "rides in the whirlwind and directs the storm"' (Shaw, 1931, IV, 283).

We quote these sentences in some detail because they underline his insistence that motion is primary and the release of latent heat by condensation, secondary; also, they provide a useful corrective to the much too facile analogy between isobars and contours which is encouraged by the use of the term 'pressure gradient'. The hills and valleys of the weather maps are so far from being everlasting that they are actually the creation of the movement of the air between them. Note that we have not yet mentioned the 'prevailing westerlies' of middle latitudes which so troubled the earlier theorists on the general circulation. We have now learned to regard them as the statistical expression of the west-to-east motion of an endless series of disturbances 'occupying the southern portion of the field of invasion of the equatorial winds by the colder polar winds' and 'expressing the result of the fluctuations of the recurrent and variable invasion' (II, 369). Shaw suspected that their generally eastward drift was related to events in the upper atmosphere, but at that time the relationship was not fully established. Elsewhere, on the westerlies he is engagingly frank—'we are not yet clear as to where they come from, the important thing is that they are on the map' (II, 253), though this is followed by the warning that 'the actual path of air over the earth's surface is hardly ever to be guessed from the single indication of a variable prevailing wind.'

The real difficulty that may face the student who has followed the argument so far is why does the polar outbreak share in this eastward motion?—and it is refreshing to find the question squarely put. 'Why does air flowing southward in the north-western area of the Atlantic turn left as it approaches temperate latitudes and join a westerly wind instead of turning right and providing an easterly wind as it ought to do?' (IV, 289). Shaw's answer that it is prevented from suffering the usual deflection 'by the coercion exercised by the distribution of pressure . . . over the North Atlantic' is clearly not consistent with his view that motion is primary (unless we have again to go back to motions in the upper air, a possibility that is by no means ruled out). With equal frankness we find him facing another difficulty that is too often glossed over: 'About discontinuity as a basis of the development of a cyclonic depression there is this difficulty, that we find it most clearly indicated in the central regions of a well-formed travelling depression. There is a general principle in dynamics that the particular condition which constitutes a cause tends to exhaust itself and disappear while the cause is producing its effect.' (IV, 300) These sentences will serve to remind us that the polar front theory does not supply us readily with *all* the answers. Expressed in a more modern idiom both these points remind us that, whilst it is comparatively easy to picture the growth and decay of an unstable wave on the polar front in three dimensions, it is much more difficult to indicate an initial cause of the feature. The important fact is that these developments occur and that they occur frequently.

This discussion has caused us to anticipate in the sense that we have become involved in circulations which might be thought to be 'local' rather than 'general' but in the middle latitudes of both hemispheres the distinction has little, if any, validity. Shaw emphasises that 'the creation of a cyclonic depression is an undertaking in the art of removing air' on a scale of the order of a billion tons and that the 'scavenging currents' (IV, 307) which make the removal possible must belong to the general circulation of the upper air.

Ferrel's initial question, the origin of the tropical highs or Horse Latitudes, is not squarely put in any passage I have been able to discover in the *Manual*. Shaw is aware, of course, that these features are cellular rather than zonal but he is inclined to regard anticyclones as comparatively passive features of the circulation, as 'dumping grounds for the air which has taken part in atmospheric action and has to be disposed of' (II, 294) rather than as 'centres of action' in Teisserenc de Bort's phrase, which is not entirely defunct. This clearly does not dispose of the question since dumping on a scale sufficient to produce the great desert girdles of the earth must be something much more than a purely random affair. Shaw pictures the dumping process as a sort of elbowing aside of a great thickness of air when the wind speed is in excess of that required by the gradient and definitely *not* as an inpouring from the top.

Nevertheless, a linkage with the high-level circulation would seem to be indicated and rising pressure implies subsidence with respect to the pressure surfaces (see p. 231).

Shaw is able to adopt this rather cavalier attitude towards the traditional tropical highs because he believes that 'the flow of the trade winds depends upon the supply of air from the polar regions' (IV, 339) and that certainly 'the most definite form of trade-wind comes from a complicated circulation that is as far from calm as the trade-winds themselves' (I, 320). He gives trajectories across the Atlantic from the northwest which would make this possible (II, 246). It should be noted that, to him, the 'real trade-winds' are 'streams of air from the north or northeast in the northern hemisphere, and from the south or southeast in the southern, *near the western shores of the continents*'[1] (II, 249). Furthermore they 'are always colder than the sea surface' (IV, 332). 'The assumed uniformity of conditions throughout a whole belt of latitude has indeed no foundation in fact . . . we want another name for the winds of any other part of the North Atlantic. We prefer the name of "intertropical flow" for the winds of the West Indies' (II, 251). At the equator the trades 'can go west or go up' and he does not exclude the possibility that a proportion of the return current moves northward near the surface on the westward side of the oceans—a possibility consistently ignored by the zonal school. Near the equator itself there is a 'line of junction of the air of the two hemispheres with completely different life histories . . . a sort of physical laboratory in which the effects of differences of physical state are disclosed' (II, 249). Here convection comes formidably into the picture and 'water-vapour is the agency by which the entropy of air is increased locally to such an extent as to produce penetrative convection', on a grand scale (IV, 344). The air thus evicted is carried away aloft by methods 'not yet determined by observation' though Shaw suspects that it is swept west and poleward till it is caught up in the grand high-level polar vortices which are maintained by the upper westerlies. That it loses heat (and entropy) during this process by long-wave radiation from its remaining small content of water vapour he considers probable. Here is a heat sink of vast extent even if of low intensity.

Shaw had 'no intention of presenting any of the major problems of meteorology as solved'. He is always most careful to distinguish between observation and speculation. We have quoted very freely because the weight and discursiveness of the original make it formidable to the general reader or to the student with heavy claims upon his time but no other meteorologist has written with such clarity or wealth of imagery[2] and our references may serve as a guide to selective reading.

Since Shaw's time extremely rapid developments in the techniques of flying have encouraged the application to the study of air motion of concepts derived initially from hydrodynamics, the analysis of motion within in-

[1] Author's italics.
[2] See especially Shaw, *The Drama of Weather*, 1940.

compressible fluids. This has deepened our appreciation of comparatively local and short-term events but it has carried with it the erection of a vast mathematical superstructure, largely inaccessible to the layman and involving a bewildering variety of assumptions most of which are plainly inapplicable to the air-ocean as a whole. We do not decry these developments but they evidently demand interpretation. The plain man may be content to leave atomic physics to the specialist since the concepts involved are remote from everyday experience, but when it is a question of the movement of a not entirely intangible (and sometimes not even invisible) substance in three dimensions he is likely to insist that he should be addressed in straightforward terms. Common sense is as effective as an 'equation of continuity' in reminding us that air which moves from one place must turn up in another and as for 'waves', 'vortices', 'shear lines' and even 'surfaces of discontinuity', the stock-in-trade of the new approach, they are encountered daily by any observant individual. He does not claim to be able to theorise about such features but he does feel that he can comprehend them to a degree that will yield satisfaction.

# Carl Rossby

Rossby has attempted such an interpretation in two papers: 'The scientific basis of modern meteorology' (1941, 599–605)[1] and 'On the nature of the general circulation of the lower atmosphere' (1949)—addressed to the intelligent but non-specialist reader. It may seem unfortunate, at first encounter, that the second of these papers involves a very considerable change of emphasis from the first, but the scientifically untidy impression that results is not necessarily a bad thing. Tidiness in science is almost always illusory—if the world were not full of loose ends, research would soon come to a standstill.

Rossby's first paper (1941) follows closely the traditional approach though expressing it in a more modern idiom. Thus it begins with the effect of the zonal distribution of radiation on convective processes but emphasises the view, only tentatively suggested by Shaw, that 'everywhere in high and middle latitudes the free atmosphere above a shallow layer of air next to the ground is constantly losing heat by radiation' (p. 604). 'In middle latitudes at 2 or 3 km above sea level, these losses would produce a cooling at fixed levels of the order of magnitude of perhaps 1 or 2°C per day. Thus, with the possible exception of equatorial regions, the free atmosphere everywhere serves as a cold source (condenser) for the circulation engine' (p. 610). He then infers, by reasoning not radically different from Ferrel's, that on a rotating globe 'the initial meridional circulation . . . necessarily must break down into at least three separate cells on each hemisphere' (p. 608). We now have, however, a distinction in type between the cells. The tropical and the arctic cells are regarded as thermally driven—the heat sources being both insolation at the

---

[1] First half reprinted in Berry *et al.*, ed. (1945), 501–29.

surface and condensation resulting from convection in the first case, and mainly condensation stimulated by uplift along the polar front in the second. In each of these 'direct' cells easterlies tend to predominate at the surface and westerlies aloft—'strong westerly winds are continually being created at high levels' (p. 611). The central cell, on the contrary is '*indirect*' or 'frictionally driven' and the passage must be quoted—'Along their boundaries with the middle cell, these strong westerly winds generate eddies with approximately vertical axes. Through the action of these eddies, the momentum of the westerlies in the upper branches of the two direct cells is diffused toward middle latitudes, and the upper air in these regions is thus dragged along eastward' (p. 611). By this arrangement, we are to believe, it is possible to generate a central cell with a marked west-to-east component of motion in both its upper and its lower limbs. These motions react upon the pattern of surface pressures giving prevailing high pressure with subsiding dry air over the tropics and prevailing low pressure along the polar front where 'cold and warm air masses are in constant battle', the conflict expressing itself through 'the formation of quasi-horizontal waves that normally progress from west to east along the front' (p. 613). Mid-latitude precipitation is therefore very different in type from that in equatorial regions.

Rossby suggests that this zonal pattern is a fair approximation to actual conditions in the southern hemisphere but that the distribution of land and sea in the northern hemisphere causes great distortion especially during the winter months. Sometimes the arctic cell remains intact but is displaced towards Siberia—yielding a strong zonal circulation in the westerlies from the Pacific to the Atlantic. Sometimes it is broken into two sectors, one lying over Asia and the other over North America. The westerlies are then weaker and with a more pronounced meridional component. He concludes: 'No adequate physical theory is available at the present time from which the fluctuations in circulation-intensity may be computed, but recent studies suggest that these fluctuations may be associated with the intermittent establishment of a direct inflow of deep moist air from the equatorial trade-wind belt into the westerlies of middle latitudes' (p. 622).

We have given but the barest outline of Rossby's treatment of the westerlies but this last sentence reveals how far he has travelled in the latter part of the paper from the comparatively simple scheme he outlines at the beginning. That scheme provides no mechanism by which trade-wind air could be transferred to the westerlies to procure this result. Nevertheless, the zonal pattern is worth brief consideration. Note that, as far as surface features are concerned, he agrees with Ferrel that westerlies are necessary to offset the drag of the tropical easterlies but omits Ferrel's point about their necessary width or strength. Furthermore he gives no clearer reason than Ferrel why air should descend at about 30° latitude and, since he brings all the equatorial air down in this region he leaves it with a shorter time to lose heat to overcome the thermal (and hence density) effects of dry-adiabatic descent. It is aloft, however, that we must look for the main points of criticism. Recalling Shaw's reference to the difficulty of relating a pressure pattern to a zonal theory we may ask, what is the pressure situation envisaged in the upper limbs of the three cells? The prevalence of westerlies in all latitudes implies a fall of pressure from equator to pole. If the gradient is comparatively uniform then all three upper limbs fuse into one and we are virtually back with Ferrel. If it is conceived as steeper in the two direct cells to accommodate itself to the strong westerly winds there, then what prevents these two gradients, both inclined polewards, from uniting together? Rossby says of the air 'dragged along eastward' that 'the excess of centrifugal force acting on these upper west winds in middle latitudes forces the air southward, but equilibrium is never reached, since the air still farther to the south, instead of piling up and thus permitting the establishment of an adequate cross-current pressure drop, cools through radiation and sinks to lower levels' (p. 611), an answer which is at least elliptical. If subsidence thus affects the pressure pattern aloft it cannot be an overload of air there which localises the subsidence as Ferrel was inclined to suggest.[1]

A much larger issue arises from the whole concept of a 'frictionally driven' indirect cell. Surprisingly enough Rossby gives no inkling of the magnitudes involved. Figure 50 shows us that a quarter of the area of the northern hemisphere lies between 30° and 50°N, a total of some 24 million square miles. The northern and southern margins of this zone have a total length of some 37 500 miles so there are 650 square miles of surface area for every mile of border section, north and south. We are thus asked to imagine that by lateral air-to-air friction aloft, and thus at comparatively low densities, enough energy is imported across these borders not only to overcome the internal viscosity of the vast mass involved but also to enable it to 'offset' the drag on the earth of the much more consistent tropical easterlies. We leave the necessary computations to the specialists, but it does seem an extremely tall order. Perhaps we could double the length of the borders by imagining them wavy or irregular but any reduction in its width will reduce its efficacy as a counterpoise to the friction of the easterlies.

Rossby was unquestionably aware of some of these difficulties and opens his second paper (1949) by remarking on 'the need for far-reaching revisions in the picture outlined above'. It is most refreshing to the climatologist then to encounter a series of five well-attested observations on tropical climates which the theorists have been far too apt to ignore. They are not, by any means, all matters of so recent discovery as Rossby is inclined to suggest. 'None so blind as he who won't see' can be as true in science as in any other human activity. We give them in detail as they are essential elements in the structure we are striving, rather laboriously, to build.

1. Actual isotherms in the upper as well as the lower air 'are very nearly horizontal . . . within a low-latitude zone of about 45 latitude degrees in

---

[1] Today indeed it is suggested that subsidence must be related to the subtropical jet stream, evidence of a *strong* poleward gradient aloft.

width', hence 'there are practically no thermal forces present to drive the meridional trade-wind circulation' (Rossby, 1949, 19). Rossby's evidence here is a north–south section through the Atlantic in August and September 1939 supplemented by mean meridional sections over America for January–February and July–August, 1941–5 but our study of solar income has already prepared us for the news.

2. Wind components in an east–west direction may *increase* with height in very low latitudes, his section actually shows values up to 40 mps (78 knots) at 13 km (8 miles) above sea-level. Ferrel had heard of these winds in connection with the spread of volcanic dust but was inclined to discount them.

3. The cloud belt over the doldrums is recognised as reducing insolation there as compared with the clear areas of the horse latitudes. The byproduct of convection is thus seen to serve 'in some measure . . . as a brake on the proposed thermal circulation' (p. 23). We can find this in Maury (1858, 210): 'The mariner . . . is surprised to find that, notwithstanding the oppressive weather of the rainy latitudes, both his thermometer and barometer stood, while in them, lower than in the clear weather on either side of them.'

4. The average variation in latitude of the doldrums over the ocean in the course of the year is 'significantly smaller than is the variation in solar declination' so that, particularly in the northern summer, 'the descending branch of the trade-wind circulation over the sea appears to be located almost right under the sun' (Rossby, 1949, 23).

5. In places the doldrum belt tends to split into 'two separate zones' between which 'descending air motion and dry-weather conditions appear to prevail' (*ibid*). It is not yet clear whether the semidesert climate of Christmas Island on 2°N is due to such motion as Rossby suggests but the duplication of the rain belt appears to be a matter of synoptic experience over a considerable extent of the equatorial girdle.

These are followed by two aerological observations which are certainly both new and unexpected: (*a*) The trades are often capped in the upper troposphere between 9 and 15 kilometres (30000–50000 ft) by 'quasi-horizontal eddies of moderate diameter' (p. 23) which move from east to west, not by counter-trades moving from west to east. (*b*) In February 1947 a marked high-pressure system with comparatively warm *easterlies* throughout the troposphere was recognised over the Arctic and 'there are good reasons to believe that similar conditions may occur with a fair degree of regularity in midwinter' (p. 26).

These facts are taken as throwing considerable doubt on the efficacy of the thermodynamic theory of the trades and of any theory which makes use of the conservation of angular momentum to explain the circulation in poleward latitudes. Rossby therefore proceeds to offer some new ideas.

He pictures the atmosphere as 'a thin spherical shell in which the distribution with latitude of the mean zonal motion is determined primarily by lateral mixing processes' (p. 26). Such processes are regarded as taking place mainly in the upper air of middle and high latitudes where absolute vorticity is greatest and their effect is to produce a field of comparatively uniform vorticity over an extensive polar cap. Towards the margin of such a cap the result would be the development of 'a sharply defined velocity maximum (jet stream), north and south of which the velocity drops quite rapidly' (p. 27). This agrees substantially with upper air observations which further confirm the impression that 'the formation of jets may be considered as a consequence of some kind of a turbulence-wave which emanates from high latitudes and gradually advances southward' (p. 31). Of course, such a 'mixing process' cannot account for the *generation* of the kinetic energy involved but basic to the whole theory is the idea that 'the zones of maximum kinetic energy do not necessarily coincide with the regions in which the rate of conversion of potential into kinetic energy reaches its maximum value' (p. 27).

The idea carries with it some interesting corollaries, thus:

1. 'Lateral mixing' will also tend to concentrate the thermal contrast between high and low latitudes at the equatorial margin of the mixed area, i.e. under the jet stream, so that 'the fronts in the free atmosphere and the jet stream would simply be different manifestations of one and the same process' (p. 32).

2. A strengthening jet aloft will develop higher pressure towards its equatorial flank and thus stimulate poleward motion near the earth's surface. An indirect meridional circulation is thus set up which corresponds to the middle-latitude cell of his earlier theory but is now regarded as covering 'not much more than 5 or 10° of latitude'. Furthermore, since the poleward thrusts at the surface may be expected to be variable, they will impart an irregular or wavelike outline to the polar front.

3. Rossby's emphasis on dynamical influences emanating from the polar regions meets with the same difficulty that Shaw experienced in his thermal approach; it is very difficult to fit the intertropical circulation into the picture. His point that there too 'dynamic, rather than thermal requirements' are paramount is well taken and this would certainly help to account 'for the remarkably small annual march of the doldrum belt'. Unfortunately this passage of his paper is unusually obscure (always a bad sign) and one has the impression that R. Fletcher's (1945) case for a twofold division of the equatorial rain-belt is magnified beyond all reasonable proportions.[1] The 'lateral mixing processes' will unquestionably involve the export of vigorous cyclonic vorticity to the very margins of the intertropical zone but what happens *in between* the high-level polar caps thus defined? Fig. 53 may well represent Rossby's guarded answer. It certainly suggests, as he does, that the high-level equatorial easterlies are also dynamically driven (how else?) and will even accommodate equatorial surface westerlies if the jet streams diverge, as they

[1] Rossby goes out of his way to disagree with Fletcher's explanation on the ground of the alleged 'permanence' of such a feature in the east-central Pacific. Sea surface temperatures are highly anomalous there. That would suffice.

may well do over the Indian Ocean in October when this anomaly is most frequently in evidence. Yet one cannot resist the thought that this approach gives but scant attention to the prime intertropical fact, the existence of five massive cells of surface trades. The ideas are stimulating but, almost certainly, he has gone too far.

It must be emphasised that Rossby's statement is by no means dogmatic. It is qualified by such phrases as 'our result, if correct' and the frank admission of an unsolved 'apparent conflict' between his views as to the origin of the jet stream on the one hand, and as to its dynamic effects (on cyclone stimulation) on the other, i.e. as to the jet-as-effect and the jet-as-cause. It is true that an

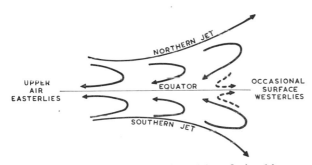

FIG. 53. A plausible interpretation of the relationship between the upper-air circulation over the trades and the subtropical jet streams. We still know relatively little about the upper-air circulation in these latitudes. The broken arrows indicate the surface equatorial westerlies experienced in the Indian Ocean particularly in October when the interval between the converging trades is unusually wide. They still present something of a mystery but are of considerable climatic importance

interpretation of the general circulation in terms of mechanical rather than of thermal processes involves inherent difficulties but it is not to be lightly dismissed simply because the full nature of those processes is not yet fully understood. The crude periodicity observed by many workers in the strength of the mean zonal flow certainly suggests the operation of some kind of mechanical effect and the pattern of the surface trades leaves a strong impression that they are *driven* towards the equator rather than *attracted* to it as the older theories would suggest. Is this a valid distinction? That may be a nice question but the author knows no other way of expressing the feel of the situation as it emerges from a close study of the facts.

The first difficulty, of course, is the comprehension of Rossby's 'mixing process'—it is quite beyond the bounds of everyday experience. In the world as we know it jets freely break down into zones of turbulence but the process is irreversible. Brunt (1941, 283) says quite positively, 'the energy of the great currents cannot be reinforced by eddy motions'. More recently Eady (1953) has asked what is it that is to be regarded as mixed? It would seem to be a property which 'far from being independent of the motion' is 'actually one aspect of it'. To produce 'organised zonal flow out of random weather systems, has the character of "unmixing"', of 'sorting out'; he suspects this may indeed occur but the mechanism was then unknown. (See Green *et. al* (1966) and Lorenz (1970).)

A second difficulty in throwing stress on air momentum is that the reserve of momentum that the air can carry is comparatively limited. Despite its great total bulk (of the order of $5.34 \times 10^{15}$ metric tons) only a proportion of it is in rapid motion at any one time. Computing the rate at which atmospheric energy is dissipated by turbulence Brunt estimated that 'if the same rate of dissipation were maintained for $10^5$ seconds, or $1\frac{1}{6}$ days, the whole kinetic energy would be destroyed' (Brunt, 1941, 286). It now appears that he underestimated the energy and overestimated its dissipation, both by a factor of two. Hence kinetic energy is, in fact, replaced about every 5 days. Even allowing for the discovery of jet streams (in comparatively thin air) it thus seems that the 'flywheel' of the atmosphere is a rather ineffective one. Atmospheric reserves lie in thermal energy from which supplies of kinetic energy must constantly be renewed.

Finally, as we have seen, the theory is weak in the tropics. No clear reason emerges why the jets should not migrate still more closely towards the equator and the steadiness of the trades seems to be regarded as quite incidental. Most modern writers seem to agree that here, at least, thermal considerations are dominant though not exclusively so. The trades pick up thermal energy via evaporation throughout their long trajectory over the sea but they can *realise it* by convection on a major scale only when they have outrun their internal inversion (a result of dynamic processes). The rain belt near the equator is thus dependent upon insolation only in a secondary sense. There is no reason whatever why it should faithfully 'follow the sun', it is drawing energy from hundreds of miles upwind. The resulting cloudbank may lower local surface temperatures but to regard it as a 'brake' on the circulation is to misplace the emphasis—it is an essential byproduct of the power plant, comparable perhaps with exhaust-pressure. When once the circulation is organised in this way, dynamic considerations shaping the pattern of release of thermal energy via the trade-wind inversion, Rossby's (1949, 23) objection to the latent heat of water vapour as the source of power for the intertropical cells would seem to disappear.

## Some current views

We turn now to a number of modern articles, often of a highly technical character. From these we select E. T. Eady's 'The cause of the general circulation of the atmosphere' (1950), and R. G. Fleagle's 'On the dynamics of the general circulation' (1957) to illustrate the theoretical approach, realising that vital links in their argument are by no means open to non-mathematical interpretation or comment. We shall then review a number of more descriptive papers in an attempt to clarify the results. All agree in throwing great emphasis on atmospheric turbulence and hence on activity in middle latitudes.

Eady's argument runs briefly as follows: On a rotating earth the meridional motion of air stimulated by thermal contrasts involves incidentally a transfer of angular momentum. In the absence of friction the exchanges thus set up would transfer much more angular momentum than is required by observed facts. The only plausible mechanism of introducing frictional stresses into the free atmosphere is by turbulence. Hence 'turbulence is . . . a necessary feature of atmospheric motion'. Gross turbulence[1] is generated in middle latitudes, (a) because here, in the zone of maximum thermal gradient and strong deflective tendencies, the strongest zonal (westerly) circulation is stimulated, and (b) because it can be demonstrated mathematically that such a zonal flow is dynamically unstable, i.e. any minor disturbance of the flow will tend to grow and produce a train of lateral wavelike meanderings which are propagated downstream (eastwards). Such 'waves' are eventually dispersed but 'regeneration begins while breakdown is still occurring'. Fleagle has added greater precision to this picture by showing that (1) this instability 'increases with latitude from the critical latitude of 37° to about 60° and then decreases to zero at the pole, that (2) growth is most rapid for disturbances of wavelength 2500 to 5000 kilometres, and that (3) the energy of the disturbances is derived from the conversion of potential into kinetic energy 'through sinking of the cold air as it moves equatorward and rising of the warm air as it moves poleward'. Eady notes two features of such waves: (a) on the average their amplitude 'increases with height up to the tropopause', and (b) their progress is 'more retarded in low than in high latitudes' (a feature exaggerated by the varying length of the parallels) so that their axes tilt or trail away 'in a general NE–SW direction in the northern hemisphere'. This enables them to transport angular momentum, especially aloft. Fleagle would add a third feature (c), that since disturbances develop most rapidly between the 50th and 70th parallels they must often 'result in formation of closed vortices'. His further suggestion that there is a 'preferred number' of such disturbances according to latitude, ranging from 8–10 at 45° to 2–3 at 80° is interesting but his treatment of the tropical zone between the critical latitudes in each hemisphere is disappointing and he almost ignores the trades.

Eady, on the other hand, struggles manfully to gear his upper waves to what is happening near the ground in a passage which is certainly not easy to follow. Briefly, since surface friction involves some motion across the isobars and since the disturbances draw upon thermal sources for their energy we must infer a general *upward* motion towards the polar limits of the westerlies (which he takes as about 60° latitude) and a general *downward* motion over the tropics. There must therefore be a corresponding *downward* motion near the

---

[1] The broad turmoil of weather systems so clearly apparent in satellite photography.

poles and an *upward* motion near the equator. He thus returns to a tricellular pattern in each hemisphere but all three cells are *driven*, 'they correspond to processes which absorb energy from outside i.e. from turbulent overturning' but 'only a fraction of the energy released by turbulent developments is used in this way'. This enables the vertical motions to take place at a comparatively leisurely speed, allowing the subsiding air to be cooled by radiation 'fast enough to account for the observed lapse rate'. The net result of this complex mechanism is not only to provide 'the observed zonal distribution of pre-cipitation' (under the rising currents) but also to strike the necessary balance between the frictional stresses of easterly and westerly surface winds. Both processes thus depend upon events aloft.

Note the emphasis of this approach. It underlines the view that 'there is little reason to regard the time or space average of the large-scale circulation as being more fundamental or more easily understood than the details of the circulation' (Fleagle, 1957, 19). The variability of climatic phenomena is no longer viewed as the product of temporary aberrations from a presumed mean pattern, it is recognised as an inherent characteristic. As Eady puts it, 'variations from mean behaviour are not only possible but are to be expected precisely because turbulent transfer is necessarily irregular'. Progress in science has often been punctuated by such subtle changes in emphasis.

It must be confessed that the concept of 'waves' in the westerlies which thus lies at the root of modern theory presents great difficulty to the non-specialist since the motion inferred is unlike anything encountered in everyday experience and can only be described in mathematical terms. Furthermore in discussing any kind of wave motion it is well known that one must *never* confuse the apparent motion of the wave system with the actual motion of any individual particle involved in it. The commonsense question—'*But what actually happens?*'—is thus far from naïve. It is fatal for the theorist to become a prisoner within his own frame of reference. There is a sharp sting in Palmén's (1951, 345) comment that 'grossturbulenz' is a 'substitute for real knowledge' and 'has therefore become rather popular' with meteorologists. Palmén insists that we must think in three dimensions, if possible on a hemispherical scale, and stresses the significance of *vertical* motion as the source of the energy involved. His careful analysis of a 'wave' over North America shows that what actually happened was a southward outburst of cold air (some $8.3 \times 10^{12}$ tons of cold air crossed the 45th parallel in 24 hours), the air subsiding strongly as it progressed along 'true trajectories' which differ widely from synoptic stream lines (Fig. 54).

The reader will recognise here a concrete picture in the Shaw tradition, a model in which mathematics is the servant and not the master. It is a valuable corrective but unfortunately it fails to tell the whole story. Palmén says that 'it is obvious that similarly large warm air-masses must replace this loss of polar air' but admits the existence of 'great difficulties in following the three-dimensional movement of tropical air parcels'. Furthermore, although he suggests that as many as five such outbreaks may be active at the same time

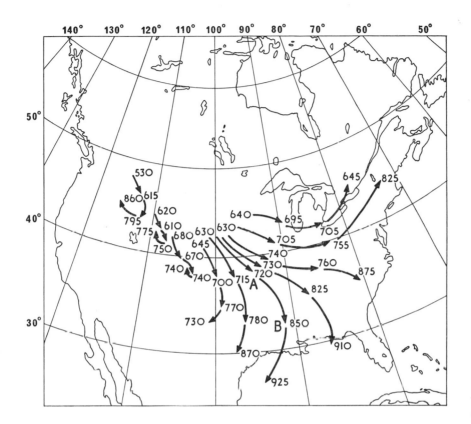

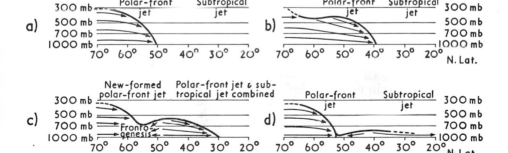

FIG. 54. The trajectories followed by a polar outbreak over North America in April 1950. The arrows show the paths followed by air parcels with a potential temperature of 290°K during a 36-hour period. Pressures within the air parcels (in mb) are given at 12-hour intervals. Note (i) the fan-shaped form of the outbreak and (ii) the increase of pressure downstream which implies widespread descent. Below, schematic downwind cross-sections of the outbreak at four successive stages of its advance. (After Palmén, 1951, pp. 350–1)

in one hemisphere it is clear that the examination of any one of them can give no clue as to why this should be so. The 'wave' school can thus quite fairly retort that it is this very *organisation* of the motion that they are striving to interpret. In the atmosphere as we know it periods of strong 'wave' activity involving strong meridional exchanges (Rossby's low 'zonal index') alternate with periods of comparative quiescence (high zonal index) when the circulation is more nearly zonal. The consequences on the weather are fundamental and they persist long enough to be a true feature of climate. The causes appear to elude us completely.[1]

It is perhaps natural to seek such causes in a sector of the atmosphere where our knowledge, although rapidly increasing, is still fragmentary—the upper air. Hence the theoretical interest of the 'jet streams,' 'ribbons' of air some hundreds of miles across moving with high velocity (often in excess of 80 knots) at levels of from 7.5 to 14 kilometres above the earth's surface. Although the term is new and dates from physical encounter with these air streams by high-flying aircraft, the fact was known in Ferrel's time from the motion of cirrus cloud and volcanic dust. The term usually implies a *westerly* motion in both hemispheres though it has also been applied (Koteswaram, 1958), with less general acceptance, to high-level easterlies over the equator. We have seen how the idea of a semipermanent jet, scarcely mentioned in his first theory, was promoted by Rossby to the very front rank in his second. Continually reinforced by his 'lateral mixing processes' it is viewed virtually as the grand driving belt of the circulation. Indeed he sees it linked, though perhaps rather obscurely, with the pulsations in zonal motion we have just described. Eady, on the contrary, dismisses the jet streams in an aside—'because large-scale disturbances are dispersive we may get energy concentrated in regions where it has not been released', and though Fleagle finds logical reasons for the existence of local jets, especially near his critical latitude (through the effect of flow-pattern on temperature distribution), the outcome is regarded as no more than incidental. 'The jet stream is a direct and rather simple consequence of baroclinic instability and its distribution with latitude' (Fleagle, 1957, 13).

The basic difficulty, of course, is that by no means all the facts about the nature of jet streams are yet to hand. Is at least one jet *always* present at some latitude or other in each hemisphere? Does it ring the earth or does it only develop in preferred localities? When encountered in widely different latitudes, is this a result of the meandering, or even bifurcation, of a single stream or are there *two* types of jet—one moving quasi-permanently above the horse latitudes and another of much more evanescent character which forms, dissipates and re-forms over the more active sections of the polar front? What is the seasonal rhythm of the system, or systems, and is there any basic difference between the patterns in the two hemispheres? Complete answers

---

[1] Rossby (1949) suggests that his mixing process has a characteristic 'relaxation period'. Starr (1951) thinks that a stable mean situation is unlikely in view of the distribution of the continents and the effect of major mountain systems. Such causes hardly appear adequate to produce the result.

to these questions must await a more searching investigation of actual conditions aloft over large sections of the earth's surface where exploration has scarcely begun. We shall give some account of the facts to hand in a later section. In the meantime let us note some of the interesting results that have emerged from laboratory experiment.

Thermal circulations have been set up in liquids contained in rotating hemispherical or paraboloid vessels (Fultz, 1951, and Fultz and Long, 1951) but, owing to difficulties with gravitational attraction, the most successful device would appear to be a flat pan where gravity is constant. The 'dishpan' thus employed is really a circular trough mounted so as to be rotated horizontally about a central axis and supplied with a heat source at the outer rim and a heat sink at the centre. Radial convection in a 'Hadley cell' is distorted by a spiral motion in the direction of the pan's rotation until a critical velocity is attained. A mobile 'waving' jet is then produced at the surface forming a 'hierarchy of general circulations' varying with the rate of heating and especially with the rate of rotation. Three principal regimes of interest have been noted. At low rotation rates a symmetrical trade-wind cell is observed with westerly jet stream adjoining the cold source. With increasing rotation, this cell breaks over into steady wave patterns . . . and finally into unsteady circulations which . . . take the form of a repeating cycle' (Riehl and Fultz, 1957–58). Two, three, four or more such 'waves' or 'petals' can be produced at will. (See Hide, 1970 Plate VI.) The authors referred to have made a very careful analysis of the three-wave model (see Fig. 55). Note the immediate inference that the jet arises, so to speak, in its own right as an essential feature of the transfer of thermal energy in a rotating system. The authors' ultimate conclusion is that it does so because the fluid strives, in difficult circumstances, to 'preserve the single vertical cell'. This becomes evident when the motion is analysed by a system of coordinates using the wave itself as a central axis.

Although such a model with a constant deflection due to steady rotation at 0.3 turns per second, a ratio of vertical to horizontal scales of 1 : 4 and with firm lateral boundaries at which heat input and output is localised would seem to offer very different conditions from those experienced by a boundless atmosphere on a spherical earth subjected to radiational gains and losses, it is found to reproduce, with fair approximation, at least some of the characteristics of middle latitudes.

1. The jet speed is about 20 per cent of that of the outer margin of the pan.
2. The 'waves' move progressively in the direction of the pan so as to make a complete circuit in about 15 pan revolutions or 'days'.
3. The jet-like motion is confined to the upper portion of the fluid, below is a system of vortices, cyclonic motion occurring beneath the poleward-moving sectors of the waves (southwesterly jets) and anticyclonic motion underlying the equatorward-moving sectors (northwesterly jets).
4. Except in the extreme upper and lower layers of the fluid, affected by the

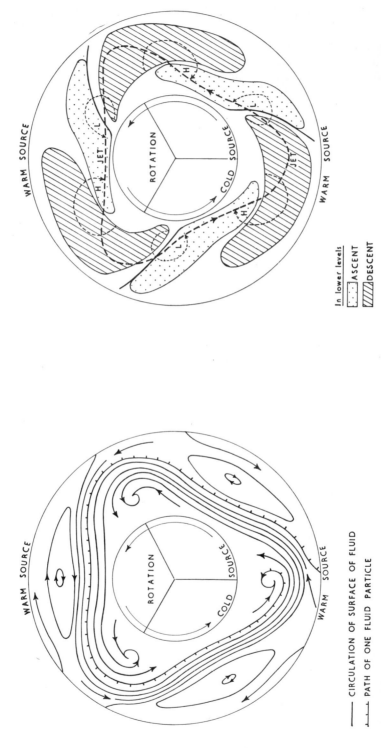

FIG. 55. A general interpretation of Fultz's rotating pan experiment (1957). (a) A regular three-wave jet. The figure illustrates how the liquid edges its way across the stream from warm source to cold source. (b) The general pattern of vertical motion in relation to each wave and the alternation of cyclonic and anticyclonic vortices which are induced beneath them

experimental conditions, temperature everywhere increases upwards and equatorwards; the analogy with the atmosphere must therefore be sought in *potential* temperature (see p. 237). There is no tropopause or stratosphere, which suggests, though it does not prove, that these features play no essential part in jet formation.

5. Fluid particles cross the jet from warm source to cold source in a path illustrated in Fig. 55. 'Hence the kinetic energy of the jet stream is built on its equatorward margin and lost on its poleward margin' (Riehl and Fultz, 1958, 402). The jet stream is thus neither 'the locus of generation of kinetic energy' nor 'a current that is "driven" by outside circulations', rather is it the manner in which the mass circulation is organised under the conditions prevailing. Note also that the jet axis must not be mistaken for either a trajectory or a streamline.

6. The pattern of vertical motion (so important in the generation of weather) is at first glance very complex. Briefly it may be said that descent is characteristic of the equatorward part of the wave (the trough) whilst ascent predominates in the poleward section (or ridge). However, the axes of major vertical motion shift diagonally across the jet axis with increasing elevation as indicated in Fig. 55. It will be seen that if we assess this motional *zonally* (i.e. along concentric circles), the area of strongest ascent will be shown to lie poleward of the area of strongest descent and 'when downward motion at the cold source and upward motion at the heat source are added to the pattern ... the classical atmospheric three-cell structure emerges'. This interpretation is not wrong, but it does arise from the zonal frame of reference which, as we have seen, has dominated our thinking since Hadley's time. Indeed it was bequeathed to us by the Greeks. Now, if the perturbations in the flow are of major scale—and the evidence of the model and of actual observations of the upper air suggests that they are—then a zonal interpretation of the facts is *artificial*. The natural frame of reference is the jet axis. Averaging on this basis we find that upward motion predominates equatorwards of the jet and downward motion poleward of it—the 'single vertical cell' already referred to. The authors' conclusion that 'the methods employed in summarising features of the flow into a general circulation picture can have a profound effect on the result' appears to be of the very first importance. Indeed, we can express the moral in much wider terms—*whenever* the climatologist is about to strike an average, however reasonable it may appear to be, he should ponder very carefully on the deed. *There are positively no exceptions* to this rule.

It will be seen that the rotating dishpan cannot answer all our questions but that it can certainly help us to visualise the type of three-dimensional motion which the general circulation involves.

*Comparison of the dishpan with the surface of the earth.* It is well to remind ourselves of some of the evident contrasts between this experiment and the actual conditions prevailing on earth.

(*a*) The most obvious of these stems from the curvature of a spherical surface and the distribution both of angular momentum and of the deflective force which it entails. It would appear that we can simulate the intertropical circulation at slow speeds of rotation and the circumpolar circulation at higher speeds but on earth both of these systems exist concurrently and clearly they must interact. It would thus appear that the dishpan can only represent *part* of a hemisphere. The ratio between its inner and outer circumferences is 42:100 which corresponds approximately to that between the Arctic Circle and the Tropic ($43\frac{1}{2}$:100) and within this range relative areas (and therefore volumes) are represented with but minor distortion.[1] Using this interpretation, it can be shown that, of the total area enclosed by the external rim of the pan (including the central core) about 65 per cent, or two-thirds, lies *outside* the jet maximum. Transferred to the earth's surface this would place the mean latitude of the jet axis at about 50° latitude. Clearly the pan margins must impose restraints upon the flow which have no counterpart in the real atmosphere. Figure 56 shows the effect of using a pan without a central core, the resemblance to atmospheric patterns is remarkable but analysis is then more complicated (Fultz, 1961).

FIG. 56. The more irregular pattern with discontinuous jets which occurs when a pan without a central core is employed. Jet areas stippled

[1] This is no longer the case if we interpret the outer rim as the equator and the inner rim as 65° latitude (ratio of lengths 100:42).

(*b*) In the pan the sources and sinks of heat are rigidly located at the two margins, whilst on the earth the main sources are distributed rather irregularly over the surface whilst the major sink covers great areas of the upper air. That this consideration does not invalidate the model is largely due to the fact that in *both* cases the circulation pattern set up by temperature contrasts is a most active agent in redistributing temperature conditions. Cause and effect are thus intimately interlocked. Of course, on the earth the pattern of sources and sinks also changes radically with the season. It would seem that on the whole the pan best represents the winter situation.

(*c*) We inspect the dishpan from *above* whilst we experience the terrestrial circulation from *below*. Far from being a criticism of the model, this is one of its great advantages. We are at once reminded of Shaw's contention that motion is primary and that pressure patterns, especially when viewed from the earth's surface, are secondary elements in the circulation. He points out that surface pressure patterns are 'a picturesque physical integration of the distribution of pressure at *all* levels' (*Manual* IV, 1931, 342) and no process of averaging can remove this characteristic. However illuminating a pressure map may be in regard to the immediate local weather situation, it is clearly *not* the place to begin an analysis of the motion of the atmosphere as a whole.

(*d*) The water in the dishpan is an incompressible fluid whilst the atmosphere is composed of gases, highly compressible and subject to the relationship between temperature and pressure which is stated by the gas laws. At first glance this would appear to be a major objection to the model. Adiabatic heating and cooling (see p. 233) imposes a brake on vertical exchange which must be overcome if the circulation is to continue on a grand scale. However, the air is supplied with a powerful weapon in overcoming this resistance in the latent heat released during the condensation of water vapour aloft and absorbed during evaporation at the earth's surface. Latent heat thus plays a notable part in stimulating vertical exchange.

(*e*) The base of the pan is uniform and flat. The surface of the earth is varied by the presence of the continents and of their major relief features. These cannot be effectively simulated, though the attempt has been made (Fultz and Long, 1951).

*Postscript.* As this work goes to press we have an authoritative restatement of the problem of the general circulation by Lorenz (1970). Note that it is still full of open questions. Like Eady and Fleagle, he places great emphasis on the activity of eddies on a scale of 'at least a few thousand kilometres' though these are 'neither as smooth as normal-mode waves nor as random as turbulence. They are an intermediate phenomenon.' They arise from baroclinic instability and are thus generated most freely in the zone of steepest temperature gradient. Averaged zonally over the year they can be shown to involve a poleward transfer of sensible heat and water-vapour, especially below the 600-millibar level (4200 metres) and a similar transfer of angular momentum, especially above the 400-millibar level (7200 metres); the latter exchange occurring in lower latitudes than the former. In this manner 'kinetic energy is converted from the eddy to the zonal form', the very process of 'unmixing' to which Eady referred. The transfer of angular momentum at great elevations is seen as having 'far-reaching ultimate effects upon the circulation at other elevations and latitudes' though why it occurs where it does is by no means clear.

Lorenz frankly admits that 'averaging over the entire year obscures many of the more interesting features of the circulation'. Notably it tends to obscure the significance of the southwest monsoon, operative over about one-fifth of the circuit of the earth. On a still smaller scale, he sees in satellite photography new evidence which may lead stress to be placed in future on the location of cloud clusters and even upon the liquid-water (or ice) content of the atmosphere. In most numerical models this is assumed to fall out very promptly but, whilst suspended, he points out that it must have major effects upon radiation balance. New roads of enquiry are thus opened.

# 7

# The general circulation: descriptive

The preceding discussion has inevitably involved some description of the general circulation as it has been seen at various periods of time though both inference and speculation have played a considerable role in some of the

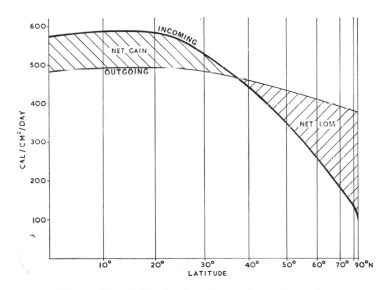

FIG. 57. The meridional distribution of incoming and outgoing radiation for the year as a whole. (Northern Hemisphere.) (After Houghton, 1954)

accounts. Yet the volume of data derived from direct observation has been accumulating at a greatly accelerating pace, particularly in recent years. It is therefore now necessary to gather this together in an orderly fashion, distinguishing as far as possible the features characteristic of each hemisphere and noting the changes in pattern which follow the recurring seasons.

We must not expect such a description to lead us directly to causes for, as Sheppard (1954) has put it, 'the atmosphere is a dynamical system with multiple feedback, almost every process reacting on every other. We cannot therefore identify an initial "cause" and proceed through to ultimate effect except in a vague and rather unprofitable way'. Nevertheless one must start somewhere! We have seen that Ferrel started with tropical heat whilst Shaw threw more emphasis on arctic cold. Yet the stress laid by Eady, Fleagle and others on the zone of maximum meridional temperature gradient has been shown to be no less logical. Let us revert once again to the fundamental facts of the earth's radiational balance.

Figure 57 (after Houghton, 1954) presents an interpretation of the relationship between incoming (short-wave) radiation and outgoing (long-wave) radiation of the earth and its atmosphere for *the year as a whole*. The two curves cross at the 38th parallel so that some six-tenths of the globe receive a net excess of radiation whilst the remaining four-tenths experience a net deficit. It is the function of the atmospheric and oceanic circulations to effect the necessary heat exchange. Note the immense width of the zone of excess (76° of meridian equals 8400 kilometres (*c.* 5250 miles)) and note furthermore that, over much of this range, the ratio of income to expenditure is almost uniform. In fact we can generalise from this diagram by taking the 25th and 50th parallels as convenient boundaries, thus establishing a *fivefold* subdivision of the globe.

1. Round the equatorial girdle from about 25°N to 25°S there is a zone of marked net radiational gain. It covers some 40 per cent of the globe's surface and within it, despite its width of nearly 5500 kilometres (*c.* 3500 miles), there is very little latitudinal contrast in the balance between incoming and outgoing radiation.

2 and 3. Over each polar cap approximately poleward of 50°N and 50°S there is an area of large radiational deficit. Each area has a diameter of 8900 kilometres (*c.* 5500 miles) but covers only about 12 per cent of the globe's surface and, within each, the deficit increases rapidly towards the poles.

4 and 5. In middle latitudes there are two transitional zones where the gradient of radiation balance is particularly steep. Each is 2775 kilometres (*c.* 1725 miles) across and embraces about 18 per cent of the globe's surface. It is across these regions that active exchange between the equatorial girdle and the polar caps must take place.

As is to be expected, this pattern is profoundly modified through the seasons on account of the changing declination of the sun and of the consequent variation in day length which occurs in high and middle latitudes. Baur and Philipps (1934–5) have made similar computations for the *northern* hemisphere in January and July[1] (Fig. 58). It will be observed that no locality within the hemisphere experiences a net radiational deficit in July and that, if the January curves can be taken as giving a broad indication of the conditions then prevailing in the *southern hemisphere*, the curves may be expected to cross at about 16°S. The area of net excess of radiation income thus still covers

[1] Also see *Met. Mag.*, **87**, 1958, 368, and *Geography*, 1961, July, 209.

rather more than six-tenths of the globe but there is now only one area of deficit and hence only one transitional zone. If the width of the latter is still taken as twenty-five degrees it will then extend from 3–28°S. The logical division of the globe's surface at the *solstices* is thus *threefold*; an area of marked net radiational gain covering some 53 per cent of the total, *one* area of large radiational deficit covering 26 per cent, and the remaining 21 per cent of the globe falling within a single transitional zone not very far from the equator. Active exchange at these periods must therefore take place between the hemispheres. (See Lorenz 1970, 20 and Johnson 1970, 125–32.)

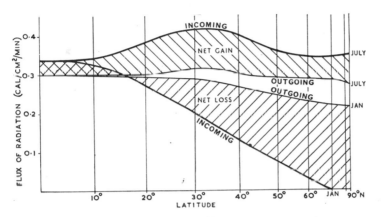

FIG. 58. The meridional distribution of incoming and outgoing radiation in the northern hemisphere in January and July. Note that in July the whole hemisphere receives a positive balance (after Baur and Philipps, 1934–5)

The exchange mechanism in the atmosphere operates via air temperature and the consequent effects on both air density and water vapour content. The picture is thus complicated by the differing responses to incoming radiation of such surfaces as cloud tops, the sea, the desert, and large areas of snow or ice. Furthermore, the air temperatures actually observed owe some of their features to the systems of circulation in both air and ocean which are thus set in train. Hence the complexities which Sheppard has described.

Figure 59 (after Shaw, *Manual* IV, 1931, 328) illustrates a feature of this exchange which is too rarely stressed since it only emerges if the data are viewed on a global scale. It shows that the variation in hemispherical solar income from solstice to solstice is followed (inversely) by the net transfer from one hemisphere to the other of not less than ten billion ($10^{13}$) tons of air. Flohn (1956, 434) has recomputed this value and expressed it as a trans-equatorial flux of $\pm 5.45 \times 10^{12}$ tons, the addition or subtraction of one-thousandth of the total mass of the atmosphere to each hemisphere. Shaw draws attention to the lag of some 27 days in this response. However, the diagram also carries a very important message with respect to the equinoxes.

Although the insolation pattern is then symmetrical about the equator, we must not expect this to be true of the circulation pattern, even if we disregard the observed 'lag'. At about these periods we shall expect the air transfer from one hemisphere to another to be *in full swing*. In the northern spring the net motion will be southwards, in the northern autumn, northwards. From the point of view of circulation therefore the two equinoxes are by no means alike. Although Shaw states that the exchange occurs 'at what level we cannot say', it would appear highly relevant that each of the five great trade-wind systems of the world attains its maximum development in the spring and its minimum in the autumn of the hemisphere in which it is situated. This periodicity is indicated both by the areas under the domination of the systems in April and October (Fig. 60) and by a frequency analysis of wind speeds observed near

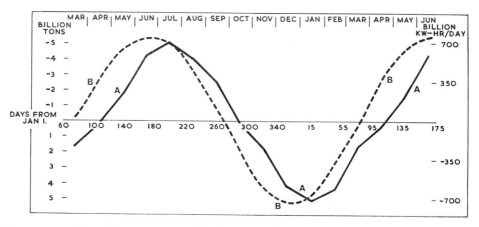

FIG. 59. Deviations from mean annual values of (a) the mass of air over the northern hemisphere and (b) daily solar income throughout the year. The air-mass scale is in $10^{12}$ tons and is inverted. The insolation scale is in $10^{12}$ kw hours per day. (After Shaw.) The two curves are out of phase by some 27 days

the core of each system over the 2-month intervals when these reached their highest and lowest mean values (Fig. 61).[1] It will be noticed in the latter diagram that wind speed in the southern trades is not quite in phase with that of the systems north of the equator. This is much less the case with the areal extent of the systems. Note, in passing, that the southeast trades of the Indian Ocean register the highest velocities of all five systems throughout the year. Indeed, when at their weakest (in January–February) they excel all the other systems at their strongest, with the very marginal exception of the Pacific northeast trades in March–April. We shall refer to this again later.

The annual temperature changes which stimulate this pulsation have been

[1] This periodicity has notable effects on the climate of the intertropical zone (see Crowe, 1951).

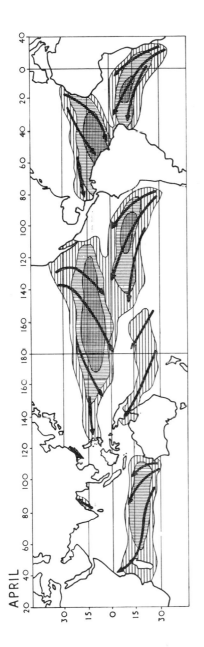

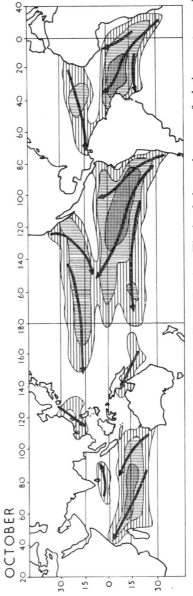

FIG. 60. The trade-wind systems of the world in April and October. The isopleths are in terms of relative constancy of wind direction and enclose shaded areas where 50, 70 and 90 per cent of all winds blew from the predominant quadrant with Beaufort force 3 or more (over 6½ knots) It is argued that the direction of lighter winds is not reliable

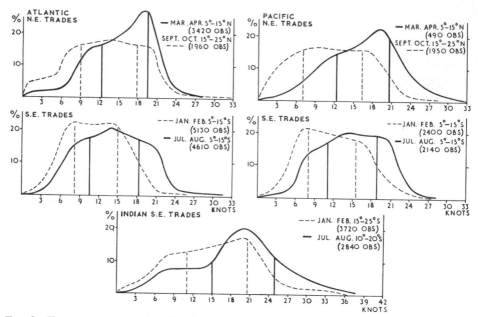

FIG. 61. Frequency-groups showing the variation of wind speed near the cores of each of the five trade-wind systems at their periods of maximum and minimum development. The method employed has taken account of any shift of the core area in latitude with the season. The upper and lower quartiles of each distribution are shown by the vertical lines

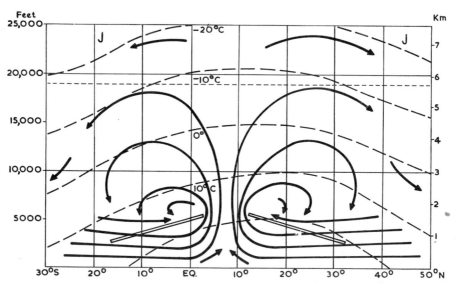

FIG. 62. A diagrammatic cross-section of the Atlantic trades from 30°S to 50°N in July. The vertical scale is highly exaggerated (×700) so as to indicate the general positions of the two trade-wind inversions and the equatorial gap between them. The upper return currents are, of course, not necessarily in the same plane as the surface winds and the circulation there is now considered to be of considerable complexity

illustrated in some detail for the northern hemisphere in Figs. 34 and 35. What we now require are typical cross-sections of the atmosphere from pole to pole showing at least the summer and winter situations. So far, really satisfactory sections are not available owing to the paucity of upper-air observations, particularly in the southern hemisphere. Even if they were, it would still be difficult to select a truly typical meridian owing to the irregular distribution of land and sea.

A diagrammatic cross-section broadly in the Atlantic region in July has been attempted in Fig. 62. Realising that the vertical scale here is about 700 times the horizontal scale, it may be concluded that from about 10°S to 35°N horizontal temperature contrasts in the middle atmosphere are almost negligible. This might well have been presumed from our studies of radiation balance (see pp. 5;15;169) but it is useful to have it confirmed. There is evidence that the same feature is present in January from about 15°N to 30°S,[1] the slight asymmetry of its seasonal swing being doubtless the result of the distribution of land surfaces over the globe.

## The intertropical circulation

The observed intertropical circulation thus occurs within a massive girdle some 45° or 4800 kilometres (*c.* 3000 miles) in width in which, except locally and near the surface, broad contrasts in temperature are completely absent. It is basically 'barotropic' in character, the surfaces of temperature, density and pressure all being very nearly horizontal. This is the feature to which Rossby gave such prominence in 1949 (see p. 181) concluding that 'there are practically no thermal forces present to drive the meridional trade-wind circulation'. This declaration was utterly revolutionary, for the consistent and persistent motion of the trades had been the fount and origin of practically all theoretical speculation about the more orderly aspects of air motion.

How was it possible to accommodate the four trade-wind systems of the Atlantic and Pacific with mean speeds ranging through the year from 12 to 16 knots, plus the even stronger system (16–20 knots) of the south Indian Ocean, within an intertropical belt thus deprived of 'thermal forces'? Little wonder that thermal explanations have been clung to in the face of some most evident facts!

Thus it has been repeatedly asserted that the trades 'follow the sun', i.e. that they are bodily displaced in latitude in broad harmony with the sun's declination. Figures 60 and 63 indicate some such motion but, if we focus attention on the general poleward limits of the systems, it is shown to be no more than 3 or 4°. (The northeast trades of the Pacific provide the one exception to this rule in July but they hold this extreme position only during that month.) This impression has been derived in fact from the development

[1] See the January and July sections near 30°W in Lamb (1961, 214–15). For more elaborate sections involving only one hemisphere see also (1) Northern hemisphere, Fig. 67 (2) southern hemisphere, Wexler (1959, 200).

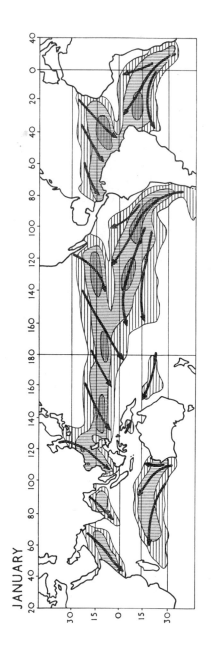

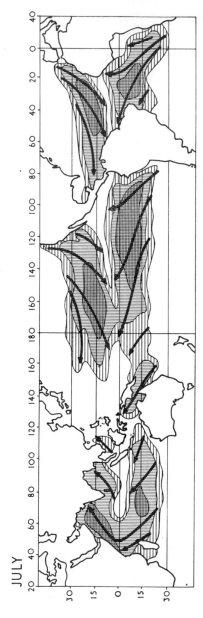

FIG. 63. The trade-wind systems of the world in January and July. Key as for Fig. 60. Note the development of the south-west monsoon in the Indian ocean in July. It persists with little change from June to September

of quite local poleward 'roots' to the systems in summer immediately off the western coasts of the continents. Within those limited areas, but there only, the summer trades may be encountered some fifteen degrees polewards of their winter limits. Much more significant is the *westward* surge of the trades in winter as compared with their position in summer, an advance and withdrawal that is clearly in evidence wherever the ocean is wide enough to accommodate it. This is the major seasonal pulsation of the trades and it is fraught with momentous consequences. Yet it would be difficult to find a purely thermal explanation for motion in a zonal direction.

The forces which move the trades must therefore be sought beyond, or at least near, the margins of the intertropical zone and particularly in the upper atmosphere. Because of the lower density of warm air as compared with cold air, the fall of pressure with height is less in the warm girdle than in the polar caps. Hence, in the upper half of the atmosphere, that is above 500 millibars, (*c.* 18000–19000 ft), and hence clear of all but the very highest terrestrial obstacles, we find that, at equivalent elevations, pressure is higher over the equatorial girdle than over the polar caps. This feature overrides all contrasts in surface pressure and would doubtless still exist in its major features even if the surface of the earth were entirely uniform. From the margins of the intertropical barotropic zone, that is from about 35°N and 10°S in the northern summer and from 15°N and 30°S in winter, the upper air can be expected therefore to attempt to slide away polewards. Here, if anywhere, would appear to lie the logical starting point for a survey of the mechanism of the general circulation.

Considerations regarding the conservation of angular momentum suggests that a poleward motion of 5 or 10° from these initial latitudes would generate theoretical *westerly* currents of the following speeds:

| | |
|---|---|
| From 35° to 40°N, 100 knots. | From 35° to 45°N, 218 knots. |
| From 10° to 15°S, 33 knots. | From 10° to 20°S, 82 knots. |
| From 15° to 20°N, 48 knots. | From 15° to 25°N, 111 knots. |
| From 30° to 35°S, 86 knots. | From 30° to 40°S, 192 knots. |

Hence realistic jet streams of between 80 and 110 knots are not improbable at about 40°N and 20°S in the northern summer and at about 23°N and 34°S in the northern winter. Such estimates can be no more than approximate for there is plenty of evidence that angular momentum is not fully conserved but it is quite clear that any considerable poleward migration beyond these limits will involve such strong accelerations that centrifugal force will offer rapidly-increasing resistance. We thus encounter again Ferrel's concept of a 'governor' to the system, clearly located in the *upper* atmosphere. Any such tendency for the development of strong zonal flow aloft in these latitudes has important effects. It offers an increasing barrier to meridional exchange and thus produces (or demands) a steepening of the poleward pressure gradient aloft. As the jet thus moves westward under balanced forces, air can escape from the intertropical zone by two routes, either aloft by the lateral shift of trajectories across the jet axis (as in the dishpan model) or by poleward motion near the ground under the impetus of the deep high-pressure cells which develop beneath it. But these cells have also an equatorial flank and so a considerable proportion of the subsident air is returned towards the equator, as the trades, to add water vapour fuel to the intertropical heat-engine.

It would thus appear that broad temperature contrasts between the inter-tropical girdle and the polar caps, the strength of the upper air circulation, the development of the subtropical high-pressure cells, and the strength of the trades are all directly interconnected but that, though the basic facts are thermal, the linkage is established by *dynamic* processes. Such a view disposes of Rossby's difficulty that the descending branch of the trade-wind circulation during the northern summer appears to be located 'almost right under the sun', that is in areas where comparatively cloud-free skies result in the highest temperatures observed on earth. It is also consistent with the observed facts not only that the trades in general are strongest in spring, when zonal temperature contrasts in the upper air are at their greatest but also with the exceptional character of the southeast trades of the Indian Ocean, a result apparently of the eccentric position of the Antarctic continent about the southern pole.

These processes generate the 'trade-wind inversion', a layer within the trade-wind air streams where temperature *increases*, instead of decreasing, with height. This layer is usually about 500 metres deep and its base is frequently encountered at between 1000 and 2000 metres above sea level (see Fig. 62). Ficker (1936) has described this feature as 'the most extensive, most uniform, and to date the least studied surface of the troposphere' and Riehl (1954, 53) has added that it is 'perhaps the most important regulating valve of the general circulation'.

This remarkable feature was first recognised by an astronomer, C. Piazzi-Smith (1858, 524–7) when he took a series of observations of temperature and relative humidity during two ascents from Orotava to Altavista (3260 m or 10 700 ft) in Tenerife in August 1856. He noted that, although the northeast wind (the trades) prevailed up to at least 9000 feet, there was 'a break or a very great anomalous deviation from the law of decrease of heat with elevation' within this air stream at around 3000 to 4000 feet accompanied by an equally sharp increase in 'the depression of the dew point'. The level at which this feature occurred was not the same on the two ascents but observers resident near the summit for two months remarked that over the sea, as far as the eye could reach, cloud tops rarely exceeded 5000 feet. Clearly, as the site for a proposed astronomical observatory, the peak had much to recommend it.

Kite observations from the research vessel *Meteor* in 1925–7 showed that the inversion could not be explained by the proximity of the Sahara for it was encountered in both of the trade-wind systems of the Atlantic at heights which increased both equatorwards and westwards. Subsequent investigation of the Pacific northeast trades by Riehl *et al.* (1951) and by M. Neiburger (1960) have confirmed that there too the westward ascent is not uniform; it follows two planes, one rising comparatively steeply from about 500 metres near the coast to 1200 metres some 500 miles (800 km) down wind and the other levelling off at 1200 to 2000 metres over the rest of the system. The actual rise in temperature with height is greatest where the inversion layer is low and an increase of 10°C is not unknown near the eastern borders of the systems. Elsewhere it is usually no more than two or three degrees but even an iso-

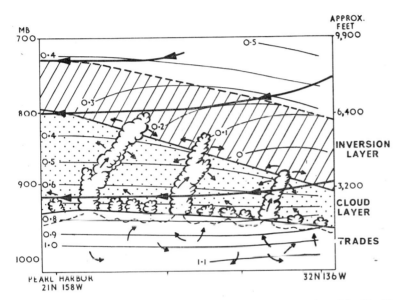

Fig. 64. A diagrammatic cross-section of the north-east trades of the Pacific from near San Francisco to Hawaii. The figure shows the variation of lapse-rate with height in °C per 100 metres (after Riehl and Malkus) and also indicates in symbolic form the generation of two main types of trade-wind cloud as the result of turbulence. The vertical exaggeration of scale is about ×770 and the horizontal scale of the clouds is also far from true. The heavy arrowed lines indicate the general subsidence within the wind system; the short arrows indicate turbulent flow

thermal layer of the order of 500 metres in depth can have a marked effect upon cloud ceilings. When the inversion level is high it is often best recognised by a sharp drop in relative humidity.

This evidence spells *subsidence* and the trades are now generally regarded as being fed for the most part by subsiding air as is suggested in a highly generalised fashion by Fig. 62.[1] Even along their eastern margins where Shaw recognised an inflow from higher latitudes (see p. 177) the air is warmed aloft by descent though the surface layers may be kept cool by the Canaries, Benguela, Californian, Peru and West Australian currents. A notable result

---

[1] It must be remembered that components of motion through the paper cannot be shown on a two-dimensional diagram.

8

of this relationship is that the stronger the trades the more rapid the subsidence and hence the better the development of their inherent inversion zone.

This has the most significant effect upon trade-wind climates. After a long and detailed survey of the rainfall records of stations within intertropical latitudes the present author concluded 'not only that the trades are normally relatively rainless winds, but that the stronger they are, the drier they are' (Crowe 1951, 67). Furthermore, this approach places the zone of equatorial rains in an entirely new light. Under comparatively clear skies (mean cloud cover about five-tenths) and in the presence of a brisk breeze, the trade-wind zones are areas of active evaporation. As Riehl *et al.* (1951, 598) put it, 'latent heat in the form of water vapour is *accumulated* by the trades'. This moisture is diffused upwards by surface eddies and thrown sporadically above condensation level to produce the characteristic 'trade wind cumulus', a scattering of bulbous clouds of limited depth. About a tenth of these may rise in 'towers' to reach, and perhaps penetrate, the inversion layer but though these may yield showers their further development is inhibited by the presence of the inversion zone (Fig. 64). Malkus has examined the complex exchanges which occur but the general effects are (*a*) a gradual lifting and weakening of the inversion down-wind and (*b*) the mass transfer of a vast amount of energy westward and equatorward until it is finally released 'in locations far removed from the site of the accumulation' (Riehl).

Maury's (1858) 'equatorial cloud ring' is this located according to the varying strength of the converging currents and not according to thermal contrasts which he had so much difficulty in finding. Satellite photographs in fact rarely show a complete ring near the equator (Fig. 65).[1] The up-draughts which occur near the equator are unquestionably stimulated by the release of latent heat aloft but Riehl (1962) has argued that most of this must

[1] Fig. 65 shows in a rather generalised form the kind of evidence now becoming available from satellite photographs. (See Barrett 1970, 25–53.) Each panel shows fortnightly means of cloud amount over Central America in June, July and August 1966, derived from NIMBUS II photo-mosaics. Areas with over 60 per cent cloud-cover are cross-hatched and areas with less than 30 per cent cloud are stippled. The main West–East axis of greatest cloud density is also indicated. On the right hand diagrams the rainfall recorded at selected stations during the *whole month* is given in centimetres.

Although valid for only a short period, the information is particularly valuable since, over much of the area, surface observations are either completely lacking or else they are profoundly influenced by relief effects.

*Note:* (*a*) The northward advance of the main cloud axis (1–3); its regeneration in (3); and its general southward drift over the Pacific towards the end of the period (4–6). (*b*) The considerable cloud-cover over the eastern flanks of both the northeast and southeast trades, particularly extensive when the southeast trades withdraw in (5) and (6). The diagram is unable to show that this cloud is probably very different in type from that found near the West–East axis. (*c*) In the Caribbean, the effect of hurricane ALMA in (1) and, less clearly, of an 'easterly wave' in (4). (*d*) Highly conjectural flow-lines have been added by the author to give the diagram greater life. It seems very probable that the comparatively clear areas are regions of divergent flow. (*e*) A close correspondence between the mean cloud amount and rainfall must not be expected; yet a general relationship is evident. It is a pity that no information was available for the narrower part of the isthmus or for Colombia.

occur in 'a few thousand large clouds' (having diameters of several kilometres) which act as 'funnels' with a total area of probably no more than one-thousandth part of the equatorial zone. He notes that clouds on this scale, towering

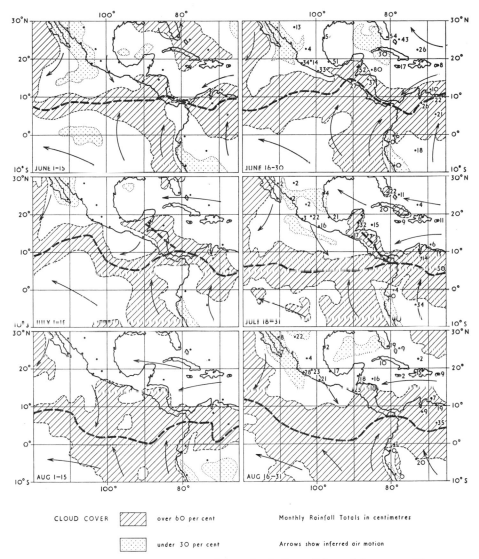

CLOUD COVER ///// over 60 per cent  Monthly Rainfall Totals in centimetres

::::: under 30 per cent  Arrows show inferred air motion

FIG. 65. Cloud distribution over Central America in June, July and August 1966 as derived from satellite photo-mosaics by E. Barrett. See note p. 206 for comments. The figures in the right-hand diagrams give the rainfall recorded at selected available stations for the whole month in question (cm)

to 14 to 15 kilometres (45 000–50 000 ft) can scarcely arise from purely convective processes. They too demand some kind of organisation such as would be provided by convergent flow. Under such circumstances the occurrence of multiple rain-belts presents no mystery and the torrential and highly variable nature of equatorial rainfall becomes much more intelligible. Intermittently and as a result of situations which are still not fully understood, the accumulated energy is released in storms of hurricane intensity. These require organisation of a still greater order of magnitude. Although rarely observed near the equator, they too occur downwind of the (autumn) trades and thus derive their fearful energy from the same source (see pp. 386-99).

Organisation of equatorial air on a far greater scale is also evident in the occurrence of summer 'monsoons', usually defined as an almost complete reversal of the more normal trade-wind circulation. The concept also carries with it implications as to heavy rainfall and it is probable that the term has been far too loosely employed in consequence. Thus Ferrel (1889, Chap. 5) recognised monsoons in Amazonia and Northern Australia in the southern hemisphere, and in Central America, West Africa and Southeast Asia in the northern. His analogy with land-and-sea breezes and his view that the cause of the wind change was predominantly thermal have proved both popular and durable, but it is rarely noted that to organise the flow he required considerable relief on the continents. Thus on p. 199: 'The monsoon influence of countries mostly level without an elevated interior, however highly they may become heated in summer or cooled in winter, is not very great.' In this way he endeavoured to dispose of the most evident anomalies in the observed facts.

If, to qualify as a 'monsoon', the summer inflow has to show features of persistence and strength at all commensurate with those of the trades which it replaces, then only the last two of Ferrel's five areas experience a monsoon circulation (see Fig. 63). In the other three regions the summer air motion is comparatively weak and no great violence is committed upon the observed facts if their summer rains are attributed to a migration of the equatorial rain-belt. Such a restriction on the use of the term may seem more reasonable when it is realised that the monsoon par excellence, the southwest monsoon of the Arabian Sea, achieves mean wind speeds in July in excess of 20 knots (in the Bay of Bengal *c.* 15 knots). Furthermore the air current is now recognised as being some 5 kilometres (16500 ft) deep, far in excess of any other known seasonal inflow and about twice the depth of any known trade. It is indeed *unique*. Given this emphasis the facts are much less amenable to a purely thermal explanation and, though a full understanding of this spectacular feature of the intertropical circulation still evades us, some interesting new facts have emerged from upper-air observations. Thus Sawyer (1947, 368), employing data for the latter part of August 1945, showed that northwest of a line from Veraval to Delhi, the southwesterly inflow was exceptionally shallow. Aloft it was capped by warm dry air from the continental interior, the border between the two air masses being marked by a strong inversion of temperature sloping upwards from about 2000 feet on the Makran coast to some 8000 to 10000 feet near its limit or 'nose' along the Veraval–Delhi line. That this was indeed a persistent feature was suggested by the widespread suppression of monsoon rain throughout the lower Indus basin (see data for Karachi p. 516). Much more recently, Ramage (1966, 145) has shown that aerological soundings taken at sea during the summers of 1963–64 reveal that the inversion extends upwards from the neighbourhood of the Somaliland coast to roof over the western half of the Arabian Sea. He attributes it to 'massive subsidence' though the effect is certainly reinforced at low levels by the upwelling of exceptionally cool water along the African coast.

Further light on the southwesterly airstream has been shed by Findlater (1969). He has confirmed the author's conclusion (Crowe, 1950, 39) that the contribution of southern hemisphere air to the summer monsoon is made predominantly *west* of the 60° east meridian and demonstrated that cross-equatorial flow often occurs in the form of one or more low-level 'jets', usually 100 to 200 miles in width and often reaching speeds of over 50 knots at heights of from 4000 to 7000 feet (1200–2100 m). They appear to be stimulated not far from the 'nose' of the upper-air inversion. Furthermore there is some evidence that heavy rains over the Indian peninsula occur a day or two later than 'jet' maxima (*ibid.*, 379). East of the 60th east meridian, cross-equatorial flow is much less active and the situation appears to be quite different. Thus the author has suggested that the comparatively low July to September rainfall at Colombo (6°56′N) may be due to the extension across the equator of the southeast trade-wind inversion (Crowe, 1951, 62). A parallel situation would appear to occur in West Africa (Accra, 5°31′N).

We may conclude this section by remarking that the northeast monsoon of the Indian Ocean is best viewed as a sixth, and temporary, trade-wind cell. It is indeed fortunate that its development is only temporary for, if it prevailed throughout the year, all the equatorial margins of southeast Asia would be desert and the Ganges valley an attenuated thread of snow-fed oases. Any linkage between this system and the winter 'Siberian high' is completely ruled out by the height of the mountain barrier to the north, except perhaps over the South China Sea where the system is certainly at its strongest. Like all the other trade-wind systems it is fed predominantly by subsiding air.

# The northern hemisphere

## Winter

The distribution of pressure in the upper or middle atmosphere can be conveniently mapped in at least two ways, (*a*) by plotting isobars appropriate to one or more fixed levels, or (*b*) by indicating in contour form the variation in height of one or more selected pressure 'surface'. Early examples of the former method of representing *mean* pressure are to be found in Shaw's *Manual* II (1936, 259–61 after Teisserenc de Bort); modern examples of the latter, again for *mean* values in the northern hemisphere, are to be found in current meteorological texts (see Fig. 66; see also the global models illustrated on Plates II-V, *Met. Mag.*, **89**, 1960, 328).

All such mean charts agree in showing over each polar cap a comparatively

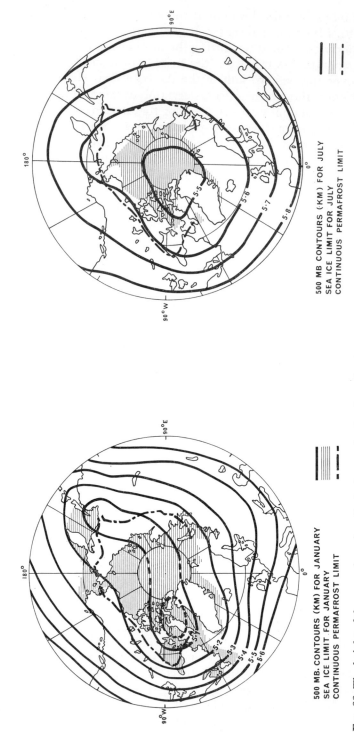

500 MB CONTOURS (KM) FOR JULY
SEA ICE LIMIT FOR JULY
CONTINUOUS PERMAFROST LIMIT

500 MB. CONTOURS (KM) FOR JANUARY
SEA ICE LIMIT FOR JANUARY
CONTINUOUS PERMAFROST LIMIT

FIG. 66. The height of the 500 mb surface (in km) over the Arctic region in January and July (after Hare, 1968, 440–41). The isopleths illustrate the seasonal change in the form and intensity of the circumpolar circulation in the upper air. The general distribution of sea ice is also indicated and, on land, the area of continuous 'permafrost' in the subsoil

uniform and more or less symmetrically disposed pressure gradient polewards from the general neighbourhood of the tropics towards their respective arctic circles. The gradient is steeper in winter than in summer, particularly in the northern hemisphere, but the dominant characteristics remain otherwise substantially unchanged throughout the year. From this evidence we are led to infer the existence over the polar caps of two great circumpolar whirls where westerly currents of great breadth and persistence move under a balance of forces derived from the pressure gradient on the one hand and the effect of the earth's rotation on the other. Indeed these whirls have been described as the twin 'flywheels' of the general circulation and the inferred motions are substantially confirmed, at least broadly as to direction, by actual observations of the wind aloft.

Yet such a picture has obvious flaws. (*a*) It is the object of a flywheel to promote *steady* motion and the winds within the polar caps are known to be far from steady at any level. Indeed constant change in both direction and force is of their very essence. (*b*) The model certainly fails to satisfy the requirements of conservation of angular momentum for the formula quoted on p. 160 yields westerlies of the order of 600 knots for poleward transfer from the 40th to the 60th parallels. Yet if poleward transfer does not occur the system is deprived of its driving force! (*c*) Charts of mean pressure, at any level, are almost universally misleading, for the 'basic current' they are designed to reveal is a figment of statistical imagination, having no reality at any one moment of time. They may show general drift but all the complex processes by which that drift takes place are lost in the statistical calculations.

It has long been known from the observation of cirrus clouds that, at times, broad belts of the upper air are in exceptionally rapid motion in a generally west-to-east direction. Seilkopf (1939) gave them the name of 'Strahlströme' or 'jet streams'. As an actual hazard to high-flying aircraft, in view of the fuel requirement to overcome exceptional headwinds, they were first encountered in 1945 by American bombers en route to Tokio. Since then it has become generally accepted that, in any cross-section of the atmosphere from equator to pole, at least one and perhaps three such streams may be encountered. They have been described as 'ribbons' of air 400 to 500 miles across but only 2 to 4 miles in depth, embedded in the general eastward drift but moving with exceptional velocity—from 80 to 200 knots or more. We are still learning rapidly about these remarkable features but already the literature is getting rather confused by the application of the term to things which may not necessarily be of similar origin. Certainly the expression '*the* jet stream' is already outdated.

Fultz's early dishpan model (see p. 190) suggested that a relatively continuous mobile jet was an innate characteristic of thermally induced flow in a rotating fluid but his later coreless pans produced a much more complex pattern (Fig. 56). Jets waxed and waned, one often replacing another at a different 'latitude'. There is every reason to expect that the circulation within a thin spherical shell would be still more complicated.

Reiter (1961; trans. 1963) in a long and authoritative work, has cleared the air a little by distinguishing between two types of tropospheric jet streams.

1. *The subtropical jet streams* (STJ) are encountered near the margins of the intertropical zone as defined above, their cores of maximum velocity usually lying near the 200 millibar level i.e. at 12 kilometres (or 40000 ft). They follow

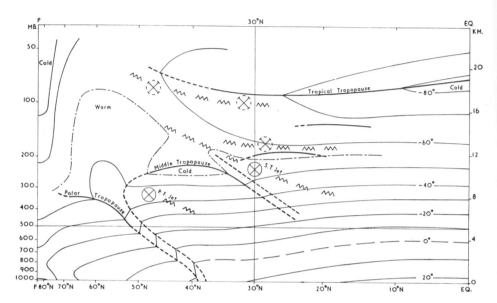

FIG. 67. A diagrammatic cross-section of the atmosphere over Central and North America in winter. Modified from the interpretation of Newton and Persson, i.e. with the horizontal scale transformed so that equal vertical columns enclose equal volumes of air. Note the multiple nature of the tropopause and the general location of the sub-tropical and polar-front jet streams (crossed full circles). The zig-zag symbols indicate shallow layers of strong winds associated with the jets. The broken circles at greater elevations represent jets in the stratosphere, less well documented. Taken overall, the vertical scale has a exaggeration of ×260

a moderately-waving course, usually between the 15th and 35th parallels, with a seasonal shift which is largely encompassed by this range. In the northern hemisphere the STJ reaches its maximum development in winter, especially over India where it may reach 150 knots. In summer it is rather difficult to distinguish this stream from more poleward jets.

2. *The polar-front jet streams* (PFJ) are more usually found with cores at the 300 millibar level, i.e. at 9 kilometres (or 30000 ft). Their general location in summer appears to be some ten degrees poleward of their position in winter but at neither season can this be expressed except in the vaguest terms as they loop and branch, dissolve and regenerate in the liveliest fashion. Wind speeds are greatest in winter and observed maxima are certainly not far short of 300 knots. In marked contrast to the STJ, the polar-front jets show an intimate relationship with surface weather.

This brings us to the major difference between the equatorial girdle and the polar caps. Throughout the year, but especially in winter, the latter display marked gradients in *surface* temperature so that density contrasts in surface air play an increasing part in moulding the character of the circulation as the poles are approached. This is illustrated in the generalised but still highly significant cross-sections of Figs. 67 and 68. The facts presented are identical but Fig. 67 is expressed in terms of temperature as normally understood whilst Fig. 68 is in terms of potential temperature (see p. 237), a more logical basis

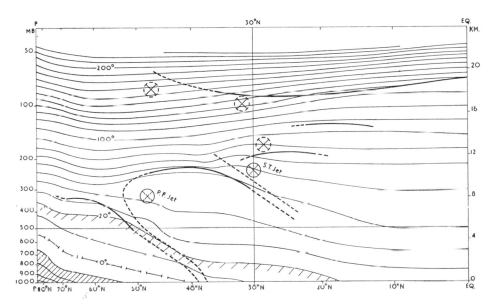

Fɪɢ. 68. The same cross-section as Fig. 67 but expressed in terms of potential temperature. This figure illustrates the general equatorward slope of the isentropic surfaces. Below 20°C (293°K) each of these is seen to meet the earth's surface. This is thus the approximate outer limit of Shaw's 'underworld' (see p. 174). Alternative interpretations are equally reasonable but the innermost core of the underworld is cross-hatched. Here the air is so seriously deficient in entropy that it can move only in close proximity to the earth's surface

for the discussion of a compressible medium like the atmosphere. Note that, as in Figs. 57 and 58, the scale along the meridian has been so drawn that equal intervals underlie equal *volumes* of air on a globe. The vertical exaggeration is of the order of 250 times.

The atmosphere of the polar caps is thus 'baroclinic', i.e. surfaces of temperature (and hence density) intersect the surfaces of pressure. Energy can thus be generated by horizontal motion. Note, particularly in Fig. 68, the existence of Shaw's 'underworld' with its central core lying north of 65°N in the region of prolonged polar night. It has no specific upper surface, *any* surface below the cross-hatched line of potential temperature 20°C (293°K) may serve to define a region of relatively imprisoned air. In the absence of surface heating, unlikely in winter especially towards the core of the region, cold, dense air can escape from this reservoir only in close proximity to the ground as a cold 'wave' or 'wedge'. As Shaw saw it, deflection of such flow should produce 'polar easterlies' thus applying a brake to further meridional exchange. Given longer to cool, the impounded air would thus become still denser, thus enhancing the surface pressure gradient. A balance of forces is thus set up but such a balance is unlikely to remain long undisturbed in the *lower* atmosphere for here conditions are affected by the varied thermal (and frictional) influences of land and sea, by the presence of major orographic barriers and by any change in the course of events aloft. This secondary governor of the circulation is thus inherently unstable, it is largely responsible for wobbles or waves within the system. Periods of polar outbreak thus alternate with more quiescent intervals when the polar reservoir is being recharged with cold air as a result of radiative losses. Such instability is particularly characteristic of the northern underworld for, owing to the large dimensions of northern Canada and Eurasia, it shows a marked tendency to become bipolar. *Two* continental cells may thus develop, extending south well into middle latitudes but linked only rather tenuously across the inner Arctic (Fig. 85). Escape may then occur from the poleward as well as the equatorward flanks of the two systems and converging currents are brought together over the winter gulf of warmth in the north Atlantic. Little wonder then that the winter weather of western Europe is bewildering in its variety! Furthermore, since disturbances are then steered far into the Arctic, the conclusion of J. Namias (1958) that the Arctic circulation differs from that of the temperate zone 'in the statistics' but not in kind has considerable justification.

Note the immediate effects of any such incursion of low-level polar air whatever its direction. It will represent an invasion of the zone normally occupied by surface air streaming poleward and westward from the western and northern flanks of the subtropical high-pressure cells. Varying widely in density, these air masses will tend to retain their identity and the zone of contact between them will be characterised by a sharp temperature discontinuity in the lower layers of the atmosphere and a steepened poleward pressure gradient aloft. Each incursion is thus accompanied by a rejuvenation of the 'polar front' in the lower layers of the atmosphere and a local acceleration of the general westerly 'jet' aloft. These two concepts thus have no independent existence; they are the joint effects rather than the causes of the flow-patterns observed. This may help us to understand why at some times and in some places both Front and Jet may be so diffuse as to be virtually non-existent.

These events occur in latitudes where the rotation of the earth imposes severe restraints upon any attempt by the air to follow a great-circle path. Large-scale turbulence is thus universal with a widely meandering jet aloft and related systems of high-pressure and low-pressure cells beneath. It has been shown by Sheppard (1958) how the meandering jet can transfer heat upwards and polewards by 'slantwise convection', thus contributing to the general thermal exchange. An asymmetry in the waves, the axes of troughs and ridges trailing from east of north to west of south, is adequate also to ensure the requisite northerly transfer of westerly momentum. As in the dishpan model (Fig. 55), poleward movement of air in the jet stream is accompanied by cyclonic motion near the surface with *ascent* predominating and equatorward movement of air in the jet is accompanied by anticyclonic motion with general *descent*. The surface pressure patterns thus tend to be carried forward in a west-to-east direction by the vigorous circulation aloft as in the dishpan model but on the actual earth they are also fundamentally affected by surface configuration, regional temperature contrasts, and the release of latent heat in areas of persistent rainfall, themselves partly a result of the circulation patterns thus set up. Polar outbreaks at the surface and their associated 'wave' developments aloft thus tend to occur on the earth in *preferred regions*—in the northern hemisphere in winter particularly over eastern America and eastern Asia—hence the semipermanent areas of high and low pressure which survive the process of long-period averaging to appear on atlas maps of mean pressure for January. But 'the jet in actual cases should not be considered as a hemispheric phenomenon without interruptions' as it appears in the early dishpan models. Rather is it a discontinuous but repeatedly regenerated feature. Each new outburst and acceleration of the eastward flow not only produces direct and immediate consequences in the weather of the locality where it occurs, it also sets up a train of indirect effects which are propagated downstream in a wavelike progression. Acceleration and deceleration of the jet are thus normal features of the flow and these, in turn, have been shown to have important effects on the pattern of vertical motion—so important in moulding surface weather contrasts. This pattern emerges from the dynamic requirements of the system and may be illustrated in an elementary fashion for the northern hemisphere as in Fig. 110.

Further complexity is added to the picture by the interaction between the circumpolar jet with that in subtropical latitudes, a relationship which is still far from fully understood. Riehl (1969, 296) has pointed out that waves in the two jet-stream systems are frequently out of phase so that 'the coupling between low and high latitudes occurs in a few narrow strips of longitude only'. Some visible evidence supporting his view that 'energy from low latitudes is injected into the temperate zone in narrow and variable areas' may be derived from a close inspection of the cloud patterns of Fig. 69. Although subject to constant change, these apparent channels are a remarkable feature on satellite mosaics of both hemispheres throughout the year.

## Summer

During high summer, surface temperatures attain remarkable levels over the interiors of the northern continents even up to the Arctic Circle. Thus mean daily maxima of the order of 20°C (68°F) are characteristic of Fort Yukon, Yellowknife, Gällivare, Turukhansk and Verkhoyansk in July though values then fall off rapidly in a poleward direction. This surface heating is diffused upwards so that the thermal gradient aloft is much weaker in summer than in winter. However the cool waters of the north Atlantic and Pacific and particularly those of the ice-laden Arctic Ocean maintain a considerable degree of baroclinicity so that the change in circulation which follows is a change in intensity rather than of kind. The Arctic reservoir of cold air is greatly reduced in area, and hence in volume, but polar outbreaks are still possible, especially in the neighbourhood of the Greenland ice-cap and less frequently across the Bering Sea. The weather over the north Atlantic and the Gulf of Alaska thus remains disturbed. Whether the distinction between the subtropical and polarfront jet streams is still maintained at this season is open to considerable doubt for the former is far enough north in summer to develop waves of considerable amplitude.

# The southern hemisphere

## Winter

Our knowledge of the circulation of the southern hemisphere is much less complete owing to the vast area of unfrequented ocean which it contains. The network of surface stations is thus very open and upper air data have become available in quantity mainly during the last decade. The evidence to hand suggests that the essential features recognised in the northern hemisphere are repeated in the south but that the broad picture is simpler since there is much less seasonal contrast. In general the circulation also appears to be more vigorous throughout the year.

Wexler (1959, 200) has presented a cross-section of the southern atmosphere near the 170th east meridian in winter. It is rather less elaborate than Fig. 67 since he has made no attempt to show the position of the polar front. Nevertheless the existence of a subtropical jet at about 25°S may be inferred and a corresponding system of surface high-pressure cells is then known to almost ring the globe. Synoptic analysis in the southern hemisphere suggests that these high-pressure cells migrate fairly steadily eastward in a fashion which has no parallel in the north (Garnier, 1958, 37). How far this is a result of wave development in the jet stream is not yet known. It may well be that the shallow cols observed are no more than distant echoes of deep lows moving in the same direction far to the south.

Southward from the tropics the section can be misleading since the process of averaging has smoothed out the poleward gradient. In his text Wexler is at pains to explain that this is due to the excellent ventilation of the whole zone by 'waves' of great vigour and amplitude. Nevertheless, even at the surface, important temperature discontinuities are known to exist. Thus, approaching the massive block of 'East Antarctica' (Gondwanan Antarctica would be a more significant and hence more appropriate title), from the eastern Atlantic or the Indian Ocean, we encounter in crudely concentric arcs: (1) the *Antarctic convergence* in about latitude 50°S (Fig. 69), beyond which warm surface waters never penetrate and the sea surface thus remains at only a few degrees above 0° C throughout the year; (2) the winter *limit of pack ice* between 56 and 58° S, usually over 3 feet thick and thus capable, in the absence of open 'leads', of generating virtually continental conditions over an area at least as large again as Antarctica itself; (3) the steep, lofty and snow-clad *margin of the continent* at 67–68° S, capable of generating winter temperatures with mean values as low as −45° C to −60° C. Off 'West Antarctica' (Cordilleran Antarctica), south of the Pacific and western Atlantic, a similar succession is found but the outlines are more irregular and thus cannot be so simply expressed in terms of latitude. Wexler's section, drawn broadly along 170° E through New Zealand to the Ross Sea, encounters (1) as far south as 60° S, (2) at about 64° S, and the edge of the great continental plateau (3), at almost its extreme poleward limit near 84° S.

Such surface contrasts must clearly generate marked density differences in the lower layers of the atmosphere. A multiple-surfaced 'underworld' is thus magnificently developed in the southern hemisphere with a core area in excess of ten million square miles in winter. In contrast to the more extensive winter Arctic core this lies exclusively in latitudes where the atmosphere is entirely cut off from any supply of direct solar energy and from the possibility of any appreciable conduction of heat from below. Furthermore about half of this area consists of lofty land-ice plateau. Hence the development of extremely low temperatures and the generation of centripetal outflows of unique persistence and violence. Katabatic, i.e. downslope, winds are even more prevalent than around Greenland. Though usually not more than a thousand feet in depth they are capable of transferring vast quantities of bitterly cold air out on to the surrounding ice-shelf and ice-pack.[1] Ball (1957) has pointed out that their burden of drifting snow (their fearsome 'blizzard' character) assists the process, indirectly by checking the rise in temperature that should follow from adiabatic compression (see p. 233) and directly by adding perhaps 1 per cent to the density of the air-snow emulsion. Hence Cape Dennison (longt. 142°40′ E) has won the unenviable reputation of being the windiest place on earth—mean wind speed over 22 months, 38 knots; for July 1913, 48 knots. Doubtless similar sites exist elsewhere but no explorer would wittingly choose such a location for a base under any temperature conditions, let alone those where, quite literally, the wind carries the sting of death.

Beyond the edge of the continent, especially off the 'eastern' sector, there is a good deal of evidence that this outflow does indeed generate 'polar easterlies' rivalling in strength and persistence the 'roaring forties' (westerly) to the north. It would be wrong, however, to regard this circulation as a relatively static one. Low-level convergence of the opposed subtropical and polar air streams continually sharpens the air temperature contrasts, thus producing polar fronts, often duplicated, anywhere between 40 and 65° S.

[1] The experimental model of Gibson and Douglas (1969) shows this occurring over the Ross Sea.

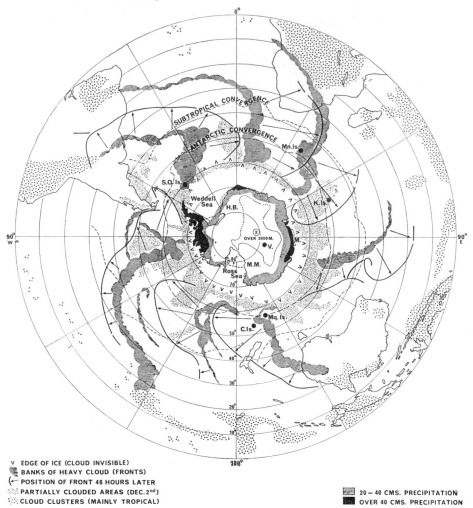

V  EDGE OF ICE (CLOUD INVISIBLE)
   BANKS OF HEAVY CLOUD (FRONTS)
   POSITION OF FRONT 48 HOURS LATER
   PARTIALLY CLOUDED AREAS (DEC. 2nd)
   CLOUD CLUSTERS (MAINLY TROPICAL)

20 – 40 CMS. PRECIPITATION
OVER 40 CMS. PRECIPITATION

FIG. 69. The Antarctic. The figure shows the areas of heavy cloud over the southern hemisphere on 2 December 1967, beyond the margin of the polar ice. Note the striking lanes often some 250 miles (400 km) in width and occasionally extending to within 20 degrees of the Equator. That these are frontal disturbances is indicated by their eastward march during the following 48 hours. Tropical cloud clusters of considerable density are also shown but the lighter veils over the trade-wind systems, so evident in Fig. 65, are much less impressive at this degree of resolution (after *Catalog of met. satellite data*, no. 5.316). On the Continent the map shows the general distribution of the area exceeding 3000 m in elevation, the area where net annual snow accumulation is less than 20 cm and the restricted localities where accumulation is known to exceed 40 cm. In the neighbourhood of point 'x' the ice cap reaches *c.* 4000 m. At sea the positions of the Antarctic Convergence and the Subtropical Convergence are indicated.

| | | | |
|---|---|---|---|
| M.M. | = McMurdo | M. | = Mirny |
| V. | = Vostok | K. Is. | = Kerguelen Island |
| Mn. Is. | = Marion Island | C. Is. | = Campbell Island |
| Mq. Is. | = Macquarie Island | S.O. Is. | = South Orkney Islands |
| H.B. | = Halley Bay | | |

With the aid of the parallel development of jet streams aloft, which evacuate the converging air, these generate an almost unceasing series of cyclonic depressions so that sea-level pressure near the margin of the continent may range from 1030 to 930 millibars over an interval of a few days—a change representing the addition or subtraction of a mass of air equivalent to one-twentieth of the normal atmospheric load. The easterlies and westerlies are

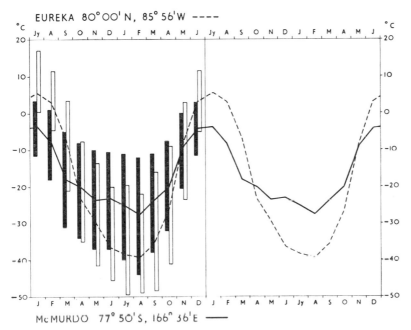

EUREKA 80° 00' N, 85° 56' W ----

McMURDO 77° 50' S, 166° 36' E ——

FIG. 70. The seasonal variation of mean monthly temperatures at McMurdo (77° 50′ S, 166° 36′ E) compared with that at Eureka (80° 00′ N, 85° 56′ W). Note that two monthly scales have been employed to facilitate comparison and that the diagram covers 24 months to make the contrasting form of the summer and winter halves of the curve at McMurdo more apparent. The 'kernlos' character of the winter at this station is evident. The vertical bars on the left-hand half of the diagram indicate the range between the mean monthly maxima and mean monthly minima at both stations. The records are for not more than a decade

thus more accurately viewed as two aspects of a single group of features, the prevailing wind direction depending on whether the centres of disturbances pass more frequently to the north or to the south. In the southern hemisphere there is much less orographical control over polar outbursts than in the north but Graham Land and the southern Andes appear to play an important part in cyclone generation. An alternative site lies off the massive coastland of Gondwanan Antarctica, particularly in the broad area between 100°E and the Ross Sea. In their lower levels such outbursts are rapidly transformed as they break out over the southern ocean but they may retain sufficient individuality to play a major part in the winter weather patterns of the three southern continents. In South America indeed 'cold waves' have occasionally been followed as far as Amazonia.

Blowing for the most part over the ocean, the upper-air polar-front jet streams of the southern hemisphere are still comparatively little known. Yet there is no doubt of their existence nor of the fact that they too develop waves of very wide amplitude. Occasionally they appear to steer depressions right across Antarctica, particularly from the Ross Sea towards the Weddell Sea, thus yielding a realistic explanation for the source of the (limited) snow supply of the interior. Hobbs' (1926, 1935) idea of a 'glacial anticyclone' which could be supplied only by rime is thus now discredited, indeed Britton and Lamb (1956, 353) write that 'the region is clearly penetrated by the very systems with which we are familiar in other latitudes and gets its precipitation through processes that are well known and well understood'. There is, in fact, another remarkable piece of evidence for the active ventilation of the upper air over Antarctica in winter. At most stations the seasonal fall of temperature is checked in April, the subsequent decline in mean monthly values being comparatively slow (see Fig. 70). German climatologists have described this as a 'kernlos' or coreless winter. Wexler (1959) attributes the feature to the advection of comparatively warm air (well below zero Fahrenheit!) aloft and the transference downwards of this (relative) heat by turbulence. Apparently there is no such effect over Siberia. Another result of this active circulation is that the difference in temperature between the warmest and coldest months aloft in all the layers from 850 to nearly 250 millibars is less than $10°C$ ($18°F$) over Antarctica, or less than one-half of the corresponding range experienced in the northern hemisphere.

## Summer

Summer in middle and high latitudes in the southern hemisphere brings light but remarkably little sensible heat. Much solar radiation is directly reflected back to space either by snow and ice surfaces or by the heavy cloud canopy which frequently extends like an aureole around the continent to the general neighbourhood of the 45th parallel. The total area with high albedo thus attains some 25 to 30 million square miles, especially in early summer. For the rest, the melting of snow and ice provides an almost inexhaustible sink and so, south of the Antarctic convergence, the change in sea-surface temperatures is no more than 2 or $3°C$, and the convergence zone shifts seasonally by scarcely more than 100 miles. Mean *daily maximum* temperatures in summer on Marion Island (46°51′S, 37°52′E) and Kerguelen (49°25′S, 69°53′E)—in the latitude of northern France![1]—reach a dismal $10°C$ ($50°F$) and though Campbell Island (53°33′S, 169°08′E), some hundreds of miles nearer the pole, achieves $13°C$ ($56°F$),[2] the record at Macquarie Island, barely a degree farther south, is even more depressing ($8°C$). Little wonder that the zone of

[1] The corresponding value for Nantes is 25° C.
[2] Yet with only some 680 hours of bright sunshine per year (Garnier, 1958,94).

maximum temperature gradient aloft shifts only a few degrees, carrying the subtropical jet with it. There is thus little poleward motion of the subtropical high-pressure zone though the warming of the three southern continents now gives this feature a more evidently cellular character, the cells being best developed over the cooler waters of the eastern sector of each ocean. In the southern Pacific this has the effect of developing a curious anomaly of climatic significance between the New Hebrides (*c.* 170°E.) and Samoa (*c.* 170°W.). From November to March the trade winds are scarcely recognisable in this area (see Fig. 63, January). The cloud pattern on the satellite photograph in Johnson (1970, figure 7) picks out the core of this area with quite remarkable clarity.

Within the southern polar cap the chief event in summer is the break-up of the ocean pack so that open water is now common to about 65–66°S off East Antarctica and to 70–71°S in the west. It is thus the presence of open water which brings 'summer' to the shores of the continent though the persisting katabatic winds and variable drift ice do much to neutralise its effects. We thus find mean daily maxima of 1°C at Cape Dennison (67°S, 142°40′E) but of no more than −4°C (25°F) at Little America (78°34′S, 163°56′W), in both cases, however, for no more than 2 months. Sharp contrasts in air temperature still remain between the continental interior and the now more open sea so that the generation of cyclonic disturbances continues with little abatement. On the whole the general pattern of the circulation appears to be rather more zonal at this season; fewer disturbances penetrate into the interior of Antarctica and fewer cold waves invade the middle latitudes. Yet the flow is still highly turbulent and an almost unceasing succession of mobile low-pressure systems pursue one another round the southern ocean along a mean path a few degrees north of the limits of the remaining pack. Drift ice is thus kept perpetually on the move to the great hazard of the whalers and explorers who visit these waters. Lamb (1959, 22) has suggested that secular changes in the vigour of this turbulent 'flywheel' and its associated jet streams might well entail worldwide readjustments which could offer a key to long-period changes in the climate of distant lands. The question is still wide open.

## The circulation in the stratosphere

If the effects of direct solar radiation in the atmosphere were confined entirely to the surface of the earth it would appear that the pattern of circulation outlined above would extend upwards to at least 75 kilometres (246000 ft or 46.5 miles). Above this level, under pressures of less than 0.05 millibars, the gases of the outer 1/20000th part of the mass of the atmosphere begin to decompose into atomic form and a new physical regime prevails. As we have seen, however, ultraviolet radiation from the sun generates ozone in the upper air. This gas is found mainly between 55 kilometres (180000 ft) and 15 kilometres (49000 ft) with maximum concentration at 35 kilometres (115000 ft). Although its total quantity is very small (averaging about 2.5 millimetres at surface temperature and pressure) its topmost fraction is sufficient to cut off all solar radiation below wavelength 0.29 microns, that is, about 5 per cent of the total radiation from the sun. Absorbing in the ultraviolet range but re-radiating as a gas according to its temperature, the upper layers of the ozone belt between 55 and 47 kilometres develop temperatures of the order of those prevailing at ground level. Direct measurement at this elevation is difficult and estimates show a wide range of variation but a mean value of about 300° K (27° C) would appear to be representative of the temperature of this warm zone and the heat thus generated is dispersed both upwards and downwards.

Above the tropopause and the isothermal zone which is often observed to succeed it, that is, in general, above 20 kilometres (124000 ft) temperature thus usually *increases* with height. Furthermore since this increase stems from the effects of radiation it may be expected to vary considerably from place to place and from time to time. A new system of circulation is thus entered of which comparatively little is yet known. Since the atmospheric load at 20 kilometres is only 55 millibars, i.e. one-twentieth of the load at sea-level, events in this system can have only limited repercussions on the pressure pattern at ground level. Nevertheless such repercussions are possible especially within the tropics where tropospheric pressure contrasts are normally weak. Note that this upper system is completely absent from the dishpan models where there is but one heat source and where the upper surface of the water imposes a quite definite and tangible limit to the circulation.

The lower stages of this system owe their character to the fact that, whereas the polar tropopause is reached at 7 to 8 kilometres (23000–26000 ft) with temperatures of the order of −45° C, the equatorial tropopause is not attained till 16 to 17 kilometres (53000–56000 ft) at temperatures of about −75° C. Since, particularly in the summer months, temperatures either remain substantially unchanged or else increase with height above this surface it follows that the thermal gradient favouring westerly winds falls off rapidly with increasing elevation until it vanishes not far from the 12 kilometres or 200-millibar level (*c.* 40000 ft). Within the next 12 kilometres of rarified atmosphere to the 30-millibar level the thermal gradient is thus favourable for an east-to-west circulation which appears to be best organised in the general neighbourhood of the equator at about 16 kilometres (52500 ft) where it overlies the generally east-to-west circulation between the converging trades. The full implications of this 'equatorial jet stream' are not yet appreciated though it is no new discovery. We have seen (p. 181) that Ferrel (1889) noted the westward transport of volcanic dust at great heights near the equator but ignored its implications. Possibly this current plays an important part in removing ascending trade-wind air from sections of the equatorial girdle and thus allows the circulation to continue in the absence of anti-trades, so long postulated and so difficult to find. To do this, of course, it must not ring the whole earth nor must it be everywhere confined closely to the neighbourhood of the equator but we have seen that all jet streams are segmented and that they grow, decay and are regenerated. In doing so they exercise potent dynamical effects upon the neighbouring air and particularly upon the air beneath them.

## The summer pattern

The equatorial jet stream, if indeed it deserves that name, is particularly a phenomenon of the *summer* hemisphere. Its axis may thus be expected to lie rather poleward of the equator when it is best developed, especially in the northern summer when the circulation is notably asymmetrical. Koteswaram (1958) has noted the presence of high-level cold easterlies over Asia about 15°N from the South China Sea to Arabia above the surface monsoon (see also the charts in Lockwood, 1963). It would appear that nowhere else in the world do the equatorial easterlies wander so far from the Equator. Koteswaram thinks that this has nothing to do with the sheer bulk of Asia but that it may well be related to the massive bulwark of the Tibet-Himalayan upland which wards off the subtropical westerlies to the north and feeds in heat to the middle troposphere (via radiation and the latent heat of condensation) thus strengthening the *southward* gradient aloft. His suggestion that there is a linkage between outbursts of rain in the monsoon current and the position and strength of the easterlies aloft is full of interest even if it cannot yet be regarded as fully established. It certainly offers a possible clue to the well-known anomaly that the monsoon 'breaks' first in Burma and advances *westward* over India, a fact that has always seemed mysterious as long as attention was confined to the surface southwesterlies. In Africa another sector of this easterly airstream has been recognised at about 10°N from West Africa out into the Atlantic. Here again is a region of notable weather anomalies. We know little of the high-level circulation round the rest of the equatorial girdle.

In the southern summer the equatorial easterlies do not appear to move far in a poleward direction. At Singapore, for instance, they are encountered aloft throughout the year. A notable anomaly at this season is the occasional occurrence of equatorial surface *westerlies* from the Bay of Bengal to at least as far east as Ocean Island, 169°36'E, (Crowe, 1951, 50, 64). Their relationship to the upper current, if any, is still unknown.

## The winter pattern

In the winter of each hemisphere the stratospherical circulation takes on a significantly different form especially above the region of the polar caps. In the absence of insolation the ozone supplies of the upper air are not renewed and the warm zone of the upper stratosphere decays. A notable result of this is the virtual disappearance of the tropopause from polar soundings in winter. Temperatures continue to decrease with height to about −75°C at 21 kilometres (50 mb) over the northern polar regions and to −90°C at the same elevation over the South Pole. In each hemisphere in winter therefore stratospheric temperatures are highest over *middle* latitudes (about 50° latitude) and from that area they decline both equatorwards and polewards. The winter hemisphere therefore develops a *westerly* flow polewards of the 50th parallel and at very high altitudes i.e. at 25 kilometres (25 mb). This may develop speeds of 50 to 100 knots and has been called the 'stratospheric polar jet' to distinguish it from the westerly circulation of lower levels. It is thought that the jet in the southern hemisphere is stronger than that in the north and that the very low temperatures registered aloft there are in part the results of this powerful whirl which excludes incursions of warmer air much more effectively than in the northern hemisphere. Wexler (1959) has presented some very interesting facts about the warming and cooling of these two cold pools in the stratosphere. The southern pool attains its lowest temperatures in August and warms 'explosively' in October when the sun returns and regenerates the ozone supply. The northern pool registers minimum temperatures in December and begins to warm up rapidly 6 weeks *before* the sun arrives. How far this picture will be changed by the accumulation of more data it is impossible to say, but there is clear evidence from the nature of the curves of 'a significant contribution to heating by dynamic processes resulting in subsidence of air and adiabatic heating'. This appears to be the main region where ozone is fed into the lower layers of the atmosphere from aloft.

It will be noted that even in the very last part of this lengthy section we have met another instance where temperature distribution is in part a *result* of air motion rather than a cause of it. Meteorology is full of these reciprocal relationships.

# 8
# The stratigraphy of the atmosphere

In meteorology the air more than 600 metres (*c.* 2000 ft) above the surface of the earth is often described as the 'free atmosphere'. The implication is that air above this level is so comparatively free from the constraints imposed by terrestrial geography that it may be assumed to be under the almost exclusive sway of the gas laws. When encountered, the term thus gives fair notice that the analysis which is to follow will be deductive in method and basically mathematical in technique.

Since climatology is concerned with the general patterns of behaviour of the atmosphere above us we cannot ignore its vertical stratification. Indeed we know this to be vital in controlling many, if not most, of the climatic events recorded on or near the ground. Our interest, however, will be descriptive rather than analytical and we shall be concerned with process only to the degree that 'genetical description' is more readily comprehensible and more intellectually satisfying than a presentation which concerns itself only with bald facts. Nevertheless we may find it useful occasionally to examine artificially simplified examples or 'models' to illustrate a point even though we know them to be only a limited portrayal of the actual facts.

Since the air-ocean is subject to large-scale circulations which keep its chemical composition virtually homogeneous, it may be difficult at first to appreciate how it can be 'stratified' in any normal sense of that term. A molecule of air that is today near sea-level might conceivably be found tomorrow, if it could be recognised, at 10000 metres. The word 'stratification' may seem to suggest a degree of stability of position which any such observation would appear to deny. Nevertheless the concept is valid for two distinct sets of reasons. In the first place, the *chances of vertical displacement* on a major scale vary widely from place to place and from time to time. Over many parts of the world vertical exchanges are so heavily damped that air motion may be considered as predominantly horizontal. Vertical contrasts are then preserved, perhaps not for very long, but at least long enough to affect the behaviour of the air mass as a whole. Because we commonly see smoke or fine dust being swept upwards we must not conclude that such motion may continue indefinitely. A flight over Britain on most days of the year will show that even the clouds have at least one fairly firm and uniform 'ceiling' above which the pilot takes his plane. Over drier and dustier lands a similar ceiling of haze is by no means unusual. Although such ceilings may be penetrated locally by vigorous upcurrents they present, over the greater part of their extent, almost as effective a barrier to vertical exchange as if they were composed of an invisible yet impermeable membrane.

Secondly, the stratification which is important in climatology is a *stratification of air characteristics*. of temperature, pressure and density, rather than of air material. The chief material contrast that occurs is in water vapour content and even this exercises its effects via yet another characteristic, relative humidity. All these features of the atmosphere are closely interrelated according to known physical laws. Furthermore, as soon as we think in terms of great volumes of air it becomes evident, on grounds of what mathematicians term 'continuity', that widespread upward motion in one sector of the atmosphere must presuppose an equivalent degree of downward motion elsewhere. We shall see shortly that, though upward motion may tend to weaken some of the contrasts between the various strata of the atmosphere, the concomitant downward motion will tend to strengthen them. Mixing and unmixing of characteristics are thus complementary processes and since they often occur over preferred regions (preferred sometimes for geographical and sometimes for dynamical reasons) *regional contrasts* in atmospheric stratification are renewed just as rapidly as they may be destroyed. They may thus remain 'real' even though any given molecule of air may pass, given adequate time, from one system to another and back.

# The standard atmosphere

The central graph in Fig. 71 represents one of the recognised 'standard atmospheres' used as a first approximation in calibrating altimeters and for other scientific and engineering purposes. It is based upon a number of assumptions, the chief of which are: (1) that the air is dry, (2) that pressure at sea level is 1013.25 millibars and temperature 15°C (288°K or 59°F), (3) that temperature decreases uniformly with height by 0.65°C for every 100 metres ascent (1°F per 280 ft) up to the level of 10 769 metres (35 332 ft), beyond which (4) temperature is taken as constant at −55°C (−67°F). In any synoptic cross-section of the atmosphere none of these conditions is likely to be completely satisfied yet the curve provides a generalised view of conditions in middle latitudes and is worth closer examination.

What are the upper and lower limits of a cross-section of the atmosphere? Even the lower limit cannot be stated with very great precision for the air enters the deepest mine or cave and plays important functions in the pore-spaces of the soil. Nevertheless the general surface of the land (conveniently written SFC) and mean sea-level provide reasonable enough limits in this direction. Aloft the problem is more difficult since the gases become increasingly rarified until molecular dissociation takes place so that the atmospheric ocean presents us with a tangible reality without a finite upper limit. Nevertheless we can weigh the column of air overhead—with a barometer—and for the standard atmosphere at sea-level this weight is equivalent to 760 milli-

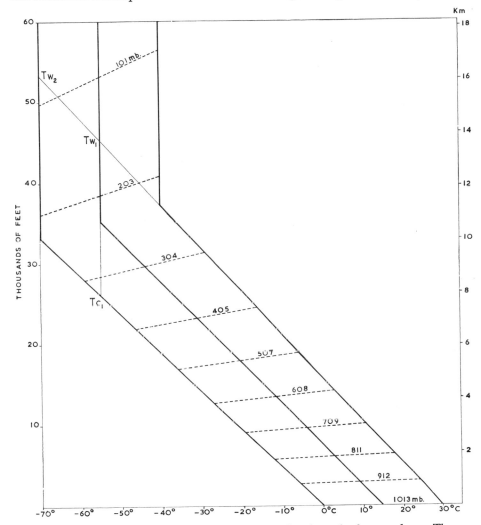

Fig. 71. A temperature-height diagram of three postulated standard atmospheres. The central graph is for the ICAN standard atmosphere with a surface temperature of 15° C (59° F) and a uniform lapse rate of 0.65° C per 100 m up to a tropopause at nearly 11 000 m. Pressure levels are shown for each tenth of the atmospheric mass—surface pressure 1013 mb. On each side is shown the effect of increasing or decreasing the temperature by 15° C uniformly throughout the whole mass. These will be referred to as standard 'warm' and 'cold' atmospheres. Alternative interpretations at tropopause level are mentioned in the text

metres (29.92 in) of mercury. Expressed in terms of force this equals 1013.25 millibars or 1.033 kg/cm² (14.7 lb/in²). One way of dealing with any measurement when a certain fuzziness is suspected in its margins is to strike off the upper and lower 1 per cent. This permits us to make the more positive statement that 98 per cent of the standard atmosphere lies between about 280 and 99 280 feet above sea-level. We can thus view the atmosphere as effectively some 30 kilometres (19 miles) thick—which is only about a half of 1 per cent of the radius of the solid earth. Events above 30 kilometres have great interest to the geophysicist and cosmonaut but have little immediate relevance to the climatologist.

An alternative treatment is to divide the atmosphere into ten layers or concentric shells of equal weight as in Table 21. Some such treatment is implicit in much modern meteorological research. It may prove convenient

TABLE 21. *The lower nine-tenths of the standard atmosphere*

| DECASPHERE | TOP OF EACH STRATUM | | | | | THICKNESS OF STRATUM | | MID-STRATUM TEMP. (°C) | MEAN DENSITY (kg/m³) |
| | PRESSURE (mb) | TEMPERATURE (°C) | (°F) | ELEVATION (km) | (ft) | (km) | (ft) | | |
|---|---|---|---|---|---|---|---|---|---|
| Ninth | 101 | −55 | −67 | 16.15 | 53 000 | 4.42 | 14 500 | −55 | 0.24 |
| Eighth | 203 | −55 | −67 | 11.73 | 38 500 | 2.66 | 8750 | −51 | 0.40 |
| Seventh | 304 | −44 | −47 | 9.07 | 29 750 | 1.98 | 6500 | −37.5 | 0.52 |
| Sixth | 405 | −31 | −24 | 7.09 | 23 250 | 1.62 | 5300 | −26 | 0.64 |
| Fifth | 507 | −20.5 | −5 | 5.47 | 17 950 | 1.37 | 4500 | −16 | 0.76 |
| Fourth | 608 | −12 | 10 | 4.10 | 13 450 | 1.19 | 3900 | −8 | 0.86 |
| Third | 709 | −4 | 25 | 2.91 | 9550 | 1.07 | 3500 | −0.5 | 0.97 |
| Second | 811 | 3 | 37 | 1.84 | 6050 | 0.96 | 3170 | 6 | 1.07 |
| First | 912 | 9.5 | 49 | 0.88 | 2880 | 0.88 | 2880 | 12 | 1.18 |
| Sea-level | 1013.25 | 15 | 59 | | | | | | 1.225 (2.7 lb) |

in the discussion which follows to coin a new word for the layers as we have defined them. We require a word free from any implications of solidity and 'decasphere'[1] presents itself as not ill-sounding and self-explanatory. Numbering these from sea-level upwards we may note that many British mountains penetrate through the first decasphere or lowest tenth of the atmosphere, that Mont Blanc extends well into the fifth decasphere, and that only the eighth, ninth and tenth decaspheres lie above the summit of Mount Everest. It is the intersection of these successive layers with the surface of the earth on high mountains that presents the mountaineer with a profound physiological challenge arising from low temperature, low pressure and the 'rarity' or low density of the air even when the weather is good, that is when these conditions are not exacerbated by the more complex factors we describe as 'exposure'. The ordinary tourist may well feel the first symptoms of distressed breathing after vigorous exercise in the third decasphere (*c.* 6000–9500 ft) where the density of the air is still about three-quarters of its value at sea-level, but experienced climbers can operate without oxygen to at least the top of the sixth

[1] An alternative, 'decisphere' is less easy on the tongue.

decasphere, where air density is about half of its sea-level value. Note that because of the fall in temperature, air density does not fall off in direct proportion with pressure and it may seem at first glance that to double the rate of respiration would be a comparatively simple response.

At these heights low pressures begin to have direct and most unpleasant results. If the rise has been rapid, as in a balloon or by mountain lift, free gases in body cavities such as the sinuses or abdomen exert a pressure which may cause acute pain and even the acclimatised mountaineer has to face a disturbed $O_2:CO_2$ balance in the blood which materially affects his efficiency. The final ceiling for the unpressurised human body is well within the tenth decasphere at 63 millibars (63000 ft) when at body temperature (37°C) the vapour pressure of body fluids will be as great as atmospheric pressure— the blood will literally boil. In fact, however, long before this height is reached the pressure of water-vapour in the lungs will so restrict the intake of unpressurised oxygen that the subject will relapse into unconsciousness.

The most striking feature of the standard atmosphere occurs at nearly 11 kilometres (*c.* 35000 ft) where the steady fall of temperature with height ceases abruptly and throughout the two uppermost decaspheres the atmosphere is regarded as isothermal. It must be emphasised that this is a highly generalised picture of the facts but a comparatively sharp change in the lapse rate, a tropopause, is usually encountered aloft in most aerial soundings. All layers below this level are known as the troposphere and above it as the stratosphere, a highly technical term which has achieved widespread currency since the development of pressurised jet aircraft. The level of the tropopause in the standard atmosphere is appropriate for middle latitudes but since it is determined in part by the activity of the troposphere it is highest within the tropics (up to 17 km; 55000 ft) and much lower towards the poles (8 km, *c.* 25000 ft). At any one place it is also higher in summer than in winter; indeed it may vary considerably in height from day to day. These changes may seem very remote from the surface of the earth but they are not without importance since the tropopause imposes a very significant ceiling upon vertical motion in the atmosphere. Most of the activity we describe as weather takes place in the troposphere; stratospheric flight is attractive largely because it is 'above the weather'. It must not be imagined, however, that conditions throughout the stratosphere are broadly uniform. Temperatures in the lower stratosphere are largely a function of tropopause height, i.e. they are lower over the tropics (often about −75°C) than over the poles (about −45°C) especially in summer. In the upper stratosphere temperature often *increases* with height owing to the absorption of ultraviolet radiation from the sun by ozone. Since this process cannot occur over the poles in winter it appears that temperature may continue to decrease to great heights at that time and no tropopause will be observed. In view of these marked contrasts in stratospheric temperatures, and hence in air density at high levels, it is not surprising that when we speak of the absence of weather in the stratosphere we do not mean the absence of strong horizontal air motion.

Returning to Fig. 71 it is an interesting exercise to consider the effect of either chilling or warming the whole column of the standard atmosphere by 15°C, it being assumed that there is no lateral transfer of air during the process so that pressure remains unchanged at sea level and at the tops of each individual stratum. It is then possible to apply Charles's Law which states that the volume of all gases over a wide range of temperatures around 0°C (273°K) varies directly with their temperature expressed on the Kelvin scale. In the present example volume will be expressed as the thickness of each stratum. The new heights of the tops of the nine lower decaspheres are given in Table 22 which also gives the difference in height above sea level of corre-

TABLE 22. *Comparison of the cold (C) and warm (W) columns represented in Figure 71 (i.e. Standard atmosphere ± 15° C at all levels).*

| DECASPHERES | HEIGHT OF TOP OF STRATUM | | | PRESSURE EXCESS (W–C) AT LEVELS GIVEN IN COLUMN C (mb) |
| | C. (km) | W. (km) | DIFFERENCE (km) | |
| --- | --- | --- | --- | --- |
| Ninth | 15.14 | 17.16 | 2.02 | 35 |
| Eighth | 11.02 | 12.44 | 1.42 | 47 |
| Seventh | 8.53 | 9.60 | 1.07 | 48 |
| Sixth | 6.68 | 7.49 | 0.81 | 45 |
| Fifth | 5.16 | 5.77 | 0.61 | 41 |
| Fourth | 3.87 | 4.32 | 0.45 | 34 |
| Third | 2.75 | 3.07 | 0.32 | 27 |
| Second | 1.75 | 1.94 | 0.19 | 19 |
| First | 0.83 | 0.92 | 0.09 | 10 |

sponding surfaces in the warm and cold columns. These isobaric surfaces bounding the various decaspheres are thus observed to slope from warm to cold at a gradient which increases with height when the temperature contrast extends through a great depth of the atmosphere as in the present imaginary example. The pressure differences represented by these gradients, however, must be affected by the upward decrease in air density. Estimates of these differences in millibars are given in the righthand column of Table 22. It will be observed that the greatest pressure contrast occurs between 6 and 12 kilometres (20 000–40 000 ft).

Now, for hydrostatic reasons, isobaric surfaces, unless otherwise disturbed, tend to take up a horizontal position; each 'finds its own level' as we say with reference to the surface of still water. If our warm and cold air masses are brought into mutual relationship, therefore, the upper layers of the warm mass will move towards the cold and the whole of the static picture we have built up undergoes profound modifications. The first effect of the shifting load aloft will be to increase pressures throughout the lower layers of the cold column and to decrease pressures in the corresponding layers of the warm. Let us pause for a moment and try to grasp the effects of this upon the stratification. A molecule of air formerly just above, say, the 709-millibar surface in the cold air will now find itself, with increasing pressure, *below* the new 709 millibar surface. Indeed, each of the lower decaspheres may be considered as having yielded some of its air substance to the stratum below and gained an equal mass from the stratum above. In a very real sense the air may be considered as *subsiding* through the pressure surfaces, it is passing from lower to higher pressure, quite irrespective of whether there is or is not any physical descent towards sea level. In the warm column, on the other hand, falling pressure will have the same effect as if the air were *rising* through the pressure surfaces. Uplift and subsidence introduce new elements into the problem which Fig. 71 is inadequate to express, but it is necessary to make it quite explicit that when these terms are used in the analysis which follows it is *motion in relation to pressure surfaces* that produces the results there described. Motion in relation to sea-level may certainly be involved but it is incidental rather than vital in this particular context.

Of course, the shifting load aloft is followed at once by other important consequences. Pressure at ground level beneath the cold column rises whilst pressure at ground level beneath the warm column falls; the change in these values is indeed a measure of the weight of air transferred. The 1013.25-millibar surface, which formerly coincided with sea-level, may now be regarded, therefore, as having been lifted above the ground within the cold column and lowered (below ground) beneath the warm column. A surface flow from cold to warm will thus develop down this pressure gradient and the complete circulation thus established as in Fig. 49 (warm to cold aloft; cold to warm below) may continue for some time until equilibrium is finally established. During this process, since air is being drawn away from the base of the cold column, a general subsidence of the upper layers is to be expected, while the warm column will be buoyed up by the arrival of colder air at its base. Physical subsidence and uplift are thus to be anticipated, but their thermo-dynamical effects must be assessed in relation to pressure change.

Before leaving Fig. 71 it should be remarked that the simple case there illustrated, where air masses differ uniformly in temperature at all pressure surfaces, is most unlikely to occur in nature. Alternative hypotheses for the uppermost layers are suggested by the fine lines.

(*a*) If the contrast in temperature is confined to the troposphere and the stratosphere is regarded as unchanged at $-55°$C, the tropopause in the cold air ($T_{c1}$) will be at 8 kilometres and in the warm air ($T_{w1}$) at nearly 14 kilometres.

(*b*) If the stratosphere over the warm air is taken down to $-70°$C the tropopause in that column ($T_{w2}$) will be found at 16 kilometres. The reader will note that $T_{c1}$ and $T_{w2}$ do not differ widely from the actual levels frequently observed in polar and tropical air respectively. These interpretations involve a considerable modification of the values given for the ninth decasphere in Table 22, and the pressure difference at the top of that layer is reduced to about 22 millibars. Even so we have here a most extreme case and it seems probable that, as a rule, the circulation in the stratosphere is only very indirectly affected by events below.

In most meteorological work pressure surfaces are taken at round intervals of 100 millibars and not at the odd values employed above. Table 23 may prove useful for general reference.

TABLE 23. *Heights of isobaric surfaces in three standard atmospheres*

| TEMPERATURE AT SEA-LEVEL | STANDARD (15°C) | | COLD (0°C) | | WARM (30°C) | |
|---|---|---|---|---|---|---|
| *Isobaric* (mb) *surface* | (m) | (ft) | (m) | (ft) | (m) | (ft) |
| 300 | 9164 | 30065 | 8631 | 28315 | 9697 | 31815 |
| 400 | 7185 | 23575 | 6777 | 22235 | 7593 | 24910 |
| 500 | 5574 | 18285 | 5263 | 17265 | 5885 | 19310 |
| 600 | 4206 | 13800 | 3975 | 13040 | 4437 | 14555 |
| 700 | 3012 | 9880 | 2849 | 9345 | 3175 | 10415 |
| 800 | 1949 | 6395 | 1845 | 6055 | 2053 | 6735 |
| 900 | 988 | 3240 | 935 | 3070 | 1041 | 3415 |
| 1000 | 111 | 365 | 105 | 345 | 117 | 385 |

## Subsidence and uplift within air masses: dry air

Although highly generalised and to some degree hypothetical, the curves of Fig. 71 are not unrelated to actual cross-sections of the atmosphere. Such curves are normally based upon observations made during a balloon or aircraft ascent, they are *environment curves* and describe the actual stratification of the atmosphere at some time and place.

We now turn to consider two other groups of curves conveniently distinguished from the above as *process* or *path curves*. These are designed to illustrate changes to be anticipated within a single sector or 'parcel' of air as it moves within the atmospheric ocean from one pressure level to another. The emphasis is thus entirely different; environment curves are concerned with observed facts, path curves are concerned with deductions from known physical laws. They are essentially a tool of the meteorological forecaster, but some familiarity with the technique is of value in the appreciation of climatic situations.

In order to apply physical laws to a real situation it is always necessary to start with a few basic assumptions which should be clearly stated at the outset and never overlooked when the time comes for a theoretical solution to be applied to the actual facts. The most significant assumption made in thermodynamic investigations relating to moving parcels of air is that the changes which occur within the parcel are 'adiabatic'. This means that, by way of a first approximation, it is assumed that heat is neither gained nor lost by the parcel via outside sources throughout the process under review. The conditions necessary to justify such an assumption are worth enumerating: (1) the parcel must be of such considerable volume that mixing processes around its margins can be regarded as having a negligible effect on the parcel as a whole; (2) the time interval under consideration must be comparatively short since radiative exchanges can never be held in check; and (3) precipitation should not fall from the parcel during the process as this will change its inherent characteristics. Fortunately for certain stages of the adiabatic process this latter provision can be ignored.

For *dry air* the relationship between pressure and temperature is comparatively simple. It is derived from a combination of Boyle's Law with Charles's Law and expressed as

$$T/P^{0.288} = \text{constant}$$

This means that $T_1/P_1{}^{0.288} = T_2/P_2{}^{0.288}$ where $T_1$ and $T_2$ represent the initial and final temperatures (in $°\text{K}$) of an air-parcel moving adiabatically from pressure $P_1$ to pressure $P_2$ expressed in any convenient units. This is a quantitative expression of the everyday experience that increasing the pressure of air (as in a bicycle pump) raises its temperature, whilst decreasing the pressure (as at the valve of a gas cylinder or car tyre) lowers it. Since in climatology it is the change in *temperature* of the moving parcel that we wish to determine—for from that we can infer its relative density as compared with that of the environmental air at the same pressure level—we can dispense with the fractional power of $P$ as long as we are given the denominators in their correct proportions. These are stated in Table 24 where the value for the 1000 millibars level has been given the abstract index of 1000 and the other values have been reduced by slide rule accordingly.

TABLE 24. *Indices for making dry adiabatic lapse rate conversions to selected pressure levels*

| (mb) | | (mb) | |
|------|------|------|------|
| 25 | 345 | 550 | 842 |
| 50 | 422 | 600 | 863 |
| 75 | 474 | 650 | 883 |
| 100 | 515 | 700 | 903 |
| 150 | 579 | 750 | 921 |
| 200 | 628 | 800 | 938 |
| 250 | 671 | 850 | 954 |
| 300 | 707 | 900 | 970 |
| 350 | 738 | 950 | 985 |
| 400 | 768 | *1000* | *1000* |
| 450 | 795 | 1050 | 1015 |
| 500 | 819 | | |

Thus, selecting three parcels of air from the lower, middle and upper parts of the standard atmosphere and lowering each dry-adiabatically by 200 millibars, we find that air of temperature 2°C (275°K) at 800 millibars will be warmed to 1000/938 × 275 = 293°K (20°C) by subsidence to the 1000-millibar level; that air of −21°C (252°K) at 500 millibars will be warmed to 903/819 × 252 = 278°K (5°C) by subsidence to the 700-millibar level; and that air of −55°C (218°K) at 200 millibars will be warmed to 768/628 × 218 = 266°K (−7°C) if brought down to the 400-millibar horizon. Over the uniform pressure range used in each of these examples the increase of temperature is thus 18, 26 and 48°C, giving rates of increase per 10 millibars of 0.9, 1.3 and 2.4°C respectively. If the exercise had been repeated for corresponding pressure levels in the cold and warm air masses shown it would emerge that the *dry adiabatic lapse rate* (DALR), as measured against pressure, is fractionally higher in the warm air than in the cold.

Now it so happens that, to a close approximation, pressure falls off logarithmically with height, furthermore, as Tables 22 and 23 clearly show, it falls off more slowly in warm air than in cold. Hence the dry adiabatic lapse rate as measured against *height* is found to be almost constant at 1°C/100 m (5.4°F/1000 ft) throughout the whole range of heights and temperatures usually found on earth. It is thus normally represented by a series of parallel straight lines on most temperature-height diagrams. It is this simple characteristic that is usually emphasised in most elementary accounts. We have deliberately adopted what may seem at first to be a more tortuous approach to underline the fact that the change of temperature is indeed a response to the change of *pressure* and that elevation above sea-level is no more than incidental.

Note also that in our examples we have spoken so far in terms of *subsidence*, though the relation between temperature and pressure is entirely reversible and the lapse rate applies equally in both directions, upwards as well as downwards. We have deliberately adopted this approach because the assumption of adiabatic change has greater validity when applied to downward movements within the atmosphere than it has to updraughts. Subsidence is usually massive, general and slow whilst uplift is often (though not invariably) local, penetrative and rapid. Furthermore uplift is often associated with precipitation whilst subsidence is not. Of the three conditions (p. 233) essential for adiabatic transformation of air masses, subsiding air is thus likely to satisfy (1) and (3) though the time factor (2) may have to be watched since the heat gained may be radiated away. Even so, heat lost in this manner, since it increases the density of the air, further encourages, rather than checks, the subsiding motion.

It is quite otherwise with rising air. Uplift often occurs in favoured localities, warmer 'bubbles' penetrating a cooler and hence denser environment and inevitably becoming mixed with it by a process now described as 'entrainment'. If the bubble is large and the process rapid the adiabatic lapse rate may still give a broad approximation to the facts but in this instance radiative losses add yet another brake to the process. As we shall see, rising *dry air* is not of much meteorological importance.

It should be clear that if, starting with our hypothetical cold and warm columns of air, we now try to visualise the thermal effects of an induced circulation between them, the subsidence within the bulk of the cold column will involve general warming whilst the relative uplift of the bulk of the warm column will involve general cooling. These effects will be nowhere more marked than within the stratosphere. Hence the observed fact that low stratospheric temperatures overlie warm and rising air and relatively high stratospheric temperatures are found above cold and subsiding air.

In general meteorological practice interest in adiabatic temperature change has a much more limited objective. It arises particularly when it is desired to test any given environmental stratification for its *internal stability*. In most incompressible fluids density is directly proportionate to temperature; their internal stability can thus be tested by a direct comparison of observed temperatures at different levels. Thus for water (above $4°C$) a stratification of warm over cold is *stable*, it resists change since the denser fluid occupies the lower position. But if a sheet of cooler water overlies warmer water—as when a lake is being chilled at the surface by radiation, evaporation or contact with cold air—the stratification is potentially *unstable*, since density then decreases downwards. Such conditions may persist for a time owing to inertia or viscosity but the situation is precarious and a minor disturbance is then sufficient to trigger off a process of widespread overturning, often generating a multicellular circulation. The gases of the atmosphere, however, are eminently compressible and in such circumstances both density and temperature are, at least in part, functions of pressure. The test for internal stability in the atmosphere must therefore be more elaborate; it is found in a comparison of 'potential temperatures'. These are values derived from observed temperature by a process of dry-adiabatic transformation, the object of the exercise being to eliminate the effect of the pressure-differential implicit in any difference in height between the layers under comparison.

The nature of the process is simply illustrated in broad terms in Fig. 72. *AB* and *CD* represent two given sections of environment curves between any selected heights $yz$ and $xz$ respectively and the parallel lines represent the DALR. It is indicated that height increases towards the top of the diagram and temperature towards the right. It is required to test the internal stability of the two air columns from *A* to *B* and from *C* to *D*. This can be done either by imagining parcels from *A* and *C* being warmed dry-adiabatically by descent to $A_1$ and $C_1$, or by imagining parcels from *B* and *D* being cooled dry-adiabatically by ascent to $B_1$ and $D_1$. Valid comparison of temperatures, in other words, is only possible *at the same pressure* and conversions to this pressure are made along the dry-adiabats. This being done it is evident that, whether we make the comparison at levels $y$ or $z$, *B* is potentially warmer, and hence relatively less dense, than *A*; whilst whether we consider levels $x$ or $z$, *D* is potentially cooler and hence relatively more dense, than *C*. The stratification between $A$ and $B$ is thus *unstable* and the stratification between $C$ and $D$ is *stable*. An alternative method of viewing this contrast which may make its importance more evident is to imagine parcels of air at $B$ and $D$ thrust upward some slight distance through the environment by an external force. The parcel from $D$ will cool at the DALR and thus find itself at once cooler and therefore denser than the environmental air at its new level, it will therefore tend to subside again as soon as the external force is released. A parcel from $B$, on the other hand, cooling at the same rate, will find itself at once warmer than the environmental air at its new level and will be buoyed up by that denser environment towards $B_1$ and beyond, its upward acceleration con-

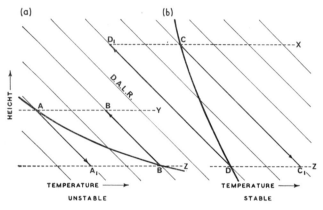

FIG. 72. The meaning of potential temperature

tinuing long after the initial disturbing force has ceased to operate. Indeed, the parcel from $B$, once triggered off, will not reach equilibrium with its new environment until it reaches some point, possibly far above level $y$, where the dry adiabat $B B_1$ is intersected by the environment curve. The presence of the stratosphere aloft ensures that this will ultimately occur. However, we must not be too dogmatic about the new point of equilibrium for two conflicting reasons. In the first place, air which has gained momentum is not readily stopped, the brakes take time to act. For this reason it is highly probable that an active air parcel will at first overrun the point of equilibrium, slow down, and then subside towards it again. On the other hand, adiabatic cooling is no more than a convenient assumption, in actual fact some considerable degree of mixing with the environment is inevitable, the buoyancy of the parcel will suffer in consequence and equilibrium may therefore be attained at levels well below that indicated by the adiabatic diagram. Which of these two considerations will bear the greatest weight will depend upon a variety of circumstances.

A grasp of the processes affecting and affected by the internal stability of air mass stratification is absolutely necessary to an understanding of climatic distributions so that the potential temperature of the various layers of the atmosphere describes a feature of the greatest significance at all levels. This comparison is simplified if, instead of making specific estimates at selected levels as in the above example, we read off potential temperature at some standard level. The 1000-millibar surface has gained international recognition for this purpose though for studies over lofty plateaus a higher surface may have local advantages. The *potential temperature* of a sample of air is thus formally defined as the temperature it would have if brought dry-adiabatically up or down to the standard pressure of 1000 millibars. It can be calculated readily by means of the indices given in Table 24, interpolating linearly when necessary, and is usually quoted on the Kelvin scale.

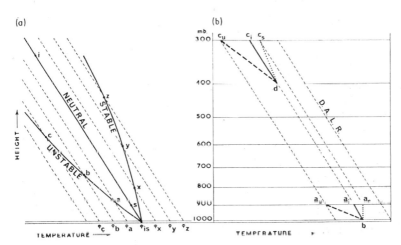

Fig. 73. (a) Stable and unstable air stratification as revealed by potential temperatures at various levels. (b) A diagrammatic illustration of Byers's Rule

Figure 73(a) illustrates the process diagrammatically. Three types of environment curve are shown. The central curve *is* coincides at all points with dry adiabat, its potential temperature at all levels is thus constant and is given by the value on the temperature scale where *is* intersects the 1000-millibar level—point $\varphi_{is}$. This column is thus in neutral equilibrium, a displaced parcel will find itself throughout the whole of its upward or downward path at the same potential temperature as its environment; its density will therefore equal that of the environment and no buoyancy forces, positive or negative, will be operating on it. Such transformations as it may undergo are known to the physicist as 'isentropic'. For the left-hand curve, the potential temperatures at points *a*, *b*, and *c* are given at $\varphi_a$, $\varphi_b$ and $\varphi_c$ so that potential temperature is observed to *decrease* with height; it should be clear from the above discussion that this is an *unstable* stratification and hence one unlikely to persist over any great length of time. The great majority of cross-sections through the atmosphere show a greater resemblance to the right-hand curve *xyz* where potential

9

temperature increases with height for this is a *stable* stratification such as we would expect to find in the atmosphere at rest. Using the data for the standard atmosphere illustrated in Fig. 71 we find that the increase in potential temperature between the bottom and the top of the third decasphere (811–709 mb) is 3.5°C and between the bottom and top of the sixth decasphere (507–405 mb) 7.6°C. The tropospheric portion of the standard atmosphere is thus not strongly stable.

In marked contrast to these values is the difference in potential temperature between the bottom and top of the ninth decasphere (203–101 mb) which is of the order of 48°C. The stratosphere thus presents a very high degree of stability and hence of resistance to vertical exchange. It forms an extremely effective ceiling to all convectional processes.

There is yet another aspect of the effects of uplift and subsidence which is worth mentioning here though its more remarkable features arise from variations in the water vapour content of the air which will be discussed later. We refer to changes in the stability pattern *within* widespread sheets or layers of the atmosphere as a result of uplift or subsidence en masse on a regional scale. We have already seen that the decaspheres, or any given fraction of them, increase in vertical thickness upwards whilst the dry adiabats may be represented by a series of parallel straight lines. We are able to show in Fig. 73(b) therefore, purely by geometrical means, what happens to the internal stability of a layer if, by some means, it suffers wholesale displacement within the atmosphere. The example employed, where we have imagined the 900- to 1000-millibar layer to have been lifted up bodily to 300 to 400 millibars, is of course a very extreme and highly improbable one, but it is the principle involved that is important. Three different internal lapse rates between 900 and 1000 millibars are represented, and the figure shows the transformations they undergo when the $a$ level has been lifted to $c$ and the $b$ level to $d$. The layer has thus increased in thickness (in feet or metres) but it still contains 100 millibars of air. In the isentropic case $a_i b$ no change in lapse rate occurs. In the other two cases the potential temperatures at all points within the layer remain unchanged and these will provide a kind of hallmark by which the layer might be recognised in its new surroundings. The stable case $a_s b$ is thus still stable at $c_s d$ and the unstable case $a_u b$ is still unstable at $c_u d$, yet the rates of change of temperature per unit height clearly approach more nearly to the dry adiabatic between 300 and 400 millibars than they did at 900 to 1000 mb. Since the diagram is equally applicable in the reverse direction we have Byers's rule (1959, 155) that 'lifting causes the lapse rate to approach the dry-adiabatic and subsidence causes it to depart more and more from the adiabatic'. He goes on to point out that divergent air flow which will have the effect of spreading any given atmospheric layer over a wider area, thus decreasing its thickness, will have an effect similar to subsidence whilst convergent air flow, decreasing the area and thus increasing the thickness, will have an effect analogous to uplift. The significance of these transformations will become clearer later. Let us note finally before leaving Fig. 73(b) that, in nature, the unstable case represented by $a_u b$ and $c_u d$ is virtually impossible since any disturbance of the layer will certainly release its internal instability. Long before it reaches 300 to 400 mb its various internal layers will have become thoroughly intermixed and its ultimate cross-section is much more likely to be isentropic, i.e. parallel to $c_i d$ and situated to the left of it. There is much less reason why this should happen to the stable case since the upper layers near $a_s$ should remain less dense than those near $b$ throughout the process, however, if some measure of mixing is imposed dynamically during the uplift

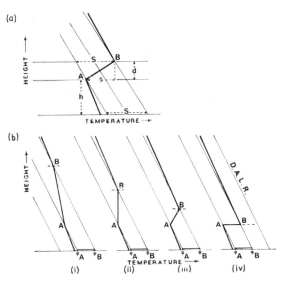

FIG. 74. 'Inversions' of temperature and their specification

its effect will be to turn $c_s d$ more closely parallel to $c_i d$. Some degree of internal mixing will thus strengthen Byers's rule for uplift and convergence. During subsidence and divergence—when indeed mixing is rather less probable—its operation will be opposed to the rule. Since, speaking in the broadest terms, the atmosphere is usually at least slightly stable we may summarise this discussion with the broad generalisation that, even for dry air; *Uplift and convergent air-flow tend to decrease stability; whilst subsidence and divergent air-flow tend to increase stability.* This rule is of the utmost importance, particularly since the effects we have just discussed are heavily reinforced by the part played by water vapour in the atmosphere.

In conclusion, it is appropriate to mention so-called 'inversions' though their visible effects are shown largely via cloud formations. When first encountered they were given this rather inappropriate name since they were recognised as usually rather narrow layers within the atmosphere where the temperature actually *increased* with height, thus providing an example of the inversion of the general rule. Shaw (*Manual* II, 1936, xxxi, xli) preferred the term 'counter-lapse' but it has not proved popular. In such a layer potential temperature will also increase with height, usually at a more rapid rate than the increase in actual temperature. As soon as this is said it is evident that inversions are no more than extreme cases of a whole family of situations involving abnormal stability.

This point may be clarified by Fig. 74. Fig. 74(a) illustrates an inversion $AB$ where $h$ is the height of the inversion base and $d$ is the depth or thickness of the inversion layer. If we ask, 'Is the inversion strong or weak or how does one measure its scale?' the true answer with respect to its stability is clearly not $s$, the difference in temperature between $B$ and $A$, but $S$, the difference between the *potential* temperatures of these two points. Fig. 74(b) carries the argument a stage further by showing four cases in which $S$ is equal. As far as convection above $B$ is concerned each of these cases will offer an equally effective barrier, the only real difference is in $d$, the depth or vertical distance over which the inversion is spread. Yet in (i) $s$ is negative and in (ii) it is zero, i.e. the layer $AB$ is isothermal. Only in (iv), an improbable case where $d$ is zero, are $s$ and $S$ equal in magnitude. Since a condition where potential temperature increases with height is the normal state of much of the atmosphere it can be argued that weak inversions are almost always present. They become notable when, for some reason or another, the increase is concentrated in one or more zones within the atmosphere.

The most widely developed inversion occurs at the tropopause and we have already mentioned some of its effects. Other inversions occur with remarkable persistence over large areas within the trade winds. Much nearer ground level they are also common in Arctic air, chilled by contact with a very cold snow surface. Yet others are found locally and temporarily in a variety of situations. Rarely can they be ignored if the climate at the earth's surface is to be properly understood. Note particularly that our general rule about stability again applies. It can be restated as follows: *Uplift and convergence tend to liquidate and disperse inversions; subsidence and divergence strengthen and sharpen them.*

There is but one noteworthy exception to this rule. It occurs when the inversion is not generated within a single air mass but is itself the product of the convergence of two air masses, the warmer overriding the cooler. Such a 'frontal inversion' is indeed a very special case.

## Uplift and subsidence in air masses: moist air

It would not be inaccurate to describe the above discussion as largely academic since, as is well known, even over the driest deserts of the world, the air contains a considerable quantity of water vapour, the presence of which we have so far ignored. What then is the effect of the addition of this gas to the mixture of nitrogen, oxygen, argon and carbon-dioxide of which dry air is composed?

The effect is twofold. One result is of minor importance though it is universal in operation, the other is only local and temporary in incidence but of the very greatest significance.

The molecular weight of water vapour is 18 whilst the value appropriate to dry air is 28.9. It follows therefore that, at the same pressure and temperature, the ratio of their densities is 18 : 28.9 or, to a very close approximation, 5 : 8. In a mixture then in which 1 molecule in 10 consisted of water vapour (i.e. one in which the vapour pressure of water comprised 10 per cent of the total pressure) the density would be $(9 \times 8) + (1 \times 5) = 77$ units as compared with 80 units for dry air at the same total pressure. Such a mixture would thus be 3.75 per cent lighter than dry air. This is an extreme case unlikely to occur in nature. A saturation vapour pressure of 101.3 millibars (one-tenth of standard pressure at sea-level) is not attained until the temperature reaches $46.1^\circ C$ $(115^\circ F)$ and though such temperatures are not unknown upon the surface of the earth it seems most unlikely that they have ever occurred in fully saturated air. The universal effect of the presence of water vapour is thus to lower the density of the air, but the result is inconsiderable—at $30^\circ C$ saturated air is only 1.7 per cent lighter than completely dry air, at $15^\circ C$ it is 0.57 per cent lighter and at $0^\circ C$ the difference is about 0.21 per cent, or so small that it almost defies computation by slide rule. It is the universal character of this result that helps to make it of so slight importance meteorologically. We can say with confidence that, in nature, completely saturated air is never confronted with (i.e. brought into meteorological relation with) completely dry air. Differences in density stemming from this cause are thus likely to be very much smaller than the figures quoted above and it is *difference* of density that is meteorologically effective. The meteorologist copes with this situation by giving his air samples a *virtual temperature*, i.e. a fictitious temperature a few degrees higher than the true value to allow for the lowered density, but this is a refinement, like his allowance for changes in the force of gravity, which we can admire and then proceed to ignore. (A table of 'Virtual temperature increment of saturated air' is given in *Smithsonian Meteorological Tables*, 298–301, Table 72.)

The second effect of the presence of water vapour stems from the well-known fact that, at atmospheric temperatures, water undergoes notable changes in molecular arrangement, 'changes of state' from gaseous to liquid, liquid to solid, and in the reverse direction. It is a commonplace of elementary physics that, during these transformations, large quantities of heat are either liberated or put into storage in 'latent' form according to the direction of the change (see p. 89). Water in its liquid, and especially in its vaporous form, thus plays the part of a deposit account in the heat bank of the earth's surface and its overlying atmosphere. Deposits are paid in by the levies made over wide areas by evaporation and the summer thaw whilst withdrawals are located wherever there is condensation or frost. The deposit account facilitates distant transfer since latent heat is not subject to the steady losses involved in radiative processes. Now, whilst condensation and freezing may well be stimulated by horizontal motion of the air from the tropics towards the poles, both are much more readily set in train by upward motion within the atmosphere. It is therefore within the great bulk of the troposphere that the release of latent heat most readily occurs and this has a major effect on its stratification and on its relative stability.

Notice that this is, to a very large degree, a one way effect. We can no longer speak of uplift and subsidence in the same breath with the implication that the result described is reversible. Subsiding air can only invest in the latent heat market if it contains cloud elements, i.e. finely divided particles of liquid water or ice held in suspension during subsidence and hence available for evaporation. It is doubtful if the total load of such material can ever exceed 2 g/m³ in air which is not actively rising; often it will be very much less. When this is exhausted the subsiding air will experience an increase of sensible heat, that is, a rise in temperature, at a rate virtually identical with that of dry air—its path curve will be along the DALR. For rising air before condensation sets in the same rule applies and up to this level the process is therefore reversible and the rules stated in the previous section remain valid. Thereafter the 'saturation adiabatic lapse rate' (SALR) takes over and if precipitation occurs this is a path along which there is no return. Note that when air is referred to as 'dry' in this context it means no more than that the air is not at saturation. This is a very different thing from the completely vapourless air we have been referring to above, but its adiabatic behaviour is virtually the same.

Since the saturation adiabatic lapse rate differs from the DALR owing to the release of latent heat at condensation, the contrast between them will clearly depend upon the amount of condensation promoted by each unit fall of pressure. We shall therefore expect this contrast to reflect differences in the moisture available; it is thus likely to be greater in warm air than in cold air and greater in the lower and middle troposphere than at extreme heights. A *variable* rate of change is therefore indicated and much computation work is saved by the use of one of the various types of diagrams now devised to ease this work.

Some idea of the scale of the contrast between the SALR and DALR can be obtained if we imagine that fully saturated parcels are taken from the 1000-, 700- and 400-millibar levels of an environment with a temperature stratification identical with that of the standard atmosphere and that each of these parcels is *raised* along the saturation adiabat by 200 millibars. The *fall* in temperature involved will then be 9.5, 15.5 and 42°C respectively, giving rates of change per 10 millibars comparable to those quoted on page 234[1] of 0.47, 0.78 and 2.10°C. In this case then the variation in lapse rate per unit change of pressure is much too wide to vanish when the scale is expressed in unit height. For the first 2 kilometres of ascent above sea level the saturated adiabatic lapse rate is thus about 0.6°C/100 m when the ground temperature is 0°C; 0.47°C/100 m when ground temperature is 15°C; and only 0.34°C/

[1] Where *descent* over the same ranges gave an *increase* of temperature at the rate per 10 millibars of 0.90, 1.30 and 2.40°C respectively.

100 m when ground temperature is as high as 30° C. Expressed in Fahrenheit degrees per 1000 feet these rates are 3.6, 2.7 and 2.0° F respectively. At higher levels all these rates gradually approach the DALR and beyond the level where temperatures of about −40° F (−40° C) are attained there is virtually no difference between the SALR and DALR.

It is clear therefore that if the dry adiabats are conveniently represented by nearly parallel straight lines, the saturated adiabats must be shown by a series

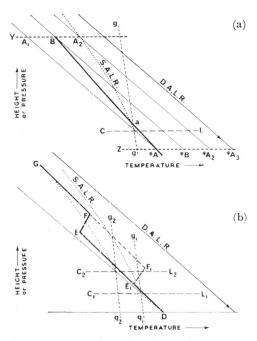

FIG. 75. The effect of the release of latent heat at condensation on (a) the stability of a normal air mass, and (b) the effectiveness of an inversion at different levels

of subparallel curves, steepening their gradient as they approach lower levels and each converging asymptotically upon an appropriate dry adiabat in the high atmosphere. This feature is illustrated in Fig. 76 to which we shall again return.

Since water vapour is normally present in greatest quantity in the lower layers of the atmosphere (temperature is higher there and the air is in closer proximity to evaporating surfaces) and since the condensation level is rarely more than a few thousand feet above the earth's surface, it is evident that the release of latent heat at condensation is greatest in the two lowest decaspheres. This has a profound effect on air stratification and particularly upon the internal stability of the air column. Water vapour has thus been described as acting as a potent 'fuel' to the atmospheric engine. To measure its contribution we therefore require some modification of the concept of potential temperature which takes account of this hidden source of heat.

Figure 75(a) illustrates this point diagrammatically in general terms. The environment curve $AB$ represents what is, on the face of it, a stable stratification (compare with Fig. 72(b)) since the potential temperature of $B$ measured at level $z$ (the level of $A$) is given by $\varphi B$, a value higher than $A$. Now suppose the air at $A$ contains $g$ grams per kilogram of precipitable water. We must add to the diagram the dew-point or mixing-ratio line $gg$ which shows at what combination of pressures and temperatures a kilogram of dry air will be just saturated by $g$ grams of water. To test for stability by the rising parcel method we must draw the path curve from $A$ along the dry adiabat $AA_1$ until this intersects $gg$ where saturation will be reached and condensation will begin. Thereafter a saturation adiabat must be followed and at level $y$ (the level of $B$), the parcel will find itself at $A_2$. The distance $A_1A_2$ measured on the temperature scale thus indicates the contribution to sensible heat made as a result of the condensation which has occurred between the condensation level $CL$ and level $y$. Note that this is enough to neutralise the initial potential-temperature difference between $B$ and $A$ and to provide an excess $A_2 - B$ (or $\varphi A_2 - \varphi B$) which will give the parcel additional buoyancy. Indeed, ever since the path curve crossed the environment curve at $a$ the parcel has been warmer than the air around it and has ceased to require any aid in its ascent from the exertion of an external force. It is evident therefore that the atmosphere represented by $AB$ is *potentially* far from stable, it is *conditionally unstable*, the requisite condition being that the inherent stability below point $a$ is reduced and finally overcome. This can be achieved in a variety of ways; by forced uplift over a relief obstacle, by the addition of either heat or of moisture to the lower layers near $A$, or by regional convergence within the atmosphere which would deepen the layer $yz$ and thus lower its initial stability. A situation analogous to that here depicted is by far the most frequent condition of the atmosphere as we find it over large areas of the earth's surface. This is one of the reasons why weather, particularly heavy cloud and precipitation, is episodic. All over the world rainfall is *intermittent;* the great regional contrasts in climate arise simply from the fact that in some places it is much more intermittent than in others. Hence the great importance of a number of 'trigger' mechanisms capable of setting weather processes in motion.

Whilst the simple rule deduced for dry air that *any* increase of potential temperature with height indicates a stable stratification thus applies equally to the immediate condition of a mass of moist air, it must require some amendment if the contingency of condensation within the displaced parcel is to be covered. To take account of the release of latent heat the desired amendment must be of the form that the increase of potential temperature with height must be *in excess of some limiting amount* which will depend on the vapour content of the parcel most likely to suffer initial displacement and the height to which it is likely to ascend. These factors cannot be covered by any simple rule, yet the stipulation necessary for the complete absence of any degree of conditional instability is readily appreciated from the diagram. It is that the plotted ascent from $A$ must lie entirely to the right of $AA_2$, or, in other words, that the environment curve must nowhere intersect the saturation adiabat appropriate to the temperature and humidity conditions prevailing at $A$. Expressed in terms of lapse-rates this means that above the condensation

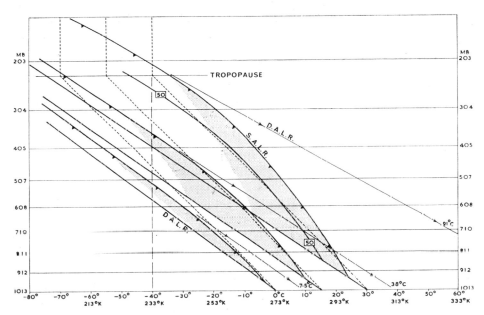

Fig. 76. Dry and saturation adiabatics for air of relative humidity 75 per cent and surface temperatures of 0°C, 15°C, and 30°C. The areas between the respective pairs of curves have been shaded only in the lower sectors to avoid overlap. The broken lines are environment curves derived from the three model atmospheres of Fig. 71. Note how the saturation adiabats approach a straight line with increasing height, particularly at temperatures below −40°C. The asymptote of each curve is thus a specific DALR line which has been indicated by an arrow leading down to the temperature scale at 1000 mb. The figures given thus express the 'equivalent potential temperature' of the surface air. A saturation adiabat for surface air of 30°C and relative humidity of only 50 per cent is also shown. It may be noted that in a temperature-height diagram covering such a wide range of temperatures, the DALR lines are not precisely parallel

level, the environmental lapse rate must be consistently less than the appropriate SALR, a feature common only in subsiding desert or arctic air.

These points are further illustrated in Fig. 76 where the broken lines are the environment curves for the standard atmosphere and for the rather artificial examples of 'cold' and 'warm' atmospheres that we have already derived from it. From the three points with surface temperatures of 0, 15 and 30° C we have drawn straight lines representing the DALR and added the curved saturation adiabats appropriate to a surface relative humidity of 75 per cent in each case. It will be evident that although (by definition) the environmental lapse rate is identical in all three atmospheres, their inherent stability with respect to air impelled upwards from the surface is by no means the same. In the 'cold' air the environment curve lies entirely to the right of the SALR at all levels. A sea-level sample of air with RH 75 per cent would thus remain cooler and hence denser than the surrounding air at all levels of a forced ascent. The cold air column is thus definitely stable and, even if surface RH rose to nearly 100 per cent, its instability would still be so slight as to be difficult to demonstrate on a diagram of this type. In the standard air there is also stability in the lower two decaspheres but from about 1500 to 6000 metres (c. 5000 to 20000 ft) the path curve for surface air of RH 75 per cent lies slightly to the right of the environment curve. Over this range, therefore, a surface sample, having overcome the initial resistance, would find itself buoyed up by a denser environment. It is clear, however, that the temperature difference between parcel and environment is never more than about 1°C, a contrast so slight that it could be easily eliminated by the slightest departure from the assumptions implicit in adiabatic theory. The balance here is delicate and an addition of water vapour to the surface air would materially increase the conditional instability though the ceiling for convection is unlikely to exceed 25000 feet.

Finally, in the 'warm' air a marked degree of conditional instability is self-evident. From about 1200 metres (c. 4000 ft) to beyond the (presumed) tropopause the humid parcel would find itself warmer than the surrounding air, the contrast attaining from 8–10°C in the sixth and seventh decaspheres. Here, although the stability of the lower layers still has to be overcome, there is a large reserve of potential energy aloft which is unlikely to be dissipated by non-adiabatic processes. If the surface relative humidity were as low as 50 per cent this reserve would virtually disappear but as values are increased above that figure the level at which breakthrough occurs is lowered and the potential energy available increases as the area enclosed by the section of the saturated adiabat lying to the right of the environment curve is enlarged. In Shaw's 'tephigram' and in numerous variants of it which have been elaborated by meteorologists, the scale of temperature and pressure are manipulated in such a way as to make the potentially available energy directly proportionate to this area.

It must be emphasised that the degree of conditional instability exhibited by an ascent is a measure only of the liability of the atmosphere to produce convectional overturning. There is nothing inevitable about this outcome. Indeed Watts (1955, 22–34) insists that 'because the vertical structure of the equatorial atmosphere nearly always favours shower development' predictions based on vertical charts are 'unreliable'. This reminds us of Shaw's statement (*Manual*, III, 1930, 285) that 'we are still a long way from understanding the convexion of warm air in the atmosphere though it has been invoked for centuries as a process which is commonly understood'. It is thus worth while to enumerate some of the obstacles that are placed in the path of convective exchange.

1. It is not entirely a simple matter to detach air in any quantity from a horizontal surface which is warming it. This is partly a result of its natural viscosity (a consequence of molecular vibration) which is remarkably high in proportion to its low density. It is also because air will only rise if it can be replaced. Expressed more precisely, the lift against gravity must arise from buoyancy which presupposes the existence, at an effective distance, of cooler and denser air able to undertake this work. This process is most efficient if the cooler air is able to move in horizontally like a shovel thus organising the convective process along a line or moving disturbance. In doing so it will have to overcome surface friction and hence the importance of the phrase 'at an effective distance'. It seems most probable that instability is often not realised in equatorial regions because no such source of cool air is available within effective range. Similarly over extensive desert flats the occurrence of mirage effects during the heat of the day shows that overheated air is being retained in close proximity to the surface. On hill slopes, where cooler air at the same elevation in the free atmosphere is never far away, this brake upon convective activity is much less effective. Hence the frequent association of convection with hill summits.

2. An alternative source of cooler air is, of course, aloft but such air will be warmed during descent just as the surface air will be cooled as it rises. Turbulent mixing must also occur. Hence, as Shaw insists (*Manual*, III, 1930, 304), the first effect of surface heating is to produce a layer of air in convective equilibrium, i.e. with a lapse rate equal to the DALR, which grows in thickness as the heating continues and as surface air is fed towards its top. At first glance this may seem to be no more than a description of the process by which the low-level stability of a conditionally unstable atmosphere is gradually eroded away but its significance goes deeper than that. In the first place, the build-up takes time and the day may begin to draw to a close before any real weather activity has been set in train. The delay thus contributes to the intermittent character of weather which we have already noted. Secondly, the surface where activity is to be looked for is now the *top* of the convective layer. If this is near the condensation level clouds may form as a product of pure chance and without any relation at all to surface inequalities. This appears to be the explanation of the characteristic clouds of the trade winds (Fig. 64). Thirdly, if a major breakthrough does occur, it is not a thin skin of surface air that provides the 'parcel' but a layer perhaps some hundreds of feet in thickness, a formidable reservoir of energy for the weather processes which ensue.

3. Finally, vigorous convection requires that a considerable volume of air must reach saturation. Saturated air is comparatively rare in any quantity near the earth's surface partly because of the mixing process just described. It is much more likely to be found near the top of a thick convective layer which takes time to build up. What happens there depends on the lapse rate

of the environment aloft but though in the case of the ascent of a single parcel or 'bubble' we may safely assume that this lapse rate remains unchanged, we certainly cannot do so if the bubbles become numerous. If the process is purely convective, for every cubic mile of air that rises, a cubic mile of the environment must somewhere descend, the rising air cooling at the SALR and the descending air warming much more rapidly at the DALR. Quite apart from the complexities introduced by entrainment and mixing, a heavy thermo-dynamic brake is thus soon clamped down on the system. Indeed it now seems quite clear that really vigorous 'convective' disturbances, such as large

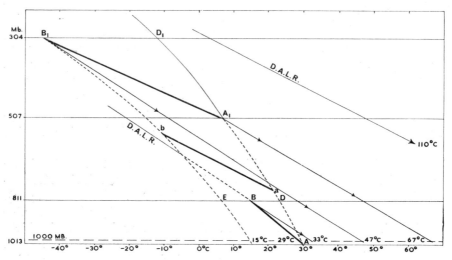

Fig. 77. Layer lability or 'convectional instability' in a compound series of air strata with relatively dry air aloft

thunderstorm systems and tropical hurricanes, cannot be maintained in the absence of a powerful wind-field aloft capable of sweeping away surplus air to some other part of the world and thus preventing it, quite literally, from 'gumming up the works'!

We have been concerned above with a rising parcel of moist air penetrating an environment that has been assumed to remain substantially unchanged, what now of the internal structure of a rising compound *layer* of the atmosphere containing differing quantities of water vapour at various levels within it? We saw on page 238 that general uplift of dry air brings its lapse rate closer to the DALR; the question now arises, what are the consequences if differing saturated adiabatic lapse rates are brought into play? Figure 77 illustrates a very extreme case. Let *AB* represent the environment curve of the two lower decaspheres of a warm atmosphere where the relative humidity at *A* is 100 per cent whilst at *B* it is only 30 per cent. The layers are initially stable since potential temperature increases with height ($\varphi_B = 33°$ C, $\varphi_A = 29°$ C). Now if these layers are lifted bodily to the 304 to 507-millibar level the trans-formations that would take place in the air initially at $A$ and $B$ are illustrated by the broken lines. The air at $A$, already saturated, will follow a saturated adiabat throughout to attain $A_1$, whilst the relatively dry air at $B$ will cool at the DALR until it reaches saturation at about the 630-millibar level and at the SALR thereafter to reach point $B_1$. The new environment curve of the layers will thus be represented by $A_1B_1$. This stratification is far from stable since potential temperature now *decreases* with height from 67°C at $A_1$ to 47°C at $B_1$. In addition to the precipitation inevitable as a direct result of uplift described we can therefore expect widespread overturning *within* the displaced layer. This may be triggered off more or less simultaneously over many tens of thousands of square miles. The condition is known as one of '*convective* or *layer instability*' and is particularly common in the south-central lowlands of the United States, where it is often associated with extensive rains and the development of families of tornadoes. Note that, given high tem-peratures and a marked contrast in relative humidity, it is not necessary for the degree of uplift to be anything like as wide as the range here postulated for the trigger action to take effect. In our present example the layer begins to become convectively unstable after rising only about 100 millibars and, with a lift of approximately 250 millibars (i.e. a rise of the base of the layer by about 2.4 km or 7750 ft) the position *ab* is attained where potential temperature decreases with height by 8°C. Clearly long before $A_1B_1$ is attained convectional over-turning within the layer thus forced upwards is likely to have been very vigorous indeed.

   Although convectional instability may thus appear to be a rather complex matter, the necessary conditions can be expressed in simple terms analogous to those used for dry air. We saw on page 237 (and Fig. 73(a)) that instability involved a decrease of potential temperature with height. In the present instance, however, we must specify the stratification of both temperature and humidity and to do this simply it is necessary to reduce them to a common term. We have already given some indication of how this can be achieved in Fig. 75(a) where $(A\varphi_2 - \varphi A)$ was seen to represent the heat liberated by con-densation up to level $y$. It is logical to continue upwards along the saturation adiabat until *all* the latent heat of the water vapour in the air sample from $A$ has been realised, that is until the saturation adiabat actually coincides with the dry adiabat to which it is approaching asymptotically. Descent along this dry adiabat to level $z$ at $\varphi A_3$ gives the *equivalent temperature* of the sample of air at $A$. Referred in the same manner to the standard pressure surface of 1000 millibars it is known as the *equivalent-potential temperature* of the sample. Precisely what is meant by this concept becomes clearer in Fig. 76. The surface water vapour content of the three atmospheres at 75 per cent relative humidity is 2.8, 8.0 and 20.5 g/kg respectively. Although condensation level is attained at a slightly greater elevation in the warm air (*c*. 945 mb) than in the cold (*c*. 960 mb) its saturation adiabat thereafter diverges much more rapidly from the dry adiabat owing to this greater water vapour reserve, yet beyond the levels where temperatures fall below $-40°C$ the SALRs and the DALRs become virtually parallel. If we project dry-adiabatically downwards from the points where the saturated adiabats reach this temperature along the arrowed lines we reach the 1000-millibar level at 7.5, 38 and 91°C. These then are the equivalent-potential temperatures of the three air samples. Were we to stop at this point the reader might well feel that he had been carried though a complex manoeuvre to very little purpose, but note the dénouement. These three values exceed the potential temperatures (at 1000 mb, not at sea-level) of the same samples by 8.5, 24 and 62°C, values which are strictly proportionate to the three mixing-ratios quoted above. In brief, we have converted the water vapour content of the three samples into excess potential temperatures at the rate of exchange of about 3°C for each gram of water per kilogram of dry air. (This rate of exchange applies only to the lower layers of the atmosphere. At 600 millibars each gram of water vapour (per kg of dry air) still available for condensation raises the potential temperature by about 3.5°C and at 300 millibars the appropriate value is 4°C. This is because the saturation mixing ratio corresponding to any given temperature increases with decreased air pressure.)

The concept of equivalent potential temperature (EPT) was first designed to produce a 'conservative air-mass property' that is, a tracer-index which remains unchanged during both dry and saturated adiabatic uplift. Hence it is analogous with, but more revealing than, potential temperature which remains constant only during dry-adiabatic transformations. In other words, whereas potential temperature specifies a unique dry-adiabat appropriate to any given point on an environment curve, the EPT specifies an equally appropriate and unique saturated adiabat though it designates that adiabat in what may appear to be an undesirably oblique fashion—i.e. in terms of the dry adiabat which is its asymptote. As soon as this is grasped we are ready to continue the analogy and declare a rule for convective instability.

*An atmosphere is convectively unstable when equivalent-potential temperatures decrease with height.*

This is clearly the case with our very extreme example in Fig. 77 where the EPT of $A$ is shown to be about 110°C whilst that of $B$ is only about 47.5°C (or only a trifle more than the potential temperature of $B_1$ for at that level practically all the vapour from $B$ has condensed out). It is clear moreover that EPT will decrease with height as long as the saturation adiabat appropriate to the upper part of the air stratum under consideration lies to the left of that appropriate to its lower portion. Neutral equilibrium, analogous to $a_1b$ in Fig. 73(b), will not be attained until the upper part of the layer in question not only has the temperature indicated by point $D$ but is also fully saturated at that temperature. Under these conditions EPT would remain constant with height, one and the same saturation adiabat being appropriate to both $A$ and $D$ and the stratification would remain in neutral equilibrium if the layer were lifted to $A_1D_1$.

In the warmer parts of the world the lapse rates of both temperature and humidity are often considerable, hence the stratification not infrequently exhibits a decrease of EPT with height. Even when the decrease is marked, however, this shows no more than that convective instability *may develop*, it gives no assurance that it *will* develop. Indeed, the uplift *en masse* necessary to bring the condition into effective reality appears to be rare in tropical and subtropical regions, the most favoured areas being the lower Mississippi valley, the Parana-Paraguay valley and parts of monsoon Asia. Byers (1959, 192) has pointed out that in the expressions 'conditional instability' and 'convective instability' the term 'instability' carries a different connotation. In the former it describes an actual state of the atmosphere, in the latter only a potential state which may or may not be realised. Even the word 'convective' is a little misleading in this context and perhaps the term 'layer lability'[1] would be more appropriate. We meet here one of the comparatively rare occasions when the English language appears to suffer from a dearth of alternative words.

Byers's interesting note that a convectively unstable layer may be transformed into an actually unstable layer by the evaporation of water into it from rain falling from higher clouds is easily illustrated on Fig. 77. Evaporation would draw heat from the dry air at *B* and as *B* moves towards saturation at *E*, the layer *AB* obviously becomes more unstable.

It may be noted that if EPT values are being used merely as indices of layer lability, the interest being merely in whether they increase or decrease with height, the necessary facts may be expressed in an alternative fashion which has much to recommend it. We have seen that we are really designating saturation adiabats. Why not do so directly by quoting the temperature at which they cut the standard 1000-millibar surface? This would be strictly analogous to the manner in which we designate dry adiabats. The advantages of this method in any graphical treatment of the data should be obvious from Fig. 76, for we are saved the difficulty of locating the appropriate asymptote towards the top of the diagram and of finding its correct designation towards the right. The new values are known as 'wet-bulb potential temperature' or, more neatly, *saturation potential temperatures* (SPT). Some degree of layer lability is thus indicated if SPTs decrease with height. In Fig. 77 therefore we have:

| | EPT (°C) | SPT (°C) |
|---|---|---|
| At point *A* | 110.0 | 29.5 |
| At point *B* | 47.5 | 15.0 |

For the physical explanation of this value the reader is referred to standard meteorological texts. We mention it here as a very handy coordinate.

The relationship between the two indices *for near-surface samples* is the simple one derived on page 250, the EPT exceeding the SPT by some 3°C for every gram per kilogram of the saturation mixing ratio at the SPT as follows:

[1] Labile = prone to undergo displacement or change; unstable.

| *SPT* | *Mixing ratio at saturation* | *EPT* |
|:---:|:---:|:---:|
| (°C) | g/kg | (°C) |
| −30 | 0.25 | −29.2 |
| −20 | 0.7 | −18 |
| −10 | 1.6 | −5 |
| 0 | 3.8 | 11.5 |
| 10 | 7.8 | 33 |
| 20 | 15.0 | 65 |
| 30 | 27.7 | 113 |

This is a convenient point to refer to the relationship between 'dry-bulb', 'wet-bulb', and 'dew-point' temperatures. Clearly comparison between these is valid only at the same pressure, normally that at station level. On an adiabatic diagram like Figs. 75(a) or 75(b) they will be represented therefore by points along either the base line or any line (pressure surface) parallel to it. For any given sample of air the dry-bulb temperature then selects the appropriate DALR line, the wet-bulb temperature defines the appropriate SALR line and the dew-point temperature indicates the appropriate mixing ratio line (*gg*). At saturation, i.e. at the condensation level, all three lines intersect at the same point, in other words the three temperature values are identical. Below condensation level however it is evident that:

$$\text{Dry-bulb } T > \text{Wet-bulb } T > \text{Dew-point } T$$

It remains for us to conclude this section with some consideration of the influence of the water vapour content of the atmosphere upon any inversions of temperature which may happen to be present. The essential facts are illustrated in Fig. 75b. Let DEFG represent an environment curve for an atmosphere which is slightly stable at all levels but contains in addition a marked inversion between $E$ and $F$. Taking a sample of air at $D$ containing $g_1$ grams of water vapour per kilogram and setting it in upward motion, we note that it reaches condensation level at $C_1L_1$, and, cooling thereafter at the saturated rate, it will be warmer than the surrounding air and thus buoyed upwards far beyond the level of $F$, the top of the inversion. The stratification is thus conditionally unstable and comparatively weak trigger actions will allow cloud to puncture the inversion with some ease. If, on the other hand, the sample contains only $g_2$/kg, condensation will not begin until level $C_2L_2$ and the saturated adiabat will then intersect the inversion between $E$ and $F$. Low level convection may still be possible but a definite cloud ceiling is likely to be established below the level of $F$. Should an inversion of precisely the same scale occur at a much lower level, such as $E_1F_1$, it is clear that it will impose a much more effective check upon updraughts. Indeed to achieve a break-through past $F_1$ even fully saturated air from $D$ would have to employ the momentum gained during the first stage of its ascent.

The following important conclusions can thus be drawn. Inversions will be *most effective*, that is, they will be least likely to suffer puncture or dissolution and hence most likely to check cloud build-up and precipitation when:

1. they are at a low level,
2. the air is cold and thus of low vapour content.
3. the air is being chilled near the earth's surface,
4. the air, though warm, is of low relative humidity.

On the contrary, inversions will be *least effective* when:

1. they are comparatively high,
2. the air is warm and thus of higher vapour content,
3. the air is being actively warmed near the surface,
4. the air, though cool, is of very high relative humidity.

These simple rules are of the utmost importance to an appreciation of cloud forms and of the distribution and character of precipitation in time and space. Most of the really violent weather phenomena build up initially beneath an inversion lid, for this prevents the gradual release of incipient instability in a steady trickling process and frees it suddenly and with almost explosive force when the resistance of the inversion is finally overcome.

# 9
# The major air masses of the world

The popular phrase 'as free as the air' is a recognition of the fact that there are no rigid barriers to the movement of air particles from pole to pole or from sea-level to the greatest heights. However fanciful it may appear, there is nothing inherently wrong in the idea that air molecules exhaled in Wigan may be subsequently inhaled in Wanganui. Yet, meteorologically, any such notion is an excellent example of misplaced emphasis. Meteorology is concerned with the *bulk* movement of air and recognises that this meets with notable physical constraints which can be overcome only by the exercise of extremely powerful forces.

A major constraint is imposed by the sheer scale of the atmosphere but this is by no means easy to visualise. We must make a conscious effort to think in three dimensions. This would be assisted if we could evolve a logical *cellular* or boxlike model of the atmosphere, but no obvious scale for the individual cells or boxes emerges from the observed facts. An evident difficulty is that even the layers (decaspheres) we have recognised increase in thickness upwards as pressure is diminished. Three possible scales suggest themselves; (a) the effective depth of the whole atmosphere—about 30 kilometres, (b) the mean height of the tropopause—some 14.5 kilometres, and (c) concentrating attention in the first place on the lower half of the atmosphere, the mean height of the 500-millibar or preferably of the 507-millibar level—about 5.5 kilometres. It may prove helpful to experiment with these. The surface of the earth has a total area of 510 100 000 km² or of 196 900 000 ml². Using scale (a) and first dividing the atmosphere crudely into cubes with this dimension (surface area 900 km²) we find that it consists of no less than 566 500 such cells, each containing some 9 200 million metric tons of air. Little wonder that these gargantuan masses offer vast inertial and frictional resistance to lateral displacement. Using scale (b) and thinking now in terms of hexagonal cells of this depth and width (surface area 182 km² or about 70 ml²) we divide the troposphere into no less than 2 800 000 cells, but this method has the disadvantage that tropopause height is known to vary widely from equator to pole. Scale (c) avoids this difficulty but its dimensions are rather small; however, by employing hexagonal cells which are rather more than three times

($\times$ 3.15)[1] as wide as they are deep (hence 17.3 km across) their surface area can be brought up to 259 km² or a round 100 ml². We thus conclude that there are nearly 2 million (1 969 000) such cells in the lower half of the atmosphere, each containing some 1334 million metric tons of air and overlain by a similar load (Fig. 78).

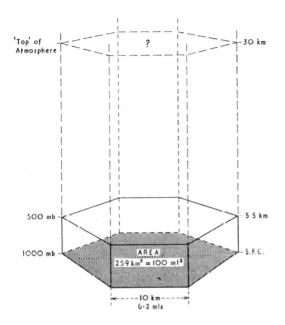

FIG. 78. A perspective sketch with true vertical scale showing an hexagonal column of the atmosphere divided into two sectors of equal mass. If the area of the base is 100 square miles (259 km²) the atmosphere can be regarded as containing nearly two million of such cells each weighing some 2668 million metric tons. Such a cellular concept may help us to appreciate the constraints affecting air-mass behaviour, particularly in the lower sectors. Each such unit of 1334 million tons is bordered by six others of approximately equal mass and capped by a similar load aloft. It derives its characteristics partly from the underlying surface, partly from any converging or diverging motion amongst its neighbours, and partly from the upper air circulation which may change the load aloft

Regarded as atmospheric units, none of these cells is 'free' in any reasonable interpretation of that word. Each cell must be conceived as having acquired, or of being in the process of acquiring, characteristics of temperature and

[1] This appears to be a reasonable proportion—see Shaw, *Manual* III (1930, 306).

humidity related to the condition of the sea, land or ice beneath it. Each is shouldered and jostled by the hydrostatic and dynamic requirements of its neighbours on all sides, and each is bowed beneath the varying load which the

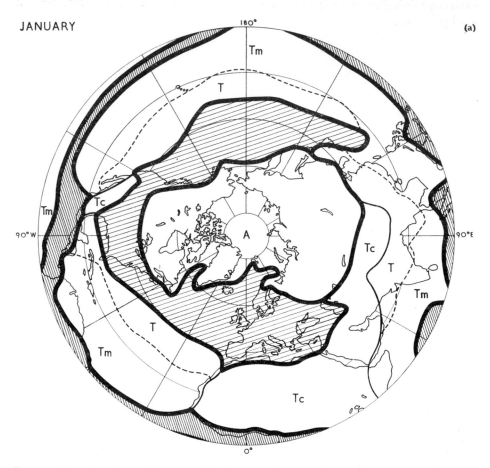

JANUARY    (a)

FIG. 79. The air masses of the world in January. Primary air masses are left blank and secondary air masses are shaded as per key (Fig. 80b). Subdivisions of the tropical masses are shown by the symbols; $T$, maritime sectors with a weak circulation, the 'horse latitudes'; $T_m$, the trade-wind systems; $T_c$, tropical continental air. Evidence is given in the text for a very real distinction between eastern and western trade-wind air but it is not yet possible to show this by a line on a map. 'Mid-latitude air' includes a group of air masses varying widely both in space and time

circulation in the upper half of the atmosphere may see fit to impose on it from above.

A fair case could be made for regarding cells of this order of magnitude as the basic units of air mass. Given information about all of them, plus a working knowledge of the upper echelon, meteorology would be well on the road towards becoming an exact science. Fortunately large areas of the earth's surface are sufficiently uniform in character to impose a strong family likeness upon thousands of adjacent cells; it is thus possible to group them together into a few broad categories or air masses. Note that this involves the introduc-

JANUARY 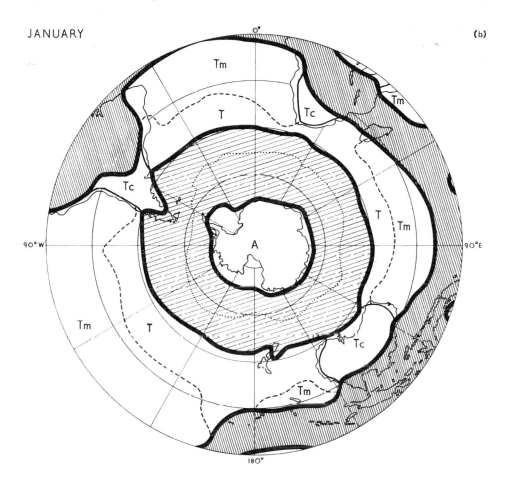 (b)

tion of a *regional* concept into the study of the atmosphere; a step that must be welcomed as giving the study a new coherence and a lively interest in place. This is atmospheric geography in a very real sense and no geographer will be surprised that it has brought with it the usual battery of difficulties—of definition, of nomenclature, of the necessity for subjective judgments, of the treatment of progressive change. Like all other regional entities, air masses are *concepts* of reality, valid only within the criteria employed to delimit them. They can be illustrated only by selected samples; they cannot be portrayed in full. Indeed, as compared with other regional concepts, geomorphological, botanical, economic or social, they offer two special difficulties: (*a*) they are much more subject to change and transformation over comparatively brief periods of time; transformation and regeneration follow one another almost continuously; (*b*) the most exciting events occur at their margins where they are thrown into relationships, the one with the other.

What are the sources of this resemblance, this similarity of structure, that develops between contiguous groups of air cells as they are moved by the general circulation across the face of the earth? Both geographical and

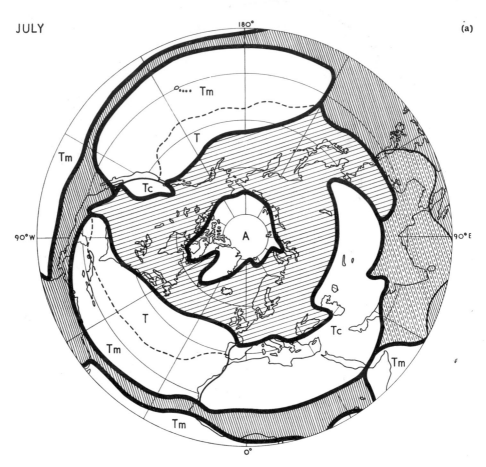

FIG. 80. The air masses of the world in July.

dynamical influences can be recognised, though of course they are not un-related. Great areas of the earth's surface, notably the tropical seas, the temperate seas, the polar caps of ice-clad sea and snow-covered land, the tropical forests, the temperate forests and their cultivated enclaves, the larger steppes and deserts, and the larger and most lofty plateaus, are suffi-ciently uniform in their response to insolation and in their available moisture supply to impose (at least for a time) a degree of widespread uniformity upon the lower layers of the air cells overlying them. Aloft, where freer motion is possible, the linkage with the ground is more remote but the air is under the sway of dynamic forces and responds to regional subsidence and regional uplift. At any one moment of time the localities where these movements occur appear to be dictated by the dynamic pattern, constantly changing, but observations make it clear that each type of motion tends to recur with unusual frequency in preferred areas.

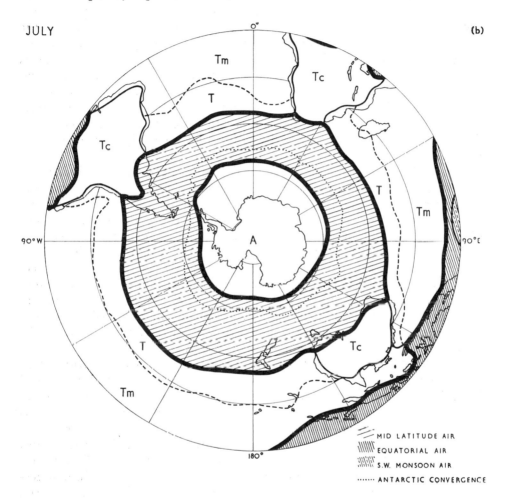

It may appear from the above paragraph that we are in some danger of equating air mass distribution with the earth's major climatic regions as distinguished by Herbertson (1905) and his successors, thus confusing cause with effect. Yet this difficulty was bound to arise since climate is the product of a complex interaction between surfaces and the atmosphere above them. Nevertheless, doubts of this nature have unquestionably raised the issue, which of these surfaces is the most fundamental, which the most clearly causative? Hence the emphasis on a major distinction between tropical and polar air, between air from latitudes with a radiational surplus and air from latitudes with a radiational deficiency. The distinction is brought to its sharpest focus in the contrast between air from over the warmer seas and air from above the polar ice-caps. The notion of a clash between a 'polar current' and an 'equatorial current' goes back to Heinrich Dove's *The Law of Storms* (1862; original in German, 1857), and the zonal concept implied has been inherited from classical Greece. The terms 'tropical' and 'polar' were crystallised into meteorological literature by J. Bjerknes about 1920 but, as the classification has been progressively elaborated, the question has arisen whether this is really as fundamental a distinction as may at first appear. We are entitled to ask; is not the distinction between dry and humid air, and above all, that between comparatively stable and comparatively unstable air, at least as valid a basis for a major classification? As an aid to forecasting, air mass analysis began in the temperate zone, but the nomenclature developed there need not necessarily be ideal for the study of the earth as a whole.

The conditions most favourable to the development of a large and relatively homogeneous air mass must clearly be found where an extensive and broadly uniform surface of the earth is found in association with high barometric pressure, low barometric gradients and thus a comparatively gentle and divergent air flow. Such conditions are well developed over the tropical seas and tropical deserts during much of the year and over the Arctic regions, especially in winter. These areas may thus be considered as the source-regions of the *primary* air masses. Elsewhere air mass characteristics are more closely related to successive transformations which occur to air in motion, especially when the flow is convergent. These may be regarded as *secondary* air masses, owing their nature as much to past weather processes as to the intrinsic qualities of a source region (Crowe, 1965).

An attempt has been made in Figs. 79 and 80 to map the broad distribution of primary and secondary air masses over the whole earth in both January and July. *Arctic* air masses are shown over compact ice at sea and over continental areas with a mean temperature for the month in question below $-10°$C ($14°$F). *Tropical maritime* masses are shown over the trade-wind systems and the areas of prevailing high pressure which lie towards their poleward margins. *Tropical continental* air masses lie in similar latitudes but in defining their limits it has been necessary to take some account of the nature of the underlying land surface; it is usually arid and, especially in central Asia, composed in part of extensive lofty plateaus. The 'northeastern monsoon' of southeast Asia is considered to be a trade-wind with continental affinities and hence is associated with both of these latter groups.

The areas covered by secondary, convergent, and more unstable air masses have been emphasised by shading. *Equatorial* air covers the 'doldrums', rather generously defined, a fairly narrow zone in the central Pacific where the trades are still active but converge together from each hemisphere, and the more humid parts of the intertropical continental interiors. It is thus shown as a continuous girdle round the earth at both periods. The south-western monsoon of Fig. 80(a) raises special problems. Towards its root off Somaliland we have interpreted it as merely a northward extension of the southern trades, but as instability increases towards India and southeast Asia it takes on characteristics much more closely related to equatorial air. It may therefore be considered as a special subsection of the latter.

We are left with an area in the middle latitudes of each hemisphere. These are the main battlefields where Dove's conflict of northern and southern air

TABLE 25. *The major air masses of the world (areas in million ml²)*

|  |  | JANUARY | JULY |
|---|---|---|---|
| N | Arctic | 21.09 | 4.72 |
|  | Mid-Latitude Group (N) | 20.95 | 28.51 |
|  | Tropical continental | 12.03 | 13.58 |
|  | Tropical maritime |  |  |
|  | Atlantic | 10.21 | 10.94 |
|  | Pacific | 15.51 | 14.03 |
|  | NE monsoon | 10.82 | — |
| Eq | SW monsoon | — | 7.15 |
|  | Equatorial | 25.45 | 18.16 |
| S | Tropical maritime |  |  |
|  | Atlantic | 10.33 | 10.12 |
|  | Pacific | 23.12 | 22.90 |
|  | Indian Ocean | 8.65 | 11.96 |
|  | Tropical continental | 4.56 | 9.51 |
|  | Mid-latitude Group (S) | 27.12 | 32.37 |
|  | Antarctic | 7.07 | 12.96 |
|  | Totals | 196.91 | 196.91 |

is fought out. Here there is much truth in the assertion that 'there is no climate, only weather'. We make no attempt to subdivide them at this stage and give them the non-committal title of *mid-latitude* air. It will be noted that these air masses form continuous girdles round the earth except in the northern hemisphere in winter where a break in the ring from the Caspian to the East China Sea appears justified by surface observations though the ring is doubt-less closed aloft. Evidence for this is offered by the fact that barometric disturbances in the upper troposphere are known to cross the Asiatic continent even in the depth of winter. It is likely indeed that mid-latitude air also covers much of the Arctic and Antarctic caps, though it is there divorced from the surface.

It must be emphasised that none of the vast subsections of the atmosphere thus delineated is entirely homogeneous, and that any line on a map must give false precision to a description of features which are themselves subject to constant change and renewal. Yet the human mind demands some anchor and, having drawn our lines, however tentatively, it is possible to pass on to useful generalisations about *areas*. The surface areas covered by the various systems are given in Table 25, due allowance having been made for the distortion produced by a polar projection.

Grouping the areas according to the hemisphere to which they owe their major allegiance and, for the time being, regarding the southwestern monsoon as an extension of Equatorial air, we reach the following summary:

|  | JANUARY (ml²) | JULY (ml²) |
|---|---|---|
| Northern air masses | 90 610 000 | 71 780 000 |
| Equatorial air masses | 25 450 000 | 25 310 000 |
| Southern air masses | 80 850 000 | 99 820 000 |

Alternatively, bisecting the equatorial air and expressing the results as percentages of the area of the whole earth, we have

|  |  | JANUARY (%) | JULY (%) |
|---|---|---|---|
| Northern system | Primary | 35.3 | 22.0 |
|  | Secondary | 17.1 | 20.9 |
|  | Total | 52.4 | 42.9 |
| Southern system | Primary | 27.3 | 34.2 |
|  | Secondary | 20.3 | 22.9 |
|  | Total | 47.6 | 57.1 |
| Whole earth | Primary | 62.6 | 56.2 |
|  | Secondary | 37.4 | 43.8 |

From these tables a number of interesting conclusions can be drawn:

1. The volume of equatorial air, thus defined, remains substantially the same at both seasons of the year. Indeed, in view of the virtual constancy of radiative income in equatorial latitudes, it would be by no means surprising to learn that it remained approximately constant throughout the year.

2. In consequence, both the northern and southern circulation systems contract from winter to summer by the same amount, i.e. by an area of nearly 19 million square miles. It is clear however that this variation does not occur symmetrically about the geographical equator.

3. If the 'circulation equator' can be assumed to bisect the equatorial air, its mean position can be inferred from a comparison of the revised area of either the northern or southern system with the areas enclosed polewards by various parallels. Its mean location is thus found to range from about 3°S in January to about 8°N in July.[1] This is considerably less than the migration of the 'heat equator' and only about a quarter of that of the 'vertical sun'.

4. The primary systems are particularly dominant during the winter season in each hemisphere when the total area involved in arctic and tropical air together *in the winter hemisphere alone* amounts to more than a third of the area of the whole earth.

5. The areas involved in secondary air masses are approximately the same during the summer of each hemisphere but this symmetry is lost in winter largely owing to exceptional conditions over the Asiatic continent.

However approximate our boundaries, it seems probable that these five points represent real features of the general circulation though we cannot expect their full implications to be immediately apparent. Even if primary air is regarded as predominantly divergent and thus subject to subsidence whilst secondary air is considered as mainly convergent and thus subject to uplift, the student cannot assume that the *areas* involved should be either similar or always present in the same proportion. This would imply assumptions as to the *speed* of the motions about which we are very ill-informed.

The geographer will notice with interest that primary air masses are generated for the most part in the 'empty quarters' of the earth. Where they prevail, although it is not completely rainless, precipitation is slight, sporadic, or clearly associated with the presence of strong relief features. Practically the whole of the habitable world, except for some oases and tropical islands, thus lies, at least for a season, under what we have called the secondary group. Here, of course, even if uplift for one reason or another gives the air mass its distinctive character, the activity varies widely from place to place and from time to time.

## Upper air analysis

The classification of air masses must clearly rest upon a three-dimensional foundation. Hence the supreme importance of the data now being provided by radio-sondes. These are ingenious, lightweight devices which are carried aloft by balloons and which radio back automatically to a receiving station (in code) a continuous record of the pressure, temperature and humidity of the various layers of the air through which they ascend. They can also be tracked by radar to yield information on wind direction and speed, level by level, long after they have disappeared from view. Such ascents are now

---

[1] If it is insisted that the southwest monsoon should be regarded as exclusively composed of southern hemisphere air, the range becomes 3°S to 10°N. Such an assumption is of questionable validity.

made, at least once a day, at many hundreds of stations in all parts of the world.

When the weather forecaster has interpreted the most recent ascent and compared it with data from other stations he can happily lay it aside and wait for the next, but during this process, an immense body of data accumulates at central forecasting offices. What is to be done with it? Where a daily weather report is printed much material is published *in extenso* but this presents a most forbidding mass of material for *climatological* investigation. A number of countries now process the data by computer and publish an upper-air summary in the form of monthly means of at least temperature and humidity for a number of selected standard pressure-levels. On a worldwide basis the most convenient source of such material is unquestionably the *Monthly Climatic Data for the World* published since 1948. The coverage and quality of material have improved steadily since that time and currently this publication gives upper-air data for over 380 land-based stations plus a dozen weather-ships. We shall use some of this information in the following discussion but a word or two is necessary as to its limitations.

1. In the lower sector of the atmosphere, which is our primary concern, the standard levels employed are at 300, 500, 700 and 850 millibars, while data are also available at station level (SFC—surface). The mean heights of these 'surfaces' may be estimated from Table 23 and it is clear that the intervals cover several thousands of feet. This keeps the problem of tabulation within manageable proportions but, as soon as the data are examined, it becomes evident that such a height range may miss the most important details. This is particularly true of the lowest layer from the surface to 850 millibars (at 1457 m, 4780 ft in the standard atmosphere and still higher within the tropics). A strong plea could be entered for the inclusion of an intermediate level, say at 925 millibars, which would still be high enough to be free from diurnal complications at most lowland stations. At stations on lofty plateaux the record at 850 millibars may be affected by the time of the ascent.

2. Some detail is inevitably lost during the averaging process. In a wide-ranging survey such a loss must be tolerated but it does mean that average ascents are likely to be of much greater significance in some parts of the world than in others. For instance, where fronts are common they are likely to be encountered at different levels on successive occasions. Averaging will spread such features over a considerable depth and may give to the mean ascent a deceptive and indeed partly spurious aspect of stability. Here we must learn from experience but if mean ascents have to be interpreted with caution, for instance over western Europe, this does not imply that they are equally suspect in lands where frontal phenomena are less common. In the following account the mean values, usually for January and July, have been further averaged over the standard 5-year period 1962–66. This does little further violence to the data but ensures that we are not dealing with a freak year. The process has also proved valuable in covering occasional gaps in the record

and exposing evident misprints, an ever-present danger in the most carefully edited tabulation.

3. Analysis and comprehension of such material is greatly facilitated if *graphical* methods are employed. An ordinary height-temperature diagram will suffice but there is much to be said for familiarising oneself with the 'tephigram' (Meteorological Office RAF Form 2810) available at small cost from H.M. Stationery Office. We shall use this method to derive a number of important air mass characteristics, notably potential temperature and mixing-ratio, from the temperature and dew-point values quoted for each standard level in *Monthly Climatic Data for the World*. Relative humidities can then be obtained with a slide-rule either by reading the saturation mixing ratio from the tephigram or by consulting standard tables.

4. Plotting ascents over the wide intervals mentioned under (1) above raises some difficulties. A purely objective approach would suggest that, since we often do not *know* the course of the curve between the points given, we should join them by straight lines. This practice has been followed everywhere above 700 millibars though evidence will be given in support of a curved path often approximately parallel with the curved saturation adiabats. The error involved here is usually negligible. This is far from true below 700 millibars and especially below 850 millibars. It can be argued with much reason that it is the function of a diagram to give a *visual* interpretation of the facts. Given limited information this is something which a purely mechanical, mindless plotting may signally fail to achieve. It is generally conceded, for instance, that in plotting isobars from spot pressures or generalised contours from spot heights, a degree of subjective interpretation is both necessary and desirable. But it must be an informed interpretation obeying reasonable rules and must do no violence to the recorded observations. This matter will be raised again with particular reference to the trade-wind inversion; suffice it to add that wherever a route other than the direct one from point to point has been employed it is clearly indicated and that, as a constraint, it has always been drawn in straight sections.

## The primary air masses of the northern hemisphere

It was first planned to present the following account air mass by air mass, drawing illustrations from both hemispheres, but it soon became apparent that such a presentation would be too diffuse, the reader's mind being carried too violently over vast distances from station to station. It has thus been decided to develop the analysis in the first place from data for the well-documented northern hemisphere, though there is no reason to regard the geographical equator as particularly sacrosanct. In the outcome, despite rigorous selection, the discussion has proved so lengthy that examples from the southern hemisphere, except for equatorial air, are presented only in tabulated form (Table 41) but notes are added which will facilitate comparison with plotted data.

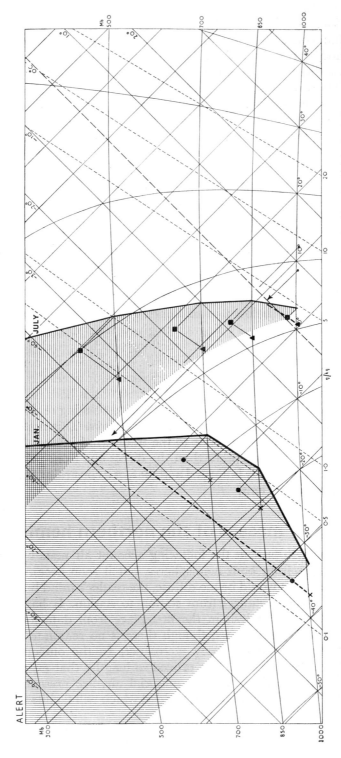

FIG. 81. Tephigram showing mean air-mass characteristics at Alert (82° 30′N) in northern Canada (Arctic Air) in both January and July. See text p. 268 for key to all tephigrams

# Arctic air

## Winter

It is hard to imagine a more uniform environment for the lower layers of the atmosphere than that presented by the ice-clad Arctic sea during the long polar night. Incoming radiation is nil, or virtually nil, for months together and the only steady source of heat is supplied by slow conduction through the 2 to 3 metres of sea ice from the unfrozen water beneath. Intermittent incursions of warmer air from lower latitudes certainly occur, but (*a*) the warmer air tends to ride aloft, (*b*) it is rapidly transformed, and (*c*) the release of latent heat at condensation is both small in amount and again concentrated for the most part in the middle layers. It would appear that 'leads' may be opened by strong winds at all seasons but, at least in winter, their total area is small and much of the heat released in this way is promptly locked away in latent form by the processes of evaporation and refreezing.[1] Yet back-radiation to space at a rate proportionate to surface temperatures on the Kelvin scale (see p. 37) continues unceasingly through the water vapour 'window'. It can be checked only by the presence of low cloud, itself evidence of the advection of more southerly air since the surface chill and the intermittent occurrence of Arctic outbreaks combine to produce subsidence aloft which precludes local cloud formation.

Over the neighbouring snow-clad continents and ice-caps much the same conditions prevail but the oceanic heat source is absent and the more varied relief stimulates air drainage from the heights so that local cold pools of even greater intensity are generated.

At the present time we know fewer details about the climate of the North Pole than of the South Pole where permanent residence is (just) possible. The most northerly station available is Alert (82°30′N, 62°20′W) near the poleward limit of the Canadian archipelago. An equally good record is also yielded by Nord (81°36′N, 16°40′W) in Greenland. The Siberian arctic is served by a number of stations but, as published, the records appear to be subject to interruptions, surprisingly enough mainly during the summer months. Fig. 81 illustrates mean ascents at Alert in both January and July and an analysis of the two curves is tabulated on the following page.

The features of these ascents are closely reproduced at other Arctic stations, notably at Barrow (71°18′N 156°47′W) and Ostrov Chetyrekhstolbovoy (70°38′N 162°24′E) both in the inner Arctic though over ten degrees of latitude farther from the pole (see Table 26). Cape Tobin (70°25′N, 21°58′W) in eastern Greenland, though not quite so intensely cold in winter (January −17°C), also falls into this category. In every case in *winter* there is a strongly developed low-level inversion of temperature extending up to about 850 millibars (at least 1200 m or 4000 ft) whilst the lapse rate thence to 700

[1] Koerner (1970, 226) makes the interesting point that fracture-and-freeze adds substantially more to Arctic ice than the thickening of old floes.

ALERT (82°30′ N, 62°20′ W) Ht 62 m

| | T (°C) | PT (°K) | EPT (°K) | SPT (°C) | MR (g/kg) | RH (%) |
|---|---|---|---|---|---|---|
| *January* | | | | | | |
| (mb) | | | | | | |
| 300 | −61 | 299 | — | — | — | — |
| 500 | −41 | 282 | — | — | — | — |
| 700 | −29 | 271 | 271 | −7.6 | 0.29 | 56 |
| 850 | −26 | 259 | 259 | −16.2 | 0.33 | 61 |
| SFC | −32 | 240 | 240 | −33.4 | 0.17 | 68 |
| (1010 mb) | LCL 945 mb | | CCL 470 mb | | P = 6 mm | |
| *July* | | | | | | |
| (mb) | | | | | | |
| 300 | −48 | 318 | — | — | — | — |
| 500 | −25 | 303 | 304 | 10.0 | 0.46 | 45 |
| 700 | −10 | 291 | 296 | 6.3 | 1.50 | 59 |
| 850 | − 2 | 284 | 291 | 4.1 | 2.55 | 67 |
| SFC | 3 | 276 | 287 | 2.1 | 4.1 | 84 |
| (1005 mb) | LCL 970 mb | | CCL 890 mb | | P = 15 mm | |

P = mean monthly precipitation, here and in tables which follow.

*Key to all tephigrams in Chapter IX.*
Each diagram shows mean ascents for the period 1962–66 during the two months named, usually July and January.

Air temperatures (°C) have been plotted at each of the standard levels: Surface, 850, 700, 500, and 300 millibars, and the points are normally joined by straight lines. The corresponding dew-point temperatures are also indicated where they are available; for January by crosses and for July by triangles. From these the mixing-ratios given in the tables have been read from the tephigram and the appropriate relative humidities determined.

The construction by which the lifting condensation level at each standard horizon is obtained is shown and from the points thus located (in January indicated by circles and in July by squares), the tabulated data for saturation potential temperature and equivalent potential temperature have been read from the original diagrams.

For surface air the convection condensation level is given by the intersection of the appropriate mixing-ratio line (reinforced) and the temperature curve. The rise in surface air temperature necessary for air to be able to reach this level dry-adiabatically is shown by the straight arrows. The saturation adiabat upwards from this point is only indicated (by curved broken arrows) where further ascent seems probable.

The area between the environment curve and the saturation adiabat appropriate to surface air is cross-hatched. This area gives a measure of the stability of the air mass.

All the diagrams have been constructed on a common framework to facilitate direct comparison but where values at all the levels shown are relatively high, the empty lefthand half of the diagram has been cut away to save space. The righthand margin is common to all.

*Note.* For some Canadian stations information was available which made it possible to adjust the standard ascents to a 50 millibar interval. The improvement in their appearance is at once evident.

Where the trade-wind inversion was clearly being obscured by the 150 millibar interval inherent in the published data a rather subjective interpretation of the ascent has been given. Such sectors are clearly shown by the use of a dotted line (see text, p. 265). The values at 850 millibars are not affected.

In the Tables, mean pressure at station level (SFC) is given in brackets.

millibars (about 2750 m) is only slightly positive—the layer is almost iso-thermal. Temperatures are so low that the water vapour content is almost negligible and the mixing-ratios would indeed have been lower yet had they been adjusted for saturation over ice rather than for water. Under these conditions, relative humidities become almost meaningless. The lifting condensation level is not high, at perhaps some 500 metres at Alert, but an

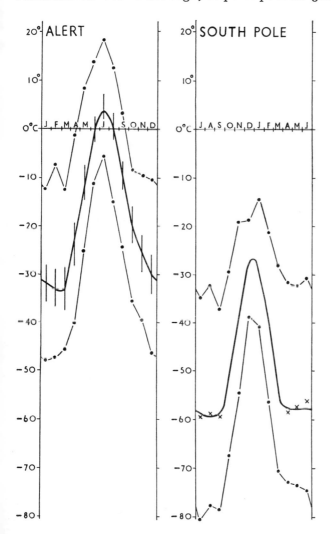

Fig. 82. The seasonal variation of temperature near the North Pole and at the South Pole. The two curves show the variation of mean monthly temperature and the absolute extremes so far recorded are indicated by dots. For Alert the range between mean daily maxima and minima is indicated by vertical bars. For comments on this statistic see text, p. 65

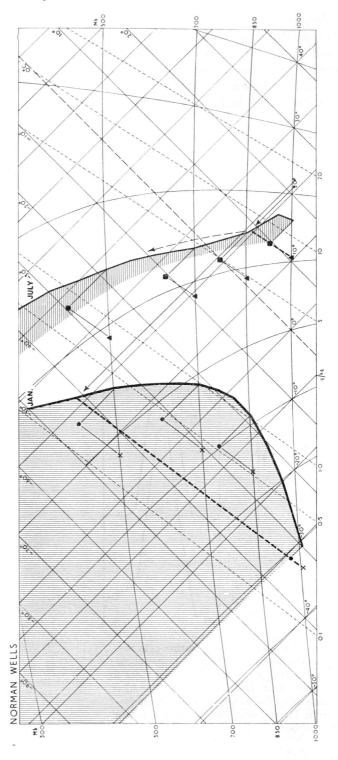

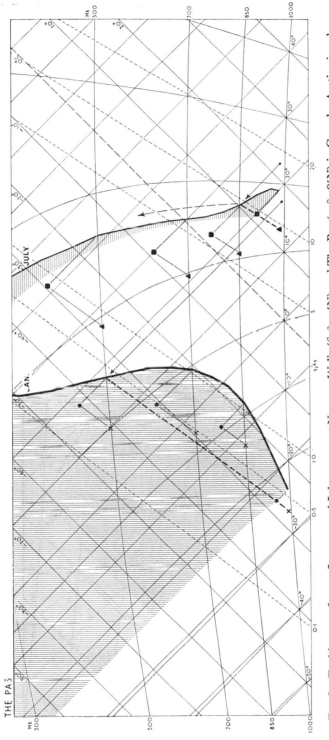

FIG. 83. Tephigrams of mean January and July ascents at Norman Wells (65° 17'N) and The Pas (53° 58'N) in Canada. Arctic air only during the winter. Note the great seasonal range

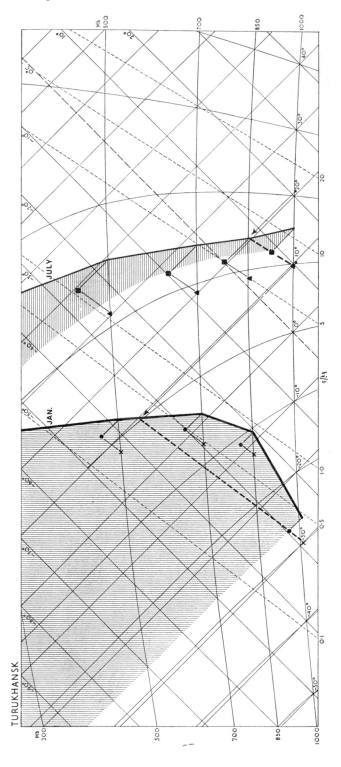

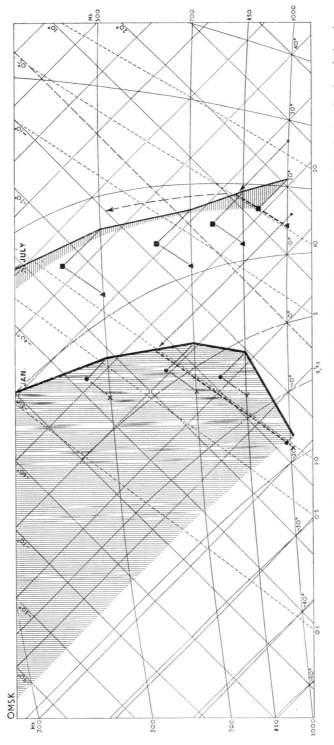

FIG. 84. Tephigrams of mean January and July ascents at Turukhansk (65° 47′N) and Omsk (54° 56′N) in Siberia. Arctic air only in winter. Note the similarity with the Canadian stations in comparable latitudes in Fig. 83

increase of potential temperature of over 30°K from the surface to 700 millibars reflects an enormous degree of stability in Arctic air masses. Convective processes are thus ruled out, indeed the convection condensation level is well above 4500 metres at all Arctic stations in winter. To rise so far from the surface, air would have to attain temperatures in excess of 10°C and, even then, it would encounter middle layers that are also stably stratified. Winter precipitation in the Arctic must thus come from an external source and, though measurements may be exaggerated by a catch of blowing snow, Alert records a mean value of only 12 millimetres in January over the years in question. In 3 of the 5 years it was very much less.

It may be of interest to note that the mean temperature of all the lower layers of air at Alert, from the surface to the 700-millibar level, in January is not far from 245°K. This may be compared with a similar estimate for tropical maritime air at Sal, in the Cape Verde islands, of 288°K for the same season. Ignoring the effect of differing water vapour content which, though slight, would be to Sal's advantage, the relative densities of these two samples are thus as 1/245 : 1/288. In other words, the Arctic air is 17 to 18 per cent more dense than the tropical maritime air in the same hemisphere.

Figure 82, showing the mean monthly regime of temperature at Alert, reminds us of the duration of the features thus described. Clearly they must prevail with little modification from November to April, and January is not in general the coldest month. We are aware, of course, that mean monthly temperatures do not convey the whole story but neither the range between mean daily maximum and minimum values nor even the absolute extreme values indicated on the diagram do much to invalidate the above generalisation. (The absolute extremes are for a short period.) At such a station the mean maximum–minimum range is far from being representative of the *true* (periodic) diurnal range. Rather is it the mean range of temperature change within a 24-hour period which is by no means the same thing (see p. 65).

We have already asserted that Arctic air extends far south over the continents in winter. This view is supported by the data quoted in Table 26(*a*) and the ascents at Norman Wells, Turukhansk, The Pas and Omsk shown in Figs. 83 and 84. Each of these stations shows similar features to those recognised above—the deep surface inversion, low mixing-ratios, and an increase of potential temperature up to the 700 millibar level of the order of 30°K. Yet the latitudes of the two latter stations are in the middle fifties. Such stations are undoubtedly affected by advection, both from the north and from the south, but winter characteristics as strong as these are surely developed on the spot and for much the same reasons as we have already given. Indeed, the absence of the Arctic ocean beneath the surface is conducive to lower temperatures on land than on the Arctic ice. Although there are insufficient data to demonstrate this point conclusively, this interpretation is certainly suggested by Fig. 85 which is based upon mean January temperatures over the period 1962–66 which we are using as standard. It carries with it a strong suggestion that the Arctic cold is bipolar and, even if the Canadian sector is not strongly developed, nobody has ever been in doubt about the severity of eastern Siberia. Here frost hollow effects (p. 443) undoubtedly contribute to the very low temperatures experienced at some stations.[1]

Verkhoyansk (67°33′N, 133°23′E) has long been regarded as the northern 'pole of cold'. Though still lower absolute minima have been recorded at Oymiakon (63°16′N 143°09′E) its January means just take second place

TABLE 26a. *Arctic air (January)*

| | WESTERN HEMISPHERE | | | | EASTERN HEMISPHERE | | | |
|---|---|---|---|---|---|---|---|---|
| Level (mb) | T (°C) | PT (°K) | SPT (°C) | MR (g/kg) | T (°C) | PT (°K) | SPT (°C) | MR (g/kg) |
| BARROW | | | | | OSTROV CHETYREKIISTOLBOVOY | | | |
| 71°18′N, 156°47′W (4 m) | | | | | 70°38′N, 162°24′E (6 m) | | | |
| 300 | −56 | 305 | — | — | −57 | 305 | 10.4 | 0.02 |
| 500 | −35 | 290 | 3.7 | 0.20 | −37 | 288 | 2.6 | 0.21 |
| 700 | −21 | 279 | −2.0 | 0.50 | −24 | 276 | −3.7 | 0.57 |
| 850 | −17 | 268 | −8.5 | 0.70 | −20 | 265 | −10.4 | 0.62 |
| SFC | −25 | 246 | −26.8 | 0.39 | −30 | 241 | −31.9 | 0.26 |
| (1021 mb) | | P = 5 mm | | | (1020 mb) | P − 9 mm | | |
| NORMAN WELLS | | | | | TURUKHANSK | | | |
| 65°17′N, 126°48′W (64 m) | | | | | 65°47′N, 87°57′E (37 m) | | | |
| 300 | −55 | 307 | — | — | −59 | 303 | 9.6 | 0.02 |
| 500 | −35 | 290 | 4.0 | 0.16 | −39 | 286 | 1.5 | 0.17 |
| 700 | −22 | 278 | −2.2 | 0.43 | −25 | 274 | −4.7 | 0.47 |
| 850 | −19 | 267 | −9.7 | 0.53 | −21 | 264 | 11.0 | 0.68 |
| SFC | −30 | 242 | −31.2 | 0.24 | −26 | 246 | −27.0 | 0.35 |
| (1016 mb) | | P = 18 mm | | | (1014 mb) | P − 23 mm | | |
| THE PAS | | | | | OMSK | | | |
| 53°58′N, 101°06′W (272 m) | | | | | 54°56′N, 73°24′E (94 m) | | | |
| 300 | −55 | 308 | — | — | −56 | 306 | 11.0 | 0.02 |
| 500 | −34 | 291 | 4.2 | 0.21 | −31 | 294 | 6.2 | 0.33 |
| 700 | −21 | 279 | −1.7 | 0.48 | −17 | 284 | 1.7 | 0.82 |
| 850 | −18 | 267 | −9.3 | 0.66 | −11 | 275 | −2.8 | 1.25 |
| SFC | −24 | 249 | −24.0 | 0.42 | −16 | 256 | −16.8 | 0.89 |
| (987 mb) | | P = 21 mm | | | (1014 mb) | P = 8 mm | | |

during our standard period and its upper-air record is rather less complete. We have thus given Verkhoyansk a diagram to itself, even though it has meant extending the tephigram well beyond its normal limits (Fig. 86). Between the surface (135 m) and the 700-millibar level, potential temperature increases by no less than 55°K. Indeed from the surface to the 850-millibar level there is an increase of *actual* mean temperature in January of 21°C (nearly 38°F).

[1] Hence also the world record sea-level pressure of 1083.8 mb (32 inches) recorded at Agata (66° 53′N, 93° 28′E, 263 m). (Giles, 1970.)

This is in close agreement with Suslov's (1961, 128) comment that 'on the average, the winter temperature increases 1°F for each 110 feet of rise' though his conclusion was apparently based on *surface* observations at lowland and upland stations in various parts of eastern Siberia. It may be mentioned in

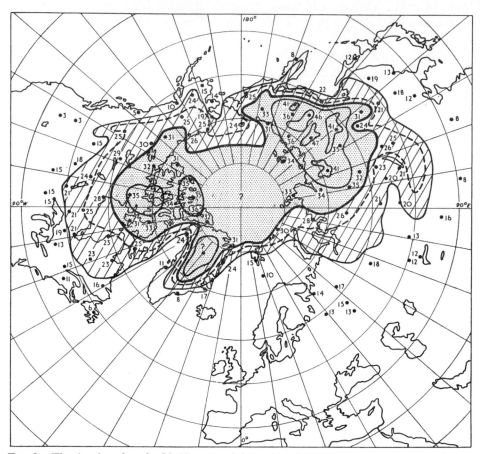

FIG. 85. The Arctic poles of cold. The actual data plotted are mean January temperatures over the period 1962–66. All the values quoted are *negative* degrees Celsius but, in view of the scale of the map, the minus signs have been omitted. The values are not reduced to sea-level. Temperatures on the Arctic ice were not known but Koerner (1970, 224) gives means of −36°C for the period December to February, 1968–9. He was then on the Canadian side of the pole at about 85°N

passing that joining the points at 700 and 850 millibars and the surface by straight lines fails to give a full picture of the intensity of the surface inversion. Unquestionably a more detailed ascent would give a curve running in a concave-upwards course through these three points. This feature has been shown in the ascent at Norman Wells where the readings for intermediate levels are known. In the lowest layers therefore the actual increase in temperature with height is even greater than the overall figure quoted above.

276

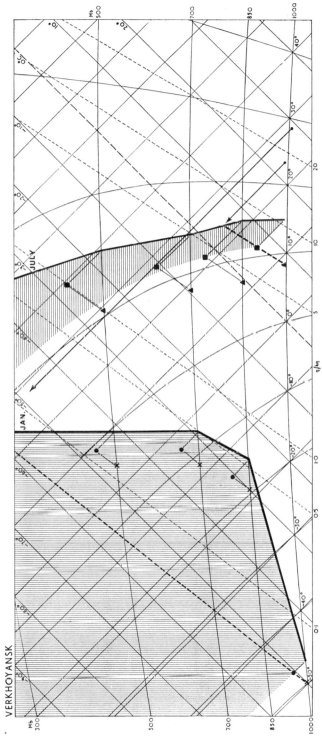

Fig. 86. Tephigrams for January and July at Verkhoyansk (67° 33′N). This illustrates the Arctic surface inversion in an extreme form since the station lies in a sheltered valley where frost-hollow effects are very strong. Note that even in July the mean ascents indicate a considerable degree of stability

It is evident that, because of its high density and indeed of the very nature of its origin, such intensely cold air can play little direct part in the general circulation, particularly in the presence of considerable physical barriers. Any general outflow from the 'Siberian high' across such a barrier in winter must thus tap only the middle and lower-middle layers of the air mass. It is an interesting observation that the January temperature curve for Okhotsk from the surface to about 600 millibars closely follows that derived by dry-adiabatic descent of the layers between 500 and 900 millibars of an air mass similar to that met with at Verkhoyansk. This would imply a descent of about 1000 metres, which is indeed the general height of the intervening plateaux. It is hardly to be expected that the mixing-ratios should confirm such a motion, since Okhotsk is a coastal station, but the values remain extremely low and, level for level, the relative humidities at Okhotsk are some 10 per cent *lower* than those recorded in the interior.

It is evident from the above discussion that there is no real need for distinguishing 'polar continental' air. Such air is Arctic in its origins, its nature and its manifestations. In winter, to use Shaw's term, it forms a single vast lenticular 'underworld' (or perhaps inner underworld) covering most of Canada and Siberia and reaching a maximum thickness of the order of 2.5 kilometres. Clearly its presence must have consequences, both direct and indirect, upon the upper-air but it is as a surface or near-surface feature that it achieves climatological significance. To the invasion of warmer air from the south it presents a positive physical barrier; such air must be directed upwards and brings with it most of the weather experienced particularly over its shallower fringes. Yet its mean borders as shown in Fig. 79 are in no sense fixed, except where contained by relief obstacles. At least at intervals, some escape must be found to relieve the growing density-differential implicit in the continuing chill but the only route open to dense cold air lies very near the surface of the ground; hence the recurrent cold waves of central North America and the Russian plains. That these two routes appear to be preferred may be to some extent an illusion since the air retains recognisably Arctic characteristics for a long time when travelling over already cold land surfaces.

What of the ready paths apparently offered by the notable re-entrants over the Norwegian Sea and the Gulf of Alaska? Escape unquestionably occurs along these avenues but two considerations have then to be kept in mind. (a) When it passes over open water, Arctic air is rapidly transformed. It reaches these islands, for instance, as 'polar maritime' air which is certainly maritime but is more likely to have come from Canada than from anywhere near the pole. However, direct outbreaks from beyond Spitzbergen do occasionally occur in mid-winter. Figure 87 illustrates a recent very striking example which brought severe weather over the whole country. Since the pressure pattern changed very little over 3 days the meridional isobars give a close approximation to the actual path or trajectory of the Arctic air. Heavy snowfall occurred as the cold front swept southward on 7 February 1969, but the two following days brought brilliant cloudless skies at least to the western

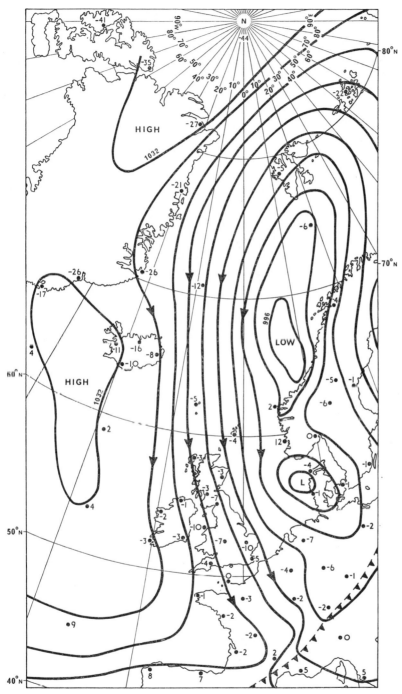

Fig. 87. The weather situation over western Europe on the morning of 8 Feb. 1969. Temperatures in °C. The course of the Arctic outbreak was virtually meridional

279

part of the country. The night of 7th/8th was the coldest February night in Manchester this century. (*b*) Escape on a major scale must be *organised*. Expressed in pressure terms this means that, to reach Britain by a direct route, an Arctic outbreak requires a situation where pressure is relatively high over the warm sea to the west and relatively low over the cold continent to the east. (In Fig. 87 the normal site of the 'Icelandic low' records pressures in excess of 1032 mb.) During the day-to-day variations of the synoptic pattern such situations can and do occur but dynamic and thermal influences are then in opposition and the pattern is unlikely to persist for long. Hence, indeed, the reason why Western Europe and Pacific North America experience the two most remarkable winter climatic anomalies on earth. Hence also why most of Britain's more prolonged cold spells come from the east rather than from the north. We then catch the tail end of an Arctic outbreak that has traversed the whole of Europe and the expression that the wind is 'straight from Moscow' may contain more than a grain of truth.

A final word on the consequences of the tendency of the inner core of the Arctic 'underworld' to develop the bipolar form illustrated in Fig. 85. This means that some air may escape from these inner regions by following a *westerly* route along their northern flanks. The actual pole is thus better ventilated than would otherwise be the case. Little wonder also that Ferrel's 'polar easterlies' have proved so hard to find in the northern hemisphere.

## Summer

If Arctic air owes its winter characteristics to the absence, or at least the inefficacy, of incoming solar radiation it follows that in summer, when insolation is virtually continuous, the term can retain validity only in a relative, rather than in an absolute, sense. The key factor at this season must be the presence of sea ice or of continental ice-fields which reduce the thermal effects of insolation to a minimum. The climatic consequences are thus a long delayed spring and an abbreviated summer during which temperatures can rise only a few degrees above freezing point. Such are the surface character-istics of a small group of stations around the shores of the Arctic sea and the Greenland ice-cap. Despite their poleward location, neither Alert nor Nord shows summer conditions at their most extreme. These are found in the inner Arctic at Barrow and Ostrov Chetyrekhstolvoboy where the atmosphere from the surface to the 850 millibar level is almost isothermal in July. This is the best evidence to hand for the persistence of a near-surface inversion throughout the summer months and the inherent stability of summer Arctic air is re-vealed by an increase of potential temperature from the surface to 700 millibars of over $20°K$ at both of these stations. It is also shown by Fig. 81 though the increase at Alert is only $15°K$. It reaches $19°K$ at Cape Tobin.

At all truly Arctic stations the summer mixing-ratios also remain low, from 2.5–4.5 g/kg even in the lower layers. Convective condensation is improbable since the level is high (it is unusually low at Alert) and it would usually require surface temperatures well in excess of 10° C. Even so, as in winter, the uplifted air would again encounter upper-air stability so that the development of cloud in depth would be inhibited. Little wonder that thunderstorms are unknown in the Arctic.

The rapid transformation which occurs in air mass characteristics as we pass from the shores of the Arctic towards the continental interiors in summer

TABLE 26b. *Arctic air (July)*

| Level (mb) | WESTERN HEMISPHERE | | | | EASTERN HEMISPHERE | | | |
|---|---|---|---|---|---|---|---|---|
| | *T* (°C) | *PT* (°K) | *SPT* (°C) | *MR* (g/kg) | *T* (°C) | *PT* (°K) | *SPT* (°C) | *MR* (g/kg) |
| BARROW 71°18′N, 156°47′W (4 m) | | | | | OSTROV CHETYREKSTOLVOBOY 70°38′N, 162°24′ E(6 m) | | | |
| 300 | −47 | 319 | — | — | −43 | 325 | 17.5 | 0.13 |
| 500 | −21 | 307 | 11.9 | 0.65 | −19 | 309 | 13.0 | 0.80 |
| 700 | −6 | 296 | 9.3 | 2.2 | −4 | 297 | 9.9 | 2.2 |
| 850 | 2 | 288 | 7.2 | 3.2 | 3 | 289 | 7.7 | 3.5 |
| SFC | 3.5 | 275 | 2.2 | 4.4 | 2 | 274 | 0.5 | 3.9 |
| (1014 mb) | | *P* = 20 mm | | | (1010 mb) | *P* = 21 mm | | |
| NORMAN WELLS[1] 65°17′N, 126°48′W (64 m) | | | | | TURUKHANSK[1] 65°47′N, 87°57′E (37 m) | | | |
| 300 | −45 | 321 | — | — | −42 | 325 | 17.6 | 0.16 |
| 500 | −18 | 310 | 13.3 | 0.78 | −17 | 313 | 14.4 | 1.1 |
| 700 | −2 | 300 | 11.6 | 2.7 | −1 | 301 | 12.1 | 2.9 |
| 850 | 8 | 294 | 11.7 | 4.9 | 7 | 293 | 11.5 | 5.0 |
| SFC | 16 | 289 | 12.8 | 8.1 | 15.5 | 28.8 | 12.1 | 7.6 |
| (1004 mb) | | *P* = 57 mm | | | (1004 mb) | *P* = 67 mm | | |
| THE PAS[1] 53°58′N, 101°06′W (272 m) | | | | | OMSK[1] 54°56′N, 73°24′E (94 m) | | | |
| 300 | −42 | 326 | — | — | −41 | 328 | 18.3 | 0.17 |
| 500 | −15 | 315 | 14.8 | 0.81 | −13 | 316 | 15.9 | 1.3 |
| 700 | 1 | 303 | 13.2 | 3.2 | 2 | 305 | 14.3 | 3.6 |
| 850 | 10 | 297 | 13.5 | 5.7 | 13 | 299 | 15.2 | 6.5 |
| SFC | 18 | 293 | 15.7 | 9.5 | 21 | 294 | 16.7 | 10.1 |
| (980 mb) | | *P* = 67 mm | | | (998 mb) | *P* = 72 mm | | |

[1] At this season, mid-latitude air.

is clearly shown by a comparison of any of the above-mentioned stations with Norman Wells, Turukhansk (Table 26(*b*)) and Verkhoyansk (Fig. 86). Within a distance of some 300–400 miles July surface temperatures are up by 12–13° C, the mixing-ratio is doubled, and the air column is much less stable. This is one type of mid-latitude air to which we shall return later.

## Tropical air

Tropical air occupies a much greater proportion of the atmosphere at any one time than Arctic air and covers a much wider variety of surfaces. Its primary characteristics apparently stem from slow dynamic descent from a very considerable height but interactions with the surface occur which generate a degree of variation within a limited range. Furthermore the patterns observed recur again and again with remarkable fidelity in widely distant parts of the earth's surface. We note four major subdivisions; (*a*) tropical continental air, (*b*) tropical maritime air (eastern), (*c*) tropical maritime air (western), and (*d*) dry monsoon air. The terms 'eastern' and 'western' refer

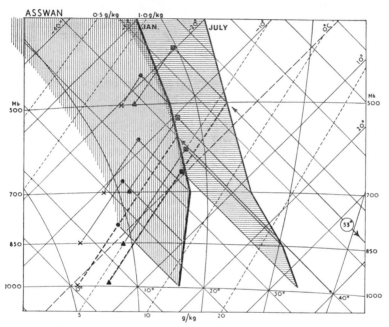

FIG. 88. Tephigrams for January and July at Aswan (23° 58′N). An excellent example of tropical continental or desert air throughout the year. The great stability indicated stems less from the thermal stratification, particularly in summer, than from the low relative humidity which raises both the lifting condensation level and the convective condensation level to extreme heights. Compare with central Australia, Table 41

broadly to the flanks of the oceans over which (*b*) and (*c*) occur. We shall see that the distinction is a real one but maritime data are still too fragmentary for us to be able to map it with any degree of precision. All of these air-masses show some degree of association with the trade-wind circulation.

(*a*) *Tropical continental air.* Standing at the opposite end of the temperature scale to Arctic air and yet also dry in depth and remarkably stable in stratification are the air masses associated with the great deserts of the earth. Such a zone extends over some 85° of longitude across the Old World from the Atlantic coast to the Indus basin and it is accompanied by a more northerly extension across the mountain divide which carries rather similar conditions far into central Asia. We illustrate its dominant features in Fig. 88 and in

Table 27. Evidently little distinction can be drawn between the Saharan and Turkestan deserts in summer but a real difference emerges during the winter months.

At the Saharan stations in winter, although potential temperature increases by some 15°K from the surface to the 700-millibar level, mean temperatures are such that the ascents follow closely the SALR up to about 400 millibars. We shall encounter a similar *thermal* stratification in regions of

TABLE 27. *Tropical continental air*

| Level (mb) | JANUARY T (°C) | PT (°K) | SPT (°C) | MR (g/kg) | RH (%) | JULY T (°C) | PT (°K) | SPT (°C) | MR (g/kg) | RH (%) |
|---|---|---|---|---|---|---|---|---|---|---|
| **FORT TRINQUET** | | | | 25°15′N, 11°37′W (359 m) (1960–64) | | | | | | |
| 300 | −39 | 330 | 19.0 | 0.07 | 16 | −30 | 342 | 22.0 | 0.18 | 17 |
| 500 | −13 | 317 | 15.2 | 0.40 | 14 | −7 | 324 | 18.3 | 1.5 | 33 |
| 700 | 3 | 306 | 12.2 | 1.2 | 17 | 14 | 318 | 17.6 | 2.5 | 17 |
| 850 | 11 | 297 | 10.0 | 2.4 | 25 | 28 | 315 | 18.2 | 4.1 | 14 |
| SFC | 17 | 292 | 11.3 | 5.3 | 42 | 33 | 309 | 20.9 | 9.7 | 28 |
| | (975 mb) | | P = 0 mm | | | (966 mb) | | P = 0 mm | | |
| **ASWAN** | | | | 23°58′N, 32°47′E (196 m) | | | | | | |
| 300 | −39 | 330 | — | — | — | −29 | 345 | — | — | — |
| 500 | −13 | 317 | — | — | . | 4 | 327 | — | — | — |
| 700 | 3 | 306 | — | — | — | 13 | 316 | 18.2 | 3.6 | 27 |
| 850 | 11 | 297 | 10.8 | 3.1 | 33 | 25 | 313 | 18.3 | 4.9 | 20 |
| SFC | 16 | 290 | 8.8 | 4.1 | 35 | 34 | 308 | 17.5 | 5.6 | 16 |
| | (994 mb) | | P = 0 mm | | | (982 mb) | | P = 0 mm | | |
| **ASHKHABAD** | | | | 37°58′N, 58°20′E (230 m) | | | | | | |
| 300 | −50 | 314 | 13.8 | 0.03 | 25 | −30 | 342 | 22.0 | 0.19 | 18 |
| 500 | −24 | 303 | 10.2 | 0.45 | 41 | −7 | 325 | 18.5 | 1.4 | 30 |
| 700 | −7 | 294 | 7.5 | 1.4 | 45 | 10 | 313 | 17.7 | 4.4 | 41 |
| 850 | 2 | 288 | 6.2 | 2.4 | 44 | 23 | 309 | 17.3 | 5.2 | 25 |
| SFC | 4 | 277 | 1.5 | 3.3 | 66 | 31 | 306 | 19.8 | 9.5 | 31 |
| | (995 mb) | | P = 22 mm | | | (978 mb) | | P = 2 mm | | |

frequent convectional overturning but here the processes at work must be very different. The key is given by the humidity data, particularly by the very low relative humidities at all levels at Fort Trinquet. This air is being warmed by subsidence from aloft, not by condensation from below—the course of the curve beyond the limits of Fig. 88 confirms this—and the correspondence with the SALR is thus essentially fortuitous. If we knew rather more than we do of the rate of radiational loss from the middle atmosphere it might indeed give us a clue to the mean rate of the subsidence. The very considerable stability of this air mass thus arises more from its low humidity than from its thermal characteristics and as long as it remains over dry lands it is unable to add much to its supply. At mean daily temperatures the lifting condensation level is at 800 millibars (about 2000 m) and convective condensation at Aswan could not normally occur below 4500 metres, requiring a (winter) surface temperature of the order of 40°C (104°F) which is in excess of the *absolute* maximum screen temperature (38°C) recorded there in January. Any development of local cloud would appear to be unlikely under these conditions and even those triggered off by favourable slopes would rapidly disperse. Little wonder that precipitation at Aswan is nil.

At Ashkhabad, on the other hand, surface inversions are much more common in January so that mean ascents appear to be virtually isothermal up to 850 millibars. This may be partly a function of latitude but Ashkhabad is also open to occasional invasions of Arctic air and obtains some winter rainfall, about 20 millimetres in January, from frontal disturbances associated with these outbreaks.

In summer, in the presence of strong surface heating, the thermal stratification of desert air is much less stable yet, despite the prevalence of mixing-ratios closely analogous to those recorded in Britain in July, relative humidities are so low that convectional overturning and precipitation are extremely rare. The lifting condensation level is thus much higher than in winter, at 650 millibars, or over 3600 metres, at Aswan and, even at Ashkhabad, the normal base of convective cloud would appear to be not below that level. At Aswan it is not far short of 6 kilometres, requiring a surface temperature of 53°C, again in excess of the July absolute maximum (51°C).

(*b*) *Tropical maritime air.* This air mass is of the greatest climatic significance for, at any one moment of time, it includes no less than a third of the whole troposphere. It is generated under dynamic conditions comparable with those producing the great deserts of the earth but, in sharp contrast to tropical continental air, it lies above vast expanses of warm sea. As we have already suggested, it gives evidence of a good deal of internal variety but it would be wrong to allow this to obscure the remarkable degree of family likeness evident wherever tropical maritime air is sampled—in both hemispheres.

The feature common to all ascents is, of course, the trade-wind inversion, developed with varying intensity but prevalent over vast areas. Equally important is active evaporation which utilises most of the incident insolation and serves, given time, to raise, weaken and eventually to penetrate, the inversion ceiling.

Here we meet the major limitation of our source material for the interval between sea-level and the 850-millibar level is so wide that, though it usually suggests the presence of the inversion, it does not permit us to illustrate it effectively by purely objective means. The point can be well made at Sal (16°44′N, 22°57′W) for which data for an intermediate level (900 mb) are available. Fig. 89 shows the features of ascents at this station for alternate months in 1961 from the surface to 600 millibars. Given data only for the surface, at 850 and at 700 millibars, an annual cycle is crudely suggested but no more. Employing the data for the 900 millibar level improves the picture a little but if the points are directly joined by straight lines we are still left with evident anomalies in the lowest layer of the atmosphere. Using reasonable assumptions about the lapse rate in proximity to the sea and noting the

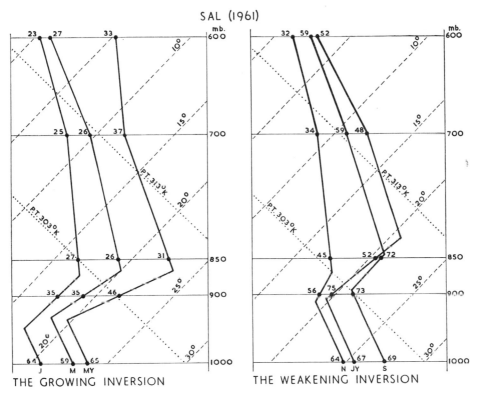

SAL (1961)

THE GROWING INVERSION          THE WEAKENING INVERSION

FIG. 89. The lower sections of mean ascents at Sal in the Cape Verde Islands (16° 44′N) for January, March, May, July, Sept. and Nov. 1961. The diagram is intended to show the difficulty of illustrating the trade-wind inversion even where data for 900 mb are to hand. The interpretation given appears to be reasonable since it shows a progressive development and decline of this important feature. The relative humidities, given in figures by the curves, provide useful corroboration

unusually high relative humidities (quoted on the figure) recorded at 850 millibars in both July and September, we find ourselves narrowly constrained to adopt the interpretation shown. This reveals a progressive development and decline in the inversion accompanied by an equally progressive change in its height. The difference in potential temperature between the top and base of the inversion-layer is thus shown to increase from a minimum of about 3°K in November to a maximum of some 10°K in May. A subjective interpretation? Yes, but not an irrational one and not without considerable support from other sources.

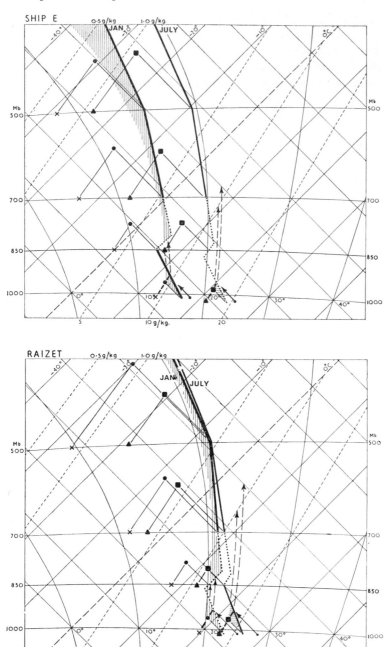

FIG. 90 Tephigrams for January and July at Weather Ship E (35° 00′ N) and Raizet (16° 16′ N) in the west Atlantic. This is tropical maritime air of the western variety, humid near the surface but capped with relatively dry air aloft. It has very limited stability in summer and even during the winter months the inversion is both high and weak

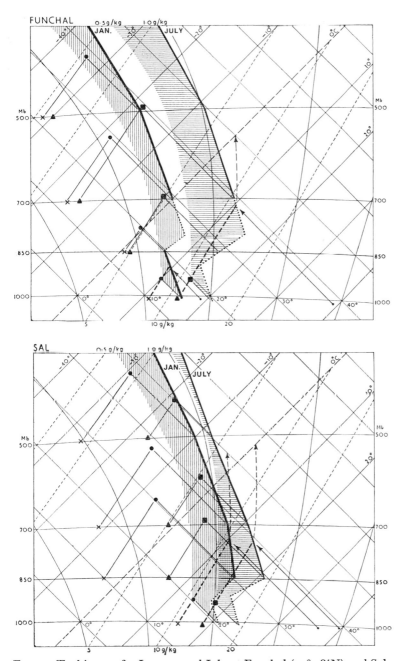

Fɪɢ. 91. Tephigrams for January and July at Funchal (32° 38′N) and Sal
(16° 44′N). This is tropical maritime air of the eastern variety with a low
and very marked trade-wind inversion, particularly during the summer.
Yet neither station is quite ideally situated with respect to the eastern trades
for Funchal lies beyond their northern limits in winter and Sal is not far from
their equatorial limit in summer

The essential characteristics of this air mass are well illustrated by a quadrilateral of stations covering a range of latitude from about 35°N to nearly 15°N in the North Atlantic (see Figs. 90 and 91 and Table 28). Although the records for these four stations are not completely without flaws, they undoubtedly represent the most complete coverage available of truly maritime data for any of the five great trade-wind cells found upon the earth's surface. They are thus worthy of careful analysis.

Common to all four stations, as might well have been anticipated, are mixing-ratios in the near-surface layers at all seasons of a magnitude considerably greater than those so far encountered. In view of the general direction of the circulation it is also not surprising that the values are highest in the southwest at Raizet (Guadeloupe), and lowest in the northeast at Funchal (Madeira), again throughout the year. In each case, however, the air from 700 millibars upwards still remains remarkably dry, a fact which suggests a normal cloud ceiling of the order of 2500 metres (*c.* 8000 ft), i.e. well below freezing level. In other respects there is plenty of evidence of internal variety in which both latitude and longitude are seen to play a significant role.

In *winter* (January) latitude has a notable effect upon recorded temperatures at all the heights quoted and the increase of potential temperature from the surface to the 700-millibar level is then greater at the more southerly stations (16–17°K) than at the northern pair (13–14°K). Raizet and Sal are within the trades but Weather Ship E and Funchal then lie in the horse latitudes, not infrequently invaded by 'polar' disturbances. Yet there is evidence for the prevalence of a weak inversion above 850 millibars at both of these latter stations so that the family likeness is still retained. Furthermore there is a notable *decrease* in saturation potential temperature at all four stations from the surface to the 850-millibar level which suggests that, given general uplift, the lower layers of this air-mass could generate 'layer lability' (see p. 251). Only at Raizet is the fall of SPT continued up to 700 millibars; elsewhere the feature is a shallow one and normally inoperative. The most important longitudinal contrast in winter is indeed between Raizet and Sal, the two genuinely trade-wind stations at this season. Note particularly the very low relative humidity at 850 millibars at the latter station in January. This is evidence of the presence of a strong low-level inversion giving a remarkably stable ascent (see Fig. 91(b)). Indeed the mean rainfall in January at Sal is negligible whilst Raizet records a mean of about 90 millimetres. Before widespread dynamic descent was generally accepted as an essential feature of the trade-wind circulation the strong inversion recognised off the coast of west Africa was usually attributed to an invasion aloft of Saharan air. That such invasions occur is shown by the distribution of observed dust-falls at sea (Fig. 92) but the ascents at Funchal, particularly in July, suggest that dynamic considerations are paramount. Indeed, since Saharan air is also subsident, this is really a distinction without a difference.

In *summer* (July) when the north-east trades occupy a slightly more northerly position, the evidence of Table 28 points much more strongly to longitudinal variations within the tropical maritime air mass. At both Funchal and Sal the increase of potential temperature from the surface to 700 millibars is now considerably greater (18–20°K) than at the more westerly stations (14°K) and a much greater proportion of this rise is achieved below 850 millibars. This indication of the strength of the low-level inversion is reinforced at both stations by the low relative humidities recorded at 850 millibars. At Weather Ship E and at Raizet the inversion is both higher and weaker and indeed

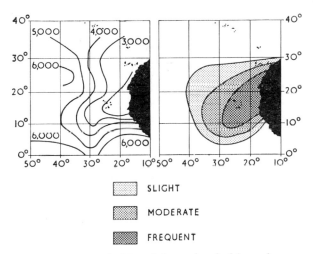

SLIGHT

MODERATE

FREQUENT

Fig. 92. The mean height of the trade-wind inversion (in feet) and the distribution of Saharan dust over the eastern Atlantic

the evidence barely justifies its inclusion in the July ascent of Fig. 90(b). Yet the sharp decrease of relative humidity between 850 and 700 millibars at both of these stations does suggest that free convection to and above the 700-millibar level is less common than a purely mechanical interpretation of Fig. 90 might imply. Even a slight modification of the 'path curve' as a result of entrainment could have marked consequences here. Doubtless the summer inversion varies in strength from time to time. Nevertheless the ascents do reveal a marked contrast in the character of trade-wind air on the eastern and western flanks of the Atlantic in July. In fact, whilst Sal records little precipitation (c. 20 mm) at this time, Raizet obtains no less than 180 milli-metres. Again, at this season, the decrease in saturation potential temperature with height is more rapid at the western stations and is continued to at least 700 millibars at both of them. The summer air over the West Indies and the Sargasso sea has thus potential layer lability throughout a very considerable depth. If this air is swept inland over the American continent and thus subjected to general uplift, the possibilities for widespread overturning are immense. There is plenty of evidence that these possibilities are much more

TABLE 28. *Maritime tropical air—North Atlantic*

### (a) JANUARY

| Level (mb) | WESTERN ATLANTIC | | | | | EASTERN ATLANTIC | | | | |
|---|---|---|---|---|---|---|---|---|---|---|
| | T (°C) | PT (°K) | SPT (°C) | MR (g/kg) | RH (%) | T (°C) | PT (°K) | SPT (°C) | MR (g/kg) | RH (%) |
| WEATHER SHIP E 35°00′N, 48°00′W | | | | | | FUNCHAL (Madeira) 32°38′N, 16°54′W (56 m) | | | | |
| 300 | −43 | 324 | — | — | — | −47 | 319 | — | — | — |
| 500 | −16 | 313 | 14.0 | 0.7 | 34 | −18 | 311 | 13.4 | 0.6 | 29 |
| 700 | −1 | 302 | 11.4 | 2.0 | 39 | 0 | 302 | 11.2 | 1.6 | 30 |
| 850 | 6 | 292 | 10.3 | 4.5 | 64 | 7 | 293 | 10.8 | 4.7 | 65 |
| SFC | 17 | 289 | 14.2 | 9.5 | 77 | 16 | 288 | 12.7 | 8.3 | 74 |
| | (1018 mb) | | | | | (1010 mb) | | P = 84 mm | | |
| RAIZET (Guadeloupe) 16°16′N, 61°31′W (8 m) | | | | | | SAL (Cape Verde Is.) 16°44′N, 22°57′W (55 m) | | | | |
| 300 | −34 | 337 | 20.8 | 0.15 | 22 | −38 | 332 | — | — | — |
| 500 | −7 | 324 | 18.0 | 0.8 | 18 | −10 | 320 | 16.7 | 0.9 | 26 |
| 700 | 8 | 311 | 16.3 | 3.6 | 38 | 8 | 311 | 15.3 | 2.4 | 25 |
| 850 | 14 | 301 | 17.2 | 8.2 | 68 | 17 | 303 | 13.9 | 3.7 | 26 |
| SFC | 23.5 | 295 | 20.3 | 14.5 | 79 | 21.5 | 294 | 17.0 | 10.9 | 67 |
| | (1016 mb) | | P = 91 mm | | | (1010 mb) | | P = 2 mm | | |

### (b) JULY

| Level (mb) | | | | | | | | | | |
|---|---|---|---|---|---|---|---|---|---|---|
| WEATHER SHIP E | | | | | | FUNCHAL (Madeira) | | | | |
| 300 | −37 | 333 | — | — | — | −37 | 332 | 19.6 | 0.1 | 18 |
| 500 | −9 | 321 | 17.3 | 1.2 | 32 | −8 | 322 | 17.2 | 0.6 | 15 |
| 700 | 6 | 309 | 15.6 | 3.7 | 44 | 9 | 312 | 15.3 | 1.9 | 18 |
| 850 | 14 | 301 | 16.6 | 7.7 | 63 | 15 | 301 | 14.4 | 4.9 | 40 |
| SFC | 25 | 295 | 21.3 | 15.8 | 82 | 21 | 292 | 16.7 | 11.0 | 71 |
| | (1025 mb) | | | | | (1013 mb) | | P = 2 mm | | |
| RAIZET (Guadeloupe) | | | | | | SAL (Cape Verde Is.) | | | | |
| 300 | −34 | 338 | 21.0 | 0.25 | 33 | −33 | 338 | 21.0 | 0.30 | 38 |
| 500 | −7 | 324 | 18.9 | 1.8 | 38 | −8 | 323 | 18.7 | 2.1 | 48 |
| 700 | 9 | 312 | 17.4 | 4.4 | 44 | 11 | 314 | 18.8 | 5.2 | 44 |
| 850 | 16 | 303 | 19.8 | 10.7 | 76 | 21 | 308 | 18.7 | 7.4 | 39 |
| SFC | 26.5 | 298 | 23.5 | 17.7 | 80 | 24 | 296 | 20.3 | 13.8 | 71 |
| | (1015 mb) | | P = 179 mm | | | (1008 mb) | | P = 18 mm | | |

frequently realised in summer than in winter. The great reserve of energy represented by this warm and very humid air mass may also be released in quite a different way, by the organisation and development of occasional tropical hurricanes (see pp. 386-99).

It would be reasonable to infer that the characteristics of tropical maritime air thus recognised in the Atlantic would be paralleled in similar latitudes in the North Pacific. Despite the risk of repetition it seems wise to check such an assumption, particularly since the trade-wind cells of the southern hemisphere leave so much open to conjecture. Data for four Pacific stations are thus presented in Table 29. In comparing these values with those of Table 28 two important considerations must be kept in mind: (*a*) There is no station in the Pacific, nor indeed in the whole world, in a position at all analogous to that of Sal. Johnston Island (a low atoll) is in the central Pacific and already has many 'western' characteristics (cp. Raizet). (*b*) The great width of the Pacific introduces features that have no close parallel in the Atlantic. Thus, at least for a season, both Weather Ship V (in winter) and Guam (in summer) may be regarded as lying beyond the western limits of tropical maritime air and their records are indeed of special interest in showing the nature of the transformation which occurs in this direction.

As in the Atlantic, all four stations show high near-surface mixing-ratios and again the values increase down-wind from Ship N, through Johnston Island, towards Guam throughout the year. From 700 millibars upwards the air is also relatively dry though, taking the year as a whole, Atlantic conditions are closely paralleled only at Ship N and Johnston Island. Here a cloud ceiling of the general order of 2500 metres is again to be expected but it is evident that in the far west (at Guam), in summer quite a new set of conditions prevails.

In *winter* (January) the effect of latitude upon the general temperature of the air is again notable but though potential temperature increases from the surface to the 700 millibar level by 15° K at the stations then within the trades (Johnston Island and Guam), as well as at Ship N beyond their northern roots, Ship V is clearly the odd man out. The very slight rise of potential temperature up to 850 mb at this location reflects a degree of low-level near-instability not paralleled in similar latitudes in the Atlantic except perhaps over the Gulf Stream. This is already mid-latitude air. As in the Atlantic there is an initial fall of SPT with height but layers involved are deep only at Johnston Island and Guam where the situation is closely analogous to that at Raizet. In winter this air is warded off the Asiatic continent by the strength and persistence of the northern 'dry monsoon' and general uplift over the Pacific can only be engineered by frontal phenomena. The long-term mean rainfall in January at Johnston Island is given as 100 millimetres (during the period 1962–66 it was only 25 mm) and though the mean fall at Guam is 120 millimetres this is one of the driest months of the year. The weather ships yield no record.

In *summer* (July), as in the Atlantic, longitude plays a relatively more important role than latitude though this may not be immediately apparent from Table 29 owing both to the westerly position of Johnston Island and to the fact that Ship V is now within the trades whilst Guam is very near or beyond their margin. As at Funchal, potential temperature increases by 19° K from the surface to 700 millibars at Ship N but evidence that the inversion is normally at a higher level than in the Atlantic is provided both

TABLE 29. *Maritime tropical air: North Pacific*

### (a) JANUARY

| | EXTREME WESTERN PACIFIC | | | | | EASTERN AND CENTRAL PACIFIC | | | | |
|---|---|---|---|---|---|---|---|---|---|---|
| Level (mb) | T (°C) | PT (°K) | SPT (°C) | MR (g/kg) | RH (%) | T (°C) | PT (°K) | SPT (°C) | MR (g/kg) | RH (%) |
| WEATHER SHIP V[1] | | | | | | WEATHER SHIP N | | | | |
| 34°N, 164°E | | | | | | 30°N, 140°W | | | | |
| 300 | −39 | 331 | — | — | — | −42 | 326 | — | — | — |
| 500 | −19 | 309 | (12.8) | (0.76) | (45) | −15 | 315 | (15.2) | (0.8) | (33) |
| 700 | −5 | 297 | 9.4 | 1.8 | 47 | 2 | 305 | 13.0 | 2.3 | 35 |
| 850 | 3 | 289 | 8.6 | 4.0 | 69 | 8 | 294 | 11.5 | 4.6 | 57 |
| SFC | 15 | 287 | 11.9 | 8.0 | 75 | 18 | 289 | 14.3 | 9.7 | 75 |
| (1011 mb) | | | | | | (1022 mb) | | | | |
| GUAM | | | | | | JOHNSTON ISLAND | | | | |
| 13°33′N, 144°50′E (110 m) | | | | | | 16°44′N, 169°31′W (5 m) | | | | |
| 300 | −31 | 341 | 21.8 | 0.27 | 30 | −33 | 339 | 21.3 | 0.20 | 24 |
| 500 | −5 | 325 | 19.0 | 1.3 | 25 | −7 | 324 | 18.2 | 1.2 | 27 |
| 700 | 10 | 313 | 17.3 | 3.5 | 31 | 9 | 312 | 15.8 | 2.7 | 27 |
| 850 | 17 | 303 | 19.2 | 9.5 | 67 | 14 | 300 | 16.5 | 7.7 | 66 |
| SFC | 25 | 298 | 22.7 | 16.9 | 82 | 25 | 297 | 20.9 | 14.6 | 73 |
| (1000 mb) | | | | $P = 118$ mm | | (1013 mb) | | | $P = 99$ mm | |

### (b) JULY

| | | | | | | | | | | |
|---|---|---|---|---|---|---|---|---|---|---|
| WEATHER SHIP V | | | | | | WEATHER SHIP N | | | | |
| 300 | −33 | 338 | 21.1 | 0.28 | 37 | −37 | 332 | (19.5) | (0.15) | (28) |
| 500 | −8 | 323 | 18.5 | 1.8 | 42 | −9 | 322 | (17.2) | (1.0) | (25) |
| 700 | 8 | 311 | 17.6 | 4.9 | 51 | 7 | 310 | 14.7 | 2.5 | 28 |
| 850 | 16 | 302 | 18.3 | 8.9 | 67 | 11 | 297 | 13.7 | 5.8 | 60 |
| SFC | 23.5 | 295 | 21.6 | 16.2 | 88 | 20.5 | 291 | 17.0 | 11.8 | 78 |
| (1016 mb) | | | | | | (1023 mb) | | | | |
| GUAM[2] | | | | | | JOHNSTON ISLAND | | | | |
| 300 | −31 | 341 | 21.8 | 0.36 | 39 | −34 | 337 | 20.7 | 0.22 | 30 |
| 500 | −6 | 326 | 20.0 | 2.7 | 54 | −7 | 324 | 18.3 | 1.3 | 29 |
| 700 | 10 | 314 | 20.2 | 7.2 | 66 | 9 | 312 | 16.3 | 3.3 | 34 |
| 850 | 19 | 305 | 21.8 | 12.5 | 78 | 15 | 302 | 18.6 | 9.4 | 72 |
| SFC | 26.5 | 300 | 23.8 | 17.8 | 80 | 27 | 299 | 23.2 | 16.9 | 75 |
| (997 mb) | | | | $P = 228$ mm | | (1014 mb) | | | $P = 33$ mm | |

[1] Mid-latitude air.
[2] Equatorial air.
Bracketed figures indicate incomplete records.

by the lower proportion of the increase which occurs by 850 mb (6°K as compared with 9°K at Funchal) and by the greater relative humidities recorded at the 850-millibar surface. It would be unwise to assume that this represents any real distinction between conditions over the two oceans for the latitude of the two sites is not identical and, in any case, the Pacific north-east trades extend into unusually high latitudes in July (see Fig. 63); the two airstreams are thus not being sampled in precisely comparable locations. In any event, the similarity between the two ascents is much more impressive than the difference. The Johnston Island record is remarkably similar to that for Raizet in July and it thus comes as something of a surprise to learn that its mean rainfall is no more than 35 millimetres as compared with 180 milli-metres at the latter station. Relief may well play some part in this contrast though occasional cloudbursts associated with the passage of hurricanes tend to raise mean values at most West Indian stations. By Guam, on the other hand, the trades have clearly lost most of their inherent stability. Relative humidity aloft is considerably higher than in any tropical air mass so far encountered and SPT values fall progressively to the 500-millibar level though the decrease in the lower layers is less rapid than at Johnston Island. This is doubtless in part a result of past weather so that the transition to equa-torial air is well under way. The mean July rainfall of Guam is some 230 millimetres, and the figure was in fact considerably higher during the 5 years of our standard period. When such air is swept into Asia during the summer months it has much to contribute to the wet monsoon.

(c) *The northeastern monsoon.* Although it prevails during only one season of the year the northeastern monsoon of southeast Asia is shown in Fig. 79 as covering an area of some 10.8 million square miles in January. So vast a system can hardly be dismissed as anomalous, and in its general space relations it is clearly associated with the tropical continental and tropical maritime air masses discussed above. In fact about a third of the area mapped lies over Asia and the remainder is superimposed upon comparatively warm seas. Considerable internal variety is thus to be anticipated. In view of the inherent climatological interest of this part of the world it has been deemed appropriate to illustrate its air mass characteristics in some detail.

The graphs of Fig. 93 and the data quoted in Table 30 present the relevant facts for both January and July at four stations near the poleward limits of this system. In this section we are concerned only with the situation in *January*. All four stations show a very considerable degree of stability in the layers from the surface to the 700-millibar level. Potential temperature increases over this range by 14–18°K and, except at Chiang Mai, nearly half of this increase is achieved by 850 millibars. Furthermore, particularly at Hong Kong, these lower layers are capped by a deep zone of stable and dry air aloft. These are the hall-marks of prevailing subsidence and the northeast monsoon is fed mainly from this source. Hence the prevailing aridity—mean January rainfall, Jodhpur 8 millimetres, Calcutta 13 millimetres, Chiang Mai 7 millimetres and Hong Kong 30 millimetres—and it is clear that, if this circula-tion was as persistent as the trades in similar latitudes, most of southeast Asia would be semidesert. The January ascent at Jodhpur indeed bears a

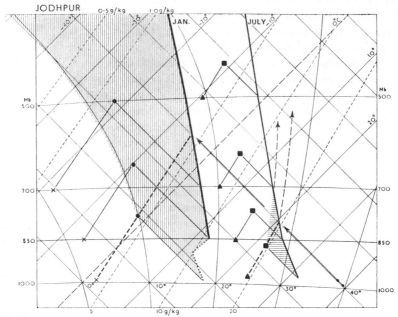

Fig. 93a

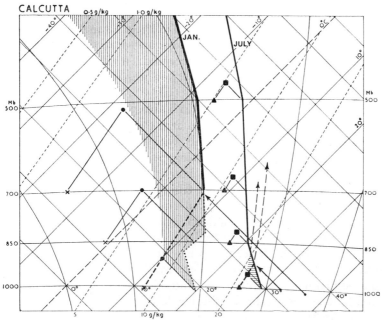

Fig. 93b

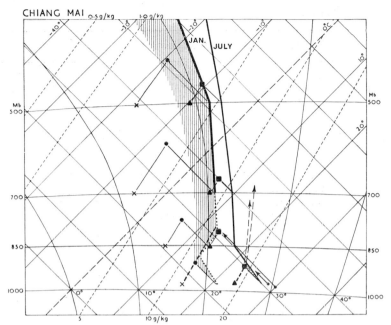

Fig. 93c

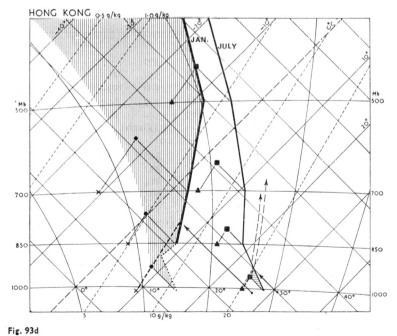

Fig. 93d

FIG. 93. Tephigrams illustrating north-east monsoon air (January) and south-west monsoon air (July) at Jodhpur, Calcutta, Chiang Mai and Hong Kong

295

TABLE 30. *Monsoon air masses—I*

### (a) THE NORTHEAST MONSOON (JANUARY)

| Level (mb) | T (°C) | PT (°K) | SPT (°C) | MR (g/kg) | RH (%) | T (°C) | PT (°K) | SPT (°C) | MR (g/kg) | RH (%) |
|---|---|---|---|---|---|---|---|---|---|---|
| **JODHPUR** | | | | | | **CALCUTTA** | | | | |
| 26°18′N, 73°01′E (224 m) | | | | | | 22°39′N, 88°27′E (6 m) | | | | |
| 300 | −35 | 336 | — | — | — | −33 | 339 | — | — | — |
| 500 | −12 | 318 | — | — | — | −8 | 323 | — | — | — |
| 700 | 3 | 306 | 12.3 | 1.4 | 19 | 6 | 309 | 13.9 | 1.9 | 23 |
| 850 | 13 | 299 | 11.3 | 2.8 | 26 | 13 | 299 | 13.0 | 4.3 | 40 |
| SFC | 18 | 291 | 9.9 | 4.4 | 34 | 20 | 292 | 14.5 | 8.7 | 59 |
| | (990 mb) | | | *P* = 8 mm | | (1014 mb) | | | *P* = 13 mm | |
| **CHIANG MAI** | | | | | | **HONG KONG** | | | | |
| 18°47′N, 98°59′E (313m) | | | | | | 22°18′N, 114°10′E (33 m) | | | | |
| 300 | −33 | 338 | — | — | — | −32 | 339 | 21.3 | 0.15 | 18 |
| 500 | −7 | 324 | 19.0 | 1.45 | 32 | −8 | 323 | 16.8 | 0.47 | 12 |
| 700 | 7 | 310 | 16.3 | 3.8 | 42 | 2 | 305 | 13.3 | 2.4 | 37 |
| 850 | 14 | 300 | 16.7 | 7.8 | 65 | 9 | 295 | 11.9 | 4.9 | 60 |
| SFC | 21 | 296 | 18.3 | 11.5 | 74 | 16 | 287 | 11.6 | 7.4 | 66 |
| | (980 mb) | | | *P* = 7 mm | | (1016 mb) | | | *P* = 30 mm | |

### (b) THE SOUTHWEST MONSOON (JULY)

| Level (mb) | T (°C) | PT (°K) | SPT (°C) | MR (g/kg) | RH (%) | T (°C) | PT (°K) | SPT (°C) | MR (g/kg) | RH (%) |
|---|---|---|---|---|---|---|---|---|---|---|
| **JODHPUR** | | | | | | **CALCUTTA** | | | | |
| 300 | −24 | 351 | — | — | — | −25 | 350 | — | — | — |
| 500 | −1 | 331 | 22.2 | 3.8 | 54 | −1 | 331 | 23.2 | 5.0 | 73 |
| 700 | 15 | 318 | 22.6 | 8.6 | 57 | 12 | 316 | 23.4 | 10.6 | 81 |
| 850 | 24 | 311 | 24.4 | 14.1 | 63 | 21 | 307 | 24.3 | 15.3 | 84 |
| SFC | 32 | 307 | 26.7 | 19.4 | 61 | 29 | 302 | 26.6 | 21.7 | 82 |
| | (973 mb) | | | *P* = 122 mm | | (998 mb) | | | *P* = 301 mm | |
| **CHIANG MAI** | | | | | | **HONG KONG** | | | | |
| 300 | −28 | 345 | 23.2 | 0.62 | 48 | −29 | 345 | 22.7 | 0.42 | 35 |
| 500 | −5 | 326 | 20.8 | 3.7 | 71 | −4 | 327 | 20.5 | 2.8 | 51 |
| 700 | 10 | 313 | 21.2 | 8.7 | 80 | 11 | 314 | 20.6 | 7.3 | 61 |
| 850 | 18 | 305 | 21.4 | 12.1 | 80 | 19 | 305 | 21.9 | 12.4 | 78 |
| SFC | 27.5 | 303 | 25.3 | 18.9 | 77 | 29 | 301 | 25.9 | 20.3 | 80 |
| | (970 mb) | | | *P* = 188 mm | | (1002 mb) | | | *P* = 286 mm | |

remarkably close resemblance to that for Aswan whilst Calcutta is more like Sal. At Chiang Mai, where the lower latitude and the surrounding uplands have combined to reduce the low-level inversion, the ascent is more reminiscent of Raizet though, not unexpectedly, the mixing-ratio of the near-surface air is considerably lower. For its latitude Hong Kong is unique in the comparatively low temperatures recorded up to at least 700 millibars. This is in harmony with the greater strength and persistence of the Chinese wing of the north-east monsoon which here, and here alone, may be fed in part from cold winter air from the Asiatic continent. Elsewhere the relief obstacles are insurmountable.

Downwind from these stations, that is, in a general equatorward direction, it is to be anticipated that this air will be warmed from below and progressively destabilized. There is some evidence that in the 'winter' of such low latitudes this occurs more rapidly over the sea than over land where night minima are comparatively low. The inversion thus roofs over almost the whole of the Indian sub-continent except perhaps its most extreme southerly tip

TABLE 31. *Monsoon air masses—II*

| Level (mb) | T (°C) | PT (°K) | SPT (°C) | MR (g/kg) | RH (%) | T (°C) | PT (°K) | SPT (°C) | MR (g/kg) | RH (%) |
|---|---|---|---|---|---|---|---|---|---|---|
| MADRAS 13°00′N, 80°11′E (16 m) | | | | | | SAIGON 10°49′N, 106°40′E (10 m) | | | | |
| (a) THE NORTHEAST MONSOON (JANUARY) | | | | | | | | | | |
| 300 | −32 | 340 | — | — | — | −32 | 340 | 21.5 | 0.20 | 23 |
| 500 | −6 | 326 | — | — | — | −6 | 326 | 18.6 | 1.15 | 23 |
| 700 | 10 | 313 | 16.7 | 3.2 | 29 | 9 | 312 | 16.6 | 3.5 | 35 |
| 850 | 16 | 303 | 17.0 | 7.1 | 52 | 15 | 302 | 17.9 | 8.8 | 69 |
| SFC | 24.5 | 296 | 20.7 | 14.4 | 74 | 25.5 | 298 | 21.2 | 14.3 | 69 |
| | (1012 mb) | | P = 24 mm | | | (1011 mb) | | P = 6 mm | | |
| (b) THE SOUTHWEST MONSOON (JULY) | | | | | | | | | | |
| 300 | −31 | 342 | — | — | — | −31 | 341 | 21.7 | 0.33 | 36 |
| 500 | 5 | 327 | — | — | — | 7 | 325 | 19.9 | 2.9 | 62 |
| 700 | 10 | 313 | 21.3 | 8.7 | 79 | 8 | 312 | 19.6 | 7.0 | 70 |
| 850 | 20 | 307 | 22.7 | 12.9 | 72 | 17 | 304 | 20.6 | 11.3 | 76 |
| SFC | 30.5 | 303 | 25.3 | 18.5 | 65 | 27.5 | 300 | 24.8 | 19.3 | 82 |
| | (1002 mb) | | P = 83 mm | | | (1007 mb) | | P = 242 mm | | |

(Trivandrum) and even at Madras (Fig. 94(a)) the mean January rainfall is no more than 24 millimetres. At Port Blair (11°40′N, 92°43′E) in the Andaman Islands, though the inversion is present, it is both higher and weaker and the January rainfall is 40 millimetres. It would appear indeed that, in contrast to most trade-wind cells, the inversion level in this air mass rises generally in a west-to-east direction, i.e. with increasing distance from the Arabian and Iranian sources of desert air. In the absence of a well-placed ascent in the Arabian Sea this feature cannot be convincingly demonstrated but the climatological evidence, e.g. from Somalia, is all in its favour. Note that at Saigon, in the east, the inversion is normally above 850 millibars, some 1500 metres, though since the mean January rainfall is only 6 millimetres it is still apparently quite effective; of course there are unquestionably relief effects here. Yet the persistence of the feature is all the more remarkable in view of the great transformation that has occurred in the lower layers since Hong Kong—near-surface temperature has increased by some 9°C and the mixing-ratio has doubled (compare Calcutta-Madras). Above the inversion level the air remains remarkably dry and the ascent (Table 31) bears a general resemblance to that at Johnston Island far to the east.

It is evident from Figs. 93 and 94 and from Tables 30 and 31 that, at all monsoon stations, aerological conditions in July differ radically from those recorded in January. We shall return to the summer or wet monsoon at a later stage but it is appropriate here to draw attention to a feature common to the ascents at Madras, Port Blair and Saigon (it is also evident at Raizet and Guam). At these stations the seasonal contrast in the *temperature* profile is confined to the layers below 750 millibars. Above that level, the air temperatures recorded in January and July differ by no more than the sampling error and closely follow the course of a saturation adiabat at least up to about 300 millibars. Yet there is a real contrast in air-mass type since the middle atmosphere is very much drier in January than in July as is very well shown by the excellent record for Saigon. (It is also suggested by the less complete records at Madras and Port Blair.) Why is it that heating by general subsidence so closely reproduces the temperature-profile characteristic of a very different process, the release of latent heat at condensation? Is it purely a matter of chance? We have met this question before (see p. 283). The author is not aware that these facts have yet attracted the attention of meteorological specialists.

Since the duration of the dry monsoon inversion is fundamental to a comprehension of the climate of India we have offered in Table 32 an analysis of ascents at four equally-spaced intervals of time at Bombay. It will be

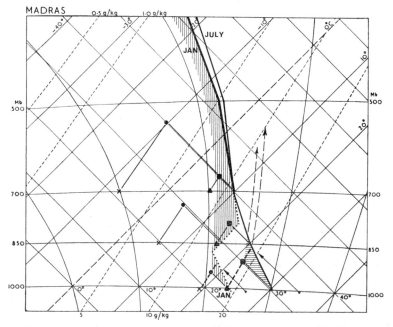

Fig 94a

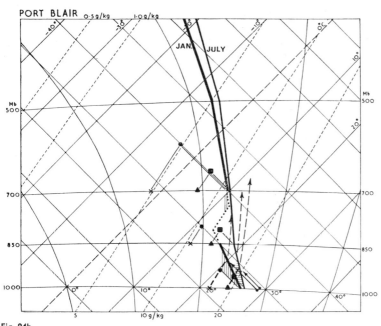

Fig. 94b

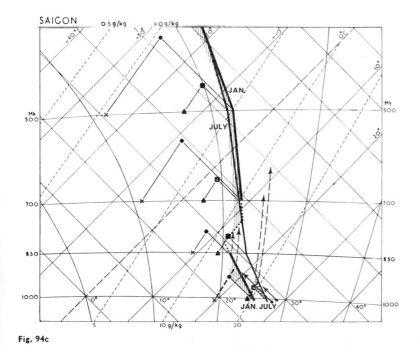

Fig. 94c

FIG. 94. Tephigrams for January and July at Madras, Port Blair and Saigon

299

recalled that this station experiences some of the sharpest seasonal rainfall contrasts known on earth (Fig. 42). Clearly the inversion is no less potent in April than in January and there are evident signs of its regeneration in October at the conclusion of the June-to-September rains. Note that saturation potential temperature decreases with height throughout the year. As with

TABLE 32. *Air masses at Bombay, 1962–66*

| Level (mb) | T (°C) | PT (°K) | SPT (°C) | MR (g/kg) | RH (%) | T (°C) | PT (°K) | SPT (°C) | MR (g/kg) | RH (%) |
|---|---|---|---|---|---|---|---|---|---|---|
| JANUARY | | | | | | APRIL | | | | |
| 500 | −7 | 324 | — | — | — | −6.5 | 325 | — | — | — |
| 700 | 8.5 | 312 | 15.1 | 2.1 | 21 | 11.5 | 315 | 19.4 | 5.6 | 46 |
| 850 | 17.5 | 304 | 15.5 | 4.8 | 32 | 23.5 | 310 | 18.5 | 6.3 | 29 |
| SFC | 24.5 | 297 | 19.9 | 13.1 | 66 | 28.5 | 301 | 23.8 | 17.2 | 69 |
| | (1012 mb) | | | P = 2 mm | | | (1008 mb) | | P = 3 mm | |
| JULY | | | | | | OCTOBER | | | | |
| 500 | −3 | 328 | — | — | — | −3.5 | 329 | — | — | — |
| 700 | 11.5 | 313 | 21.6 | 8.4 | 69 | 10 | 315 | 18.7 | 5.5 | 50 |
| 850 | 18.5 | 305 | 22.7 | 13.8 | 86 | 21 | 308 | 20.3 | 9.2 | 48 |
| SFC | 27.5 | 300 | 25.7 | 20.4 | 86 | 29 | 301 | 25.0 | 19.2 | 75 |
| | (1002 mb) | | | P = 709 mm | | | (1008 mb) | | P = 88 mm | |

most tropical maritime air masses, the air is thus open to layer lability at all times but the regional uplift necessary to release this potential can only occur in the presence of a westerly or southwesterly circulation. It should not surprise us that the fall of SPT from the surface to 700 millibars is rather less in July than during the other 3 months quoted since this may well be evidence that overturning has already occurred. It is greatest in October, a month of rare but violent rainstorms (Fig. 48).

## The secondary air masses

It will have been noted that all the various types of air mass discussed above fall into coherent and comparatively well defined systems where divergence predominates and a close causal connection is built up between air mass characteristics and the nature of the terrestrial surface beneath. Having mapped them, however approximately, we are left with three large sectors of the atmosphere, one in the neighbourhood of the equator and one in the middle latitudes of each hemisphere, where this relationship, though by no means completely lost, is usually of a much less immediate nature. If these are 'source regions' the term is here used in a rather different sense. Air converges upon these three zones from their northern and southern flanks and brings with it characteristics generated within its primary source. At first recognisably 'modified Arctic' or 'modified tropical' air of one variety or another, it is subjected to a complex series of reactions or weather processes which lead in general to some loss of stability, much intermixture and frequent and widespread precipitation. The nature of the air then found at any given place or time thus depends much more on the past weather processes in which it has been involved than on the intrinsic qualities of the place where it is being observed. It is in this sense only that we use the term 'secondary'. Here the stream of causation is *downwards* rather than upwards from the earth's surface. In consequence the contribution made by such air-masses to the habitability of the earth, both by plants and man, is of the very greatest significance.

## Equatorial air

As tropical maritime air moves onwards towards the equator the temperature and humidity of its lower layers increase until the trade-wind inversion is undermined. The active convection which is then triggered off builds up a deep cloud-cover, distributing high humidities throughout much of the troposphere and imposing a firm check on further increase of surface temperatures. Furthermore the convergence from the two hemispheres of deep zones of warm, humid and rather unstable air may produce heavy precipitation at almost any point round the complete circuit of the earth and particularly where they invade the neighbouring continents. We have already emphasised that the actual location of this zone of equatorial air has only the loosest possible connection with the 'vertical sun'—it is rather the result of a delicate balance of forces operating from each hemisphere. Such forces are by nature variable, hence equatorial lands experience considerable variety of weather even where the climatic pattern suggests year-long monotony. Indeed this monotony has often been grossly exaggerated in the past and it is with no little surprise that one learns that Ocean Island (0°52′S, 169°36′E) in the west-central Pacific has perhaps the most variable rainfall on earth (Crowe, 1951, 47, 50).

Table 25 shows that equatorial air covers over 25 million square miles in January, that is rather more than an eighth of the total surface of the earth. This is probably a conservative estimate for the equatorial borders of the trades are extremely ill-defined. Figure 79 suggests that, at this season, only about 25 per cent of the area lies *north* of the equator. In July, on the other hand (Fig. 80), of the comparable area—including the southwestern monsoon—no more than 15 per cent lies *south* of the equator. These proportions give some measure of the seasonal oscillation which occurs. It is thus evident that less than half of the equatorial zone experiences truly equatorial conditions throughout the year. Indeed it is quite difficult to find a station where mean ascents in January and July do not differ in some important particular. Temperatures at all heights may be so similar that it is not easy to show separate curves on the scale of our diagrams but there is often a significant contrast in upper-air humidity or else, as in the Indian Ocean, there is a well-known change in the general direction of the circulation. This may result in

TABLE 33. *Stations near the geographical equator*

| | | JANUARY | | | | | JULY | | | |
|---|---|---|---|---|---|---|---|---|---|---|
| *Level* (mb) | *T* (°C) | *PT* (°K) | *SPT* (°C) | *MR* (g/kg) | *RH* (%) | *T* (°C) | *PT* (°K) | *SPT* (°C) | *MR* (g/kg) | *RH* (%) |
| CANTON ISLAND 2°46′S, 171°43′W (3 m) | | | | | | | | | | |
| 300 | −30.5 | 342 | 22.1 | 0.30 | 29 | −31.5 | 340.5 | 21.7 | 0.21 | 23 |
| 500 | −4.5 | 327 | 19.3 | 1.5 | 28 | −5.5 | 326 | 18.8 | 1.3 | 25 |
| 700 | 10.5 | 313.5 | 17.6 | 3.9 | 35 | 9.5 | 313 | 17.3 | 3.8 | 35 |
| 850 | 17.0 | 304 | 19.2 | 9.5 | 65 | 17.5 | 304 | 18.8 | 9.0 | 61 |
| SFC | 28.0 | 300.5 | 23.8 | 17.5 | 71 | 28.0 | 300.5 | 24.2 | 18.0 | 73 |
| | (1008 mb) | | | | | | (1009 mb) | | | |
| SINGAPORE 1°21′N, 103°54′E (18 m) | | | | | | | | | | |
| 300 | −31.5 | 340.5 | 21.8 | 0.46 | 51 | −32.5 | 339 | 21.3 | 0.40 | 49 |
| 500 | −6.4 | 324.5 | 19.8 | 2.8 | 61 | −7.0 | 324 | 19.5 | 2.7 | 60 |
| 700 | 8.0 | 311.5 | 19.2 | 6.7 | 68 | 8.5 | 311.5 | 19.1 | 6.5 | 66 |
| 850 | 16.5 | 303 | 20.0 | 10.7 | 76 | 17.5 | 304 | 20.3 | 10.9 | 76 |
| SFC | 25.5 | 298 | 23.1 | 17.6 | 84 | 27.0 | 299 | 24.3 | 18.9 | 83 |
| | (1008 mb) | | *P* = 285 mm | | | | (1007 mb) | | *P* = 163 mm | |
| NAIROBI 1°18′S, 36°45′E (1798 m) | | | | | | | | | | |
| 300 | −31.5 | 341 | 21.7 | 0.23 | 25 | −32.0 | 340 | 21.5 | 0.17 | 19 |
| 500 | −6.0 | 325 | 18.4 | 1.8 | 38 | −6.0 | 325 | 18.7 | 1.4 | 29 |
| 700 | 9.0 | 311.5 | 19.7 | 7.1 | 70 | 7.0 | 310 | 19.6 | 7.5 | 82 |
| SFC | 17.5 | 307 | 21.6 | 11.6 | 75 | 15.5 | 304.5 | 20.3 | 10.6 | 79 |
| | (822 mb) | | *P* = 45 mm | | | | (823 mb) | | *P* = 19 mm | |

| SINGAPORE | *T* | *PT* | *SPT* | *RH* | *Mix. Ratio* |
|---|---|---|---|---|---|
| May–August (southerly air-stream) | | | | | |
| (mb) | (°C) | (°K) | (°C) | (%) | (g/kg) |
| 500 | −5.3 | 327 | — | — | — |
| 600 | 2.6 | 319 | 20.3 | 71 | 5.5 |
| 700 | 9.5 | 313 | 20.1 | 70 | 7.2 |
| 800 | 15.5 | 308 | 20.9 | 74 | 10.5 |
| 900 | 21.2 | 303 | 22.6 | 79 | 14.2 |
| 1000 | 25.1 | 298 | 23.9 | 91 | 18.9 |
| November–February (northerly air-stream) | | | | | |
| (mb) | (°C) | (°K) | (°C) | (%) | (g/kg) |
| 500 | −6.6 | 325 | — | — | — |
| 600 | 1.6 | 318 | 19.8 | 75 | 5.5 |
| 700 | 8.7 | 312 | 19.8 | 73 | 7.5 |
| 800 | 14.3 | 306 | 20.4 | 78 | 10.2 |
| 900 | 20.3 | 302 | 21.8 | 80 | 13.4 |
| 1000 | 23.6 | 297 | 23.0 | 91 | 17.2 |

The convective condensation level at both periods is at about 300 metres (*c.* 1000 ft).

part from the absence of satisfactory records from such places as Amazonia, the central Congo and Borneo but ground-based records also show that rainfall 'evenly distributed throughout the year' is the exception rather than the rule. There is usually a well-marked seasonal regime even where the fall in all months is sufficient to support genuine equatorial forest.

It would not seem unreasonable to suppose that ascents from stations as near as possible to the geographical equator would give the clearest indication of the inherent characteristics of equatorial air. In fact only three such stations are at present available and two of them present rather anomalous features. Data for Canton Island (lat. 2°46′S) an atoll in the west-central Pacific, for Singapore (lat. 1°21′N) in an insular location but to some degree affected by the monsoons, and for Nairobi (lat. 1°18′S) on a lofty African plateau, are presented in Table 33. It is clear at a glance that the temperature records have much in common but a closer inspection reveals notable contrasts in the humidity of the upper air; the surface rainfall records are also known to differ widely. At Canton Island the sharp drop in relative humidity between the 850 and 700 millibars levels in both January and July suggests the presence of a fairly persistent inversion. This is, in fact, really a trade-wind station and the air mass belongs to the western tropical maritime category. It is near the western limit of the trades during the southern summer and, like Ocean Island some 1300 miles farther west, its rainfall is then extremely variable. Thus the January fall in 1966 was 530 millimetres, while in January 1967 only 7 millimetres was recorded. Under such conditions mean values for a few years are quite valueless.

At Nairobi the cross-section of the atmosphere is curtailed since the station level is already at about 820 millibars. Here too the upper air is relatively dry in both January and July. The chief rains fall in fact between March and May and the April record has therefore been analysed. At 500 millibars the mixing-ratio is then 2.9 g/kg and the relative humidity 58 per cent, but for much of the year this station appears to be far from typical.

We are thus left with the Singapore record to give us a first key to the innate qualities of equatorial air. Here at least, despite the proximity of the monsoons, there is little contrast between January and July and one gets the impression of year-long uniformity. In fact it is quite difficult to separate the two curves in Fig. 95. A study by I. G. John (1949, 24–5), based upon meteorological flights from December 1946 to July 1948, yields the more detailed record for the two periods, May to August and November to February, given in the lower section of Table 33. Although a southerly air stream prevails during the former period and a northerly air stream is characteristic of the latter, both are rather weak and there is little distinction between them.

Like western tropical maritime air (Raizet, Johnston Island), equatorial air thus has the two characteristics:

1. High temperatures (about 25°C, 77°F) and high mixing-ratios (over 16 g/kg) near sea level.

2. An ascent curve not far from the SALR at least from the 850-millibar level to about 300 millibars.

In addition, note:

3. The very limited degree of near-surface stability and the absence of any evidence of persistent low-level inversions.
4. The very considerable water vapour content present in great depth. Typical mixing-ratio values appear to be in excess of 6 g/kg at 700 millibars and of 2 g/kg at 500 millibars.
5. These features are associated with considerable rainfall, usually in excess of 100 millimetres per month. Indeed, although, as in tropical maritime air, saturation potential temperature at first decreases with height, the decrease is less marked than in TM air simply because high-reaching convection has already occurred.

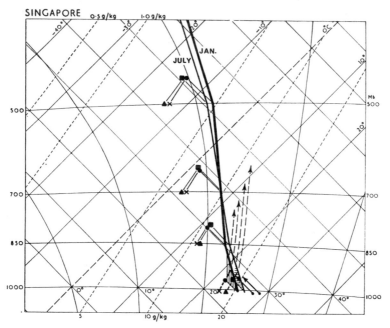

FIG. 95. Equatorial air at Singapore in January and July

In such air instability showers are easily set off by surface heating over land during the day and by orographic lifting at any time. Yet rainless spells of considerable duration undoubtedly occur. Watts (1955, 32–4) has made the important point that, though near the equator the air is 'invariably moist and conditionally unstable' so that 'clouds of great vertical extent might be predicted daily' from the thermodynamic diagram, yet very considerable contrasts in day-to-day weather patterns are the rule rather than the exception. Though 'much towering cloud does develop . . . the amount varies greatly . . . without correlation with any changes in the tephigram'.[1] He thus concludes that the forecaster must supplement the ascent data by a study of regional patterns of air flow. It seems therefore that, even in this basically unstable air mass, the most potent trigger mechanism for the release of weather activity is provided by regional convergence.

We may now use the five characteristics of equatorial air outlined above as a key or touchstone for the recognition of this air mass when it prevails, perhaps only for a season, at stations some degrees removed from the geographical equator.

TABLE 34. *The equatorial west Pacific*

| Level (mb) | T (°C) | PT (°K) | SPT (°C) | MR (g/kg) | RH (%) | T (°C) | PT (°K) | SPT (°C) | MR (g/kg) | RH (%) |
|---|---|---|---|---|---|---|---|---|---|---|
| | | JANUARY | | | | | JULY | | | |
| | | (Winter) | | | | | (Summer) | | | |
| TRUK 7°28'N, 151°51'E (2 m) | | | | | | | | | | |
| 300 | −30.0 | 342.5 | 22.1 | 0.28 | 27 | −30.5 | 341.5 | 22.0 | 0.36 | 36 |
| 500 | −4.5 | 327 | 20.0 | 2.4 | 44 | −5.5 | 326 | 20.2 | 2.7 | 53 |
| 700 | 10.5 | 314 | 18.2 | 4.6 | 40 | 9.5 | 313 | 19.9 | 7.0 | 65 |
| 850 | 18.0 | 304.5 | 20.1 | 10.3 | 68 | 18.0 | 305 | 21.1 | 11.5 | 73 |
| SFC | 27.0 | 299.5 | 24.3 | 18.7 | 81 | 27.0 | 299 | 24.5 | 19.2 | 85 |
| | (1009 mb) | | | P = 213 mm | | (1009 mb) | | | P = 913 mm | |
| | | (Summer) | | | | | (Winter) | | | |
| LAE 6°44'S, 147°00'E (8 m) | | | | | | | | | | |
| 300 | −31.0 | 341.5 | — | — | — | −32. | 340 | — | — | — |
| 500 | −6.0 | 325.5 | 20.0 | 2.75 | 55 | −6.5 | 324.5 | 19.7 | 2.65 | 57 |
| 700 | 9.5 | 313 | 19.8 | 6.8 | 62 | 8.0 | 311 | 19.1 | 6.5 | 66 |
| 850 | 19.0 | 305.5 | 20.8 | 11.0 | 67 | 15.5 | 302.5 | 19.7 | 10.6 | 80 |
| SFC | 27.5 | 300 | 24.2 | 18.2 | 77 | 24.5 | 297 | 22.5 | 17.0 | 86 |
| | (1006 mb) | | | P = 252 mm | | (1010 mb) | | | P = 529 mm | |

The western Pacific with the greatest accumulation of warm ocean water on earth (Crowe, 1951, 24) provides a logical starting-point for our review. We have already suggested that the ascent at Guam (lat. 13°33'N) in July (Table 29) is of this type and we now supplement this with the data for two stations, Truk (lat. 7°28'N) and Lae (lat. 6°44'S), lying at approximately equal distances astride of the equator (Table 34). At Truk in January, although the upper air is more humid than at Guam, there is still some evidence of an inversion at about 800 millibars though Lae is then clearly within equatorial air. In July, on the other hand, Guam, Truk and Lae are all submerged beneath an ocean of equatorial air which thus covers a range of

[1] Johnson (1970, 122–3) makes a similar point with reference to Gan Island (0° 41'S, 73° 09'E) as between 'wet' and 'other' days during the Julys of 1960–64.

not less than 20° of latitude. It is also evident that the border between this air-mass and the surrounding western tropical maritime air is rather vague and ill-defined. The data for Lae suggest that, in this part of the world, equatorial air remains more persistently south of the equator than north of it though its northward extension in July is remarkable.

TABLE 35. *The equatorial Atlantic*

| | JANUARY | | | | | JULY | | | | |
|---|---|---|---|---|---|---|---|---|---|---|
| Level (mb) | T (°C) | PT (°K) | SPT (°C) | MR (g/kg) | RH (%) | T (°C) | PT (°K) | SPG (°C) | MR (g/kg) | RH (%) |
| DOUALA 4°01′N, 9°43′E (13 m) | | | | | | | | | | |
| 300 | −31.5 | 341 | 21.7 | 0.19 | 21 | −31.5 | 341 | 21.8 | 0.27 | 30 |
| 500 | −6.5 | 324.5 | 18.3 | 1.2 | 26 | −6.5 | 325 | 19.6 | 2.4 | 50 |
| 700 | 8.5 | 312 | 17.3 | 4.3 | 43 | 8.5 | 312 | 19.5 | 7.0 | 70 |
| 850 | 17.5 | 304.5 | 20.5 | 11.2 | 76 | 16.0 | 303 | 20.3 | 11.3 | 81 |
| SFC | 27.5 | 299.5 | 24.6 | 19.0 | 81 | 25.0 | 297 | 22.8 | 17.5 | 87 |
| | (1008 mb) | | | P = 61 mm | | (1011 mb) | | | P = 710 mm | |
| LUANDA 8°15′S, 13°14′E (70 m) | | | | | | | | | | |
| 300 | −33.0 | 339 | 21.5 | 0.45 | 56 | −34.5 | 336.5 | 20.6 | 0.20 | 29[1] |
| 500 | −6.5 | 324.5 | 20.0 | 3.1 | 68 | −6.5 | 325 | 18.2 | 0.92 | 19 |
| 700 | 9.5 | 312.5 | 19.8 | 7.0 | 65 | 8.0 | 311.5 | 17.1 | 4.2 | 42 |
| 850 | 17.5 | 304 | 20.5 | 11.1 | 75 | 18.0 | 305 | 18.1 | 7.6 | 48 |
| SFC | 26.0 | 299 | 23.4 | 17.5 | 80 | 21.0 | 293.5 | 19.1 | 13.4 | 84 |
| | (1003 mb) | | | P = 26 mm | | (1007 mb) | | | P = 0 mm | |
| ALBROOK (Panama) 8°58′N, 79°33′W (6 m) | | | | | | | | | | |
| 300 | −34.0 | 337 | 20.8 | 0.35 | 49[2] | −32.0 | 339.5 | 21.6 | 0.46 | 54 |
| 500 | −5.5 | 326 | 19.3 | 1.7 | 33 | −6.0 | 325 | 20.1 | 3.0 | 62 |
| 700 | 10.0 | 313.5 | 17.8 | 4.2 | 38 | 10.0 | 313 | 20.2 | 7.1 | 64 |
| 850 | 17.0 | 304 | 19.7 | 9.9 | 67 | 18.0 | 304.5 | 21.7 | 12.5 | 82 |
| SFC | 27.0 | 299 | 22.6 | 16.2 | 71 | 27.0 | 299 | 23.2 | 17.0 | 75 |
| | (1010 mb) | | | | | (1009 mb) | | | | |

[1] Tropical maritime (eastern).
[2] Tropical maritime (western).

In the Atlantic, with its very different configuration, the pattern is broadly reversed. The more permanent sources of equatorial air lie north of the equator, in the Gulf of Guinea and the seas flanking the Central American isthmus. They are thus limited in area. Furthermore, owing to the great persistence of the southeast trades, the southward extension in January occurs mainly over the South American and African continents. The data for Douala (lat. 4°01′N) and Luanda (lat. 8°51′S) are useful here. They are given in Table 35 along with mean ascents at Albrook, Panama (lat. 8°58′N) though none of these stations is quite ideally situated. Douala, at the apex of the great re-entrant angle of the Atlantic coast of Africa, is one of the wettest places on that continent. Although within 300 miles of the equator, its rainfall nevertheless exhibits a strong seasonal regime and the summer maximum is usually attributed to the 'West African monsoon'. If this term implies a reversal of near-surface winds the evidence from Douala is not convincing for, in the afternoon (1400 hours) a 9- to 10-knot sea breeze from the southwest or west prevails with remarkable consistency, on over 80 per cent of occasions, throughout the year. It is true that soon after sunrise (at 0700 hours) winds from the northeast and east are recorded on 37 per cent of occasions in January as compared with only 15 per cent in July but the circulation is then weak and there is a very high proportion of calms (MO 492, 1949, 162). Surprisingly enough, the upper-air record is equally equivocal. The July record fulfils all five of our specifications for equatorial air but in January at Douala, as at other West African stations e.g. Abidjan, the air from 700 millibars upwards is unusually dry. This is usually attributed to the 'harmattan' but if there is an inversion of temperature between the two air masses it has been lost in the 850 to 700 millibar interval or has been obscured in the process of averaging. On either count it is hard to believe that it can be more than two or three degrees, which illustrates the delicate balance of forces in equatorial lands. On the evidence to hand it would be dishonest not to draw attention to the fact that, whether seen from the point of view of potential temperature or of saturation potential temperature, the January ascent at Douala is apparently less stable than that for July. This is the only tropical record which the author has analysed where such a difficulty has been encountered though Abidjan shares some of its features. It may well be that Watts's emphasis on the importance of streamline analysis for day-to-day forecasting in equatorial lands has indeed a longer-term climatic relevance in this particular part of the world.

The record for Luanda, on the other hand, is comparatively straightforward. In July there is an inversion of nearly 10° C below 850 millibars, little wonder that the rainfall is nil. This is clearly eastern tropical maritime air. In January, however, the air has several equatorial characteristics though a reasonable interpretation of the ascent would still show a weak inversion and precipitation is far from heavy. The station then lies near the margin of the southeast trades and it is possible that the high water vapour content in the upper air is derived from processes operating in the interior of the continent (see Broken Hill, Fig. 96 and Table 36).

Albrook, Panama, is unfortunately the only near-equatorial station available in the Americas for Bogota is too lofty to be much help to us. Despite its latitude there appears to be little doubt that the air is equatorial in July though a slight inversion above 850 mb may still be present. In January this feature is stronger and the ascent then bears a close resemblance to the Raizet record in July.

For the transition between equatorial and tropical continental air we must rely mainly upon the records of African stations where some difficulty arises from the elevation of the station sites. Darwin in northern Australia is also useful here. Ascents along an approximately north–south section across the equator from Fort Lamy (lat. 12°08′N) through Bangui (lat. 4°23′N) to Broken Hill (lat. 14°27′S) are illustrated in Fig. 96 and the Darwin record is shown in Fig. 97. The relevant analyses are given in numerical form in Table 36.

The remarkable fact which emerges from these observations is that, even where a strong seasonal contrast in air mass type is self-evident, it is not revealed by the temperature record of the middle and upper atmosphere. Humidity data are therefore essential. If low humidities may be accepted as evidence of general subsidence, these stations thus provide additional illustration of the tendency of that process to produce environment curves closely parallel to the SALR (see pp. 283; 298) in all but the very lowest layers of the atmosphere. At Fort Lamy, for instance, air temperatures—and hence also potential temperatures—are virtually identical in January and July at all levels from 850 millibars upwards, yet the two sets of ascents differ fundamentally in their inherent stability. In January the dryness of the whole column, the low-level inversion, and the great heights of the lifting condensation level (740 mb or some 2700 metres) and the convective condensation level (530 mb or 5300 metres) point clearly to air of Saharan origin and it is scarcely surprising that the rainfall is nil. In July, on the other hand, although the mean data do not completely satisfy all our postulated requirements for equatorial air, it seems evident that such air is frequently present. Occasional summer incursions of Saharan air aloft are by no means improbable at such a

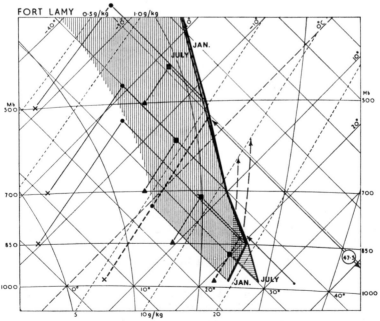

Fig. 96a

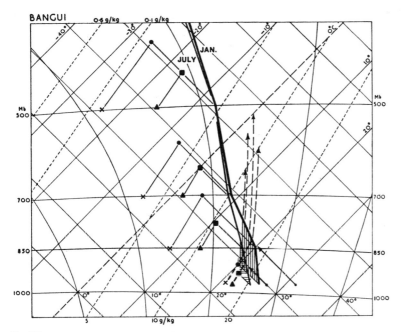

Fig. 96b

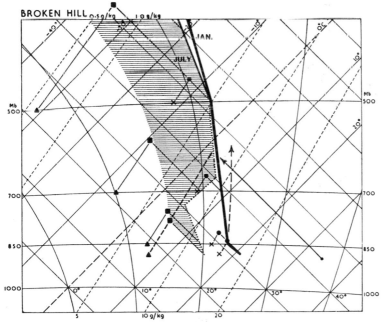

Fig. 96c

FIG. 96. Air mass contrasts astride the equator in central Africa. The tephigrams show mean ascents in January and July at Fort Lamy (12° 08′N), Bangui (4° 23′N) and Broken Hill (14° 27′S)

station and hence the quite exceptional degree of layer lability revealed by the fall of saturation potential temperature of 6.3°C between the surface and the 700-millibar level. A similar feature has been encountered at Douala in January and indeed it appears to be widespread along the West African coast

TABLE 36. *The transition from tropical continental to equatorial air*

| Level (mb) | JANUARY | | | | | JULY | | | | |
|---|---|---|---|---|---|---|---|---|---|---|
| | *T* (°C) | *PT* (°K) | *SPT* (°C) | *MR* (g/kg) | *RH* (%) | *T* (°C) | *PT* (°K) | *SPT* (°C) | *MR* (g/kg) | *RH* (%) |
| FORT LAMY 12°08′N, 15°02′E (300 m) | | | | | | | | | | |
| | | *(Winter)* | | | | | *(Summer)* | | | |
| 300 | −31.5 | 341 | 21.7 | 0.13 | 14 | −31.5 | 341 | 21.8 | 0.30 | 32 |
| 500 | −6.5 | 325 | 17.7 | 0.57 | 12 | −7.0 | 324 | 19.3 | 2.3 | 51 |
| 700 | 9.5 | 313 | 15.0 | 1.5 | 14 | 10.0 | 313 | 17.9 | 4.6 | 42 |
| 850 | 20.5 | 308 | 13.7 | 2.05 | 11 | 21.0 | 308 | 19.7 | 8.7 | 46 |
| SFC | 23.5 | 298 | 14.0 | 5.5 | 29 | 28.0 | 303 | 24.2 | 16.5 | 65 |
| | (975 mb) | | | P = 0 mm | | (975 mb) | | | P = 156 mm | |
| BANGUI 4°23′N, 18°34′E (386 m) | | | | | | | | | | |
| 300 | −32.0 | 340 | 21.5 | 0.18 | 20 | −32.5 | 339 | 21.4 | 0.33 | 40 |
| 500 | −7.0 | 324.5 | 18.3 | 1.25 | 27 | −7.0 | 324.5 | 19.4 | 2.4 | 51 |
| 700 | 8.5 | 312 | 17.2 | 4.0 | 39 | 8.5 | 311.5 | 18.8 | 6.1 | 62 |
| 850 | 20.0 | 307 | 18.7 | 7.6 | 42 | 18.5 | 305 | 20.3 | 10.3 | 65 |
| SFC | 26.0 | 302 | 23.7 | 16.3 | 73 | 25.0 | 300.5 | 23.9 | 17.6 | 84 |
| | (965 mb) | | | P = 21 mm | | (969 mb) | | | P = 184 mm | |
| BROKEN HILL 14°27′S, 28°28′E (1206 m) | | | | | | | | | | |
| | | *(Summer)* | | | | | *(Winter)* | | | |
| 300 | −31.0 | 341.5 | 22.2 | 0.52 | 53 | −32.5 | 339 | 21.3 | 0.10 | 12 |
| 500 | −6.0 | 325.5 | 20.3 | 3.2 | 65 | −6.0 | 325.5 | 18.1 | 0.59 | 12 |
| 700 | 9.0 | 312 | 20.4 | 8.0 | 78 | 6.0 | 308.5 | 15.3 | 3.3 | 40 |
| 850 | 18.0 | 305 | 22.1 | 12.9 | 85 | 13.5 | 300 | 15.7 | 6.7 | 58 |
| SFC | 21.0 | 305 | 23.3 | 14.8 | 82 | 16.0 | 299 | 15.8 | 7.3 | 56 |
| | (880 mb) | | | P = 226 mm | | (885 mb) | | | P = 0 mm | |
| DARWIN 12°26′S, 130°52′E (27 m) | | | | | | | | | | |
| 300 | −30.0 | 342.5 | — | — | — | −31.5 | 340.5 | — | — | — |
| 500 | −5.0 | 326 | 20.3 | 2.9 | 55 | −6.0 | 325 | 18.2 | 0.79 | 16 |
| 700 | 10.0 | 313 | 19.7 | 6.7 | 61 | 9.5 | 313 | 14.9 | 1.6 | 15 |
| 850 | 19.0 | 306 | 21.3 | 11.7 | 70 | 14.0 | 300 | 14.6 | 5.4 | 45 |
| SFC | 28.0 | 301 | 25.3 | 19.5 | 79 | 24.0 | 296 | 16.9 | 9.6 | 50 |
| | (1003 mb) | | | P = 341 mm | | (1010 mb) | | | P = 2 mm | |

(e.g. Abidjan) at that season when the zone of Equatorial air lies farther south. The ascents at Bangui in January have much the same character though precipitation is light. In July, however, the air is unmistakably equatorial, evidence that there is no difficulty in generating such an air mass over a continental interior. This view is further reinforced by the January ascents at

Broken Hill, about a thousand miles south of the equator. Here, although the surface temperature (and mixing ratio) is depressed by the elevation of the station, the humidity content of the air from 850 millibars upwards is the highest so far encountered. In the southern winter (July), on the contrary, this

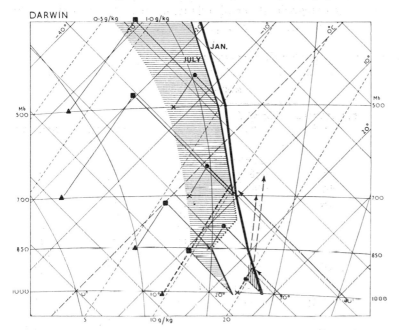

FIG. 97. Tephigrams for January and July at Darwin, Australia (12° 26′ S)

station is under tropical continental air with an inversion of the order of of 9–10°C at about the 700-millibar level. Little wonder that the rainfall is nil. Not surprisingly the level of the winter inversion is clearly affected by the height of the ground for air can descend no farther. The Darwin record repeats much the same features at a station near sea-level.

## Monsoon air

If, after this extensive survey, we now turn back to the July data for Monsoon stations (Figs. 93 and 94 and Tables 30, 31 and 32), it is evident that, from the point of view of air mass characteristics alone, a differentiation between south-west monsoon and equatorial air is a matter of degree rather than of kind. The real difference is in the system of circulation in which it is involved, the air is in active and consistent motion whereas most equatorial air is relatively quiescent. At the more southerly stations, Trivandrum, Madras, Port Blair and Saigon, ascents in southwest monsoon air are virtually identical with those in equatorial air but as the airstream moves northward of the fifteenth parallel its temperature is raised a few degrees throughout a great depth of the atmosphere and the humidity content shows a significant increase to the limits of the available records. These features are particularly well developed in the Indian realm both at coastal stations (Bombay and Calcutta) and in the interior (Jodhpur, Nagpur, Allahabad).

It is useful to underline this development by grouping the *July* records at a number of stations into categories determined primarily by the data themselves as in Table 37. The equatorial values represent means derived from the ten

TABLE 37. *Comparison of equatorial and monsoon air (July) 1962–66*

| | | MONSOON | | | |
|---|---|---|---|---|---|
| LEVEL (mb) | EQUATORIAL 10 stns[1] | *Group I* 3 stns[1] | *Group II* 5 stns[1] | CALCUTTA | BAHREIN |
| | | *Temperature* (°C) | | | |
| 100 | −75 | −78 | −73 | −73 | −77 |
| 150 | −67 | −66 | −60 | −60 | −62 |
| 200 | −54 | −52 | −46 | −46 | −48 |
| 300 | −32 | −29 | −25 | −25 | −27 |
| 500 | −6 | −5 | −2 | −1 | −4 |
| 700 | 9 | 10 | 13 | 12 | 17 |
| 850 | 17 | 18 | 21 | 21 | 30 |
| SFC | 26 | 28 | 30 | 29 | 34 |
| | | *Mixing-ratio* (g/kg) | | | |
| 300 | 0.34 | 0.46 | — | — | 0.3 |
| 500 | 2.5 | 3.1 | 4.2 | 5.0 | 1.6 |
| 700 | 6.7 | 7.7 | 9.6 | 10.6 | 3.1 |
| 850 | 11.3 | 11.9 | 14.7 | 15.3 | 4.5 |
| SFC | 18.0 | 19.5 | 20.1 | 21.7 | 20.2 |

[1] See text for stations in each group.

stations, Majuro, Guam, Truk, Lae, Singapore, Minicoy, Bangui, Douala, Abidjan and Panama, all (except Lae) north of the equator. Grouping the *January* figures for five stations south of the equator, Lae, Darwin, Dar es Salaam, Broken Hill (ignoring the surface record) and Luanda, produces mean values within 1°C of this series except at 100 millibars, where the figure is −81°C. Although data for the highest levels are a little fragmentary it does appear that this represents a real asymmetry between the hemispheres. The monsoon records fall naturally into two groups; those in southeast Asia— Saigon, Chiang Mai and Hong Kong (group I); and those in northern India—Bombay, Jodhpur, Nagpur, Allahabad and Calcutta (group II). The southeast Asian monsoon is only slightly warmer and more humid than equatorial air but the transformation over northern India is certainly significant. As compared with the equatorial record, temperatures are up by 4°C to the 500-millibar level and by 7°C in the three succeeding layers. At the same time the mixing-ratio is increased by some 10 per cent near the surface to over 50 per cent before the record gives out. What are the causes of this change? A comparison of the 1962–66 records for Calcutta and Bahrein in July may help us here. Below 600 millibars the former station records the most humid air and the latter the warmest air encountered anywhere in this survey. In the lower levels temperatures may be raised by insolation as summer days lengthen with increasing distance from the equator. Bahrein gives some measure of this effect under virtually cloudless skies but the monsoon brings much cloud to Calcutta and mean temperatures in July are in fact lower than in April or May. In the higher levels air temperature can be raised by (a) advection from a warmer source-region, (b) dynamic descent, and (c) the release of latent heat at condensation. Advection towards Calcutta in the layers from 500 to 150 millibars appears to be ruled out since no such warmer source region is known to exist—unless it is over the Tibetan plateau. Bahrein surely gives some measure of the effect of descent and the fact that, above 600 millibars, the Calcutta air is still warmer than at this station is at least a pointer in favour of the third solution. Positive proof lies with the humidity record but, though this does its best, it fails us above 500 millibars.

It thus seems highly probable that the transformation of equatorial air into southwest monsoon air is mainly the result of the release of latent heat at condensation, that is to processes operating within the air mass itself. Equatorial air achieves most of its characteristics by the same processes so that the difference is indeed one of degree.

In view of the exceptional humidity of the air at Calcutta, closely paralleled by that at Allahabad, it is not surprising that exceptionally heavy precipitation is possible in the Himalayas even at elevations in excess of 20 000 feet.

## Mid-latitude air

At this stage of our discussion we find a considerable proportion (roughly a quarter) of the atmosphere of the northern hemisphere still unclassified. This lies between the divergent air masses of the tropical girdle and the arctic cap and shares some of the characteristics of each. Though both of the primary air masses are decidedly stable, there is a wide difference in temperature, and hence in density, between them (see p. 274) and hence convergent flow produces an *interaction* between the two masses which is revealed by widespread frontal disturbances. Under these circumstances, whilst some ascents may be made wholly within (modified) arctic or tropical air, others must undoubtedly transect the sloping surface between them, that is they will be of a composite character. Mean values are therefore much less reliable in this area than anywhere else on earth. Yet both arctic and tropical air masses are progressively modified as they move outwards from their source regions and the frontal interaction produces further changes, notably via evaporation from falling precipitation. There is hence ample scope for another 'secondary' group of air masses even if it is indeed rather variable and ill-defined. We describe this group broadly as mid-latitude air. Such a category is clearly desirable in a world-wide survey even though air mass analysis as first devised for forecasting in Europe and America was mainly concerned with internal distinctions *within* this complex system.

Many of these complexities are indeed confined to the northern hemisphere where the distribution of continents and oceans produces sharp regional and seasonal contrasts. The reality of mid-latitude air is much less likely to be called in question in the southern hemisphere where it occupies a continuous zone from 1500 to 2250 miles wide, situated almost exclusively over the open sea (Figs. 79 and 80). There it produces some of the most depressing climates known on earth at a few islands lost in the great waste of water.

Before turning to the mean ascents at a selected few of the great wealth of northern hemisphere stations it may be useful to outline the features they have in common.

1. Everywhere convergent flow-patterns are of relatively frequent occurrence. This arises from the position between the divergent systems of the Arctic and the tropics and must tend materially to reduce the inherent stability of air from each (pp. 239-40).

2. As arctic air moves towards middle latitudes it is progressively warmed from below, especially over the open sea and over the continents in summer. Its low-level inversion is thus gradually eroded though its initial scale is such that it is not easily eliminated. Tropical air, on the contrary, becomes in-increasingly stable as a result of surface cooling during poleward motion, particularly in winter. Mean ascents in middle latitudes thus show a persistent degree of residual stability even over the open ocean and, on this evidence, the student may wonder why it ever rains at all! The problem is the reverse of that encountered in equatorial air where ascents suggest that it should rain every day but here, as there, the solution lies in the pattern of air flow.

3. In middle latitudes convergent motion of two air masses of different densities produces *confrontation*, a self-explanatory term much to be preferred to the ugly hybrid 'frontogenesis'. The warmer air mass then moves aloft over the cooler mass and a whole sequence of events known as 'frontal phenomena' is thus set in train (see pp. 351–63). The mid-latitude area is thus a field of action, a battlefield of atmospheric conflict, and it is to this, rather than to any quasi-homogeneity, that it owes its real identity. Rainfall is possible, indeed probable, from such an encounter regardless of the initial stability of the two contenders. It falls from the uplifted humid tropical air and may well be enhanced by the release of the latent layer lability (decrease of SPT with height) which we have already recognised as one of its more consistent features.

4. The mid-latitude zone is thus, *par excellence*, one of 'overhead weather'. The nature of the earth's surface still has its effects but, for the most part, they are secondary to those of mobile disturbances steered mainly from above. As these systems move in a generally eastward direction they bring alternating patterns of low and high pressure, of uplift and subsidence, so that the keynote

of local climates is wide variability. It is true that such effects are not unknown in the tropics, nor are they completely absent above the dense air of the arctic caps, but here they are deep and all-pervading. Their ultimate result is to produce a new atmospheric stratification which it is now our task to examine.

Ranging widely in latitude in the northern hemisphere, locally from about 25°N to at least 75°N at both seasons of the year, mid-latitude air may be expected to show a variety of transitions and transformations. Stations have

TABLE 38. *Mediterranean climates*

| LEVEL (mb) | JANUARY | | | | | JULY | | | | |
|---|---|---|---|---|---|---|---|---|---|---|
| | $T$ (°C) | $PT$ (°K) | $SPT$ (°C) | $MR$ (g/kg) | $RH$ (%) | $T$ (°C) | $PT$ (°K) | $SPT$ (°C) | $MR$ (g/kg) | $RH$ (%) |
| OAKLAND 37°44′N, 122°12′W (2 m) | | | | | | | | | | |
| 300 | −47 | 319 | — | — | — | −37 | 332 | 19.6 | 0.14 | 28 |
| 500 | −20 | 309 | 12.3 | 0.57 | 36 | −9 | 321 | 17.1 | 0.93 | 25 |
| 700 | −3 | 299 | 9.7 | 1.45 | 33 | 9 | 312 | 15.6 | 2.5 | 24 |
| 850 | 4 | 290 | 7.0 | 2.4 | 37 | 18 | 305 | 13.8 | 3.1 | 20 |
| SFC | 10 | 281 | 6.5 | 5.7 | 76 | 18 | 289 | 13.8 | 8.9 | 69 |
| | (1022 mb) | | | $P = 98$ mm | | (1015 mb) | | | $P = 0$ mm | |
| GIBRALTAR 36°09′N, 5°21′W (3 m) | | | | | | | | | | |
| 300 | −46 | 320 | — | — | — | 37 | 333 | — | — | — |
| 500 | −19 | 309 | 12.6 | 0.58 | 34 | −9 | 321 | 16.9 | 0.85 | 23 |
| 700 | −3 | 300 | 10.2 | 1.6 | 36 | 10 | 314 | 16.0 | 2.3 | 20 |
| 850 | 5 | 291 | 9.0 | 3.7 | 57 | 19 | 306 | 16.2 | 5.0 | 30 |
| SFC | 14 | 285 | 10.6 | 7.6 | 79 | 24 | 295 | 19.2 | 12.7 | 68 |
| | (1020 mb) | | | $P = 154$ mm | | (1016 mb) | | | $P = 1$ mm | |
| ATHENS 37°58′N, 23°43′E (107 m) | | | | | | | | | | |
| 300 | −51 | 314 | — | — | — | −36 | 335 | 20.3 | 0.18 | 29 |
| 500 | −25 | 303 | 10.2 | 0.50 | 47 | −10 | 321 | 17.1 | 1.10 | 30 |
| 700 | −8 | 293 | 7.5 | 1.6 | 54 | 7 | 310 | 15.9 | 3.5 | 39 |
| 850 | 0 | 286 | 6.5 | 3.4 | 72 | 17 | 303 | 17.3 | 7.3 | 51 |
| SFC | 10 | 283 | 7.4 | 5.6 | 73 | 27 | 300 | 19.1 | 10.7 | 45 |
| | (1001 mb) | | | $P = 53$ mm | | (999 mb) | | $P = 4$ mm | | |

been selected from both the Old and the New World to illustrate three of these features, namely from 'Mediterranean' lands where mid-latitude conditions prevail during the winter months, from continental interiors where the most significant events occur in summer, and from coastal or oceanic areas where arctic air takes on the traditional 'polar' features as it passes out over open sea at all seasons of the year.

The Mediterranean transition is well shown at Oakland, California (lat. 37°44′N) and Gibraltar (lat. 36°09′N) and, in a rather different form, at Athens (lat. 37°58′N) (Fig. 98 and Table 38). In July it is evident that the two former stations lie under what is unmistakably tropical maritime air of the 'eastern' variety. The trade-wind inversion is clearly present at both stations and indeed at Oakland, where potential temperature increases by no less than 16° from the surface to the 850-millibar level (inversion about 13° K), it reaches quite extreme proportions. That this is partly the result of the cool waters of the California current is evident from the low surface temperature in July. Indeed anyone who has had the misfortune to visit the Golden Gate in summer is unlikely to forget the experience. This may also be the reason why this station also fails to show an initial fall of SPT with height, a usual feature of tropical air ascents. That the inversion is very much an oceanic feature is underlined by the July ascent at Athens. Here we have clearly suggested that the summer drought of the eastern Mediterranean is due rather to the presence of tropical continental air, warm and yet relatively dry even near the surface. The lifting condensation level is thus at no less than 1700 metres (compare with Ashkhabad) and, even so, the ascent still shows stability aloft. Although we have prepared the reader for some degree of apparent stability in mid-latitude air, it may nevertheless come as something of a surprise that the contrast between January and July ascents at these stations is not more marked. Certainly the inversion is much weaker in January but all three mean ascents remain comparatively stable. This may be partly the result of the averaging process and we shall encounter the same feature in western Europe. As we have indicated above, the likelihood of frontal rain cannot be gauged from the tephigram which presents a rather static picture.

In the interiors of the great American and Eurasian continents, however,

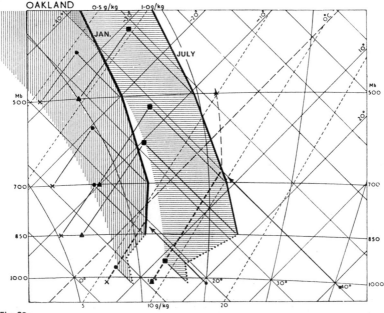

Fig. 98a

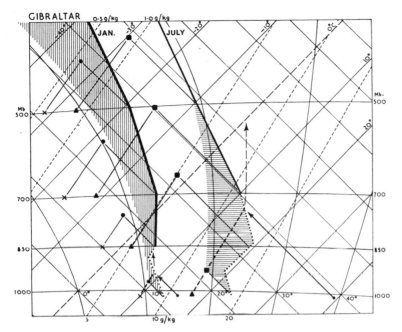

Fig. 98b

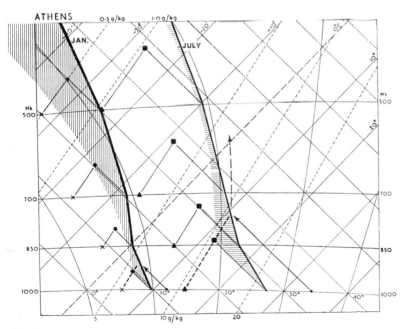

Fig. 98c

FIG. 98. Air mass contrasts at three 'Mediterranean' stations. The tephi-grams show mean ascents in January and July at Oakland, California (37° 44′ N), Gibraltar (36° 09′ N) and Athens (37° 58′ N)

317

the tephigram represents most effectively the immense seasonal variation in air mass type which occurs. We have already encountered the contrast in Canada (Norman Wells and The Pas) and Siberia (Turukhansk and Omsk) where clearly arctic air in winter is replaced by air of a very different quality in summer (Figs. 83, 84 and 86). At all these stations there is a remarkable increase in temperature throughout the lowest half of the atmosphere,

TABLE 39a. *Varieties of mid-latitude air (January)*

| Level (mb) | T (°C) | PT (°K) | SPT (°C) | MR (g/kg) | RH (%) | T (°C) | PT (°K) | SPT (°C) | MR (g/kg) | RH (%) |
|---|---|---|---|---|---|---|---|---|---|---|
| | | | | | (JANUARY) | | | | | |

NORTH PLATTE
41°08′N, 100°41′W (849 m)

ALMA ATA
(43°14′N, 76°56′E (851 m)

| Level | T | PT | SPT | MR | RH | T | PT | SPT | MR | RH |
|---|---|---|---|---|---|---|---|---|---|---|
| 300 | −50 | 315 | — | — | — | −53 | 310 | 12.5 | 0.02 | 31 |
| 500 | −25 | 303 | 9.8 | 0.42 | 40 | −27 | 300 | 8.6 | 0.30 | 35 |
| 700 | −9 | 292 | 6.4 | 1.30 | 47 | −10 | 292 | 5.7 | 1.0 | 38 |
| 850 | −4 | 282 | 1.6 | 1.65 | 50 | −2 | 283 | 2.7 | 1.7 | 45 |
| SFC | −6 | 273 | −2.8 | 1.8 | 69 | −6 | 273 | −3.0 | 1.9 | 71 |
| | (919 mb) | | | P = 11 mm | | (929 mb) | | | P = 26 mm | |

DAYTON
39°54′N, 84°12′W (306 m)

BUDAPEST
47°26′N, 19°11′E (140 m)

| 300 | −48 | 316 | — | — | — | −51 | 313 | 10.5 | 0.04 | 32 |
|---|---|---|---|---|---|---|---|---|---|---|
| 500 | −24 | 303 | 10.1 | 0.46 | 42 | −26 | 301 | 9.1 | 0.47 | 53 |
| 700 | −10 | 291 | 5.7 | 1.20 | 48 | −11 | 290 | 5.5 | 1.4 | 58 |
| 850 | −5 | 280 | 0.7 | 1.6 | 56 | −5 | 281 | 1.9 | 2.2 | 71 |
| SFC | −4 | 271 | −4.0 | 2.2 | 77 | −3 | 269 | −4.5 | 2.5 | 84 |
| | (984 mb) | | | P = 81 mm | | (1007 mb) | | | P = 41 mm | |

CHARLESTON
32°54′N, 80°02′W (15 m)

KAGOSHIMA
31°38′N, 130°36′E (283 m)

| 300 | −42 | 325 | (17.6) | (0.16) | (52) | −39 | 330 | — | — | — |
|---|---|---|---|---|---|---|---|---|---|---|
| 500 | −16 | 313 | 14.1 | 0.75 | 35 | −22 | 306 | 11.5 | 0.43 | 31 |
| 700 | −1 | 301 | 11.1 | 2.0 | 40 | −8 | 293 | 7.0 | 1.24 | 41 |
| 850 | 5 | 291 | 8.3 | 3.1 | 49 | −1 | 284 | 4.5 | 2.6 | 66 |
| SFC | 8 | 280 | 5.0 | 4.9 | 73 | 6 | 278 | 3.5 | 4.4 | 75 |
| | (1019 mb) | | | P = 65 mm | | (1020 mb) | | | P = 75 mm | |

ranging indeed from the order of 20° C near the 500-millibar level to 40° C or more near the surface. At the same time the humidity content, as given by the mixing-ratio, shows broadly a fivefold to twentyfold increase over the same height range. The extreme stability of the January ascents is thus almost completely eliminated and convectional development becomes possible with a rise of surface temperatures of not more than 4–6° C above mean July values— a not infrequent occurrence in lands with a considerable diurnal variation of temperature. Note also that at The Pas and Omsk (Table 26b) saturation potential temperature decreases with height in the lowest 3000 metres of the atmosphere in July; frontal uplift may therefore release widespread layer lability.

South of these stations with a genuine arctic winter there is an almost embarrassing wealth of data over the continents though the plateau lands of the American West and Asian East are ill served. To show the nature of the transitions which occur in the mid-latitude zone we have had to exercise a rigorous selection but three pairs of stations will suffice to illustrate the major features (Tables 39a, b). At North Platte (lat. 41°08′N) and Alma Ata (lat.

TABLE 39b. *Varieties of mid-latitude air (July)*

| Level (mb) | T (°C) | PT (°K) | SPT (°C) | MR (g/kg) | RH (%) | T (°C) | PT (°K) | SPT (°C) | MR (g/kg) | RH (%) |
|---|---|---|---|---|---|---|---|---|---|---|
| NORTH PLATTE | | | | | | ALMA ATA | | | | |
| 41°08′N, 100°41′W (849 m) | | | | | | 43°14′N, 76°56′E (851 m) | | | | |
| 300 | −34 | 337 | 20.7 | 0.20 | 29 | −36 | 334 | 20.0 | 0.24 | 42 |
| 500 | −9 | 322 | 18.1 | 1.63 | 40 | −11 | 320 | 17.3 | 1.7 | 51 |
| 700 | 11 | 314 | 18.9 | 5.2 | 44 | 8 | 312 | 17.5 | 4.7 | 47 |
| 850 | 20 | 307 | 20.4 | 9.6 | 53 | 21 | 308 | 17.9 | 6.2 | 32 |
| SFC | 24 | 305 | 22.3 | 13.2 | 62 | 24 | 305 | 19.7 | 9.5 | 45 |
| | (915 mb) | | | P − 64 mm | | (905 mb) | | | P = 35 mm | |
| DAYTON | | | | | | BUDAPEST | | | | |
| 39°54′N, 84°12′W (306 m) | | | | | | 47°26′N, 19°11′E (140 m) | | | | |
| 300 | −35 | 335 | 20.3 | 0.17 | 27 | −40 | 329 | 18.7 | 0.19 | 47 |
| 500 | −9 | 322 | 17.3 | 1.00 | 26 | −13 | 316 | 15.9 | 1.30 | 47 |
| 700 | 6 | 309 | 15.5 | 3.4 | 40 | 2 | 304 | 14.5 | 3.9 | 62 |
| 850 | 15 | 301 | 16.7 | 7.6 | 61 | 12 | 298 | 15.2 | 7.0 | 69 |
| SFC | 23 | 298 | 19.5 | 12.0 | 63 | 21 | 294 | 16.4 | 10.0 | 63 |
| | (980 mb) | | | P = 90 mm | | (1000 mb) | | | P = 53 mm | |
| CHARLESTON | | | | | | KAGOSHIMA[1] | | | | |
| 32°54′N, 80°02′W (15 m) | | | | | | 31°38′N, 130°36′E (283 m) | | | | |
| 300 | −33 | 338 | 21.1 | 0.24 | 32 | −29 | 345 | 22.9 | 0.52 | 41 |
| 500 | −7 | 324 | 18.7 | 1.72 | 39 | −5 | 327 | 20.0 | 2.30 | 42 |
| 700 | 7 | 310 | 17.5 | 5.0 | 54 | 11 | 314 | 19.7 | 6.0 | 50 |
| 850 | 17 | 303 | 19.2 | 9.6 | 68 | 18 | 305 | 21.7 | 12.1 | 76 |
| SFC | 26 | 298 | 22.3 | 16.0 | 75 | 27 | 299 | 24.2 | 18.4 | 79 |
| | (1015 mb) | | | P = 196 mm | | (1010 mb) | | | P = 343 mm | |

[1] Virtually equatorial.

43°14′N) (Fig. 99) we have two stations of considerable elevation on the approaches to the arid interior of the two continents. January temperatures are low for the latitude though well above arctic levels, yet there is still evidence of a near-surface inversion and both ascents show a very stable stratification. Clearly this air has strong arctic affinities, the higher mean temperatures being probably the joint result of (*a*) warming in transit, and (*b*) occasional incursions of warmer air from elsewhere. The water vapour

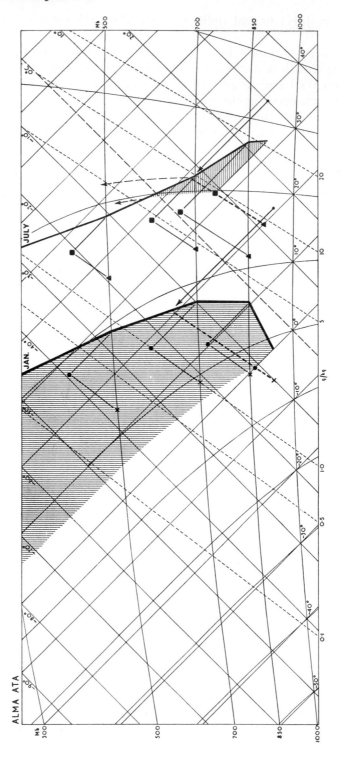

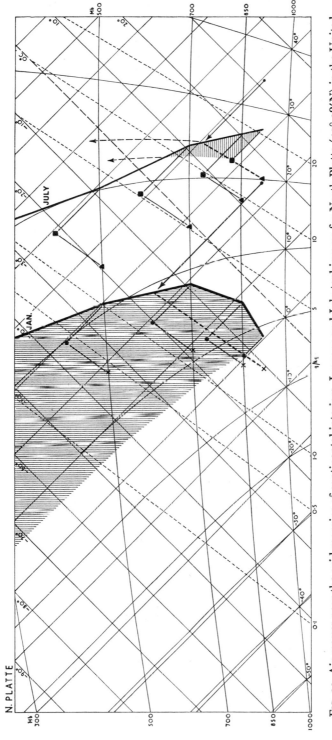

FIG. 99. Air masses near the arid margins of continental interiors. January and July tephigrams for North Platte (41° 08′ N) in the United States and for Alma Ata (43° 14′ N) in Russian Turkestan

content, though still low, is also well above arctic levels. It too probably represents the combined results of precipitation through frontal surfaces and direct advection from more humid regions. One has a distinct impression, however, that the latter process must be rather rare, particularly at Alma Ata where all probable sources are very far away. In July the position is transformed, the great increase in surface mixing-ratios (fivefold at Alma Ata and more than sevenfold at North Platte) being particularly significant at stations so far from the sea. Both ascents are then conditionally unstable but because relative humidities are low, particularly at Alma Ata, there is still a considerable degree of low-level stability that has first to be overcome. Active convection is possible, on this evidence, if surface temperatures rise some 6–7° C above the monthly mean. This is by no means out of the question but the cloud base indicated is no less than 1800 metres above ground-level at North Platte and 2400 metres above Alma Ata. The initial decrease of SPT with height suggests that, at both stations, the summer air has tropical affinities. Under suitable frontal conditions this could assist in the generation of summer rain, meagre at Alma Ata for this is tropical continental, rather than mid-latitude air (compare Ashkhabad, Table 27).

At first glance Dayton, Ohio (lat. 39°54′N) and Budapest (lat. 47°26′N) would appear to be an odd pair of stations but they both lie in comparatively humid continental interiors and, in fact, may be taken as representing the 'corn belts' of America and Europe. At both stations the mean ascents for January appear to be no less stable than those at North Platte and Alma Ata, potential temperature increasing by some 20° K from the surface to the 700-millibar level. The quite considerable winter precipitation (largely snowfall) of these stations must therefore be primarily a result of frontal activity and the humidity of the modified arctic air has been correspondingly increased. In July, however, both ascents are much less stable than those at the former pair of stations and, even in the absence of frontal activity, convectional processes are free to operate when surface temperatures exceed the mean values by no more than about 4° C. Indeed, the fact that from about 850 millibars (the mean convectional cloud base—some 1500 metres above sea level) to nearly 300 millibars the two ascents approximate closely to the SALR suggests that such processes are by no means uncommon. Yet the two July ascents are by no means identical. At Dayton the air at 700 millibars remains relatively dry so that saturation potential temperature decreases sharply with height. Furthermore the tropopause is much higher than at Budapest. Both of these facts suggest not infrequent visitations of air of tropical maritime origin over Dayton and the layer lability implied undoubtedly contributes to occasional widespread thunderstorm activity in that area and indeed over most of the eastern States.

By Charleston (lat. 32°54′N) and Kagoshima (lat. 31°38′N) the transformations noted above have progressed another stage (Fig. 100 and Table 39). Despite the coastal location of Charleston and the insular location of Kagoshima (data for Chinese stations are not available), the January ascents

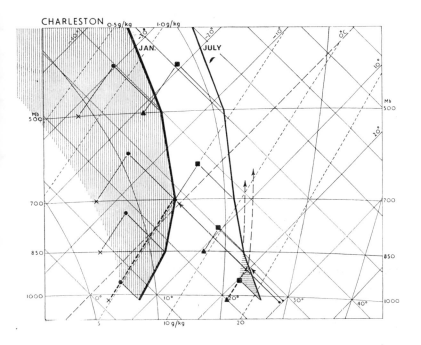

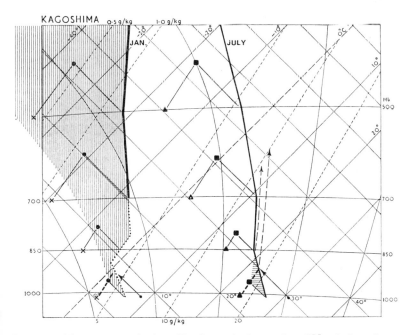

Fig. 100. Air masses on the humid sub-tropical margins of North America and Asia. The data are for Charleston S.C. (32° 54′N) and Kagoshima (31° 34′N) in southern Japan. Unfortunately no ascents are available for the coast of continental China

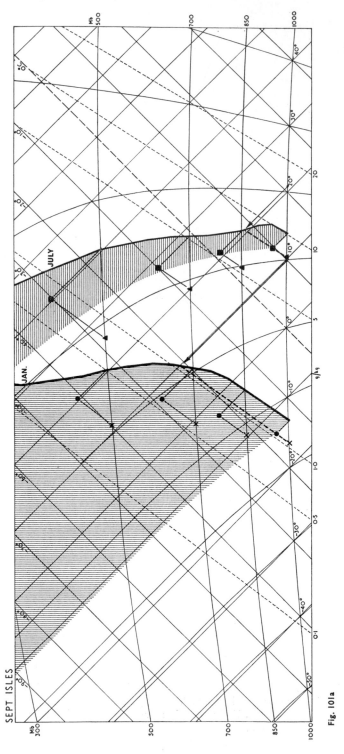

Fig. 101a

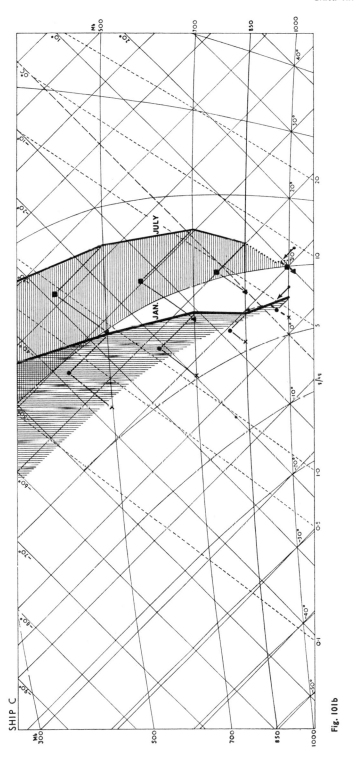

Fig. 101b

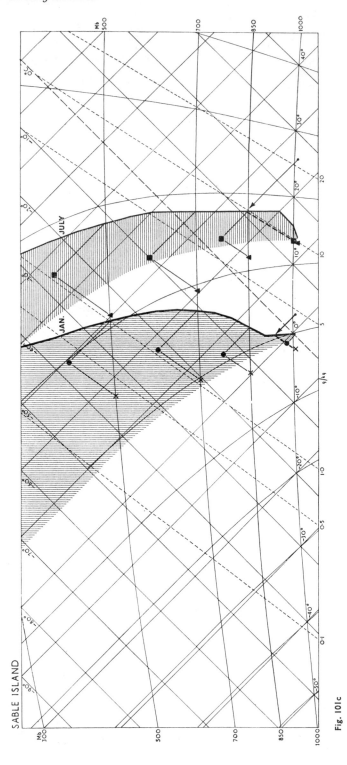

Fig. 101c

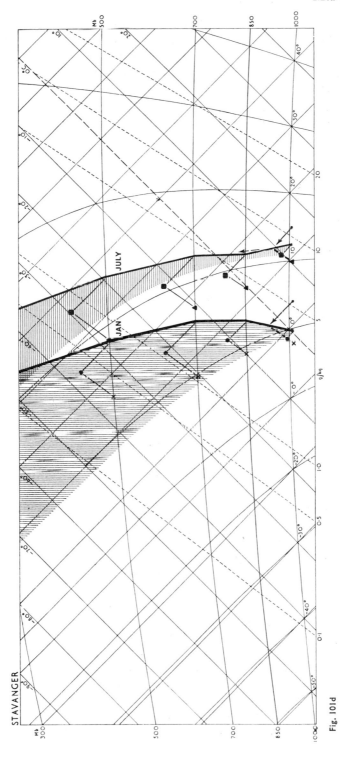

FIG. 101. Mid-latitude air at a group of stations astride the North Atlantic. The tephigrams are for Sept Iles (50° 13′N, 66° 16′W), Sable Island (43° 56′N, 60° 02′W), Weather Ship C (57° 00′N, 35° 30′W) and Stavanger (58° 53′N, 5° 38′E)

Fig. 101d

are still so stable and the air is still so dry that an ultimate Arctic origin is to be suspected. True that a modification in depth has been achieved in transit but the considerable winter precipitation of both these stations must rest exclusively upon frontal activity. In summer, however, though it is not suggested that such activities come to an end, both ascents are conditionally unstable and the requisite trigger for more local overturnings may be supplied by relief features in excess of 600 metres (2000 ft) or by day temperatures some 3–4°C above the overall mean values. The source of this air is unquestionably tropical maritime but the trade-wind inversion has been outrun or overcome and hence the higher mixing-ratios recorded at the 700-millibar level (compare Raizet and Charleston). Indeed the July ascent at Kagoshima, over 2150 miles north of the equator, very nearly fulfils all the requirements postulated for equatorial air (see p. 304 and Table 37), an interesting commentary on the unique character of the climate of southern Japan.

It is in oceanic locations, particularly in the higher latitudes, that mean ascents become of the least value as climatic indicators. Although rainfall is known to be well-distributed throughout the year the overall pattern, as it emerges from the means for both January and July, is one of very considerable stability in atmospheric stratification. One must frankly admit that this came as a considerable surprise. At first it was thought that the effect must be spurious, that it must arise from the high frequency of 'composite' ascents which transect frontal surfaces, yet the result is so persistent and universal in its distribution over the temperate oceans that other solutions must be canvassed. (*a*) The rainfall is pre-eminently frontal and frontal rain is possible even when both of the air-masses concerned have a considerable degree of initial internal stability. (*b*) Even in the wettest regions of middle latitudes it rains during only a fraction of the time (see p. 141) so that an overall view must be heavily weighted by rain-free conditions. (*c*) Such regions are frequently overrun by air of arctic origin. Despite its transformation over the sea into 'polar maritime air' which so frequently sweeps over these islands at all seasons of the year, the initial stability of arctic air is so immense that it may well dominate the mean picture. (*d*) Frontal activity is now known to involve active *descent* as well as ascent (p. 363) so that stability is readily restored.

Despite these reservations, the reader is entitled to be shown the results of applying to this more difficult situation the same method as has been used over the rest of the world. Figure 101 illustrates mean ascents in January and July at four stations in the North Atlantic and in Tables 40a, b, these are compared with four Pacific stations.

In January the air at Sept Iles and Kharbarovsk has near-Arctic characteristics. Surface temperatures are low, there is an increase in potential temperature of the order of 20°K in the first 300 millibars of the ascents, the content of water vapour is slight, and saturation potential temperatures increase very rapidly with height. Such air is extremely stable and winter snows, heavy at Sept Iles but light at Kharbarovsk, can come only from aloft through frontal surfaces. By the time such air is brought by northeasterly winds to Sable Island, and particularly to Sapporo, the qualities of its lower layers are much transformed by passage over the open sea. This appears to be the main factor at work at Sapporo where northeasterlies are common and the change above 700 millibars is slight. At Sable Island the greater depth of the apparent transformation may well owe something to more frequent visitations of southerly air as a result of a much more disturbed circulation. Yet the impression remains at both stations of a dominant stability and the considerable winter precipitation must be entirely of frontal origin. At

TABLE 40a. *Mid-latitude air* (*January*)

| | ATLANTIC | | | | | PACIFIC | | | | |
|---|---|---|---|---|---|---|---|---|---|---|
| LEVEL (mb) | T (°C) | PT (°K) | SPT (°C) | MR (g/kg) | RH (%) | T (°C) | PT (°K) | SPT (°C) | MR (g/kg) | RH (%) |
| **SEPT ILES** 50°13′N, 66°16′W (58 m) | | | | | | **KHARBAROVSK** 48°31′N, 135°10′E (72 m) | | | | |
| 300 | −52 | 311 | — | — | — | −55 | 307 | 11.2 | 0.02 | 62 |
| 500 | −32 | 293 | 5.6 | 0.24 | 47 | −37 | 288 | 2.9 | 0.22 | 67 |
| 700 | −19 | 280 | −0.7 | 0.59 | 49 | −23 | 287 | −3.0 | 0.58 | 69 |
| 850 | −15 | 270 | −6.9 | 0.80 | 57 | −19 | 267 | −9.3 | 0.77 | 73 |
| SFC | −13 | 260 | −14.2 | 1.07 | 77 | −19 | 253 | −20.3 | 0.65 | 79 |
| (1005 mb) | | | P = 100 mm | | | (1010 mb) | | P = 9 mm | | |
| **SABLE ISLAND** 43°56′N, 60°02′W (9 m) | | | | | | **SAPPORO** 43°03′N, 141°20′E (18 m) | | | | |
| 300 | −48 | 317 | — | — | — | −51 | 313 | — | — | — |
| 500 | −25 | 302 | 9.7 | 0.38 | 38 | −35 | 291 | — | — | — |
| 700 | −11 | 290 | 5.4 | 1.1 | 46 | −20 | 280 | −0.6 | 0.75 | 65 |
| 850 | −6 | 280 | 0.9 | 1.7 | 57 | −12 | 273 | −3.5 | 1.45 | 80 |
| SFC | 0 | 272 | −1.3 | 3.25 | 84 | −4 | 268 | −5.5 | 2.2 | 76 |
| (1010 mb) | | | P = 125 mm | | | (1008 mb) | | P = 111 mm | | |
| **SHIP C** 57°00′N, 35°30′W | | | | | | **SHIP P** 50°00′N, 145°00′W | | | | |
| 300 | −51 | 314 | — | — | — | −48 | 317 | — | — | — |
| 500 | −27 | 300 | 8.7 | 0.34 | 40 | −25 | 303 | 10.0 | 0.42 | 40 |
| 700 | −11 | 290 | 5.4 | 1.20 | 50 | −9 | 292 | 6.4 | 1.22 | 44 |
| 850 | −3 | 282 | 3.4 | 2.5 | 73 | −2 | 284 | 4.3 | 2.6 | 65 |
| SFC | 6 | 278 | 4.0 | 4.6 | 81 | 6 | 278 | 4.3 | 5.0 | 87 |
| (1003 mb) | | | | | | (1009 mb) | | | | |
| **STAVANGER** 58°53′N, 5°38′E (13 m) | | | | | | **YAKUTAT** 59°31′N, 139°40′W (9 m) | | | | |
| 300 | −52 | 311 | — | — | — | −51 | 313 | — | — | — |
| 500 | −28 | 298 | 7.9 | 0.36 | 48 | −30 | 296 | 6.7 | −0.31 | 52 |
| 700 | −13 | 288 | 4.3 | 1.10 | 54 | −15 | 286 | 3.2 | 0.98 | 55 |
| 850 | −5 | 280 | 1.8 | 2.2 | 71 | −7 | 279 | 0.5 | 1.8 | 65 |
| SFC | 0 | 272 | −1.3 | 3.4 | 87 | −4 | 268 | −5.2 | 2.5 | 89 |
| (1014 mb) | | | P = 88 mm | | | (1008 mb) | | P = 276 mm | | |

TABLE 40b. *Mid-latitude air (July)*

| | ATLANTIC | | | | | PACIFIC | | | | |
|---|---|---|---|---|---|---|---|---|---|---|
| LEVEL (mb) | T (°C) | PT (°K) | SPT (°C) | MR (g/kg) | RH (%) | T (°C) | PT (°K) | SPT (°C) | MR (g/kg) | RH (%) |
| **SEPT ILES** 50°13′N, 66°16′W (58 m) | | | | | | **KHARBAROVSK** 48°31′N, 135°10′E (72 m) | | | | |
| 300 | −42 | 326 | — | — | — | −36 | 335 | 20.4 | 0.31 | 50 |
| 500 | −16 | 313 | 14.4 | 0.77 | 34 | −10 | 320 | 17.5 | 1.70 | 48 |
| 700 | −1 | 301 | 12.3 | 2.9 | 56 | 5 | 307 | 16.1 | 4.6 | 61 |
| 850 | 8 | 294 | 12.1 | 5.4 | 69 | 14 | 300 | 17.1 | 8.4 | 72 |
| SFC | 14 | 287 | 11.9 | 8.0 | 79 | 21 | 294 | 18.2 | 12.1 | 77 |
| | (1003 mb) | | P = 102 mm | | | (996 mb) | | P = 102 mm | | |
| **SABLE ISLAND** 43°56′N, 60°02′W (9 m) | | | | | | **SAPPORO** 43°03′N, 141°20′E (18 m) | | | | |
| 300 | −37 | 332 | 19.6 | 0.19 | 37 | −34 | 337 | — | — | — |
| 500 | −11 | 319 | 16.5 | 1.10 | 34 | −9 | 322 | 17.9 | 1.50 | 37 |
| 700 | 4 | 306 | 14.2 | 3.0 | 43 | 6 | 309 | 16.6 | 4.75 | 57 |
| 850 | 11 | 298 | 14.4 | 6.2 | 62 | 13 | 300 | 17.3 | 8.7 | 75 |
| SFC | 15 | 286 | 13.4 | 9.8 | 94 | 19 | 292 | 16.8 | 11.4 | 81 |
| | (1014 mb) | | P = 74 mm | | | (1006 mb) | | P = 100 mm | | |
| **SHIP C** 57°00′N, 35°30′W | | | | | | **SHIP P** 50°00′N, 145°00′W | | | | |
| 300 | −41 | 328 | (18.8) | (0.10) | (23) | −41 | 326 | (18.2) | (0.17) | (48) |
| 500 | −14 | 315 | 15.2 | 0.89 | 36 | −16 | 314 | 14.6 | 0.88 | 38 |
| 700 | 1 | 303 | 12.4 | 2.2 | 38 | −1 | 301 | 11.9 | 2.5 | 48 |
| 850 | 7 | 293 | 10.6 | 4.4 | 60 | 6 | 292 | 10.2 | 4.6 | 69 |
| SFC | 12 | 283 | 9.9 | 7.6 | 89 | 11 | 283 | 9.7 | 7.7 | 92 |
| | (1018 mb) | | | | | (1020 mb) | | | | |
| **STAVANGER** 58°53′N, 5°38′E (13 m) | | | | | | **YAKUTAT** 59°31′N, 139°40′W (9 m) | | | | |
| 300 | −44 | 323 | — | — | — | −45 | 322 | — | — | — |
| 500 | −19 | 309 | 12.9 | 0.77 | 46 | −19 | 310 | 13.3 | 0.85 | 48 |
| 700 | −3 | 298 | 10.6 | 2.4 | 56 | −3 | 299 | 11.3 | 2.75 | 61 |
| 850 | 5 | 291 | 9.6 | 4.4 | 70 | 5 | 291 | 10.0 | 4.7 | 73 |
| SFC | 13 | 285 | 10.9 | 7.8 | 83 | 11 | 283 | 9.7 | 7.3 | 86 |
| | (1010 mb) | | P = 89 mm | | | (1016 mb) | | P = 214 mm | | |

Weather Ships C (57°00′N, 35°30′W) and P (50°00′N, 145°00′W), in the winter storm belts of the two oceans, the apparently inherent stability of mid-latitude maritime air remains though it is considerably reduced. Potential temperature here increases upwards by only 12–14°K in the lowest 300 millibars and the change in saturation potential temperature is slight. Note that owing to the comparatively low winter temperatures, the content of water vapour is, by world standards, no more than moderate though relative humidities above 80 per cent prevail. At the eastward landfall of this oceanic air, at Stavanger and Yakutat, the reader may be surprised to observe that the ascents are again more stable despite winter rains that are exceptionally heavy at Yakutat and by no means negligible at Stavanger. This is partly an effect of latitude, the Arctic outbreaks are fresher and some indeed may come from the interior of the two continents. Yet it does suggest that in these latitudes the effect of orographic barriers operates more through the modification of frontal processes (see Fig. 139) than via the release of latent instability in any single air mass.

In July, though the situation is very different, a number of interesting features remain. Kharbarovsk is virtually continental, though the seasonal reversal of winds so characteristic of eastern Asia now brings in a deep layer of warm moist air to a degree not experienced in eastern Canada. This air has some layer lability and penetrative convection is also possible when surface temperatures reach some 5°C above the overall mean. Elsewhere a considerable degree of long-term stability is again the order of the day. In the western Atlantic, at Sept Iles and Sable Island, this is primarily the result of a near-surface inversion over cool inshore waters, associated of course with the summer fogs which persist for long periods over the Newfoundland seas. Yet, in mid-ocean at Weather Ships C and P, the July ascents are even more stable than those of January. At first one might suspect that this has some relation to the poleward migration of the Azores and Californian high-pressure cells but a closer examination of the data suggests that (modified) arctic air is dominant. Thus the humidity content is scarcely tropical and, as in winter, the SPT still increases with height. In view of the restricted area of the source region of genuinely arctic air in summer the persistence of vestigial remnants of its unique qualities over the middle-latitude oceans during this season may be surprising, particularly since the transformation over land has proved to be so rapid. Yet evidence from still more northerly stations, e.g. Jan Mayen, not reproduced here, certainly suggests that this feature comes out of the north. At its eastern landfall, at Stavanger and Yakutat, this air retains much the same character though its mean stability is somewhat decreased, possibly owing to occasional incursions from the continents, and the ascents are thus less stable in summer than in winter. Yet frontal rainfall must be dominant.

In conclusion it should be remarked that, in middle latitudes, day-to-day contrasts in air mass type are much more marked than anywhere else on earth. So also is the relative frequency of air mass occurrence from year to year bringing with it aperiodic variations from abnormally warm to abnormally cold, and unusually wet to unusually dry, seasons. Our method is not appropriate to the analysis of these variations—an alternative approach is discussed later (p. 338)—but the long-term view is not without importance. It would be rash indeed to dismiss the above features, shared in common by a wide variety of stations, as no more than the product of the averaging process or of pure chance. Indeed, many of them were encountered during preliminary work on individual years preceding the standard period 1962–66 which we have now employed. For this very reason it seemed wise to put the analysis on a broader foundation.

## The air masses of the southern hemisphere

The question naturally arises how far the features thus recognised in the northern hemisphere have their parallels in the south. There the records, particularly during the earlier years, are more fragmentary and widely spaced. There are no weather ships but island stations give some indication of events over the vast extent of ocean. Furthermore some of the more interesting features of the northern hemisphere do not recur or appear only in very attenuated form. There is no monsoon, at least in any strict interpretation of the term, the limited areas of Mediterranean climate are not reinforced by a Mediterranean Sea, the mid-latitude continental area is much restricted and Australia, to its great distress, offers the only example of a desert of truly continental proportions. In climatic variety the surface of the hemisphere is thus relatively impoverished, yet much remains and zonal contrasts reach their full development.

To save space and avoid tedious repetition it has been decided to present some of the facts in tabular form only, with appropriate annotation (Table 41). Mean temperatures and humidities for July (winter) and January (summer) over the standard period are thus given below and comparisons with northern stations are suggested. In order to facilitate such comparison, humidities are expressed in terms of mixing-ratio instead of as dew-point temperatures as they appear in the original document. Should it be so desired it is just as easy to plot these figures on a tephigram after a little practice and, in the author's opinion tabulated data are more readily appreciated in this form. The transformation has involved plotting the ascents but that was necessary as a basis of selection. It is hoped, of course, that the serious reader will feel disposed to make his own direct approach to *Monthly Climatic Data for the World* or to similar published material—including, of course, the results of individual ascents at some stations of particular interest to him.

TABLE 41. *Air masses of the southern hemisphere*

| LEVEL (mb) | JULY $T(°C)$ | $MR$(g/kg) | JANUARY $T(°C)$ | $MR$(g/kg) | LEVEL (mb) | JULY $T(°C)$ | $MR$(g/kg) | JANUARY $T(°C)$ | $MR$(g/kg) |
|---|---|---|---|---|---|---|---|---|---|
| 1. SOUTH POLE (2800 m) 90°00′S | | | | | 2. HALLEY BAY (30 m) 75°31′S, 26°36′W | | | | |
| 100 | −82.3 | — | −38.8 | — | 100 | −80.7 | — | −43.2 | — |
| 200 | −74.8 | — | −43.6 | — | 200 | −73.2 | — | −46.8 | — |
| 300 | −65.6 | — | −52.7 | — | 300 | −62.3 | — | −51.7 | — |
| 500 | −44.6 | — | −36.2 | 0.20 | 500 | −39.9 | — | −31.0 | 0.25 |
| SFC | −59.5 | — | −29.2 | (0.19) | 700 | −25.2 | 0.33 | −16.5 | 0.73 |
| | (685 mb) | | (690 mb) | | 850 | −21.1 | 0.45 | −9.8 | 1.4 |
| | | | | | SFC | −27.6 | 0.28 | −5.2 | 2.0 |
| | | | | | | (989 mb) | | (985 mb) | |

| LEVEL | JULY | | JANUARY | |
|---|---|---|---|---|
| (mb) | $T(°C)$ | $MR$(g/kg) | $T(°C)$ | $MR$(g/kg) |

**3. MIRNY OBSERVATORY** (30 m) 66°33′S, 93°00′E

| | | | | |
|---|---|---|---|---|
| 100 | −74.6 | — | −41.0 | — |
| 200 | −69.9 | 0.005 | −44.1 | 0.11 |
| 300 | −59.7 | 0.012 | −50.2 | 0.04 |
| 500 | −36.1 | 0.18 | −29.1 | 0.37 |
| 700 | −21.8 | 0.50 | −14.6 | 0.91 |
| 850 | −19.0 | 0.51 | −8.3 | 1.6 |
| SFC | −16.7 | 0.80 | −2.4 | 2.3 |
| | (985 mb) | | (981 mb) | |

**4. ARGENTINE ISLAND** (10 m) 65°15′S, 64°16′W

| | | | | |
|---|---|---|---|---|
| 100 | (−71.0) | — | −43.7 | — |
| 200 | −69.1 | — | −45.9 | — |
| 300 | −58.4 | — | −50.1 | — |
| 500 | −33.8 | 0.24 | −27.9 | 0.41 |
| 700 | −17.9 | 0.75 | −12.7 | 1.40 |
| 850 | −11.2 | 1.30 | −5.2 | 2.3 |
| SFC | −10.0 | 1.4 | −0.6 | 3.35 |
| | (992 mb) | | (989 mb) | |

**5. D.N. ORCADES** (4 m) 60°45′S, 44°43′W

| | | | | |
|---|---|---|---|---|
| 100 | −69.3 | — | −43.0 | — |
| 200 | −66.1 | — | −45.8 | — |
| 300 | −57.0 | 0.016 | −48.3 | 0.05 |
| 500 | −33.8 | 0.25 | −25.9 | 0.52 |
| 700 | −17.8 | 0.80 | −10.8 | 1.65 |
| 850 | −11.4 | 1.35 | −4.2 | 2.7 |
| SFC | −9.8 | 1.65 | 0.8 | 3.5 |
| | (995 mb) | | (991 mb) | |
| | | | $P = 33$ mm | $P = 35$ mm |

**6. USHUAIA** (6 m) 54°48′S, 68°19′W

| | | | | |
|---|---|---|---|---|
| 100 | −63.5 | — | −47.1 | — |
| 200 | −63.1 | — | −47.8 | — |
| 300 | −53.9 | 0.027 | −47.0 | 0.12 |
| 500 | −29.3 | 0.42 | −24.0 | 0.72 |
| 700 | −12.2 | 1.5 | −8.5 | 2.23 |
| 850 | −3.9 | 2.7 | 0.3 | 3.7 |
| SFC | 2.0 | 3.3 | 9.3 | 5.1 |
| | (1002 mb) | | (996 mb) | |
| | | | $P = 38$ mm | $P = 55$ mm |

**7. MACQUARIE ISLAND** (6 m) 54°30′S, 158°57′E

| | | | | |
|---|---|---|---|---|
| 100 | −56.5 | — | −50.9 | — |
| 200 | −58.9 | — | −50.3 | — |
| 300 | −54.9 | — | −45.7 | — |
| 500 | −29.2 | 0.26 | −22.5 | 0.68 |
| 700 | −11.5 | 1.10 | −7.4 | 1.80 |
| 850 | −3.9 | 2.25 | −1.1 | 3.2 |
| SFC | 3.1 | 4.35 | 6.6 | 5.5 |
| | (1003 mb) | | (1000 mb) | |
| | | | $P = 68$ mm | $P = 99$ mm |

**8. MARION ISLAND** (23 m) 46°53′S, 37°52′E

| | | | | |
|---|---|---|---|---|
| 100 | −57.3 | — | −52.8 | — |
| 200 | −58.6 | — | −51.2 | — |
| 300 | −49.8 | 0.03 | −42.8 | 0.12 |
| 500 | −25.8 | 0.39 | −19.6 | 0.66 |
| 700 | −10.0 | 1.22 | −5.6 | 1.8 |
| 850 | −2.2 | 2.4 | 0.7 | 3.3 |
| SFC | 4.0 | 4.4 | 6.8 | 5.15 |
| | (1005 mb) | | (1003 mb) | |
| | | | $P = 228$ mm | $P = 194$ mm |

**9. INVERCARGILL** (1 m) 46°25′S, 168°19′E

| | | | | |
|---|---|---|---|---|
| 100 | −52.4 | — | −55.4 | — |
| 200 | −54.9 | — | −55.0 | — |
| 300 | −52.3 | — | −42.9 | — |
| 500 | −26.2 | 0.4 | −17.2 | 0.9 |
| 700 | −8.8 | 1.2 | −1.4 | 2.5 |
| 850 | −0.4 | 2.7 | 6.0 | 4.6 |
| SFC | 5.0 | 4.6 | 13.6 | 7.5 |
| | (1011 mb) | | (1010 mb) | |
| | | | $P = 75$ mm | $P = 99$ mm |

**10. GOUGH ISLAND** (40 m) 40°21′S, 9°53′W

| | | | | |
|---|---|---|---|---|
| 100 | −56.8 | — | −58.0 | — |
| 200 | −57.0 | — | −54.4 | — |
| 300 | −47.4 | 0.037 | −39.1 | 0.14 |
| 500 | −22.8 | 0.46 | 14.1 | 0.76 |
| 700 | −7.7 | 1.90 | 0.8 | 1.32 |
| 850 | 0.1 | 3.1 | 7.2 | 4.25 |
| SFC | 8.6 | 5.8 | 13.7 | 7.9 |
| | (1008 mb) | | (1011 mb) | |
| | | | $P = 313$ mm | $P = 226$ mm |

**11. ILE NOUVELLE AMSTERDAM** (28 m) 37°50′S, 77°34′E

| | | | | |
|---|---|---|---|---|
| 100 | −56.2 | — | −61.4 | — |
| 200 | −53.5 | — | −54.2 | — |
| 300 | 46.5 | — | −36.0 | 0.18 |
| 500 | −21.7 | 0.39 | −11.4 | 1.03 |
| 700 | −5.8 | 1.40 | 2.9 | 2.30 |
| 850 | 0.8 | 2.9 | 8.4 | 5.1 |
| SFC | 10.5 | 6.3 | 16.1 | 9.4 |
| | (1016 mb) | | (1011 mb) | |
| | | | $P = 123$ mm | $P = 86$ mm |

**12. QUINTERO** (2 m) 32°47′S, 71°32′W

| | | | | |
|---|---|---|---|---|
| 100 | −61.6 | — | −68.0 | — |
| 200 | −56.6 | — | −53.5 | — |
| 300 | −45.8 | — | −38.0 | 0.06 |
| 500 | −18.3 | 0.63 | −10.6 | 0.59 |
| 700 | −0.3 | 1.55 | 8.0 | 1.5 |
| 850 | 9.1 | 2.8 | 15.7 | 4.6 |
| SFC | 11.5 | 7.1 | 17.5 | 9.6 |
| | (1018 mb) | | (1015 mb) | |
| | | | $P = 87$ mm[1] | $P = 0$ mm[1] |

[1] At Valparaiso.

T ABLE 41. *Air masses of the Southern hemisphere*—continued.

| LEVEL (mb) | JULY T(°C) | JULY MR(g/kg) | JANUARY T(°C) | JANUARY MR(g/kg) |
|---|---|---|---|---|
| **13. CAPE TOWN (49 m) 33°58′S, 18°36′E** | | | | |
| 100 | −61.1 | — | −66.9 | — |
| 200 | −55.2 | — | −54.2 | — |
| 300 | −43.4 | 0.035 | −36.4 | 0.13 |
| 500 | −17.3 | 0.41 | −9.4 | 0.79 |
| 700 | −1.2 | 1.5 | 6.9 | 2.15 |
| 850 | 6.3 | 3.65 | 14.5 | 5.2 |
| SFC | 11.7 | 7.1 | 20.8 | 10.9 |
| | (1015 mb) | | (1007 mb) | |
| | | | P = 83 mm | P = 11 mm |
| **14. GUILDFORD (15 m) 31°56′S, 115°57′E** | | | | |
| 100 | −62.7 | — | −71.7 | — |
| 200 | −51.5 | — | −54.4 | — |
| 300 | −42.4 | — | −36.9 | — |
| 500 | −19.4 | 0.38 | −9.9 | 1.0 |
| 700 | −3.3 | 1.5 | 7.4 | 2.8 |
| 850 | 3.8 | 3.9 | 16.6 | 5.4 |
| SFC | 13.1 | 7.2 | 25.1 | 9.6 |
| | (1016 mb) | | (1011 mb) | |
| **15. NEUQUEN (270 m) 38°57′S, 68°07′W** | | | | |
| 100 | −58.3 | — | −61.7 | — |
| 200 | −57.2 | — | −54.4 | — |
| 300 | −47.4 | 0.045 | −40.4 | 0.12 |
| 500 | −21.8 | 0.55 | −14.5 | 0.83 |
| 700 | −4.5 | 1.52 | 2.7 | 2.7 |
| 850 | 3.5 | 2.8 | 12.8 | 4.7 |
| SFC | 4.9 | 3.8 | 23.4 | 6.5 |
| | (986 mb) | | (977 mb) | |
| | | | P = 10 mm | P = 12 mm |
| **16. RESISTENCIA (52 m) 27°27′S, 59°03′W** | | | | |
| 100 | −65.8 | — | −72.7 | — |
| 200 | −55.2 | — | −51.8 | — |
| 300 | −40.6 | 0.05 | −32.9 | 0.24 |
| 500 | −13.1 | 0.68 | −7.6 | 1.28 |
| 700 | 3.2 | 2.5 | 8.5 | 4.5 |
| 850 | 10.1 | 4.8 | 18.3 | 7.75 |
| SFC | 14.2 | 8.0 | 26.7 | 15.3 |
| | (1013 mb) | | (1004 mb) | |
| | | | P = 45 mm | P = 118 mm |
| **17. WINDHOEK (1728 m) 22°34′S, 17°06′E** | | | | |
| 100 | −69.8 | — | −75.5 | — |
| 200 | −52.3 | — | −51.8 | — |
| 300 | −35.2 | 0.037 | −30.9 | 0.19 |
| 500 | −8.8 | 0.41 | −6.0 | 2.1 |
| 700 | 5.4 | 1.9 | 12.6 | 5.2 |
| SFC | 13.0[1] | 3.2 | 23.0[2] | 9.2 |
| | | | P = 1 mm | P = 77 mm |

[1] SFC at 835 mb. [2] SFC at 830 mb.

| LEVEL (mb) | JULY T(°C) | JULY MR(g/kg) | JANUARY T(°C) | JANUARY MR(g/kg) |
|---|---|---|---|---|
| **18. PRETORIA (1368 m) 25°45′S, 28°14′E** | | | | |
| 100 | −68.3 | — | −73.9 | — |
| 200 | −52.2 | — | −52.3 | — |
| 300 | −37.3 | 0.029 | −32.1 | 0.23 |
| 500 | −11.5 | 0.33 | −6.3 | 1.9 |
| 700 | 3.0 | 1.6 | 10.3 | 7.3 |
| SFC | 10.0[1] | 4.7 | 21.7[2] | 12.6 |
| | P = 10 mm | | P = 125 mm | |

[1] SFC at 870 mb. [2] SFC at 865 mb.

| LEVEL (mb) | JULY T(°C) | JULY MR(g/kg) | JANUARY T(°C) | JANUARY MR(g/kg) |
|---|---|---|---|---|
| **19. GILES (514 m) 25°02′S, 128°18′E** | | | | |
| 100 | −71.4 | — | −77.8 | — |
| 200 | −53.9 | — | −52.7 | — |
| 300 | −35.6 | — | −33.0 | — |
| 500 | −12.4 | 0.57 | −7.4 | 1.4 |
| 700 | 1.8 | 1.6 | 9.8 | 4.4 |
| 850 | 9.0 | 3.1 | 21.6 | 6.1 |
| SFC | 13.8 | 4.7 | 30.1 | 7.6 |
| | (961 mb) | | (948 mb) | |
| **20. CLONCURRY (188 m) 20°40′S, 140°30′E** | | | | |
| 100 | −75.3 | — | −79.7 | — |
| 200 | −54.3 | — | −53.0 | — |
| 300 | −32.9 | — | −32.1 | — |
| 500 | −8.5 | 0.57 | −5.8 | 1.4 |
| 700 | 4.7 | 1.23 | 8.9 | 5.2 |
| 850 | 10.8 | 3.2 | 20.4 | 9.2 |
| SFC | 17.8 | 4.7 | 31.9 | 13.6 |
| | (995 mb) | | (984 mb) | |
| | | | P = 9 mm | P = 112 mm |
| **21. LIMA (34 m) 12°01′S, 77°07′W** | | | | |
| 100 | −74.8 | — | −81.1 | — |
| 200 | −55.1 | — | −53.8 | — |
| 300 | −33.2 | 0.14 | −31.6 | 0.4 |
| 500 | −7.3 | 1.63 | −5.6 | 2.55 |
| 700 | 10.5 | 2.45 | 10.0 | 5.0 |
| 850 | 17.5 | 3.2 | 16.8 | 9.8 |
| SFC | 16.2 | 10.0 | 21.5 | 13.7 |
| | (1010 mb) | | (1007 mb) | |
| | | | P = 6 mm | P = 1 mm |
| **22. TAHITI (2 m) 17°33′S, 149°37′W** | | | | |
| 100 | −71.4 | — | −72.6 | — |
| 200 | −51.6 | — | −51.7 | — |
| 300 | −32.3 | 0.105 | −30.9 | 0.29 |
| 500 | −6.8 | 0.70 | −6.6 | 2.0 |
| 700 | 8.3 | 2.5 | 8.5 | 5.3 |
| 850 | 14.9 | 8.7 | 16.4 | 10.9 |
| SFC | 24.4 | 15.2 | 26.5 | 17.5 |
| | (1013 mb) | | (1011 mb) | |
| | | | P = 70 mm | P = 423 mm |

| LEVEL (mb) | JULY T(°C) | MR(g/kg) | JANUARY T(°C) | MR(g/kg) |
|---|---|---|---|---|

**23. NANDI (Fiji Is.) (16 m) 17°45′S, 177°27′E**

| LEVEL (mb) | JULY T(°C) | MR(g/kg) | JANUARY T(°C) | MR(g/kg) |
|---|---|---|---|---|
| 100 | −76.1 | — | −80.6 | — |
| 200 | −54.7 | — | −53.5 | — |
| 300 | −32.7 | — | −30.8 | 0.4 |
| 500 | −7.3 | 1.50 | −5.0 | 2.4 |
| 700 | 8.6 | 3.9 | 9.9 | 6.1 |
| 850 | 14.8 | 8.3 | 18.0 | 11.9 |
| SFC | 22.4 | 13.4 | 26.2 | 18.2 |
| | (1009 mb) | | (1005 mb) | |

**24. RAOUL ISLAND (49 m) 29°15′S, 177°55′W**

| LEVEL (mb) | JULY T(°C) | MR(g/kg) | JANUARY T(°C) | MR(g/kg) |
|---|---|---|---|---|
| 100 | −64.0 | — | −69.7 | — |
| 200 | −52.3 | — | −54.2 | — |
| 300 | −38.3 | 0.22 | −35.9 | 0.24 |
| 500 | −15.8 | 0.75 | −9.2 | 0.97 |
| 700 | −0.5 | 2.05 | 6.8 | 2.7 |
| 850 | 6.2 | 4.9 | 12.3 | 7.2 |
| SFC | 16.3 | 9.2 | 21.5 | 13.5 |
| | (1010 mb) | | (1009 mb) | |
| | | | P = 170 mm | P = 82 mm |

**25. LORD HOWE ISLAND (11 m) 31°31′S, 159°04′E**

| LEVEL (mb) | JULY T(°C) | MR(g/kg) | JANUARY T(°C) | MR(g/kg) |
|---|---|---|---|---|
| 100 | −60.6 | — | −68.6 | — |
| 200 | −50.7 | — | −55.1 | — |
| 300 | −43.0 | — | −37.9 | — |
| 500 | −20.5 | 0.44 | −10.6 | 0.89 |
| 700 | −3.8 | 1.70 | 5.3 | 2.7 |
| 850 | 4.2 | 4.4 | 11.2 | 7.1 |
| SFC | 15.9 | 8.6 | 21.8 | 12.7 |
| | (1013 mb) | | (1013 mb) | |
| | | | P = 166 mm | P = 105 mm |

**26. WILLIS ISLAND (8 m)[1] 16°18′S, 149°59′E[1]**

| LEVEL (mb) | JULY T(°C) | MR(g/kg) | JANUARY T(°C) | MR(g/kg) |
|---|---|---|---|---|
| 100 | −77.1 | — | −82.1 | — |
| 200 | −54.4 | — | −52.9 | — |
| 300 | −31.9 | — | −30.2 | — |
| 500 | −6.1 | 1.4 | −5.2 | 2.1 |
| 700 | 8.6 | 3.5 | 9.7 | 6.0 |
| 850 | 11.7 | 6.0 | 17.6 | 10.1 |
| SFC | 23.7 | 12.6 | 28.5 | 19.8 |
| | (1014 mb) | | (1007 mb) | |

[1] 1963–67.

**27. CARNARVON (3 m) 24°53′S, 113°39′E**

| LEVEL (mb) | JULY T(°C) | MR(g/kg) | JANUARY T(°C) | MR(g/kg) |
|---|---|---|---|---|
| 100 | −71.8 | — | −78.3 | — |
| 200 | −53.7 | — | −52.8 | — |
| 300 | −34.7 | — | −33.6 | — |
| 500 | −12.0 | 0.62 | −7.6 | 1.40 |
| 700 | 2.4 | 1.55 | 10.1 | 4.3 |
| 850 | 8.3 | 4.6 | 22.0 | 5.8 |
| SFC | 16.5 | 7.7 | 27.3 | 14.2 |
| | (1019 mb) | | (1008 mb) | |
| | P = 52 mm | | P = 5 mm | |

**28. COCOS ISLAND (5 m) 12°05′S, 96°53′E**

| LEVEL (mb) | JULY T(°C) | MR(g/kg) | JANUARY T(°C) | MR(g/kg) |
|---|---|---|---|---|
| 100 | −77.1 | — | −81.3 | — |
| 200 | −55.1 | — | −54.7 | — |
| 300 | −32.7 | — | 32.2 | — |
| 500 | −6.2 | 1.4 | −5.7 | 1.6 |
| 700 | 8.1 | 4.6 | 9.1 | 4.8 |
| 850 | 15.2 | 9.5 | 16.4 | 10.6 |
| SFC | 25.4 | 16.6 | 26.9 | 17.7 |
| | (1012 mb) | | (1010 mb) | |

**29. MAURITIUS (55 m) 20°18′S, 57°30′E**

| LEVEL (mb) | JULY T(°C) | MR(g/kg) | JANUARY T(°C) | MR(g/kg) |
|---|---|---|---|---|
| 100 | −74.1 | — | −77.3 | — |
| 200 | −52.9 | — | −53.9 | — |
| 300 | −32.5 | 0.10 | −32.5 | 0.23 |
| 500 | −8.4 | 0.69 | −6.7 | 1.3 |
| 700 | 5.4 | 2.4 | 9.0 | 4.6 |
| 850 | 11.4 | 7.3 | 16.7 | 10.2 |
| SFC | 20.8 | 11.6 | 26.4 | 18.1 |
| | (1015 mb) | | (1006 mb) | |

NOTES TO TABLE 41—STATIONS IN THE SOUTHERN HEMISPHERE

1. *Amundsen-Scott Station at the South Pole* (2800 m). During July (winter) outgoing radiation from the elevated surface of Antarctica is producing a shallow layer of uniquely cold air in the fourth decasphere, the 700- to 600-millibar horizon. Note that during the same month incoming radiation is producing anomalously high day temperatures at the same level over Tibet. Even in January (summer), despite continuous insolation, the surface inversion, though weaker, is sufficient to bring down temperatures to a level comparable with those encountered in the free air at the same level over Verkhoyansk—then in the depth of winter. The water vapour content is extremely low throughout the year, though data are limited.

2. *Halley Bay*—in the Weddell Sea south of the Atlantic. Clearly arctic air at both seasons and

6–8° C colder than Alert in summer throughout the lowest half of the atmosphere, despite a lower latitude.

3.  *Mirny Observatory*—on the comparatively open coast of the massive plateau of Eastern (Gondwana) Antarctica south of the Indian Ocean. Arctic air, though surface characteristics are modified by occasional incursions from the pack-laden sea.

4.  *Argentine Island*—on the west coast of Graham Land, south of Cape Horn. Exposure to mid-latitude frontal storms decreases seasonal contrast and increases mixing ratios but Arctic characteristics predominate.

5.  *D. N. Orcades (South Orkneys)*—east of Drake Strait. Similar to (4) and still heavily under the influence of the Antarctic pack-ice throughout the year.

6.  *Ushuaia*—in southern Chile near Cape Horn. Mid-latitude air throughout the year with some traces of continental heating in summer.

7 & 8.  *Macquarie Island*—southwest of New Zealand and *Marion Island*, southeast of Cape Town. Although in latitudes broadly comparable with those of Manchester and Nantes these islands experience a uniquely unpleasant climate with such a slight seasonal range that they are virtually both summerless and winterless.

9.  *Invercargill*—in the south of New Zealand. Mid-latitude air throughout the year.

10.  *Gough Island*—in the South Atlantic near Tristan da Cunha. Mid-latitude air in winter (July). In January the mean ascent appears to be rather more stable but the precipitation record suggests that this is an illusion comparable to that recognised over the western approaches to Europe. Lively frontal activity.

11.  *New Amsterdam*—in the south of the Indian Ocean. On the border between tropical and mid-latitude air especially in summer (January) when an inversion between 900 and 800 millibars may be inferred when the temperature and humidity records are plotted.

12.  *Quintero, near Valparaiso*—Tropical air over cool coastal waters. A strong low-level inversion characteristic throughout the year reaching a potential temperature range of some 15° K in January. Relative humidity at 850 millibars rarely exceeds 33 per cent.

13 & 14.  *Cape Town and Guildford (near Perth, Australia)*—The difference between summer and winter at these southern 'Mediterranean' stations appears to lie less in the strength of the inversion than in the height of the Lifting Condensation Level—highest in January (summer).

15.  *Neuquen*—northern Patagonia. Southern tropical-continental air reminiscent of Alma Ata but less extreme.

16.  *Resistencia*—north Argentina. Tropical continental in July but almost equatorial air in January (summer).

17.  *Windhoek*—southwest Africa. Southern semi-desert at considerable elevation (1728 m).

18.  *Pretoria*—Transvaal. Southern tropical-continental air, approaching high equatorial air in January (summer).

19.  *Giles*—central Western Australia. Southern desert.

20.  *Cloncurry*—central Queensland. Southern tropical continental air. Much less stable in January but with a high lifting condensation level.

336

21. *Lima*—Peru. Coastal desert with a low inversion throughout the year but attaining a range of potential temperature of about 20° K in July.

22 & 23. *Tahiti*—south central Pacific, and *Nandi*, Fiji Islands. Tropical maritime air with little seasonal range. The trade-wind inversion is relatively high (about 850 mb) and never strong, though best developed in July (winter). Compare these stations with Raizet.

24 & 25. *Raoul Island and Lord Howe Island*. Both towards the poleward margins of tropical maritime air with the trade-wind inversion best developed in January (summer) during the southward movement of the wind systems.

26. *Willis Island*—in the Coral Sea off the Queensland coast. With the development of the Coral Sea trades there is a marked inversion between 850 and 700 millibars in July (winter) but in January the air is almost equatorial.

27. *Carnarvon*—coast of Western Australia. Stable air near the eastern margin of the southeast trades of the Indian Ocean with a marked seasonal temperature range owing to the presence of the Australian continent.

28. *Cocos Island*—Indian Ocean. Southeast trades but never far from their equatorial margin—hence little seasonal range and comparatively weak inversion.

29. *Mauritius*—western sector of Indian Ocean trades. Inversion most marked in July (winter) and air comparatively unstable in January, the hurricane season. Again compare with Raizet.

*Temperatures and saturation mixing-ratios of three model standard saturated atmospheres*

| LEVEL (mb) | COLD[1] | | STANDARD | | WARM[1] | |
|---|---|---|---|---|---|---|
| | $T$ (°C) | $MR$ (g/kg) | $T$ (°C) | $MR$ (g/kg) | $T$ (°C) | $MR$ (g/kg) |
| 300 | −60 | 0.024 | −45 | 0.23 | −30 | 1.0 |
| 500 | −36 | 0.35 | −21 | 1.4 | −6 | 4.8 |
| 700 | −20 | 1.10 | −5 | 3.9 | 10 | 11.4 |
| 850 | −9.5 | 2.2 | 5.5 | 6.7 | 20.5 | 18.2 |
| 1013.25 SFC | 0 | 3.8 | 15 | 10.7 | 30 | 27.3 |

[1] See text pp. 227; 230; 232.

It has been found useful to compare these *fully-saturated models* with the values quoted in Table 41. They are, of course, never experienced in nature.

# 10
# Other approaches to air-mass classification

## Air masses over the British Isles

In any area where variety is a major feature of the weather it is self-evident that monthly means derived from surface observations of any of the elements of climate must involve a considerable degree of abstraction, of unreality. This is particularly true of the features shown by mean ascents since the root cause of short-period variations must rest in the atmospheric circulation which imports air characteristics generated elsewhere. Throughout the whole year the weather of the British Archipelago depends heavily upon imported air though its two outstanding characteristics, moistness and moderation, may be regarded as at least partially 'home-brewed' since they arise from the nature of the surrounding seas. The fault of averages taken over some arbitrary period of time is that they must necessarily combine observations made during air invasions from different source regions. The result is therefore a confused amalgam which illustrates neither the inherent contrasts between the invading masses nor the relative frequency of their occurrence.

In such circumstances generalisation must follow a more onerous route. A logical solution is to classify the observed air samples according to their inferred origin and trajectory and to derive mean values from each group, combining like with like. Such a process should reveal the inherent characteristics of each category and yield its relative frequency as a useful by-product. It has been argued indeed that this is the *only way* in which the innate variability of mid-latitude climates can be made amenable to valid and useful generalisation but it is not without difficulties. Clearly such an approach must start with a comparatively restricted area as focus and it is only possible where a long series of reliable weather-maps is available. Furthermore, since there are no hard and fast distinctions in nature, any classification must involve some element of subjective judgment. This may have only a slight effect upon the *means* for the various groups but it cannot be eliminated from any statement of relative *frequencies*. Finally, since the intellectual appreciation of any classification becomes more and more difficult as the number of categories increases, some considerable degree of generalisation is certainly desirable. Very few of us can comprehend more than seven or eight groups *at one and the same time* and this is what this exercise demands if its results are to be fused into a single climatic picture.

An experiment along these lines was made with British data by Belasco (1952). It yielded a table of mean monthly frequencies of various air masses at Kew, Scilly and Stornoway as recorded each day at 1800 hours GMT over the 12-year period 1938 to 1949. In all, Belasco found it desirable to distinguish twenty-three types of air over Britain, of which some ten to eleven had originated in a northern or cold source-region whilst five to six had arrived from southern or warm sources.[1] Of the remainder, five types were associated with comparatively gentle air motion in high-pressure areas and the last two were so closely involved with an active front or depression that separate classification was deemed advisable. By combining pairs of related groups he reduced this number to seventeen categories for his final tabulation Belasco (1952, 33–4). In the following summary we have generalised still further by combining the records for the 4-month periods, winter (November to February) and summer (May to August) at Stornoway and Scilly and reducing the number of categories to eight.

I. 'Northern air' includes Belasco's four polar categories $P_1$ to $P_4$ as well as the rare occurrences (in winter) of cold air from Russia ($A_1$ and $A_2$). Broadly speaking, it includes all air from north of 60°N.

II. 'Central North Atlantic air' includes categories $P_5$ and $P_6$, that is, air reaching Britain mainly from the west, i.e. from an ocean source broadly between 60 and 45°N.

III. 'Air with northern affinities' includes Belasco's categories $P_7$ and $P_R$, i.e. polar air which has reached our shores only after a very considerable detour to the south and has undergone modification in the process.

IV. 'Southeast European air' is Belasco's group C. It is given separately not because of any intrinsic importance but so that it may be combined with northern air in winter and southern air in summer if that seems reasonable.

V. 'Southern air' includes Belasco's two tropical categories $T_1$ and $T_2$, i.e. air from the more tropical parts of the North Atlantic south of 45°N.

VI. 'Air with southern affinities' ($T_3$, $T_4$ and $T_Q$) includes air from around the Iberian peninsula or the Mediterranean.

VII. 'Highs' includes all five of Belasco's categories of air closely associated with anticyclonic systems. The essential point is that the pressure system itself imposes some uniformity upon the air stratification whilst the gentle motion enables it to take on 'home-brewed' characteristics.

VIII. 'Fronts and lows' (F and D) includes air in frontal zones where the vertical stratification may involve two distinct air masses and 'polar

[1] The numbers are given in this way since his group 'C', continental air from east of 10°E and south of 50°N, is cold in winter and warm in summer.

air moving eastwards and southwards near the central region of depressions' where the structure of the atmosphere is undergoing profound modification through the medium of active weather processes.

TABLE 42. *Summary Table of air mass frequencies over the British Isles 1938–49* (*percentage of days*)

| STORNOWAY (58°11′N) | WINTER | | | SUMMER | | |
|---|---|---|---|---|---|---|
| I. Northern air | 25 | | | 27 | | |
| II. Central North Atlantic air | 18 | = 63 | | 18.5 | = 60 | |
| III. Air with northern affinities | 20 | | | 14.5 | | |
| IV. Southeast European air | 6 | | 6 | 3 | | 3 |
| V. Southern air | 6 | = 11 | | 3.5 | = 6 | |
| VI. Air with southern affinities | 5 | | | 2.5 | | |
| VII. Air in high pressure areas | 10 | = 20 | | 15.5 | = 31 | |
| VIII. Air near fronts and lows | 10 | | | 15.5 | | |
| SCILLY (49°56′N) | | | | | | |
| I. Northern air | 15 | | | 20 | | |
| II. Central North Atlantic air | 19 | = 44 | | 19 | = 47 | |
| III. Air with northern affinities | 10 | | | 8 | | |
| IV. Southeast European air | 8 | | 8 | 5 | | 5 |
| V. Southern air | 15 | = 21.5 | | 9 | = 12 | |
| VI. Air with southern affinities | 6.5 | | | 3 | | |
| VII. Air in high pressure areas | 15.5 | = 26.5 | | 22 | = 36 | |
| VIII. Air near fronts and lows | 11 | | | 14 | | |

In comparing these two stations (which are about 575 miles apart along approximately the same meridian, *c.* 6°20′ W), by traditional methods we could quote their mean seasonal temperatures,

Stornoway, winter 42°F (5.6°C) summer 53°F (11.7°C)
Scilly,      winter 47°F (8.3°C) summer 58°F (14.4°C)

thus concluding that there was a mean difference of 5°F (2.8°C) between them for most of the year. These figures could be supplemented by a statement of the means of monthly extremes:

| | WINTER | | SUMMER | |
|---|---|---|---|---|
| | *Max.* | *Min.* | *Max.* | *Min.* |
| Stornoway | 52°F (11.1°C) | 27°F (−2.8°C) | 67°F (19.4°C) | 39°F (3.9°C) |
| Scilly | 55°F (12.8°C) | 37°F (2.8°C) | 69°F (20.6°C) | 48°F (8.9°C) |

which are a good deal more revealing. It is open to debate whether the above table of frequencies adds much to this account. Certainly the relatively greater prevalence of Northern air (I) at the more northerly station and the relatively greater prevalence of Southern air (V) at the more southerly station might well have been anticipated. More interesting perhaps are the higher values in group VII at Scilly since they give some indication of the contribution which the Azores high makes to the climate of the southwest during both seasons of the year. The obverse of this is presented by the greater values under III at Stornoway since these are evidence of widespread vigour in the circulation.

Nevertheless the table does reveal some interesting facts:

1. Note that air that has freshly arrived from some distant part of the northern, central or southern sectors of the North Atlantic is recognised in winter on 59 (44 + 15) and in summer on 56 (47 + 9) per cent of the days at Scilly and respectively on 69 (63 + 6) to 63.5 (60 + 3.5) per cent of the days at Stornoway. This is a much more acceptable statement of the dominance of Atlantic air-streams throughout the year than emphasis on the 'prevailing westerlies'.

2. Notice also that Northern air rather narrowly defined (group I) is more frequently encountered at both stations in summer than in winter and that the contrast is greatest at the more southerly station. That this is no mere chance is suggested by the parallel fact that Southern air (group V) is more common at both stations in winter than in summer. Although the differences do not exceed from 3 to 6 per cent of occasions, these remarkable facts must go some way towards explaining the low seasonal range of temperatures along our western shores.

3. Groups VII and VIII may perhaps be thought of as consisting of air undergoing local transformation as a result of dynamic processes. That group VII should be more frequent in summer at both stations is not surprising but would one have expected group VIII to follow its lead?

4. It follows from (1) and (3) that air from the European continent must be comparatively rare at these westerly stations and that it must be more common in winter than in summer despite the more powerful cyclonic westerly circulation at that period. Combining the values for groups IV and VI we obtain winter frequencies of 11 per cent at Stornoway and 14.5 per cent at Scilly but these would become 13 per cent and 17.5 per cent respectively if Belasco's 'arctic' category were detached from group I and added here. That air with a recognisably transcontinental path may reach the west of these islands on from a sixth to an eighth of all winter days is a fact worthy of note though we must recall that its nature is rapidly modified during its short traverse of the Narrow Seas.

It is, however, the *characteristics* of the various types of air embraced within the mid-latitude category that particularly interest us in this chapter. Some idea of the usual range of contrasts encountered over Britain may be obtained by comparing mean ascents in 'tropical maritime air with an anticyclonic path from the Azores' (Belasco's $T_2$) with 'polar maritime air with a cyclonic track from north and northeast of Iceland' (Belasco's $P_1$). Mean ascents for

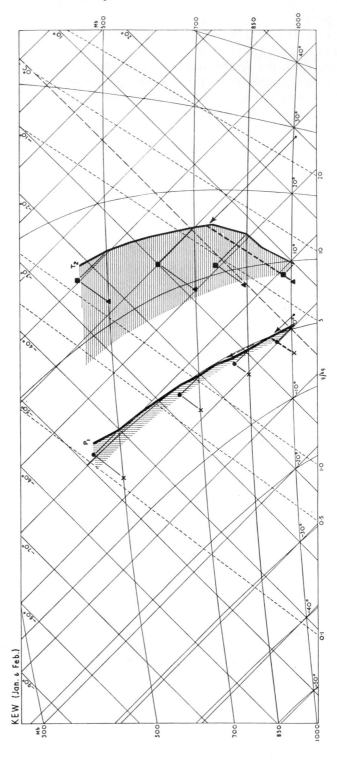

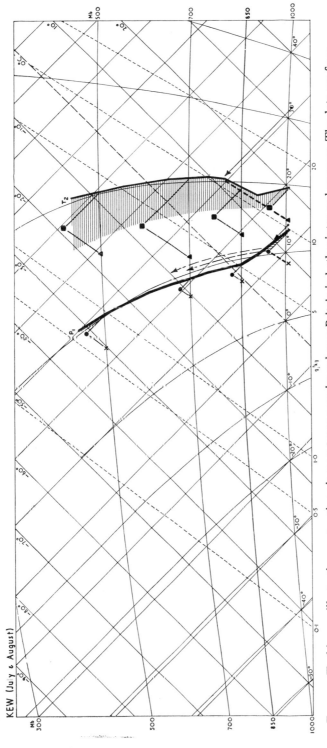

FIG. 102. Tephigrams illustrating contrasting air mass types experienced over Britain in both winter and summer. The data are from Belasco (1952). $P_1$ represents a common type of 'polar' air and $T_2$ a not infrequent type of 'tropical' air

January–February and July–August in these two subtypes are illustrated in Fig. 102 and Belasco's material is summarised in a more modern format below.

TABLE 43. *Tropical and polar air masses over Britain (after Belasco)* [1]

| LEVEL (mb) | WINTER (JANUARY–FEBRUARY) | | | | | SUMMER (JULY–AUGUST) | | | | |
|---|---|---|---|---|---|---|---|---|---|---|
| | $T$ (°C) | $PT$ (°K) | $SPT$ (°C) | $MR$ (g/kg) | $RH$ (%) | $T$ (°C) | $PT$ (°K) | $SPT$ (°C) | $MR$ (g/kg) | $RH$ (%) |
| *Type $P_1$ (polar)* | | | | | | | | | | |
| 500 | −40.5 | 283 | 0.3 | 0.12 | 52 | −25.5 | 301.5 | 9.7 | 0.62 | 66 |
| 700 | −20.5 | 280 | −0.8 | 0.71 | 64 | −8 | 294 | 8.7 | 2.5 | 80 |
| 850 | −9 | 276.5 | −1.4 | 1.7 | 75 | 2 | 288 | 8.3 | 4.3 | 84 |
| SFC[2] | 1 | 273 | −1.2 | 3.0 | 74 | 14 | 286 | 10.3 | 7.1 | 72 |
| *Type $T_2$ (tropical)* | | | | | | | | | | |
| 500 | −16 | 313.5 | 15.0 | 1.24 | 54 | −8.5 | 322.5 | 18.4 | 2.0 | 49 |
| 700 | 1 | 303 | 13.0 | 2.95 | 51 | 7 | 310 | 16.2 | 3.8 | 43 |
| 850 | 7.5 | 293.5 | 11.0 | 4.45 | 59 | 13 | 299.5 | 15.8 | 7.0 | 62 |
| SFC[2] | 10 | 282 | 8.1 | 6.4 | 83 | 20 | 292 | 16.5 | 10.9 | 74 |

[1] In Fig. 102 mean soundings at 950 mb have been ignored in the graphs since only early morning records were selected at this level. They do not match the surface record.
[2] Surface values are those given by Belasco for Kew.

The data thus presented illustrate the wide range of air mass types which occur over Britain, most of the other types recognised by Belasco falling between these extremes. It is evident, as we have already suspected, that mean ascents must be used with caution in these latitudes. Note that both air-masses are comparatively moist so that the lifting condensation level is rarely above 1850 ft, i.e. well within the range of the British uplands on which 'hill fog' is thus extremely common. The stability of *tropical* air moving northwards is well brought out. Low relative humidities in the middle layers suggest that this is partly the result of continued subsidence but this is reinforced in winter by surface chill. At this season potential temperature increases by no less than 21°K from the surface to the 700-millibar level and penetrative convection is out of the question. In summer, on the other hand, the drier air aloft induces a limited degree of layer lability up to about 800 mb though penetrative convection would then require surface temperatures to exceed 32°C (90°F), rare at Kew. The *polar* air has clearly been profoundly modified en route from its arctic sources. It retains a limited degree of stability in winter though the curve follows a saturation adiabat with remarkable fidelity. In summer, on the other hand, the stratification is conditionally unstable and a very moderate trigger action would be sufficient to set off deep thunderstorm activity.

The dominant feature of the mid-latitude zone however, is that air streams as different as these—termed 'equatorial' and 'polar' currents by Dove (1862, 75) may, as he says, 'flow alongside of each other in channels which are constantly changing'. Today we would put it that they are brought into dynamic relationship or confronted with each other and the result will clearly depend upon their relative densities. Air density varies inversely as virtual temperature (temperature adjusted for water-vapour content, see p. 241) expressed in degrees Kelvin and the following results may be of interest.

| (mb) | WINTER | | | SUMMER | | |
|---|---|---|---|---|---|---|
| | *Virtual temperature* (°K) | | *Relative density* | *Virtual temperature* (°K) | | *Relative density* |
| | $T_2$ | $P_1$ | (%) | $T_2$ | $P_1$ | (%) |
| 500 | 257.5 | 232.5 | 90.5 | 265 | 247.5 | 93 |
| 700 | 274.5 | 253 | 92 | 280.5 | 265.5 | 94.5 |
| 850 | 281.5 | 264 | 94 | 287.5 | 275.5 | 96 |
| S.F.C. | 284 | 274.5 | 96.5 | 295 | 288 | 97.5 |

The percentages give the relative density of tropical as compared with polar air at each of the levels stated. As might well have been anticipated, the density contrasts and hence the vigour of the reactions during confrontation, are considerably greater in winter than in summer.

# The synoptic climatology of Central Europe

An alternative appproach to this kind of problem which has proved popular on the Continent is to examine a long series of daily weather charts with the object of grouping them into a number of characteristic patterns and thus establishing the relative frequency of various 'weather types'. This method has the attraction of beginning with real situations and the exercise, though arduous, is certainly stimulating. In the author's experience, however, there is always some sense of anticlimax when the results are presented to those who have not been through the drill. Sadly enough, relative frequency is also something of an abstraction so that 'reality' has again eluded us!

A practical difficulty is that a significant pressure-pattern must relate to an area of at least sub-continental scale but to relate this to weather we must focus attention upon a much more limited region, preferably not far from the middle of the chart. The grouping of patterns also involves some degree of subjective judgement, particularly with reference to marginal or awkward cases. There is always a temptation to over-elaborate the classification with the two-fold risk of increasing the play of pure chance and of making the scheme difficult to comprehend as a whole.

As an example we may quote the work of F. Baur (referred to at length in Blüthgen 1964, pp. 311–25). Baur considered daily weather charts for the whole of Europe from 1881 to 1943 and classified them from the point of view

345

of Central European weather into 18 main weather types (grosswetterlagen), some of which he further subdivided. We give a highly-condensed summary of his results below, designed to illustrate at least the main features which emerged. The figures give the quarterly frequency of seven groups of types expressed as a percentage of the total number of days in a year.

| Type | I | II | III | IV | V | VI | VII | |
|---|---|---|---|---|---|---|---|---|
| December–February | 7.5 | 6.0 | 1.5 | 2.0 | 4.0 | 2.0 | 2.0 | 25.0 |
| March–May | 6.0 | 4.5 | 3.5 | 3.0 | 2.0 | 1.5 | 4.5 | 25.0 |
| June–August | 8.5 | 7.0 | 3.5 | 3.0 | 0.5 | 0.5 | 2.0 | 25.0 |
| September–November | 8.5 | 6.5 | 2.0 | 2.0 | 3.0 | 1.0 | 2.0 | 25.0 |
| | 30.5 | 24.0 | 10.5 | 10.0 | 9.5 | 5.0 | 10.5 | 100 |

*Group I* includes days with an area of high pressure centred over the Atlantic south-west of Britain but extending far eastward over most of Central Europe. Pressure falls steadily but rather slowly polewards and light westerly or north-westerly winds thus prevail over Central Europe. In summer and early autumn this pattern brings warm and sunny spells, in winter the weather is usually cold, often with mist or fog. The seasonal variation in the frequency of this type is inconsiderable.

*Group II* sees the passage of depressions in a west-to-east direction across the British Isles towards Baltic Russia whilst relatively high pressure is maintained to the south. The poleward gradient in Central Europe is thus increased and the westerly air-stream is accelerated, fronts bringing variable weather with cloud and precipitation over most of the area. This type is rather more common in summer though the seasonal contrast is not very marked. Indeed this summer maximum is also exhibited by some intermediate cases (Baur's 'HW' type) which have been classed with group I. If these were shared between groups I and II, the mean frequency of these two groups would be about equal, yielding the impression that on no less than 55 per cent of all occasions the weather of Central Europe showed an alternation between types I and II. It seems that much the same could be said of south-eastern England.

The remaining five types are much less frequent but they show some interesting contrasts in seasonal distribution.

*Group III* includes occasions when a high pressure over the north Atlantic extends over the western coasts of Europe whilst low pressure predominates towards Russia. A cool and rather unstable air-stream from the north thus floods over Central Europe. Though infrequent, this occurs most commonly in spring and summer.

*Group IV* comprises all cases when a low is situated squarely over Central Europe. Winds are variable and instability showers are common. Like group III this is mainly a summer phenomenon.

*Group V*, with high pressure towards the Black Sea and light south-easterly to southerly winds over Central Europe, is rare in summer. In autumn and winter it brings cold weather though föhns may be experienced in suitable localities.

*Group VI* is relatively rare. Broadly the pattern is the reverse of that experienced with groups I and II, the high pressure being over Scandinavia and relatively low pressure towards the south. The weather is variable with easterly winds.

*Group VII* includes a well-marked number of occasions particularly distinguished by Baur. An extensive low pressure over the Atlantic extends also over much of western Europe, bringing a southerly air-stream and widespread precipitation over our focal region. Unlike any of our other groups it shows a marked spring maximum and may also be associated with föhns.

The general picture which thus emerges is thus one of very considerable variety. Whether this approach adds much to the account furnished by traditional climatological description the reader must judge for himself. The difficulty, of course, is that a statement of mean frequencies gives no information as to either the *duration* of individual 'spells' or to the *sequence* in which the various types of situations occur. To answer these questions the whole mass of material must be processed afresh. This procedure yields evidence of some value in long-range forecasting.

## The relative classification of air masses

The transformations which both tropical and arctic air masses undergo as they invade middle latitudes are so considerable that the paths they have followed may be no less significant than their original source regions. To a native of Cornwall, for instance, 'returning mP air' may have so many of the qualities of 'mT air' that he may wonder whether the reference to pole and tropic is not irrelevant. Besides, mixing as a result of past weather between two such very different constituents can produce a wide spectrum of sub-types, and active weather is just as likely to follow from the encounter of subtype with subtype as it is from the clash of the parent air streams. Indeed this appears to be by far the most usual occurrence.

Bergeron (1928, 50) met this difficulty when he proposed a relative classification into 'warm' and 'cold' air masses. A 'warm' air mass is one which transfers net heat to the surface over which it is passing whilst a 'cold' air mass receives a net transfer of heat from that surface. Emphasis is thus placed upon the local response to the invading air mass rather than upon its inherent characteristics or its source expressed in geographical terms. This is a useful qualitative distinction though it will be noticed that one and the same air current may display 'warm' features in one context and 'cold' features in another. The distinction is most clearly defined when the air current is moving over the sea where the mutual relationships are not complicated by the diurnal variation of temperature.

'Warm' characteristics are likely to be found whenever (*a*) air is moving

347

in a generally poleward direction at any period of the year, (*b*) air moves from over the open sea towards pack or fast ice, (*c*) air moves in winter from a comparatively warm ocean to cross a colder continent, and (*d*) when air moves in summer from a continental interior out across some part of the sea kept cool by upwelling or the presence of a polar current.

The general effect of the surface chilling that follows is the production of a low-level inversion which may be associated with haze or fog if condensation level is reached. Often surface turbulence distributes the chilling through the lowest two or three thousand feet of the atmosphere and the inversion may be raised to produce a more or less continuous stratus deck below which mixing has brought the lapse rate very near the DALR. Such inversions are much less impressive than those characteristic of Arctic air in its source region and are therefore much more easily dissipated. Contributing to this effect are (*a*) long-wave radiation aloft which is likely to reduce air temperature above the inversion level by 1 to 2°C per day, (*b*) the customary veering of poleward winds with height which may bring in cooler air in the upper layers, and above all, (*c*) the combined effects of convergence and uplift, commonly experienced by warm air and potent forces in liquidating its newfound stability.

The types of weather thus associated with incursions of 'warm' air masses are thus:

1. Stratiform cloud, often low and covering the whole sky in a vast unbroken sheet.
2. Poor visibility involving haze even if it does not reach densities normally classified as fog.
3. High humidity involving low rates of evaporation by day and the deposition of dew at night.
4. Little other precipitation except perhaps some very fine drizzle or 'Scots mist' which may or may not evaporate before it reaches the surface of the ground. Whether hilly areas will record much more precipitation than the lowlands will depend a good deal on the strength of the inversion but such airstreams are often able to find their way round orographic obstacles and uplands experience little more than a wet hill fog.[1]

'Cold' characteristics, on the other hand, are likely to be betrayed when (*a*) air is moving in a generally equatorward direction at any period of the year, (*b*) air moves off sea ice towards open water, (*c*) air moves in winter from a cold continent out over the open sea, and (*d*) when air moves in summer from a warm sea towards a still warmer continental interior.

The general effect of warming any air column at its base is to decrease its stability. If the 'cold' air is already near neutral stability, active convection is likely to occur with the build-up of cumulus cloud and the possible outbreak of scattered showers. If, on the other hand, it is initially rather stable a lofty inversion may cap the lower turbulent layers thus providing a lid to cumulus growth as it does so frequently in the Trades. Summer cumulus may then

[1] Though see Pedgley (1970).

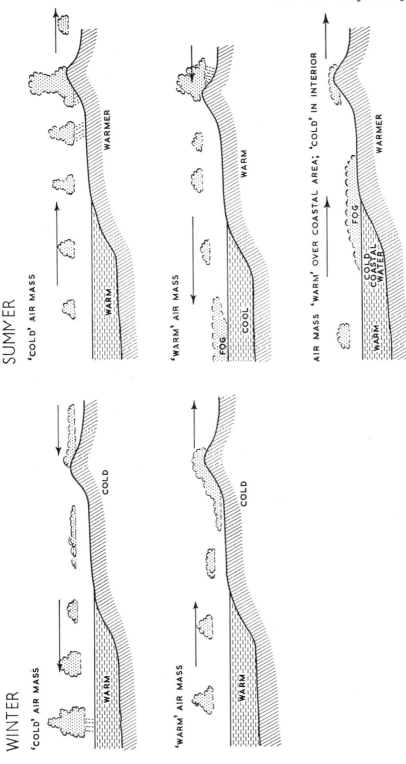

Fig. 103. An elementary yet vivid impression of the impact of 'cold' and 'warm' air masses along a continental coast in winter and summer (after Petterssen, 1941)

349

provide a bright and interesting sky without the risk of rain. The further development of the situation will depend on (*a*) long-wave radiation aloft which now acts in favour of increasing the instability, (*b*) the tendency of equatorward winds to back with height which may bring in warmer air aloft and thus operates in the opposite sense, and (*c*) the degree to which the cold air is subject to divergence and subsidence which will also renew its failing stability.

In such circumstances the weather features to be anticipated are thus:

1. Cumulus cloud forms of one variety or another according to local circumstance. With a high inversion lid their development will be in breadth rather than height and the proportion of clear sky may be small. Elsewhere they may become tall shower clouds but the accompanying 'bright intervals' are likely to offer some compensation.

2. Visibility is normally good and often excellent. Atmospheric impurities are partly washed out of the air by the showers and partly distributed through such a deep layer of the air that their concentration near the surface is much reduced.

3. Humidities are usually comparatively low (say, below 75 per cent) and evaporation is thus brisk both from the ground surface and from the occasional showers of rain. The water vapour thus added to the lower layers is rapidly dispersed aloft and so the process is able to continue unchecked.

4. Although showers are characteristic and local falls may be quite heavy, perhaps involving hail, cold air-streams, of themselves, are unlikely to produce general rain though they respond readily to orographic obstacles and are responsible for a large proportion of the summer rainfall over hilly ground. Fig. 103 illustrates some of these features pictorially.

It will be evident, even from this brief account, that the relative classification of air masses brings us into immediate touch with reality in a much more stimulating and interesting fashion than the more fundamental but more remote geographical classification is able to do. This is indeed because it is essentially local in application and it is only local climatology that any of us is able to perceive by direct observation. 'Visit Loch Lomond when the wind is in the northwest' was the advice given to the author by a park-keeper many years ago, and nobody acquainted with western Britain can fail to recall the brilliant days of unparalleled beauty which punctuate the long spells of grey murk from the southwest. Yet, clearly, the full story has not yet been revealed. What of the prolonged rains, the gales, the sudden changes of wind, the violent outbursts that accompany thunder and hail? These phenomena rarely occur within any single air mass, they are mainly the results of outbursts of energy along the margin between two conflicting masses.

# 11
# Fronts and weather disturbances

The concept of air masses implies that large sectors of the atmospheric ocean have intrinsic characteristics which endow each sector with some considerable degree of individuality. But individuality in turn implies the existence of finite limits both in space and time. This remains true even when the concept of individuality is applied to a generalised group like a species of plant or animal which achieves continuity of existence by a continuous replacement of its constituent parts. We have seen that air masses are in a state of more or less continuous regeneration and dissipation. What then is the nature of their limits and how can they be recognised?

Arguing from first principles it can be asserted that the only sharp discontinuities on the earth arise from its basic geography—above all from the broad distribution of continent and ocean, supplemented perhaps by some of the really major relief features. That these discontinuities may find their reflection in the nature of the superincumbent air, especially in its lower layers, seems not unreasonable, and it is possible to conceive how, given time, the contrasts may develop in depth. Are air mass contrasts therefore found only in close relation with oceanic coasts? There is, of course, another source of contrast, latitude and the radiative balance that it implies, but changes in latitude are *progressive* and it is hard to see how sharp discontinuities can arise from such a source. Yet comparatively sharp discontinuities *can* be recognised in the atmosphere, even over the open sea. There can be only one cause for such features, the system of circulation both general and local, which brings air streams from widely separated regions into close interrelationship along converging paths. We have called such a process 'confrontation' and since about 1917 such discontinuities have been known as 'fronts'.

It is interesting to note that the term 'front' was not derived in this way, however happy such a derivation would have been. It was, in fact, borrowed by analogy from the Western Front in France during the First World War. A military front implies the existence of *two* opposing forces, both potential sources of warlike activity even when the front is actually 'all quiet'. Outbursts of such activity, if prolonged, are known as a 'battle' which, with its early, middle and later phases has its own identity symbolised by its being given a name. The actual battle may involve only a portion of the front but, as it develops in intensity, resources from deep within the opposing systems may be brought into play. Furthermore, even if the battle never involves more than a fraction of the opposing forces, upon its outcome the stability of the whole line may ultimately depend. If we replace the opposing armies by two air masses differing in depth in temperature (and hence in density) and thrown into mutual relationships by a converging circulation, then the release of their potential energy, the battle, is represented by the development of a weather system along the front and the analogy is seen to be remarkably apt. There is a difference, of course, the weather system organises itself, for better or for worse, without the benefit of General Staff! The important point however is that the development, like a battle, achieves its own identity, it grows and decays and, whilst it is in existence, it is quasi-independent of the two air masses from which its energies are ultimately drawn. We meet again an example of action and re-action and the difficulty in meteorology of pointing to a prime cause.

Such weather systems have been known in general terms successively as 'storms', 'cyclones', 'lows', 'depressions', 'troughs' and 'disturbances', the term varying according to the angle of view and the implications which it is intended to carry. Shaw (*Manual* II, 1936, 345) reminds us that 'Clement Wragge, in Australia, used to give (proper) names to those that appeared on his maps' and this happy idea, which fell into disfavour when meteorology became scientific, has been revived by the less inhibited Americans and by television weathermen. This is not high-spirited tomfoolery; a 'thing', however intangible, which has a life of some days and which may carry within it the seeds of widespread destruction may well be considered to deserve a proper name, to help establish its identity.[1]

What is the nature of the identity of a weather disturbance? Unlike an air mass it cannot be simply expressed in terms of the usual physical scales, of temperature, humidity or lapse rate. A disturbance is essentially an active relationship between two air masses, it is thus recognised as a pattern of events or, more specifically, as *a pattern of air motion*. The remarkable fact is that this motion succeeds in sharpening the contrast between the two air masses by narrowing the zone of transition from one to the other. It is thus virtually impossible to discuss fronts without becoming involved in a consideration of the disturbances which develop along them. A quiescent front can be imagined but it will be broad and indeterminate and its chief interest to the observer will be the chance that the situation may so develop that it will become active i.e. sharpened and disturbed.

The history of the concept of the weather disturbance is an interesting illustration of the circuitous manner in which scientific truth is often attained. The first notion that the vagaries of weather, especially in the mid-latitude zone, were not random events but followed, rather vaguely perhaps, some kind

---

[1] The American practice is to begin the alphabet afresh each hurricane season, thus 'Hazel' was the eighth Atlantic hurricane observed in 1954, 'Janet' was the tenth in 1955, and Camilla the third in 1969. See Dunn and Miller (1964).

of system was derived indeed from the observation of air motion. Over a century ago Dove (1862, 11–12) propounded his Law of Gyration, namely that 'in the northern hemisphere, when polar and equatorial currents succeed each other, the wind veers in the general direction, S, W, N, E, S, round the compass'. He realised that this was not an infallible rule but thought that

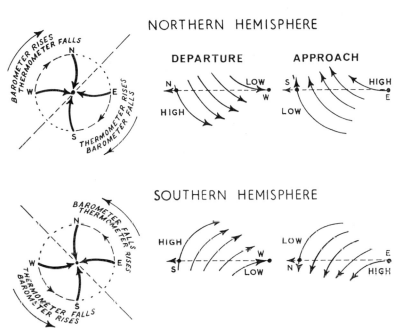

FIG. 104. Dove's observations on the relationship between changes in wind-direction, air temperature and pressure in the cyclonic zones of both hemispheres, expressed in diagrammatic form. In the righthand diagrams, to allow for the westward drift of the disturbance, the point of observation must be imagined to move to the left along the broken arrows as the 'cyclone' approaches and departs

exceptions were least common 'between W and N' and 'between E and S'. Furthermore the rule was reversed in the southern hemisphere. We have expressed his concepts in diagrammatic form in Fig. 104 showing, on the left, the swing of the wind and concomitant pressure and temperature changes as experienced at a fixed point and, on the right, a rough interpretation of how these changes may occur in two quadrants. In the latter, to express the generally westward drift of the air streams we must imagine the point of observation to move progressively from right to left.

There was nothing particularly new in this since seamen had been aware that the wind veers 'with the sun' much more frequently than it backs 'against the sun' since the days of Aristotle. Dove gave the observation fresh meaning and, as he developed his theme, he added comments that have a decidedly modern tone even if the wording is rather old-fashioned. 'There are really only two winds, viz. northerly and southerly, inasmuch as all other winds are merely occasional deviations from these two directions' (p. 77). 'The falling side of the (barometric) wave belongs to a different current from that which produces the rising side' (p. 269). 'The southerly current displaces the northerly in the upper strata of the atmosphere before it does so in the lower strata; while the displacement of the southerly by the northerly takes place at first in the lower and afterwards in the upper strata' (p. 83). 'The warm and light air is displaced by that which is cold and heavy on the west side, more rapidly than the cold and heavy air by that which is warm and light on the east side' (p. 87). And we must not omit, 'Most winds are liars,[1] as they do not come from the regions from which they appear to flow' (p. 79)—a sound observation still far too frequently overlooked.

The advent of the electric telegraph made possible the construction of synoptic weather maps—maps of wind and weather at some specific moment of time as recorded at a widespread network of stations. The new technique seemed to make all earlier work archaic and, since wind observations at land stations are liable to orographic distortion, students of weather maps became obsessed with the barometer. In brief, Dove's 'storms' became 'lows'. It is never good science to forsake reality for abstraction and for a whole generation meteorology built up a body of myth which is still only slowly being purged from its system. There were advances, of course, especially in forecasting technique. It was observed from successive synoptic maps that pressure patterns 'moved' and that often they moved systematically, particularly from west to east. But two real intellectual difficulties had been introduced.

A. What was it that 'moved'? The barometer has no cognisance of lateral motion, it can only go up or down.
B. The concept of movement was being derived from a sequence of static snapshots. Could such a technique give a true picture of the interaction of the moving parts?

These difficulties were all the more insidious in that they were not at first recognised. Point A found a ready answer in the analogy with water vortices familiar to everyone, surely it was an air vortex which moved. It was easy to show that a vortical pattern would impress its stamp upon the barometric record to produce the observed closed isobars. Hence the introduction of the term 'cyclone' and a great deal of confusion between intertropical and extratropical cyclones which Dove had clearly regarded as quite distinct phenomena. The network of synoptic stations had to be brought to a very much finer mesh before it became evident that the observed temperature pattern within most extratropical cyclones was much too asymmetrical to support the notion of simple vortical motion. Point B was largely lost in the elaboration of techniques devoted to the analysis of the synoptic record as it was presented to the observer. Although full of interesting and important information such a

---

[1] 'Deceivers' or 'impostors' would probably be a better translation.

chart is closely analogous to a single frame of a cinematograph film. The prints displayed in the cinema vestibule may give us some idea of the setting of the film but the closest study of them will reveal little of its action or of its dramatic quality. A snapshot of a hurdler in full flight is not be be taken as evidence that the force of gravity has been suspended. Yet mistakes analogous to this

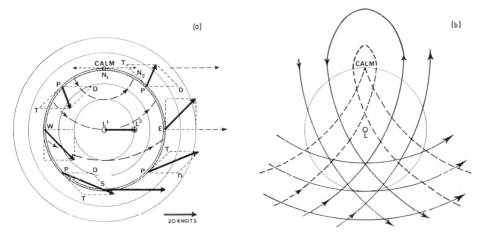

Fɪɢ. 105. Wind directions in a moving vortex. Shaw deplored the 'unfortunate habit' of regarding the travel of a cyclone as 'something which could be treated independently of the cyclone itself.' (a) This is an instantaneous representation of a ring of air (stippled) rotating at 20 knots round a centre $L^1$ which is itself moving forward to $L^2$ at the same speed. The instantaneous wind-direction at any point on the ring is then given by the heavy arrows derived from the parallelograms shown. It will be noted that at S the wind is westerly at 40 knots whilst at $N_1$ there is a calm since the two motions are directly opposed. In the forward half of the ring the motion is obliquely *away* from the centre; in the rear half it is obliquely *towards* it. In fact the air in the ring is behaving as if it were rotating, not about $L^1$, but about $N_1$ (along the dashed concentric circles). (b) Here we abandon the instantaneous approach and try to visualise the motion through time i.e. the trajectories of various air particles or samples. These are generated by regarding the ring as a wheel rolling along $N_1$–$N_2$. It is then evident (i) that the rotatory motion is largely an illusion, (ii) that the 'ring' can only be maintained by the import of air from the rear and a corresponding evacuation of air in the forward sector, (iii) that air-motion to the south of the centre L is vigorously from west to east, (iv) that between L and N southerly winds are abruptly succeeded by northerly winds as the system passes overhead, and, finally, (v) that north of N the net motion is from east to west but at velocities much lower than in the westerlies to the south. Shaw used such a model to warn his readers of the danger of considering a pressure-pattern as an identity. Yet 'moving lows' were on the pressure maps for all to see!

have certainly been made. Thus it has been noted that surface wind direction is at a wider angle to the isobars at the rear of a fast-moving depression than at the front where, indeed, the wind motion may sometimes appear to be *up* the barometric gradient. A simple graphical exercise in combining rotational with translational motion (Fig. 105) will show that there is no mystery about this. It is a glimpse of reality but by no means the whole story.

The chief defect of the new emphasis upon barometric pattern was that it distracted attention from the *discontinuities* inherent within most mid-latitude weather systems. The facts were known, at least in part, but they were brushed aside and a period of scientific blindness prevailed. Thus 'line squalls' had no place in the theory although they were feared by sailors and were by no means confined to the sea. Indeed Dove (1862, 233) had quoted a vivid account of the disastrous effects of a sudden incursion of cold air over central Russia in December 1850—'a thaw had preceded the violent NW gale; but the first gust caused the thermometer to fall about 40°F below the freezing point, so that persons caught out of doors fell dead'. Indeed Ferrel (1889), must carry heavy responsibility for this neglect. His observations were impeccable thus—'as cold and warm portions of air . . . tend to keep apart, there is often a long line several hundred miles in length in the central or perhaps rather easterly part of the cyclone where there is a short but sharp temperature and pressure gradient and a very sudden change of wind' (p. 290). Yet, obsessed with the notion that all cyclones were convectional in origin, he dismissed these facts as having 'nothing to do with the temperature disturbance upon which the origination of the cyclone depends' (p. 285). Rarely have observed facts been so completely cast aside in the pursuit of a theoretical model!

With the installation of autographic apparatus at Kew in 1867 there was no longer any need to rely upon eyewitness accounts or subjective impressions. Shaw (*Manual* II, 1936, 374) was able to quote occasions when the new records showed 'a sudden change simultaneously in pressure, temperature, wind-direction and force' but notes that these 'important details' went disregarded and the publication of the traces he reproduces as an example (*ibid*, 375) was discontinued in the very year when hundreds of young lives were lost in the foundering of the *Eurydice* off Portsmouth (24 March 1878) in a line-squall of unusual violence. Shaw's comment that the decision to replace the traces by the publication of five-day means of hourly values was 'rather pathetic' might well be added to examples of typically British understatement! The decision, made in the interests of economy, undoubtedly set back meteorology for nearly a generation.

Indeed, it took a man of Shaw's calibre to find a way out of the intellectual quagmire which the worship of the pressure pattern had produced. The steps by which he convinced himself of the importance of what we would now call air mass discontinuities in the production of weather are worth enumerating, both because of their intrinsic importance and because they illustrate the volume of evidence and argument necessary to unseat an established theory. We bring these points together from scattered portions of his *Manual* and should remember that even these potent arguments, developed years before his great work was published, failed to gain widespread acceptance until the Norwegian School invented a vocabulary to bring the new ideas into focus.

1. The study of the actual distribution of cloud and rainfall in relation to the centres of a number of depressions showed a wide variation of pattern (Shaw

and Lempfert 1906, 1955, 40, 52, 91; also *Manual* II, 1936, 386). Shaw comments that in any attempt to generalise from these patterns to portray a 'typical cyclone', 'the most noteworthy characteristics of each cyclone would be lost'. 'There was no symmetry with respect to the centre for any of the meteorological elements with the exception of pressure and, to a certain extent, of the winds' (*Manual* IV, 1931, 251).

2. 'Perhaps the most notable feature of the embroidery of the barogram is the sudden but permanent rise of pressure that sweeps as a well-marked line across the country and is the prelude to a sudden veer of wind and fall of temperature. These phenomena belong exclusively to the right-hand side of the path of a depression' (II, 1936, 379). Or, as he expresses it in more modern terminology, 'the cold front is the dominant element of the partnership' (IV, 1936, 286).

3. In his attempt to plot the actual trajectories of air samples from successive synoptic charts (there are numerous examples in Shaw, 1955) he showed that 'the pressure distributions are a guide to the instantaneous flow but not to the long treks. Speaking generally for northward-moving air (equatorial air) centres of depressions are goals to be arrived at by passing from high pressure to low; for southward-moving air (polar air) they are marks to be passed and the transition is from low pressure to high' (IV, 1931, 282). This was perhaps his most notable contribution.

4. Implicit in this view was Shaw's conviction that motion is primary. 'It is really the motion of the air which is developing those entities (moving surface highs and lows) on its flanks as it travels' (II, 1936, 400). Or, in a passage already quoted in the chapter on the General Circulation, 'the pressure distribution is regarded as the "banking" required for the maintenance of the air-currents' (IV, 1931, viii).

5. Furthermore Shaw was the first meteorologist to realise that 'the formation of a cyclonic depression involves the removal of some hundreds of thousands of millions of tons of air' (*ibid*, 307). He could find no mechanism for the eviction process in the simple vortex and, with great foresight, states that 'the only form of disposal that suggests itself is delivery into some passing current in the upper air which acts the part of scavenger' (*ibid*, 307). His suggestion that the scavenging current probably belongs 'to the primary cyclonic vortex of the hemisphere' is an early hint of the importance of the jet stream, then unknown.

6. Finally, in a still wider context, Shaw's development (after Helmholtz) of the concept of entropy as applied to the atmosphere implied that atmospheric stratification was indeed a physical reality. He demonstrated that such strata must dip systematically from the poles towards the equator (III, 1930, 252–3). It was therefore a comparatively simple step for the Norwegians to seize upon one surface in the series and to regard its intersection with the earth's surface as the 'polar front'. But Shaw recognised clearly that what happens at a front depends to a very great extent upon conditions to the rear or, as he put it with great simplicity, 'we may read a great deal about fronts but we find little or nothing about their backs' (II, 1936, 395).

Each of these points finds a place in the Polar Front theory of mid-latitude disturbances as developed during the same period by the Norwegian School under V. Bjerknes and his son, J. Bjerknes. They achieved a synthesis by what can only be called a magnificent piece of over-generalisation. In effect they grouped Shaw's isentropic strata (see Fig. 67 and *Manual* II, 1936, 116 or III, 1930, 252–3) into three stratigraphical systems, two closed or local systems, one over each pole, rather like piles of inverted saucers, and a universal system enveloping the whole earth though in contact with the ground at its base only between the tropics since it rides over the closed systems towards the poles. The closed systems they called Polar Air and the open system Tropical Air, thus giving fresh life to Dove's dichotomy. It was then a comparatively simple step to compute mean air temperatures within adjacent samples of polar and tropical air up to some selected elevation above sea level and to regard the density contrast inferred as occurring abruptly at an imaginary 'surface' between the two air masses. This 'surface' was required by the mathematical analysts and the computations showed (*a*) that, on a rotating earth it must have a calculable mean slope downwards from the pole, which would vary directly with the sine of the latitude and with wind shear (the contrast in wind-speed within any pair of systems) and inversely with the contrast in temperature (and hence density); (*b*) that, in the presence of wind shear the surface was likely to develop wavelike ridges and troughs which would normally progress round the margin of the Polar air from west to east; and (*c*) that many of these waves would be 'unstable', i.e. that their amplitude would change with time, giving each wave a recognisable life-cycle from initiation to decay. The intersection of the inferred surface with the ground was called the Polar Front and, theoretically at least, such an intersection should ring each pole as a continuous though sinuous line. Viewed in three dimensions, however, the surface becomes the top of a comparatively shallow dome of polar air, subject to frequent contortions. It is indeed the top of Shaw's 'underworld'. The major contribution of the Norwegian mathematical model was the light it shed upon the *dynamics* of such a system. Identifying the two wind discontinuities revealed in Shaw's 1911 diagram (Fig. 106) as two active sectors of the Polar Front and calling one a 'warm front' and the other a 'cold front' from the nature of the air they introduced, the Norwegians showed that their model fitted many of the observed facts with a remarkable degree of faithfulness and could thus be used in forecasting. Furthermore, they introduced a useful vocabulary and showed that their basic ideas could be effectively illustrated in diagrammatic form (Fig. 107).

It should be apparent from the above account that it is by no means impertinent to put the questions: But does the Polar Front really exist? Is it no more than a convenient mathematical abstraction? Even if the atmosphere can be thought of as broadly stratified, on what grounds are we justified in selecting any one single member of the series—far less any one 'surface' between adjacent strata—and exalting it in importance above all others? In the poleward gradient of surface air temperatures from about 40°C to

(a)

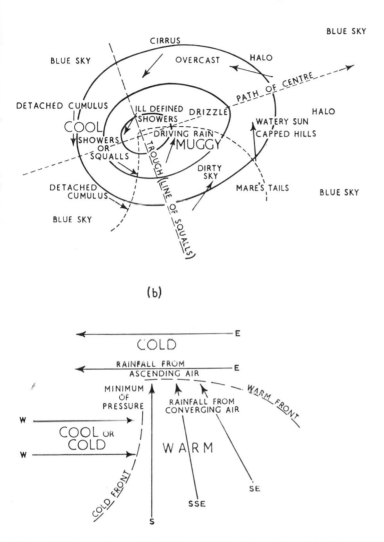

(b)

FIG. 106. Two early models of the mid-latitude weather disturbance or 'cyclone'. (a) Abercromby's diagram of the distribution of weather in such a system in the northern hemisphere. (b) Shaw's 1911 figure emphasising the wind-discontinuities characteristic of such systems—'fronts' have been added

—40° C there is no single point which is *critical for air density*. True that, half way between these extremes, at 0° C, a critical point occurs for *water* which has momentous consequences but, initially at least, the Polar Front theorists did not tie their concept to the presence of sea ice or frozen ground. It was much later than a comparatively shallow 'Arctic Front' was introduced well behind

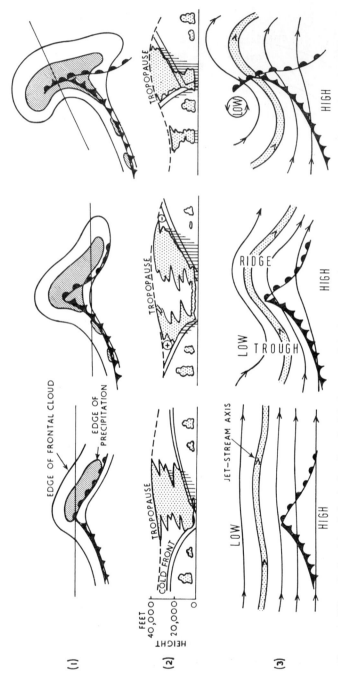

FIG. 107. The relationship of the 'polar front' with the core of the jet stream in a developing mid-latitude disturbance. The diagram is for the northern hemisphere and illustrates many features of such systems (after *A Course in Elementary Meteorology*, HMSO, 1962, p. 154)

the Polar Front and in a position suggesting some relation to the presence of sea ice. Again, the mean isentropic surfaces (Fig. 68) begin to take on a distinct upward tilt polewards from 10° latitude in winter and 30° latitude in summer but the Polar Front is rarely regarded as lying so near to the equator as this. Besides, the same tilt is characteristic of *all* of the lower layers encountered between these parallels and the region of lowest surface temperatures. Throughout this wide range of latitudes the atmosphere is baroclinic, i.e. the gradients of temperature, air density and entropy are propitious for the formation of fronts but an active discontinuity can only arise where these contrasts are somehow brought to a focus. Above all, this must be the work of converging air-flow so that, once again, we are brought back to Shaw's dictum that 'motion is primary'. In these circumstances it is perfectly natural that fronts should be (*a*) of varying degrees of development, (*b*) frequently discontinuous, and (*c*) sometimes multiple rather than single. Much of the air taking part in these developments will be mid-latitude air and, though there will always be a warm and a cold side to a front, the terms 'tropical' and 'polar' as applied to these air bodies may sometimes be more confusing than helpful. In brief, we are left with the concept of fronts, but the notion of a single unique Polar Front gradually fades away as we approach it.

In an attempt to give some precision to these rather elusive ideas, Bergeron (1928, 67–73) suggested as a general working rule that an *air mass* should have a diameter of from 500 to 5000 kilometres within the bulk of which the horizontal temperature gradient should not exceed 1°C per 100 kilometres. (This would involve the subdivision of the major air masses discussed in Chapter 9.) It follows then that the length of fronts, the active, ribbon-like margins which may develop between two such masses, must fall within a similar range. It is however the comparatively narrow width of such a margin that gives it dynamical significance and justifies its title as a *front* and Bergeron suggested as appropriate dimensions a width of from 5 to 50 kilometres as measured in a horizontal direction. Over this range, he thought, the temperature gradient might run at from 1°C per kilometre to 1°C in 10 kilometres (fronts narrower than 5 kilometres and gradients in excess of 1°C per km seem never to occur owing to turbulent mixing). Since frontal analysis is not concerned with merely diurnal variations, he also considered that, to deserve the name, both air masses and fronts should have a vertical development in excess of 1 kilometre (roughly 3000 ft).

It is clear that these two definitions must exclude a considerable body of undifferentiated or transitional air where horizontal temperature gradients of 1°C in 10 to 100 kilometres are spread over distances ranging from 50 to 500 kilometres. These sectors of the atmosphere may be regarded as a kind of reserve from which new masses or new fronts are developed out and into which old air masses or old fronts may pass as they gain or lose their distinctive qualities. The presence of such a reserve helps to remind us that the distinctions drawn in air mass classification are distinctions of degree and not of substance.

Even with the help of these definitions, the indication of a front on a synoptic chart is still, in part, a matter of interpretation. Fronts were not shown on British charts until 1934 and, five years later, Brunt (1941, 242) remarked that an examination of the charts showing fronts issued by different European meteorological services showed differences that were 'astonishing, not merely in the details of the fronts, but in their main outline'. Although improved observations in the upper air have reduced this disparity in interpretation it has also revealed unexpected features of unquestionable significance.

Since 1950 carefully organised flights through fronts by instrumented air-craft have revealed the following facts (see Sawyer, 1955, and Freeman, 1961):

1. Over Britain at least, practically all fronts are *dual* in character. Far from being a single surface they are observed to present two approximately parallel surfaces of discontinuity. The mean *horizontal* distance between such surfaces measured by the Meteorological Research Flights was 130 miles (*c* 210 km) and the range observed was from 50 to 200 miles (*c*. 80–320 km). The *frontal zone* is thus an observed reality and its breadth is considerably in excess of that envisaged in Bergeron's definition though the usual horizontal tempera-ture contrast, about 5° C (9° F), is very close to his estimate. The *vertical* distance across the frontal zone is observed to be of the order of 100 millibars, so that the frontal zone is often about a mile in depth. Within this belt the decrease of temperature with height is much less than usual.

2. On both flanks of the frontal zone, in both the warm air and the cold, the horizontal temperature gradient is observed to continue in the same sense though at a reduced rate so that it is possible to speak of a *frontal region* with a total width of from 500 to 600 miles (*c*. 800–1000 km) across which the gross difference in temperature is of the order of 8–9° C (14–16°F). This confirms the suggestion that fronts are best thought of as a local steepening of the contrasts observed in Bergeron's zone of transition and reminds us of Shaw's comment on the importance of 'backs'. Petterssen (1956, i, 191) makes the same point when he says that 'the importance of fronts depends more upon the over-all density contrast and the slope of the frontal surface than upon the sharpness of the apparent discontinuities'. On the battle-front analogy employed above this is no more than a reminder of the significance of readily-available reserves.

3. Although, along such paired fronts, it is the first surface encountered on moving from warm air to cold that seems to be most important in the immediate production of weather, it was found over *warm* fronts that the inclination of the sloping cloud base was often considerably steeper than that of the front itself (Fig. 108). Indeed, in the upper levels, the cloud mass was often observed to lie completely within the warmer air, and the heaviest rainfall was apparently generated in this sector. This has helped to clear up a difficulty that had troubled thoughtful meteorologists for some years, namely that ascent at known wind speeds up a gradient of the order of 1 in 200 is quite inadequate to produce even moderately heavy rainfall. Clearly the weather produced is in

large measure a function of conditional instability within the warm mass itself and the front may be providing little more than the necessary trigger mechanism which sets this instability in motion. The same process is recognised, often with still greater clarity, along advancing *cold* fronts. Here most of the rainfall is often observed in advance of the actual intersection of the first frontal surface with the ground and its arrival may then be marked by an immediate clearing of the sky (see Fig. 114).

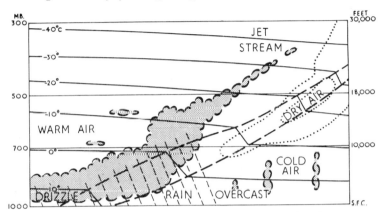

FIG. 108. A cross section through a warm front over Britain. Temperatures are in °C. The diagram illustrates the 'frontal zone', the typical cloud and rainfall distribution and the tongue of dry air aloft discovered by the Meteorological Research Flights. Vertical exaggeration some ×65

4. Still more remarkable was the observation that, particularly in warm fronts, there was often a wedge of extremely dry air between the bounding surfaces of the frontal zone above 800 mb (2000 m). Relative humidities as low as 5 per cent have been recorded in this region. The only possible inference to be drawn is that the air in this sector, far from experiencing uplift along the frontal surfaces, has indeed *subsided* as it has been drawn into the system perhaps even from the lower stratosphere. A corresponding dry pocket was also sometimes observed above the nose of a cold front, sharpening the transition from torrential rain to clear skies. This feature is not yet fully understood though it can be argued that vigorous ascent in the warm air must be accompanied by descent somewhere and where better than over the cold air immediately adjacent to the updraught?

5. For the organisation of this flow and, indeed, for the evacuation of the air aloft which enables the surface barometric depression to persist, we must look to the nature of the air flow *above* the depression and its associated fronts.[1] In other words we must explore the jet stream, first positively identified in 1945.

[1] For informative case-studies of warm and cold fronts over the British Isles, see Browning and Harrold (1969, 1970).

For cold fronts in the southern hemisphere, see Garnier (1958) and Gentilli (1969).

# The jet streams

Since cold air is denser than warm air it is clear that pressure must fall off more rapidly with height in a cold air column than in a warm one (Fig. 49). Hence it has long been recognised that, as we pass upwards through the atmosphere over middle latitudes we must encounter a 'thermal wind' blowing from a generally western direction around the high-level polar trough and increasing steadily in vigour up to the level of tropopause. It would seem a logical consequence that, where surface temperature contrasts were being sharpened—as along a developing front—the poleward gradient would likewise steepen aloft and a corresponding ribbon-like acceleration of the westerly circulation would therefore occur. Palmén (1951, 602) has expressed this as follows: 'Any pronounced front with an inclination not too small must be accompanied by a strong upper-level wind concentrated into a relatively narrow band'. He hastens to add that these must be regarded as 'parallel' developments; they are by no means necessarily related as cause and effect.

High altitude investigation has confirmed this presumption and the jet stream is now a recognised feature of the circulation pattern. It appears on vertical cross-sections of the atmosphere as a comparatively narrow band (or bands) of fast-moving air embedded in the upper westerlies. When well developed it may have a vertical thickness of about 8 kilometres (usually between 5500 and 13500 m aloft) and a horizontal width of from 300 to 400 miles. Air speeds in its core (at about 10 km) may run from about 60 knots to over 100 knots, the velocity falling off more rapidly towards the polar flank than towards the tropics. The jet is often accompanied by a discontinuity in the tropopause which is often 100 millibars lower on the polar side. A common position for the core of the jet is vertically above the intersection of a sloping front with the 500-millibar surface. Projected on to the ground this will often bring the core some 400 to 600 miles in advance of a warm front and some 200 to 300 miles behind a cold front but the width of the jet is such that, in fact, it often overlies most of the frontal zone. Southward outbreaks of polar air are accompanied by a tongue-like extension or 'trough' of low pressure aloft around which the jet stream swings in a southward loop; northward penetrations of tropical air are accompanied by a tongue-like 'ridge' of high pressure aloft around which the jet must make a northward loop. There is thus maintained a broad parallelism between the polar front and the jet stream (Fig. 107) and just as the former cannot properly be regarded as continuous, so the latter varies widely in intensity from time to time and from place to place.

So far so good, but what is the real nature of the relationship, if any, between these parallel phenomena? Do they interact, in any way, the one upon the other and what is the nature of their common cause? This raises some difficult questions of a four-dimensional character which are still being actively explored by meteorologists. Thus Green (*et al.*, 1966), reviving Shaw's emphasis on motion along isentropic surfaces, has shown that, in a polar outbreak such as that illustrated in Fig. 54, the poleward return current is concentrated within a comparatively narrow band over its eastward flank. Here moisture initially raised into the air above the western trades by cellular convexion is swept polewards to be condensed in the middle layers of the atmosphere as a result of 'slantwise convexion' along a lengthy path. The energy thus released serves to accelerate the rising current and thus energy is fed into the right flank of the polar-front jet. Certainly satellite cloud photographs give a most vivid impression of the presence of such active channels (see Fig. 69). Newton (1970) has also analysed the function of 'synoptic scale eddies' in middle latitudes and Walker (1970) concludes that we are thus offered 'a plausible interaction mechanism between tropical and extratropical circulations' so that they can be seen as operating as a unity. He also points out that, since the initial moistening occurs and the long ascent begins in layers well above the earth's surface, the incipient development of a mid-latitude disturbance may not, at first, be revealed by surface observations.

Fultz's model (p. 190) suggests that a meandering jet is indeed a necessary feature of a thermally driven circulation in a rotating fluid. In the atmosphere, as in the model, the speed of air particles in the jet is much greater than the speed of propagation of the 'waves' and it seems certain also that air is transferred laterally across the jet in much the same way as was observed in the rotating dishpan. The description of the jet stream as a 'meandering river' of air is thus more picturesque than accurate and the analogy must be used with caution.

## Long 'waves' in the upper westerlies

The recognition of long wave-like sinuosities in the upper air stream, at first suggested by 5-day means but now, as upper-air records have improved, demonstrable synoptically (Sawyer, 1957, 26–9), has led meteorologists to look aloft for the atmospheric 'higher command'. Can we find there some system of organised motion which, though clearly not in control of all the vicissitudes of the surface battle, may at least affect the nature of their general pattern? If such waves were found to be at all regular in scale and rate of progress we might expect to find some periodicity, however crude, in the events below. Even if they are gradually damped out and then regenerated it would be a great advantage if it could be shown that regeneration usually occurs in preferred regions; the appearance of each new wave could then be expected to set in train a series of events downstream, the broad sequence of which could be forecast, taking into account, of course, all the other factors which might be brought into play.

In view of the innate complexity of the atmospheric circulation, this was expecting rather too much. In the southern hemisphere the westerlies blow for the most part over an empty sea. Little is therefore known about their circulation aloft but it would appear that there can be no more than one likely preferred region, that east of the Graham Land–Andean barrier which is broken by the 500-mile gap of Drake Strait. In the northern hemisphere, on the other hand, the westerly current is subjected alternately to continental and oceanic influences and the massive barrier of the Rockies and the still wider though lower uplands of eastern Siberia appear to supply two points of anchorage for westerly 'waves'. Mathematical analysis has shown that such waves may be generated over a very wide range of wavelengths and how are we to recognise one from another in the earlier stages of their formation? Furthermore, it has been shown that between wavelengths 2000 to 5000 kilometres (*c.* 1500–3000 miles) such waves are inherently 'unstable', that is, their amplitude increases with time.[1] We are thus involved not merely with the regular process of a recognisable feature of upper air flow, but with a phenomenon which has a life-history of its own. Each wave may become distorted to a different degree before its energy is dispersed amongst a host of minor eddies and forecasting events downstream thus becomes much more difficult. No wonder that the attempt to discover periodicities in the weather, particularly of western Europe, has proved a source of perennial heartbreak.

Particular difficulty is encountered at periods of the initiation and dispersal of what has become known as 'blocking'[2]. The amplitude of the wave then exceeds its wavelength and cutoffs may be produced at both of its extremities. We are then faced with a cyclonic circulation round a 'cold pool' in the upper air to the south and an anticyclonic circulation round an upper air 'warm pool' to the north with momentous consequences upon the weather beneath (Fig. 109). Such a pattern involves meridional transfer of air on a massive scale, it appears to impose a check upon wave development *upwind*, and a prolonged spell of cold or warm weather may result. Attempts have been made to show that the 'index cycle', the alternation of periods of predominantly zonal and predominantly meridional circulation, has a crude periodicity of 6 weeks but, since the basic cause of the phenomenon is virtually unknown, the changeover, which is usually abrupt, has an exasperating knack of taking the forecaster by surprise.

To the climatologist, not interested in forecasting but seeking only to understand what is shaping the broad pattern of events, the long waves in the jet stream are still of interest. The southward sweep of a 'wave' aloft is observed to follow in broad outline the margin of a tongue of cold air also moving southward at or near the surface. The greater density of the cold air in the lower half of the atmosphere involves the development of a 'trough' of low pressure in the upper atmosphere and the jet is accelerated cyclonically around the margin of this trough. But, as Fultz's model has shown, the jet stream is a 'stream' only in a rather limited sense. The system can tap energy from the atmosphere only if it also involves *vertical* motion and this, in turn, involves a degree of transverse exchange across the jet axis. On the average, descending motion predominates on the west side of the upper trough (the

[1] Not to be confused with stratigraphical instability.

[2] For an analysis of 'blocking' over western Europe, see Sumner (1954).

rear portion of the wave), and ascending motion predominates on the east side (the front portion of the wave). This motion stimulates the development of surface anticyclones under the former and surface cyclones under the latter.

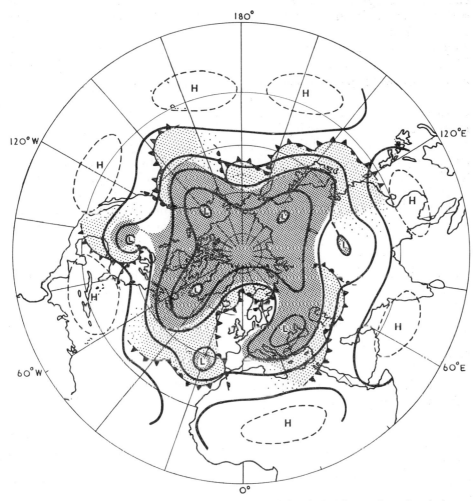

FIG. 109. Schematic circumpolar chart for the 500 mb level over the northern hemisphere. The area of polar air at 500 mb is heavily shaded and discontinuous areas of lighter shading show polar outbreaks near the surface. Note the 'waves' at the 500 mb level with 'cold pools' over the USA and the Azores and a 'warm pool' over north-western Europe (modified from Palmén, 1951, p. 607). Modern cloud-photograph mosaics from satellites vividly confirm the general accuracy of this interpretation and enable us to follow the motion of the systems from day to day

However the relationship between 'wave' and cyclones is rarely as simple as this may suggest. Wave distances along the polar front near sea-level are often of much shorter wavelength than the upper waves aloft. The poleward arm of the upper wave may then be associated with the growth of a whole 'family' of cyclonic depressions instead of a single very large disturbance. Since the air in the jet stream moves much more rapidly than the 'wave' itself, since it blows *through* the wave, it provides the 'scavenging current' sought by Shaw to sweep away surplus air that would otherwise check the further growth of the surface lows. J. Bjerknes (1951, 579, and 1954, 156) has expressed the same thought in more technical terms: 'Deepening of the surface centre will occur only if this upper-air divergence overcompensates the low-level convergence.' In carrying out this vital function the jet stream has a powerful influence in steering lows upon their predominantly eastward path.

More recently it has been shown that a jet stream aloft may have still more subtle effects upon the circulation beneath. Even if the stream is regarded as a more or less continuous waving band which circuits the earth it is unquestionably stronger in some localities than in others. The stronger sectors may therefore be regarded as having an 'entrance' where the flow is accelerating and an 'exit' where the flow is slowing down. Theoretical considerations suggest that such changes in acceleration must have important dynamical consequences and Murray and Daniels (1953) have confirmed this view from observations of jet streams over the British Isles. At the entrance to a jet stream in the northern hemisphere, where the air is accelerating, they demonstrated the existence of a crossflow component of the order of 10 knots at the level of the jet axis, the direction of flow being 'from right to left, looking downstream'. In the southern hemisphere the crossflow would be from left to right but we can combine both instances by saying that the crossflow at jet entrances is from high to low pressure, i.e. *down* the barometric gradient at the jet stream level. At the exit of a jet, on the other hand, they found a crossflow component of equal velocity in the opposite direction, that is, from left to right in the northern hemisphere, from low to high pressure, or *up* the barometric gradient at jet level. This exchange of air across the axis of a jet must have a profound effect upon the air beneath. Expressing it in terms applicable to either hemisphere, it implies *ascent* in the lower atmosphere on the tropical flank of a jet entrance and on the poleward flank of a jet exit and *descent* in the lower atmosphere on the poleward flank of a jet entrance and on the tropical flank of a jet exit. In view of the important effects of such vertical motions, however slow, upon atmospheric stability (see p. 239) we are by no means surprised to learn (Johnson and Daniels, 1954, 215) that 'in the entrance, twice as much rain occurred to the right of the axis as to the left' and that 'in the exit, there was twice as much rain to the left as to the right of the axis' (Fig. 110). This observation was made over the British Isles, i.e. in the northern hemisphere. In the southern hemisphere we would expect the rule to be reversed. It may be added in passing that the association of the jet stream with rainfall is not surprising since it normally lies above a front and that in the centre portion of the jet precipitation was found to occur fairly uniformly over a zone extending for a distance of the order of 500 miles on either flank of the jet axis.

Both the wavelike meanderings and the acceleration and deceleration of the jet stream may thus be expected to impose some dynamic control over the apparently random distribution of surface weather systems, since both involve a shifting of the atmospheric load aloft. But the lower layers of the atmosphere do not respond in purely passive fashion to this direction from above since it is within these layers that thermal energy is fed into the system via the varying thermal responses of land and sea and especially via the release of latent heat

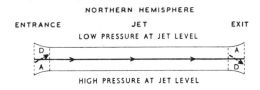

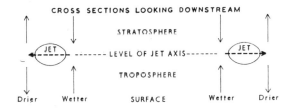

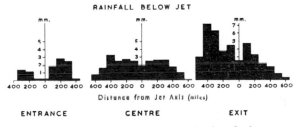

FIG. 110. Airflow at the entrance and exist of a jet stream and the observed effects of such circulations on precipitation in Britain (after Murray and Daniels, 1953 and Johnson and Daniels, 1954)

when condensation occurs. Multiple feedback occurs in a system of highly elaborate exchanges—cause is effect and effect is cause—and the whole complex apparatus that is mid-latitude air operates in an environment determined by the thermal and dynamic requirements of the great pools of arctic and tropical air on either flank. Little wonder that mid-latitude weather is consistent only in its innate inconstancy.

## Anticyclones

It is evident from the above discussion that the circulation within mid-latitude air is characterised by turbulence in a very broadly horizontal plane on a scale and intensity unequalled in any of the other major air masses of the world. Since all the major active 'vortices' in this turbulent system are constrained, as a direct consequence of the rotation of the earth, to spin cyclonically (i.e. anticlockwise in the northern hemisphere and clockwise in the southern), it follows that somewhere in the field between them the motion must appear to be in the reverse sense. A synoptic chart of the whole hemisphere thus always shows an apparently random distribution of 'highs' and 'lows' and the question naturally arises, what is the relation between them? Both types of system seem to preserve their identity for days together and both are seen to 'move' from day to day though not necessarily at the same speeds. Are we faced then with two distinct types of meteorological entities requiring separate analysis? The fact that the weather patterns associated with each are very different has certainly encouraged meteorologists to think so.

The anticyclone as a system of closed isobars surrounding an area of high pressure was first recognised by Francis Galton (1863, 385) who described it as an 'area of barometric elevation . . . usually a locus of dense descending currents . . . plunging from the higher regions upon the surface of the earth . . . flowing away radially on all sides'. Obsessed, as were all the students of the new synoptic charts, with the idea of the 'pressure gradient', he was under the impression that the air actually moved spirally outwards from a high in a clockwise direction (in the northern hemisphere) so as to be able to continue its journey via an anticlockwise spiral towards the centre of a neighbouring low. The movement, though circuitous, was thus considered to be 'down the gradient' throughout. This apparently simple picture of the circulation dies hard, but a moment's reflection will convince the reader that at best it can only be half the story. If the systems are to be *maintained* over any reasonable span of time both the pressure gradient and the systems of rotation must be reversed aloft and the three-dimensional model which thus emerges has some of the spurious characteristics of perpetual motion. Furthermore, it should be added that even the first most tentative observations of motion in the upper air, based upon cloud motion or sporadic balloon ascents, never gave a shade of support to the notion that the circulation there was even approximately a mirror-image of that at the surface. Clearly then, the pattern must be much more complex.

The declared intention of Shaw's famous *Life History of Surface Air Currents* (1906, 1955) was 'to find typical illustrations of spiral curves connecting areas of high and low pressure and thus indicating the lines along which the air was travelling between those areas'. Yet it stands as a major landmark in the literature of the subject mainly because it produced unassailable evidence that even the surface flow was quite different from what Galton had imagined. We have already seen that in cyclones, instead of spiral flow, Shaw found wind discontinuity and a most eccentric distribution of the area of major rainfall. In anticyclones his conclusions were not less remarkable, thus 'we have failed to identify the central areas of well-marked anticyclones as the regions of origin of surface air currents'. His trajectories led back rather to 'the shoulders or protuberances of anticyclones' or to 'cols'. Still more remarkable, however, were his observations (*a*) 'that there is very strong presumption in favour of the replacement of surface air by air *descending* from near the minimum of pressure', and (*b*) that 'there are many cases in which air flows along the surface to regions of increased pressure'. Air *descending* in a cyclone and actually moving *up* the pressure gradient!—it must have seemed rank heresy at the time! His broad conclusion that anticyclones were 'masses of air which for some reason is not taking part in the circulation going on around it' and that consequently 'the moving currents may be regarded as maintaining the anticyclone quite as truly as being maintained by it' were subsequently elaborated in his *Manual*.

Later discussions of anticyclones, notably those by C. E. P. Brooks (1932) and Brunt (1941, ch. 18) stress the distinction between the 'cold' and 'warm' varieties. Cold anticyclones owe their high pressure to the presence of a dense cold mass of air in their lower layers. They are comparatively easy to understand and, since they are shallow features, the pressure pattern is rapidly transformed with increasing height. Warm anticyclones, on the other hand, frequently extend upwards through the greater part of the troposphere and must owe their exceptional atmospheric load to the circulation at near-stratospheric levels. For a long time, therefore, their origin and mechanism have been shrouded in mystery, indeed it is still one of the chief aspects of the unsolved problem of the general circulation. The polar anticyclones fall clearly into the cold category and the tropical anticyclones as clearly belong to the warm. Both types may be encountered in middle latitudes but there is little evidence that they are generated there. Often they can be recognised as extensions or invasions from the high pressure areas on its poleward or equatorward flanks and, even when apparently self-contained, they can be often traced back to cut-off segments of the arctic or tropical systems. We have already discussed how this can occur as a result of 'polar outbreaks' near the surface and of wide meandering in the jet stream aloft. Such severed systems will normally undergo progressive transformation by subsidence (perhaps of the order of 200–500 m/day) and by the surge of the cyclonic systems round their flanks. Occasionally, however, they are preserved for some time by a propitious arrangement of the circulation both aloft and below and bring prolonged periods of comparatively settled weather. Their abrupt eventual collapse is apt to catch the forecaster off his guard.

Sutcliffe (1954) has posed the question, why does the synoptic meteorologist form the impression that cyclones preponderate over anticyclones in middle latitudes? He reaches the conclusion that it is 'because they develop a vigorous circulation more quickly and, as each system moves on, generally to merge with a pre-existing cyclone to the east, the intervening ridge is squeezed out before it has time to develop'. The liberation of latent heat in ascending cloudy air is a potent force in cyclone development and an almost unbroken series of cyclones is particularly characteristic of the southern hemisphere. In the northern hemisphere, on the other hand, both thermal and dynamic effects contribute to a greater frequency of anticyclones. In the first place, 'over the winter continents . . . radiative cooling counteracts the warming by subsidence in the anticyclonic development' and therefore 'the preferred tracks of anticyclones feature importantly in Russian synoptic methods'. And, secondly, the alternate distribution of warm seas and cold continents in this hemisphere favours the occasional development of 'blocking' in the long waves of the jet stream so that 'mature anticyclones may build up across the westerlies' especially over northwest Europe. The factors contributing to this event are so complex that 'the occurrence of these situations is, beyond a period of a day or two, effectively fortuitous'.

In conclusion it is perhaps wise to remind ourselves that, from the point of view of vorticity, the opposition implied by the terms cyclone and anticyclone is only relative; the absolute vorticity of the two systems is in the same sense since it is derived from the rotation of the earth (see pp. 164–67). What we actually observe is a difference in degree of rotation between the mobile air and the solid earth beneath it. There is no real difference in kind despite the weather contrast. Any process which stretches the vortex tubes of an anticyclone can generate an apparent cyclone and any process which shrinks the vortex tubes of a cyclone can slow down the motion until it becomes apparently an anticyclonic system. It is the accompanying uplift or subsidence which generates the contrast in weather pattern. The important distinction is that, in the presence of an appropriate 'scavenging current', uplifted air may be rapidly evacuated from the area so that the stretching process may be carried to great extremes, hence the fearful torque of the tornado. Subsiding air, on the contrary, meets with two constraints, (*a*) that imposed by the presence of the surface beneath, and (*b*) that arising when zero absolute vorticity is attained, the 'anticyclonic' rotation would then be equal to the comparatively slow turntable rotation of the earth beneath. Tight anticyclonic spin can, therefore, only arise very locally as a result of frictional turbulence or wind shear. Anticyclones on weather maps always appear as systems of open isobars with weak pressure gradients.

# 12
# Violent convective phenomena

Sir Napier Shaw (1914, 1955, 168) defined convection as 'the descent of colder air in contiguity with air relatively warmer' and explained that he placed the emphasis thus 'advisedly ... because the driving power of the convective circulation comes from the excess of density of the descending portion'. The study of the more violent forms of convective circulation—thunderstorms, tornadoes and tropical hurricanes—has fully justified this rather unusual view. Although, in each case, the release of latent heat as a result of uplift and condensation is known to play a vital part in the energetics of the system, the almost explosive release of this energy over a limited area has been found to depend upon the development of an organised downdraught which localises and reinvigorates the upward motion. This fact also helps to explain the relative infrequency of these phenomena for, whereas 'bubble' convection is likely to arise over any area where the air is heated a degree or two above the temperature of its surroundings, the organisation of such vertical motion to storm dimensions is the exception rather than the rule.

## Thunderstorms

C. E. P. Brooks (1934) has estimated that about 16 million thunderstorms occur each year over the earth as a whole. Assuming that their average duration is about an hour this means that, at any one moment of time, it should be possible to locate about 1825 storms in some stage of development or decay.[1] It may seem, at first, that to describe such a phenomenon as 'exceptional' is hardly appropriate, but these figures must be interpreted in relation to the vast dimensions of the air ocean and to the very moderate area affected by each storm. Even allotting the generous estimate of 100 square miles per storm, we derive a mean frequency for the earth as a whole of only eight storms per year at any one point of observation. A process which is in operation for only 8 of the 8760 hours of the year can hardly be regarded as normal! In actual fact, of course, thunderstorms are far from evenly distributed over the surface of the earth. Their observed frequency varies

[1] Blüthgen (1964, 262) says more modern observations have raised this figure to about 3000 storms per hour.

from virtually zero in the polar regions (for reasons which should be evident from Chapter 9) to a mean value of some 80–90 storms per station per year in Florida where they are triggered off by the converging sea breezes from opposite sides of the low peninsula. In mountainous areas within the tropics their frequency is probably much greater but records are comparatively brief. Lumb (1970) reports 200 thunderstorm days per year near Lake Victoria. In middle latitudes they are a summer phenomenon of continental interiors and stations in Ohio, for instance, record some fifty storms per year.

Table 44 shows the seasonal variation in thunderstorm frequency at four stations in America.

TABLE 44. *Mean Number of Thunderstorm Days 1904–1943*

| STATION | JAN. | FEB. | MAR. | APRIL | MAY | JUNE | JULY | AUG. | SEPT. | OCT. | NOV. | DEC. | TOTAL |
|---|---|---|---|---|---|---|---|---|---|---|---|---|---|
| Tampa | 1.2 | 1.5 | 2.6 | 3.4 | 8.0 | 16.3 | 21.7 | 20.0 | 12.9 | 3.1 | 0.5 | 0.7 | 91.9 |
| Santa Fé | 0.1 | 0.3 | 1.1 | 2.4 | 6.9 | 9.6 | 19.1 | 16.4 | 7.5 | 2.5 | 0.4 | 0 | 65.3 |
| Omaha | 0.1 | 0.2 | 1.3 | 3.4 | 7.0 | 9.6 | 8.6 | 8.6 | 5.8 | 2.5 | 0.6 | 0.1 | 47.8 |
| Seattle | 0.1 | 0.1 | 0.3 | 0.4 | 0.8 | 1.1 | 0.7 | 0.8 | 0.8 | 0.6 | 0.3 | 0.2 | 6.2 |

*Source:* Byers and Braham (1949).

Although thunderstorms, by definition, involve 'thunder heard' and the concomitant display of lightning, it must be emphasised that these pyrotechnics are as purely incidental to the real processes at work as the sparks from a grindstone or the roar of an exhaust. Lightning is of interest to the geophysicist and must not be forgotten by the architect, but there is no evidence whatever that it plays any really significant part in the physical processes of the weather.

The real distinction between a thunderhead and an ordinary cumulus cloud is to be found only in the vigour and depth of the convective process; yet this distinction in degree is so marked as to be virtually a distinction in kind. Following a suggestion by Köppen, Ferrel (1889, 461) demonstrated that the vital feature was the development of an active *downdraught* in the rain area beneath the growing cloud. He was also aware of the cool squall which precedes the storm and discusses the advance across country of groups of thunderstorms along a crescentic front (*ibid*, 454). Our knowledge of the precise nature of these events has been much improved in recent years. Direct observation of cloud interiors by instrumented aircraft is now almost routine, while the power of the ground observer has been greatly increased by the development of 10 centimetre radar techniques.

The American 'Thunderstorm Project' 1946–49 did much to clarify the life history of the typical thunderstorm. Three distinct stages of development were recognised—(a) the cumulus or development stage during which a great depth of cloud is built up; (b) the mature or active stage characterised by heavy rainfall and lightning discharge; and (c) a final or dissipating stage during which the activity gradually subsides and the major portion of the cloud-system disappears (Fig. 111).

(*a*) *The development stage.* The initial cumulus cloud is built up by the ascent of successive 'bubbles' of warm air rising either directly from the heated ground, especially from slopes of favourable aspect, or, more usually, from the top of a layer in convective equilibrium (see p. 237) which has been developed and deepened by surface turbulence. Significant 'bubbles' appear to range in diameter from about 300 to 1500 kilometres (1000–5000 ft) and their lift through the environmental air is unlikely to exceed twice this distance since they are eroded around their margins and leave behind them a 'wake' that is more humid than the undisturbed environment. Each successive bubble, following the trail blazed by its predecessor, thus suffers less loss of lift so that later bubbles overtake and overtop the pioneers. The summit of the cloud thus builds up by the development of successive turrets and gives the impression on a time lapse film of a vigorous boiling action. Around the flanks of the cloud, on the other hand, the evaporation of cloud droplets lowers the air temperature and results in a slow and mildly turbulent descent.

Figure 111 (a) is a highly generalised representation of this stage and no attempt has been made to indicate the separate initial cells. The cloud is 1–2° C warmer than its environment throughout and general uplift is the rule. Updraughts range from between 1.5 and 3 m/sec in the lower part of the cloud to from 5–8 m/sec aloft. The accompanying table may help the reader to convert these speeds into more familiar units.

| m/sec | | ft/min | | mph | | knots |
|---|---|---|---|---|---|---|
| 1 | = | 197 | = | 2.2 | = | 1.9 |
| 5 | = | 984 | = | 11.2 | = | 9.7 |
| 30 | = | 5906 | = | 67.1 | = | 58.3 |

The large cloud shown would contain about 60 cubic miles of air. About half of this will have risen from below whilst the other half has been entrained in the process. Clearly growth to this stage not only takes time, normally from ten to fifteen minutes, but it also implies the prior existence of a large *reservoir* of warm air near the surface, awaiting eviction. Hence the importance of Shaw's insistence that 'it is by the growth in thickness of the isentropic layer that the proper destination of warmed air is reached' (*Manual*, III, 1930, 304). Note that there is so far no precipitation below the base of the cloud though it now contains large quantities of liquid water in tiny droplets. Perhaps half of these are supercooled and there may be traces of snow in the upper levels.

(*b*) *The mature stage.* This is illustrated by Fig. 111(b). The mixture of supercooled droplets and ice crystals in the central portion of the cloud is now producing raindrops, and rain or hail is observed below cloud base. Even if this never reaches the earth's surface it sets off an active *downdraught* within and below the cloud, partly as a result of the frictional drag as it attains its terminal velocity and partly because it chills the surrounding air, directly and

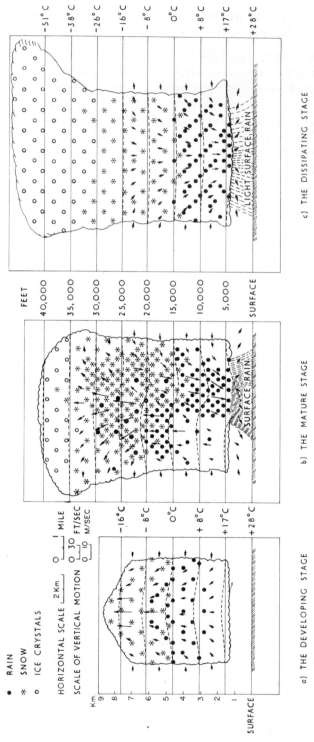

Fig. 111. The life history of a large continental thunderstorm. The diagram shows the three stages of youth, maturity and senility which usually are passed through over an interval of some 30 minutes (after Byers and Braham, 1949)

as a consequence of evaporation. This downdraught often reaches 4–8 m/sec towards cloud base and two remarkable developments follow. (1) Within the cloud, especially in its lower layers, horizontal contrasts of temperature may now reach from 4–5° C so that the updraught is accelerated, in extreme cases reaching 30 m/sec. The cloud therefore builds up by another 10000 ft or more, increasing its volume in our example to over 100 cubic miles and bringing its

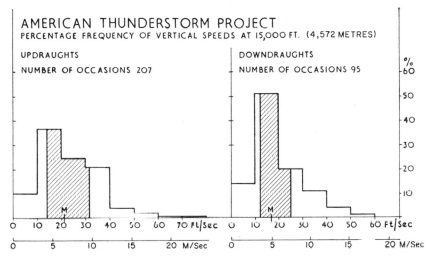

Fig. 112. Frequency-groups of the upward and downward velocities actually observed during the American Thunderstorm Project

upper levels well within the range of complete glaciation. (2) At ground level, on the other hand, the cold downdraught must spread outward, forming a micro-cold-front that is often slipperlike in shape. The toe of this downflow, advancing in violent gusts often against the surface wind, frequently outruns the storm and plays the part of an active scoop which virtually shovels renewed supplies of warm surface air into the tumult aloft. The process has therefore become self-stimulating and is likely to continue until the supply of unstable air is exhausted. Neighbouring storms may be active competitors. This phase normally lasts from 15 to 30 minutes and is marked by dangerously bumpy conditions in the cloud as well as by the characteristic electrical phenomena.

Figure 112 illustrates the range of vertical motions in mature thunderstorms actually measured at 15000 feet during the course of the American Thunderstorm Project.

(*c*) *The dissipating stage.* This is reached when the updraughts cease and the cloud takes on much the same temperature as the environmental air. The upper part of the cloud is completely glaciated and the fine ice particles may be carried away for miles in the upper wind as a long veil of altostratus. Lower down the cloud is still being drained of its finer raindrops, delayed in their descent by their comparatively low fall-speeds. Indeed, many of these evaporate on the way down, chilling the air and producing a slow descent (of between 2 and 4 m/sec) over the whole area of the cloud base. Light rain continues but within 30 minutes or so of the cessation of the active updraught the storm is likely to degenerate into thin layers of stratified cloud which gradually disperse.

If such a large storm brought no more than a quarter of an inch of rain to an area no larger than the cloud itself, say about 25 square miles, this would represent no less than 400000 tons of rain and hail. The release of latent heat involved in condensing this mass of water amounts to approximately $25 \times 10^{13}$ calories, that is, the energy of about a dozen Hiroshima bombs. Yet this is only a small fraction, perhaps one-tenth, of the total energy of the storm, the rest having been lost through mixing with the environmental air and the evaporation of cloud and rain droplets.

*Hail* is the characteristic type of precipitation produced by thunderstorms even when none actually reaches the ground in the frozen state. Hail production appears to coincide with the reinvigoration of the system following the scooping action of the downdraught as it advances along the ground ahead of the storm. The increase in vertical velocities which follows may carry small raindrops far into the freezing layers to produce comparatively large ice pellets. In falling through the cloud again these collect both ice crystals and supercooled water, the presence of the ice helping to dispose of the heat of fusion as the water freezes on. The layered structure often encountered in large hailstones is now thought to arise from variations in the relative proportions of ice crystals and supercooled water droplets encountered at various parts of the fall-path. Newton (1966) considers that a fairly strong wind-shear with height is necessary to the production of hail large enough to do serious damage to growing crops. Ludlam sees some justification for the belief held by farmers on the southern flanks of the Alps that damaging hail can be broken up by large rockets. In such terrain the really powerful updraughts are located along the mountain slopes and the shock wave from a rocket may well break up the larger aggregates within a range of the order of 100 metres. The method is expensive but the damage to ripening grapes can be disastrous.

Although all thunderstorms apparently reach the freezing level the production of hail is by no means inevitable. Appleman (1960) has noted that maps of the mean frequency of thunderstorms and of the occurrence of hail in the United States are far from indicating a similar distribution. Furthermore hail is most common in April and May whilst thunderstorms reach their peak in June, July and August. He argues that hail is produced by clouds with a relatively low droplet concentration and with a freezing level not far above cloud base. When the droplet concentration is great and the freezing level well above cloud base he considers that the larger droplets may grow to a size adequate to achieve fallout before penetrating far above the freezing level. Hail production is then 'suppressed' though the smaller cloud elements may be carried far aloft to produce graupel or ice crystals which may act as a further seeding mechanism.

The generation of the powerful electrical potentials accompanying thunderstorms has proved a lively field for speculation amongst physicists. Active precipitation appears to be a prerequisite even if the elements are re-evaporated before reaching the ground. There also appears to be a close relation between the generation of the charge and the occurrence of the icing phase. Many possible explanations fail because they can be shown to be incapable of producing the necessary potential gradient in the limited time available and

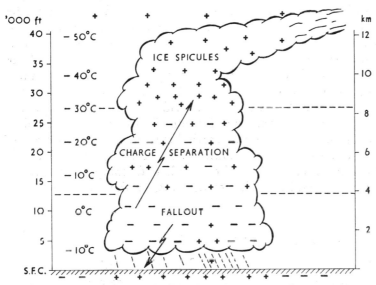

FIG. 113. The typical charge distribution in a thunder cloud. The normal electrical field in the atmosphere is positive in the ionosphere and negative at the earth's surface. In a thundercloud a charge-separation occurs broadly between 4 and 8 km, primarily as a complex result of the freezing process. Ice spicules carrying a positive charge are swept aloft whilst the fallout, consisting of hail or melting ice-aggregates, transfers a corresponding negative charge towards the base of the cloud. A local positive charge is thus induced at the earth's surface beneath the storm. A small area with a positive charge is frequently observed near cloud base over heavy rain. Mason suggests that this may arise from intercepted point-discharges from the ground. When the massive electrical resistance of the atmosphere is overcome, lightning discharges along the paths shown are possible. The electrical reserves of the charge-separation zone are usually sufficient to permit such discharges at frequent intervals

Mason favours the splintering of the ice shell of freezing droplets when the water inside freezes and expands, supplemented perhaps by the collision of soft hail (graupel) with supercooled droplets. Both these processes appear to be particularly effective between the $-15$ and $0°C$ levels, and in each case the larger element carries a negative charge downwards as it falls whilst a corresponding positive charge is swept upwards towards the cloud top on the finer elements.

It is the astonishing length of the discharge path that gives lightning its fearful character and puts it quite beyond the experience even of those familiar with high-tension laboratories. Strokes from cloud to earth may be well over a mile in length whilst strokes from cloud top to cloud base or from cloud to cloud may exceed 3 miles. Voltages of the order of tens of millions are required to overcome these gaps and the current may reach a maximum of 20000 amps. Since the duration of the flash, however, is to be measured in fractions of a thousandth of a second the actual charge transferred is not large: it averages about 20 coulombs.[1] A stroke of lightning has been described as an avalanche of free ions, the speed of the avalanche averaging about one-sixth of the speed of light. Along its path, ranging from about 2 to 9 inches in width, the air is abruptly raised to a temperature of the order of 15000°C. The noise of the discharge arises from a shock wave produced by explosive expansion and it is heard as a thump, a tearing crack or a rumbling roar according to the position of the observer and the number of individual discharges comprising the stroke. Figure 113 shows the usual distribution of the charge in a thundercloud and the induced charge on the earth beneath. The small area of positive charge near the base of the cloud is an interesting anomaly. It is thought to be due to the interception by heavy rain of discharges from prominent points, e.g. tree tops and lightning conductors, on the ground (Mason, 1962b).

## Thunderstorm Systems—Squall Lines

Isolated thunderstorms may occur over uplands or hilly islands with sufficient frequency to suggest that the outburst has some close connection with unequal heating of the earth's surface, i.e. that they are of a 'thermal' type. But, as Ferrel (1889) realised, many of the most violent thunderstorms occur in groups or lines and advance across country in much the same manner as frontal disturbances; they would therefore appear to be related to a pattern of circulation on a regional or, at least, sub-regional scale. Some degree of regional convergence will clearly assist in overcoming the normal stability of the atmosphere which presents an initial barrier to large-scale convectional overturning. Even in England, Ludlam has noted the association of thunderstorms with the occurrence of weak sea-breeze fronts and we have already mentioned the importance of this effect in Florida where two coastal zones are opposed.

The dominant trigger mechanism for the release of whole families of thunderstorms in middle latitudes, however, is undoubtedly the advance of a cold front. Particularly over the central plains of the United States this sometimes stimulates the development of a 'squall line' several hundreds of miles in length and containing a score or more of active thunderstorm centres. The normal direction of advance is then from northwest to southeast and the

[1] A coulomb is the quantity of electricity transferred by a current of one ampere per second.

activity is unusually prolonged. From crop-insurance claims in the Dakotas Frisby (1963)[1] has plotted fierce hail swaths from 5 to 15 miles wide and from 50 to 200 miles long, the direction of the swath being broadly parallel to the wind at 500 millibars, though often diverging at a slight angle to the right of it. The basic cause of the instability is unquestionably the confrontation of warm moist air from the Gulf of Mexico with cooler, and often very dry, air from the north-west, but the detailed structure of the squall line, under these conditions, is related to the organisation of the thunderstorm downdraught. As shown in Fig. 114, thundery outbreaks often begin at the cold front itself but the presence

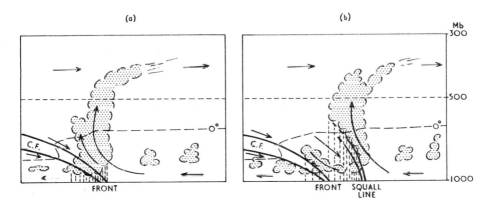

FIG. 114. The development of a squall-line ahead of a cold front. (a) Shows the cold front with its initial cloud pattern. (b) Shows a stage in the advance of the squall-line and the vertical circulation which has given rise to it (after Petterssen, 1956)

of the cold mass in the rear of the storm gives additional forward impetus to the downdraughts so that they may outrun the advancing front by 100 to 200 miles in a 'pseudo cold front' which is the scene of most violent activity (see Petterssen, 1956, ii, 174–87, and Fulks, 1951, 647–8).

Browning and Ludlam (1962) analysed a parallel situation over England which gave rise to the famous Wokingham hailstorm of 9 July 1959. Here the storm came up with the mid-tropospheric wind from the south-west and cut a hail swath across south-eastern England from 10 to 15 miles wide and 130 miles long. The duration of this storm is thought to indicate that it contained a circulation 'in a virtually steady state', for 'in the presence of wind shear (the south-westerly wind increasing in strength with height) the up- and down-draughts can be maintained continuously . . . from opposite sides of the storm'. Such storms are practically independent of surface heating effects.

The Wokingham storm was kept under continuous observation by radar but we can give here only some of the inferences derived therefrom. Seen in plan the storm was far from symmetrical. The downdraught spread mainly to the *right* of the path and here the gust front was succeeded within a few

[1] See also Stout *et al.* (1960, 317–2).

minutes by an abrupt and very heavy fall of large hail, followed by heavy rain gradually decreasing in intensity. It is this distribution of the downdraught that causes the storm to work its way slowly across the upper wind in the manner noted in America by Frisby. To the *left* of the path, on the other hand, the gust front was weaker though rain fell earlier from the 'overhang', increased in intensity (mingled with small hail) under the centre and then gradually decreased as the storm passed on. Figure 115 shows the storm in cross-section along its direction of travel at the stage of maximum intensity.

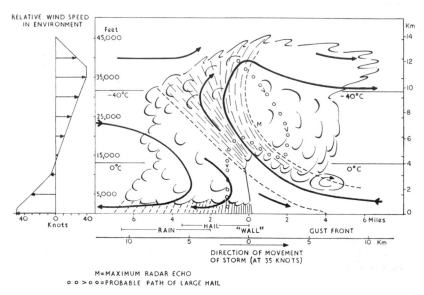

FIG. 115. A cross-section of the Wokingham storm with a true vertical scale

At this period the cloud top lost its discrete turrets and towered to 12–13.5 kilometres (40000–45000 ft), that is, to well within the stratosphere, for a considerable period of time. Immediately beneath the lofty central peak is a dense fall of large hail which produces a strong radar echo sharpening to a radar 'wall' below 4000 metres (13000 ft). A clear downdraught advances 4 miles in front of this 'wall' to produce the 'gust front' where the surface wind is abruptly reversed in direction. Since the storm is advancing at 35 knots against a north–north-easterly wind of about 15 knots along a front of some 7 to 8 miles, the lowest 2450 metres (8000 ft) of the warm moist air ahead is being engulfed into the system at the rate of at least 10 cubic miles per minute! This produces the immense overhang ahead of the 'wall' where updraughts of the order of 2000 ft per minute (10 m/sec) toss back the great majority of the falling precipitation elements. The leading edge of the overhang is usually marked by a turbulent roll or arch of cloud whilst towards the 'wall' the vigorous updraught maintains a cloud-free 'vault'. Towards the −40° C level the updraught must have reached 30 m/sec in order to support the 2-inch hail characteristic of this particular storm. Note that air thus raised is then swept on *ahead* of the storm by the strong winds of the upper troposphere. Behind the 'wall' the cloud base is low and precipitation is widespread. Yet much of the lower cloud is in *descending* air, the flow being fed from air above 4000 metres which is overtaking the storm and being chilled by the partial evaporation of the falling raindrops. The evicted warm air is also partly replaced by flow from the left flank of the storm. These complex exchanges transform the characteristically intermittent thunderstorm phenomena into a virtually continuous process.

Further analysis of intense storms suggests that we really require a three-layer model, the wind in each successive layer veering strongly in relation to the layer beneath it. Warm moist air near the surface provides energy to the system; a drier stratum above it is chiefly responsible for the evaporation from fallout and hence, in the view of K. Browning (1968), yields the bulk of the cool downdraught; whilst a strong upper current broadly steers the storm by providing the necessary scavenging mechanism. Carlson and Ludlam (1968) further suggest that broad geographical features may play a significant role in setting the stage for some intense storms in high summer. In western Europe they have found evidence that the dry Meseta of Spain may raise the temperature of the 1 to 2 kilometre layer of an already warm and dry air stream from the Sahara, thus producing a 'plume' of potentially very warm air over France. Under the inversion thus generated only small-scale convection is possible and the surface layers build up their water vapour content until the plume is outrun. In Texas and Oklahoma they suggest that a similar 'lid' to invading trade-wind air from the Gulf may be provided by warm and dry middle layers off the elevated surface of northern Mexico. The release mechanism which makes deep overturning possible is provided in each instance by the line along which the plume is replaced by an upper current from a more westerly direction. In Europe this is Atlantic air, in Texas it is tropical continental air, but the general effect is similar and the storms move northeast, sometimes accompanied by tornadolike phenomena, fortunately much less common in Britain than in the United States.

It may be noted that the cold front hailstorms of the Mid-West which normally travel in a south-westerly direction must fall into a distinct category. The same is probably true of the severe hailstorms of the pampas of Argentina vividly described by W. H. Hudson in *Far Away and Long Ago* (1918, ch. 5). These move towards the north-east but it must be recalled that we are there dealing with the southern hemisphere.

In tropical latitudes, particularly where very moist and very dry air masses are brought into interrelationship as in West Africa, rather similar squall-line thunderstorms develop though their motion is normally in an east-to-west direction. The precise mechanism of the miscalled West African 'tornadoes' is not fully known but the humidity contrast is responsible for layer lability and since the dry air is brought down to the ground after the passage of the storm, it is playing the part of the 'cold' element in the reaction.

# Tornadoes

The name of these extremely violent but fortunately very local storms appears to be derived from 'tronada', the Spanish word for a thunderstorm. Indeed, they are commonly associated with thunderstorm activity and may well be triggered off by the rapid descent of air heavily laden with hail and rain. The partial evaporation of the precipitation elements counteracts the rise of temperature that should follow from adiabatic compression and the tornado is often followed by a marked fall in temperature. Their essential feature, however, is the development of cyclonic spin to a degree which baffles analysis and almost defies description. Hence the appropriateness of the American vernacular term, 'twisters'.

From numerous accounts of tornadoes it is possible to gather that they usually develop from a low cloud base, heavy with mammatus forms, and begin as a rotating pendant which narrows downwards and often generates considerable noise (from wind shear) even when it has little effect at ground level. Furthermore, even after striking the ground, the tornado is often observed to lift and strike again, a skipping action that certainly suggests that the root cause of the phenomenon must be sought aloft, though pockets of cool air near the surface may contribute to the effect.

In American experience, the damage following a strike is usually confined to a swath about a sixth of a mile wide and from 2 to 4 miles long but the devastation is so complete that instrumental observations are out of the question. Freak effects abound—straws driven through planks, wool stripped from sheep, heavy farm gear carried for miles and large trees uprooted and tossed about like weeds—indeed, tornadoes have a literature of their own scarcely inferior to the wild fantasies of Baron Münchhausen. From the nature of the damage it is possible to infer the conditions in the heart of the storm. Damage arises from three sources: (*a*) the sheer pressure of winds of perhaps 200 to 400 mph. Ferrel (1889) estimated that a surface wind of 50 mph at normal temperatures exerts a pressure of about 7.5 lb/ft² and argues that the value increases as the square of wind speed.[1] Pressures of the order of 100 to 500 lb/ft² seem not improbable though it must be remembered that the real damage is effected by the shock waves in gusts. (*b*) The sudden fall in atmospheric pressure in the heart of the vortex. It is now considered that this is of the order of 25 millibars, but it may well be considerably more. Such a pressure fall, occurring abruptly, has an explosive power of over 50 lb/ft², more than enough to lift a roof and blow walls out in all directions. (*c*) Finally, there is a powerful upward suction related to vertical velocities which must be far in

---

[1] The accepted formula is $p = kv^2$ where $v$ is wind speed in mph and $p$ is wind pressure in lb/ft². $k$ varies with air density and thus depends on both temperature and pressure. It is evident that Ferrel was using $k = 0.003$, but the engineer J. Smeaton (1724–92) drew up a table based upon $k = 0.005$ which, one gathers, is still in favour among practical engineers. The higher values provide a useful safety factor. Experiments on the windy summit of Mount Washington, USA, pointed to $k = 0.004$ but it is known that the pressure per unit area is lower for large surfaces than for a relatively small plate (see Fleming, 1930).

excess of those occurring in the ordinary thunderstorm. E. M. Brooks (1951, 675) quotes an instance 'when a tornado got ahead of its cumulonimbus cloud and immediately created a new cumulonimbus cloud which shot up to an elevation of 35000 ft in 1 minute'. Even allowing for the evident crudities of this estimate, it does suggest that vertical velocities are of much the same order as the horizontal wind so that the values quoted under (a) give us some idea of the weights which the updraught would be capable of supporting. Most of the debris thus lifted by the rotating column is rapidly centrifuged out and distributed at random over the surrounding countryside.

In a frictionless vortex it is not difficult to show that as the distance from the centre $(r)$ varies, the tangential velocity (wind speed) varies as $1/r$, the speed of rotation (spin) as $1/r^2$, and the pressure gradient as $1/r^3$, at least to within a short distance from the centre where all of these values become theoretically infinite. But although this mathematical model gives a first approximation to the orders of magnitude involved, it yields no key to the origin of the vortex itself. Wegener (1928) suggested that spin was first generated along a horizontal axis in the cloud arch which precedes the updraught and that a tornado was formed when one end of this vortex was brought down towards the ground. This idea faces the difficulty that, whereas the end on the right flank of the storm would yield the observed cyclonic spin, that on the left flank would give spin in the opposite direction. As long as the descent was regarded as a purely random affair Wegener's idea appeared to be untenable but Browning and Ludlam's (1962) demonstration of the asymmetrical nature of a squall-line storm increases its acceptability though it does not dismiss all doubts. Certainly the preferred area for tornado generation is near the toe of the downdraught on the right flank where both the weight of hail and the horizontal wind shear at the surface are at a maximum. Furthermore we know that strong vortices abhor a free end and will extend themselves towards the nearest available 'surface', i.e. downwards to the ground and upwards perhaps to the tropopause. Tornadoes usually arise within a small 'low', perhaps 5 to 10 miles in diameter, itself embedded within a large trough with strong cyclonic wind shear. A rich source of cyclonic vorticity is therefore available which it is the task of the generating processes to amplify into alarming proportions.

Nevertheless the energy of the system must be derived from atmospheric instability and the ascending air which stretches the vortex 'tubes', increasing their rate of spin, must be evicted aloft to maintain the low pressure and prolong the life of the system. Hence again the importance of a marked wind shear with height and the significance of the jet stream aloft on the general trend of the disturbance. As we have seen above (Fig. 110), situations beneath the right entrance or left exits of accelerating jets will be most favourable to the development of deep instability in the lower atmosphere.

Tornadoes are most frequent and most powerful in the central part of the United States, especially from Kansas to Indiana. On the average about 150 are recorded each year though on one noteworthy occasion, 19 February

1884, no less than 57 occurred on one day.[1] In this central area high frequencies between 1400 and 2200 hours suggest that surface heating certainly plays its part but elsewhere there is much less evidence of a diurnal variation (see Petterssen, 1956, ii, 191). Early summer, April, May and June, when convective instability is widespread, is the favourite period but tornadoes can be set off locally under a wide variety of conditions.

Violent whirlwinds of tornado type though usually on a much reduced scale are not unusual during squall-line thunderstorms. In Britain they number about a dozen a year but the effects are usually very local indeed. C. E. P. Brooks (1954, 39; see also Lamb, 1964, 36) has listed and mapped the strongest of these and shows that here too the usual course of travel is from south-west to north-east.

Waterspouts appear to be mild tornadoes developing over the sea. Although feared in the days of sail it is now known that the visible pendant is merely cloud or light precipitation and sea spray and there is no real danger to shipping. Surprisingly the circulation can be either clockwise or counter-clockwise. This suggests that lateral wind shear is a significant factor. 'Dust devils' behave in similar fashion (Woodward, 1960).

## Tropical rotating storms

Tropical storms present some of the most intriguing and difficult problems in meteorology. Superficially, with their broadly radial symmetry they appear to be related to the simple air vortices such as tornadoes and waterspouts. Yet their subregional scale, comparatively long life and fairly consistent tracks suggest analogies with mid-latitude disturbances—a relationship that seems to be reinforced when it is observed that occasionally hurricanes leave the tropics and join the eastward procession of polar front depressions. Rapidly expanding its sphere of influence and drawing into its system of circulation air from widely contrasting sources, such a disturbance is soon in no way distinguishable from its new-found neighbours.

Tropical rotating storms are essentially a maritime phenomenon—or, as Maury (1858, 330) picturesquely puts it 'hurricanes prefer to place their feet in warm water'. They are therefore rarely observed in the formative stage. Too large to be fully comprehended by a single observer and yet small enough to slip through the very wide mesh of permanent stations available in the tropics, they appear in full fury at irregular intervals along tropical coastlands and their story is thus one of intermittent catastrophe. Fortunately storms of hurricane intensity are comparatively rare—there are some 40 to 50 such storms in the tropical half of the air ocean each year—but undoubtedly they offer the most awe-inspiring examples of energy exchange in the atmosphere and are capable of devastation and loss of life on a scale paralleled only by volcanic eruptions and major earthquakes. There is a vast literature on hurricanes and certain elements of a broad pattern have emerged. It is known that their incidence is related to the withdrawal in autumn of the trade-wind

---

[1] A similar event occurred in the Mississippi delta region in 1971, again in February.

systems from the western portions of the tropical oceans and their imminent arrival can be gauged from a number of portents in both sky and sea. Yet, even with modern methods of air reconnaissance and radar tracking, we are far from a fully convincing explanation of their origin and mechanism.

Tropical storms arise over seas with a surface temperature of 80–82°F (27–28°C)[1] overlain by a deep layer of warm moist air. The trade-wind inversion, if still present, is thus at 1800 to 3000 metres and comparatively weak. Yet they do not appear to occur within a zone some 5° from the equator, presumably because the air there is too deficient in vorticity to develop into a cyclonic system. Initiated perhaps as a group of thunderstorms, the system takes on its unique character when it organises itself into cyclonic rotation and from then onwards it may decay or develop according to its response to a number of external stimuli which we are beginning to understand. It becomes a full hurricane when the wind strength exceeds 65 knots (75 mph) and, since the experience of the *Charles Heddle* which 'running before the wind' made five complete circuits of a slow-moving storm in five days off Mauritius in 1845, it has been regarded as established that the circulation is essentially rotary in character. Needless to say, 'running before the wind' thereafter became unpopular in the tropics and masters were issued with a number of rule-of-thumb sailing directions calculated to help them from repeating this devastating experience. On a modern synoptic chart a hurricane appears as a group of almost circular isobars, usually from 100 to 400 miles across, round a central low with a pressure of the order of 950 millibars (28 in of mercury). Indeed values lower than 900 millibars have been recorded but it is clear that the central low is not the *cause* of the storm, it is at least as much the result of the vortical circulation. Note, however, that this vortex is very different in shape from the tornado, it is biscuit-like, many times wider than it is deep. A depth of about 11 to 13 km or 36000 to 42500 feet appears to be common. This important fact is too often obscured by cross-sections drawn with a greatly exaggerated vertical scale.

In spite of fierce winds, sometimes exceeding 130 knots (150 mph), the forward motion of a hurricane is usually comparatively slow (from 8–15 mph). Yet it may have a life of nearly a week. Clearly it can maintain this stately progress only by drawing on reserves of energy from far beyond its limited area and by reacting, in turn, upon the weather of lands and seas many miles from its actual path. The sheer ferocity of the disturbance is thus one of the factors contributing to its rarity. Bergeron (1954, 150) has suggested that a hurricane 'propagates by steps . . . a secondary centre . . . gradually taking over the role of the main centre'. This certainly fits in with some of the observations made and serves to remind us that even when a weather feature is graced with a feminine name (a familiar American practice) its continuous identity may be nonetheless suspect. Within the tropics the usual direction of

---

[1] Perlroth (1969) has stressed the great importance of the *depth* of the warm layer. Storms reach hurricane intensity only where the downward temperature gradient in the sea is less than some 4°C in the first 70 metres. This is because the wind itself produces strong vertical mixing in the sea.

progress is westward with a slight poleward bias. Early accounts laid great stress on the tendency of such storms to follow a parabolic path, 'recurving' into the westerlies at about the 30th parallel. It is now known that this is the exception rather than the rule. Forecasting the track of an approaching hurricane is of great practical importance if adequate precautions are to be taken to reduce loss of life, but it is still a task beset with difficulties.

The unique and characteristic feature of the true hurricane is the 'eye of the storm', a central area of light winds and comparatively cloudless skies though filled over the ocean with mountainous and confused seas. At ground level the eye may range from 5 to 25 miles or more in diameter and the sudden lull which it brings, followed later by the equally abrupt return of the gale from the opposite direction, is a fearful experience. In August 1951 R. H. Simpson (1952) flew on two occasions into an unusually large eye and photographed its magnificent cloud walls extending up to 35 000 feet. He describes it as a 'coliseum' 40 miles across, and floored with broken cloud at about the 8000-foot level. Within it he observed a temperature of 16°C at 500 millibars (17 000 ft), the highest ever recorded at that level. An earlier sounding made at Tampa in 1944 (Riehl, 1948, 194) is summarised below:

| Pressure level (mb) | Temperature within eye (°C) | RH within eye (°C) | Normal temperature in hurricane rain areas (°C) |
|---|---|---|---|
| 100 | −74 | (20) est | −77 |
| 300 | −19 | 23 | −30 |
| 500 | 4 | 53 | −5 |
| 700 | 15 | 86 | 10 |
| 900 | 20 | 93 | 20 |

Malkus considers this to be fairly representative of the normal hurricane eye and infers that such a combination of high temperatures and considerable relative humidities above 700 millibars can be explained by subsidence of the order of 20–30 ft/min combined with the evaporation of cloud particles from anvil cirrus and some admixture, in reasonable proportions, with fragments of the cloud wall. The subsiding air may take about a day in its descent within the eye and is ejected into the surrounding updraught near the ground so that the process is continuous (Malkus, 1958a). The body of exceptionally warm air in the eye above 700 millibars plays an important part in maintaining the deep low in the centre of the hurricane; indeed it is hard to see how the system could exist without it. Yet the mechanism of its formation is unknown and we can say little more than that it appears to arise as a necessary by-product of the vortical circulation. What Malkus describes as 'one of the most amazing features of hurricane eyes', the weakness of the air-motion there in comparison with the turmoil outside, seems to be a result of the prevailing subsidence. The air comes from the upper parts of the system where there is little spin and most of the cyclonic vorticity which it gains as a result of admixture with the wall cloud is dispersed by the lateral spreading that occurs below 850 millibars as it approaches the earth's surface.

Ringing the eye, particularly in the earlier stages of hurricane development, is a funnel-shaped region where warm moist air circulates with extreme velocity. It is fed from the surface layers (perhaps up to 1000 metres) which rapidly gain in angular velocity as the radius of rotation decreases. Here is experienced the extreme fury of the storm. The centrifugal force thus generated soon balances the pressure gradient and the surface air is then thrust upwards; 'it is finally lifted in an orderly fashion nearly simultaneously within a big area of a ringlike shape' (Bergeron, 1954, 141).[1] Condensation follows and the massive release of latent heat provides energy in amounts considerably in excess of the immediate dynamic requirements of the storm. Local updraughts of the order of 4–5 m/sec (800–1000 ft/min) appear to be probable, i.e. of much the same scale as those observed along a squall line, but the masses of air involved are larger and the system is much more prolonged. Radar observations suggest that it is incorrect to view this process as smoothly continuous; the outbursts occur spasmodically but at frequent intervals and the major updraught may be first in one part of the 'ring' and then in another. Although this is the major rain-making part of the storm the updraughts are so strong that comparatively little rain falls within it.

The major rain belt thus lies outside the region of maximum wind—perhaps from 25 to 60 or more miles from the centre according to the size and stage of development of the system. Again falls of 0.5–1.0 in/hr are not greatly in excess of the intensities observed in a thunder-squall but if the system moves slowly such intensities may continue for hours. A yield of 5 to 10 inches of rain is by no means unusual and falls in excess of 24 in/day are not unknown. Bergeron considers that such falls must be associated with heavy *downdraughts*, especially below 4 to 5 kilometres (13000–16500 ft) and Simpson certainly noted a series of squall-lines almost parallel to the surface wind as he approached the typhoon 'Marge'. In this way the region of active rain is carried outwards from the central 'ring' in a series of spiral bands.[2] Each of these has a life of an hour or two but fresh bands are generated with each new outburst of activity in the central zone. As the hurricane ages more and more of its rainfall comes from these marginal developments and the rain area is often displaced towards the poleward sector of the system. It is in this ragged outward margin of the rain zone that electrical phenomena most usually occur. In Bergeron's view the rain downdraughts also react upon the updraught in the central core, evaporation from rain increasing the humidity of the inflowing air whilst the warm broken sea-surface rapidly corrects the loss of temperature that such evaporation entails (Fig. 116). Certainly the updraught within a

---

[1] This may be an over-generalisation. Later workers insist that most of the uplift occurs in discrete 'hot towers' but within the ring these are closely-spaced.

[2] In satellite photographs the appearance closely resembles a spiral nebula. Astronomers ascribe the banded formation to wavelike developments in a rotating medium.

hurricane appears to occur on a massive scale without the loss of lift which follows from the entrainment of environmental air in all other convectional disturbances. The lapse rate in the rising air thus follows very closely the saturated adiabat appropriate to a surface air sample.

The outer ring of the hurricane lies under a great pall of high cloud diverging from the centre and carrying both moisture and sensible heat to the upper layers over great distances. This gradually cools by radiation and subsides.

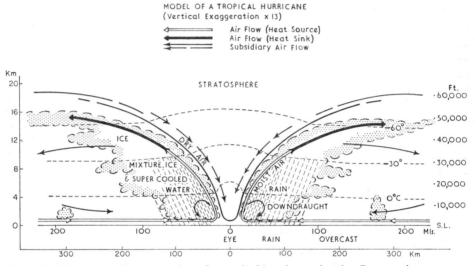

Fig. 116. Diagrammatic cross-section of a tropical hurricane showing Bergeron's interpretation of the vertical circulation. Vertical exaggeration about ×13

The intense convection of the storm is thus balanced by air descent in three distinct ways—within the eye, in the rain downdraughts and by slow regional subsidence over many thousands of square miles. The greater the storm the less likely an immediate successor.

Estimates of the amount of water vapour passing through such a storm reach values of the order of $15 \times 10^9$ tons/day and it is not uncommon for half of this to fall as rain. The energy of a well-developed storm has been estimated as equivalent to four hundred 20-megaton hydrogen bombs per day. About 3 per cent of this is converted into energy of winds (J. Simpson, 1967, 95).

## Hurricane zones

It is well known that hurricanes are the scourge of particular areas of the tropical seas, especially from July to October in the northern hemisphere and from December to March in the southern hemisphere. In the Indian Seas rather similar disturbances accompany the changeover from the northern to the southern monsoon in May–June and October–November. It is difficult to give more than a very general picture of their relative frequency in different areas, since data are not available for a uniform period, since some storms may never be recorded, and since a rigid definition has only been agreed upon in recent years. Ramage (1959, 233) has attempted to express their annual frequency in more satisfactory terms by comparing the mean number of all tropical storms (wind strength in excess of 40 knots) and of true hurricanes (wind strength in excess of 65 knots) with the varying areas of sea over which they may be anticipated. At first he included all sea areas beyond the fifth parallels with a surface temperature above 26°C (79°F) during the appropriate season in each hemisphere. The total area thus involved was about

TABLE 45. *Estimated areas of the chief hurricane zones*

| Area | Million sq. miles |
|---|---|
| *Northern hemisphere* | |
| Atlantic west of 70° W | 1.7 |
| Atlantic east of 70° W | 2.2 |
| Arabian Sea | 1.1 |
| Bay of Bengal | 1.3 |
| South China Sea | 1.3 |
| Pacific west of 170° E | 4.5 |
| Pacific east of 170° E | 3.2 |
| *Southern hemisphere* | |
| Indian Ocean west of 90° E | 3.3 |
| Indian Ocean east of 90° E | 1.4 |
| Pacific west of 150° W | 3.8 |
| Total | 23.8 |

39 million square miles but this embraced some 3.5 million square miles in the South Atlantic and eastern South Pacific where, because of the unceasing domination of the trades, such storms are never observed. It thus appeared that the net had been thrown rather too wide. By introducing another criterion involving conditions in the upper air, Ramage reduced the hurricane zones of the northern hemisphere to the areas quoted below. We have ventured to make estimates for the southern hemisphere to round off the table.

Generalising for the world as a whole and using the totals given in Table 46 we can thus derive a mean frequency of all tropical storms of 2.2 per million square miles and of hurricanes of 1.6 per million square miles. The areas swept by individual hurricanes varies widely but if we take 100000 square miles as not too unreasonable[1], the mean chance of any one point being involved would appear to be only 0.16 times per year.

In fact Table 46 shows that hurricane frequency varies widely from place

[1] Satellite data now to hand suggest that the typhoons of the western Pacific are both larger in scale and longer in life-span than the Atlantic hurricanes. This estimate may well prove much too small in that region.

to place. In the western North Pacific it is two and a half times the world average. Tropical storms, on the other hand, are as frequent, area for area, in the Bay of Bengal as they are in the western North Pacific. Authorities vary widely on the appropriate figure for the western South Pacific and the value given by Ramage may well be too low. The figures for hurricanes in the southern hemisphere are estimates based upon the assumption that the proportion of hurricanes to storms is much the same as in the corresponding areas to the north.

TABLE 46. *Mean number of tropical storms and hurricanes*

| | Atlantic | | | Indian Ocean | | | West Pacific | | | East Pacific | | | Total in Hemisphere |
|---|---|---|---|---|---|---|---|---|---|---|---|---|---|
| | | *a* | *b* | | *a* | *b* | | *a* | *b* | | *a* | *b* | |
| **ALL TROPICAL STORMS** | | | | | | | | | | | | | |
| *Northern hemisphere* | | | | | | | | | | | | | |
| West | | 3.6 | *2.1* | Arabian Sea | 1.3 | *1.2* | China Sea | 3.4 | *2.7* | | | | |
| East | | 3.7 | *1.7* | B. of Bengal | 5.4 | *4.1* | Pacific | 18.6 | *4.2* | Pacific | 5.7 | *1.8* | |
| Total | | 7.3 | *(1.9)* | | 6.7 | *(2.8)* | | 22.0 | *(3.8)* | | 5.7 | *(1.8)* | 41.7 |
| *Southern hemisphere* | | | | | | | | | | | | | |
| | West | | | | 4.7 | *1.4* | Pacific | 4.0 | *1.0* | | | | |
| — — | East | | | | 2.1 | *1.5* | | | | | — | — | |
| Total | | 0 | | | 6.8 | *(1.5)* | | 4.0 | *(1.0)* | | 0 | | 10.8 |
| **STORMS OF HURRICANE INTENSITY** | | | | | | | | | | | | | |
| *Northern hemisphere* | | | | | | | | | | | | | |
| West | | 2.3 | *1.4* | Arabian Sea | 0.7 | *0.7* | China Sea | 1.9 | *1.5* | | | | |
| East | | 2.5 | *1.1* | B. of Bengal | 2.0 | *1.5* | Pacific | 17.5 | *3.9* | Pacific | 2.2 | *0.7* | |
| Total | | 4.8 | *(1.2)* | | 2.7 | *(1.1)* | | 19.4 | *(3.4)* | | 2.2 | *(0.7)* | 29.1 |
| *Southern hemisphere* | | | | | | | | | | | | | |
| | West | | | | 3.5 | *1.1* | Pacific | 3.0 | *0.8* | | | | |
| — — | East | | | | 1.6 | *1.1* | | | | | — | — | |
| Total | | 0 | | | 5.1 | *(1.1)* | | 3.0 | *(0.8)* | | 0 | | 8.1 |

*Note.* Tropical storms = winds over 40 knots; hurricanes = winds over 65 knots.
*a* = Mean number of tropical storms or hurricanes per year in each sea area.
*b* = Mean number of tropical storms or hurricanes per million square miles of sea with temperatures above 26° C beyond 5° latitude from the equator. The bracketed values must not therefore be confused with the totals.
After Ramage (1959, 233).

## The origin of tropical storms

Why are tropical rotating storms so rare? The necessary prerequisite on which all theorists agree, the presence of a pre-existing shallow tropical low embedded within a deep layer of potentially unstable air, appears to be a very common phenomenon over the intertropical seas. Why is it that the cyclonic spin thus generated is so infrequently whipped up to storm proportions? The question is not dissimilar to that already encountered in regard to other violent weather phenomena: why does the instability find local and intermittent relief instead of being frittered away in minor and scattered overturnings? But we are faced here with disturbances of quite a different order of magnitude. What *organises* the processes at work on such a titanic scale?

Answers to these questions have been sought by various authors at different times by emphasis upon three major processes—convection, confrontation and superimposition. These must not be regarded as mutually exclusive but they provide a key to three fairly distinct schools of thought.

The *convectional* school argues that, under the essentially barotropic conditions prevailing within the intertropical zone (conditions of virtually horizontal isentropic surfaces), convection offers the only method for the release of energy in the atmosphere. It points to the towering clouds and torrential rainfall as clear evidence that convection must be operating to an extreme degree in tropical storms and regards the organisation of the airflow into a vortex-like system as an inevitable but secondary consequence of inflow into a large updraught upon a rotating earth. Ferrel elaborated this view in a rather heavy chapter covering some 120 pages of his *Popular Treatise on the Winds* (1889), though he confuses the issue by attempting to force extratropical cyclones into the same framework in the face of clear evidence that the temperature field within them was very different indeed. His approach is essentially deductive. Arguing from the premises that 'without ascending motion there can be no clouds' (p. 238), that uplift occurs locally whilst descent is 'a very gradual settling down' (p. 228), and that latent heat provides 'a continuous source of energy as long as moist air is being drawn in from all sides' (p. 230), he concludes that the deflection of air-currents upon a rotating earth must produce a cyclonic 'gyration around this centre' (p. 242). Furthermore this must be supplemented by a 'gyration in the opposite sense' aloft if the system is to have a continuous existence.

Unfortunately he introduced here the illustration of 'the behaviour of water in a shallow basin, where it is allowed to run out through a hole in the centre' (p. 242), a false analogy that has continued to gurgle through the literature until the present time. It may be true that 'the principal difference in the two cases is that the water runs down through the hole and disappears while the air runs upward over the central area and flows away above' (p. 243) but there is all the difference in the world between 'disappearing' and 'flowing away above'. The water is *withdrawn* from the system to play no further part in its activity but the air is *superimposed* upon the surface system and still continues to make its contribution to the total atmospheric load. Even by introducing the effects of surface friction and centrifugal force upon such a disturbance it is hard to explain more than a comparatively shallow surface low. The progressive motion of such cyclones Ferrel attributed to 'the general motion of the atmosphere . . . at high altitudes where the centre of energy is' or 'where the condensation of the aqueous vapour occurs' (pp. 275–7). Their translation was thus quite passive like 'small whirling eddies' carried along by 'a stream of water.' Here again analogy is dangerous and Shaw,

393

much concerned whether 'the conditions and arrangements' of experimental vortices 'have their counterpart in the atmosphere', devotes much of Chapters 9 and 10 of his *Manual*, Vol. IV to showing that, on the contrary, progressive motion in relation to the ambient air is *essential* to the maintenance of an apparent atmospheric vortex (see Fig. 105). Indeed he adds, 'as soon as the stage of revolving fluid is arrived at . . . death follows' (IV, 1931, 285). The difficulty of course was in making 'satisfactory arrangements for the delivery of the air when it has reached the limit of its convection' (p. 278), and 'it is much more easy to explain convection along the core as the *effect* of an existing circulation above than vice versa' (p. 277).

Recently Bergeron (1954) has injected new life into the convectional school. Conceding that other factors may play their part in initiating and steering the disturbance, he sees in the organisation of the rain downdraught the distinctive characteristic of the tropical hurricane. Reviewing other types of convective systems he notes that 'the cool rainfall-wind' spreads out on reaching the surface until it produces a broad cool 'shield' which soon 'quenches' a local system. Squall lines can maintain their existence for some hours only because of their rapid forward motion. Hurricanes, on the other hand, can maintain a violent circulation for days because the rainfall-wind circulation is 'inverted', in the sense that it is *turned in* upon the system, reinvigorating it and producing an approximation to flow in a 'steady state'. Indeed the system is not truly a tropical storm until this 'inversion' has occurred. Such a development is rare because it must occur within an appropriate pressure pattern and the updraught must be strong enough to throw the great bulk of the rainfall outwards beyond the updraught core. Over a land surface this would not be possible for the rain would chill the surface, but over a warm sea the cool hurricane rain sinks down through the warm sea-water and the air is promptly reheated by contact with the rough sea surface to supply fresh fuel for the updraught. Indeed, evaporation from falling rain under these conditions will increase the saturation potential temperature (the entropy) of the inflowing current and 'in this way the opposing action of the rainfall-wind will be transformed into a supporting action' (p. 147). When once this 'inversion' has occurred there appears to be no limit to the further development of the disturbance apart from the exhaustion of its available energy supply. Bergeron considers that similar developments may help to explain the sudden deepening of secondaries occasionally observed in middle latitudes. His theory is not a complete explanation of the tropical hurricane but he hopes that it demonstrates 'what kind of atmospheric machinery we are dealing with and its general manner of functioning' (p. 133). It contributes little to our knowledge of the travel of the system though we have already noted that he considers that it 'seemingly propagates by steps'. As with other convective systems, however, its decay will occur when it 'has acquired a cold core and its motion is now automatically braked' (p. 162).

*Confrontation.* Ever since isobaric patterns began to dominate meteorological thinking, the superficial analogies that can be drawn between tropical and

\extratropical cyclones have proved almost irresistible. Just as Ferrel had tried to force extratropical disturbances into a convectional pattern against the weight of the evidence, so the exponents of the new Norwegian theories soon set out with equal enthusiasm, and less encouragement from the facts, to discover new types of fronts which might provide a key to the organisation and translation of tropical storms. Bergeron himself confesses to initiating this attempt in 1928.

Two distinct types of convergence offered promising avenues of enquiry.

1. The most obvious was along the zone where the northern and southern trade-wind systems appeared to 'clash' from opposite hemispheres. Swinging 'with the sun' though lagging some two months behind it, this zone was known to reach its most poleward location in the autumn, the season of greatest hurricane frequency. At this period fresh autumn trade-wind air would encounter old spring trade-wind air from the other hemisphere, 'old' because it had been subjected to the experience of crossing the Line. Would this not create a front situated far enough from the equator for cyclonic spin to be engendered along it? The intertropical front thus deduced has singularly failed to reveal itself to direct observation and, even on theoretical grounds, it has proved very difficult to specify its nature. Even where the trades do approach each other they rarely exhibit such contrasts as could promote a frontal structure. Indeed Riehl (1954, 238) asserts that the equatorial trough is 'placed so that northern and southern trades reach it after complete equalisation of their thermodynamic properties'. It had been overlooked that, in their respective hemispheres, the trades are much stronger and more stable in spring than in autumn; at a rendezvous some degrees beyond the equator they thus meet on approximately equal terms. Besides, the act of crossing the Line generates relative *anticyclonic* spin. Cyclonic development, if it is to occur, must be expected in the local or autumn air stream and, if so, it is by no means clear what part the alleged front is supposed to play. The observation that tropical storms are much more common in the western parts of the oceans where the trades, far from 'clashing', are always separated by a broad zone of doldrum or equatorial air led Garbell (1947, ch. 7) to propose the term 'Tropical Front'. This involves a relationship between an active trade wind and the relatively inert but potentially unstable equatorial mass. There is no doubt that the seasonal readvance of the trade-wind air stream into the doldrums produces widespread rainfall but no clear evidence that cyclones are generated in this way. Indeed the readvance of the trades in a westward and equatorward direction is the signal for the close of the hurricane season!

2. Garbell also develops the idea of trade-wind or 'temperate fronts' (Ch. 5). These are viewed as embedded within each trade-wind air stream and arise from the periodic reinvigoration of the system. The fresh or 'polar' air lies to the west and it is regarded as forming a wedge up which old trade-wind air, already tending to move away from the equator, is forced to rise. In the northern hemisphere such fronts are presumed to run from north-east to

south-west and to slope upwards towards the west. That such discontinuities occur in the trades is more than probable, indeed they are likely to be far too common to give a useful key to storm generation. Furthermore any disturbance along such a 'front' will be damped out unless it moves north-east (south-east in the southern hemisphere) a most unusual track except in the western South Pacific.

Not to be defeated Deppermann (1936) combined both these ideas in his concept of a 'triple point'. This was virtually the intersection of a temperate front with the intertropical front but, since he was working in the Philippines, the three air masses involved were the northern monsoon in the west, old north-eastern trade-wind air in the east, and the southern monsoon in the south. Triple points were regarded as very likely regions for hurricane generation, the system working its way northwards along the margin of the northern monsoon air. It must be recalled that most of the typhoons of the Philippine seas are hurricanes in a late stage of development and no one denies that, under such conditions, both here and in the eastern United States, such storms can draw into their orbit air from widely different regions. It is at this stage, when they are about to 'recurve' that fronts appear to have their most potent influence on the system. The rainfall distribution becomes increasingly asymmetrical—it is concentrated more and more within the forward sector—and as the system expands and accelerates it takes on to an increasing degree the attributes of a severe extratropical disturbance.

It is nevertheless difficult to believe that fronts have much to do with the *origin* of tropical storms. Not only are they hard to find or to specify in the tropics but even if they were present in the formative stage, they would be rapidly 'wound round' the system so that it would become virtually an occlusion. How can we explain an occlusion which not only continues to exist but actually *grows* for days together?

*Superimposition.* Hence the modern interest in theories involving super-imposition or a two-storey approach to the problem. The idea that the upper air has a part to play is by no means new. Even in 1857 Dove suggested that 'the primary cause of tropical cyclones' was to be found in sporadic outbursts of upper easterlies which 'interfere with the free passage of the upper or counter-trade-wind on its way to the tropics, and force it downwards into the lower or direct trade-wind' (Dove, 1862, 188). This 'must necessarily generate a rotary motion in a direction opposite to that of the hands of a watch' in the northern hemisphere and the disturbance 'must advance at the same rate as the upper cross-current which produces it'. This may sound a little archaic but note that 'the diminution of barometrical pressure is not the cause of the violent disturbance of the air but rather a secondary effect of it' (*ibid*, 198). Ferrel (1889), struggling with his purely convectional model, was also much concerned with the high-level anticyclone which had to evict air from the system. He thus faced but did not solve what Byers (1959, 379) calls 'unquestionably the most important . . . dynamic requirement . . . that upper air divergence must exceed low-level convergence' and that, furthermore, this must be 'maintained over a considerable period of time on a *proper scale*'.

The chief modern exponent of the two-storey approach is unquestionably Riehl (1954) who devotes nearly 150 pages to its analysis in detail. Briefly, he divides the tropical atmosphere into two main layers, one below 300 millibars or about 9 kilometres, and the other above. In the *lower* layer, dominated by a more or less steady easterly current (the trades), he has recognised the occurrence of wavelike disturbances with a wavelength of about 1000 miles. These are most marked in the central portions of this layer between 3 and 6 kilometres and they are propagated westward with a speed of about 13 knots (15 mph). At any given point, therefore, especially during the summer months, one such disturbance may be expected every 3 days. The mean speed of the trades, especially in their lower layers, is greater than the forward motion of the 'waves' so that the air enters the disturbance from the rear (the east) and passes through it towards the west. Since the waveform is damped down as one passes along its axis towards the tropical highs, the air streams must converge on the upwind, rear or eastern flank and diverge again as they leave the downwind, forward or western flank of the disturbance. This has a profound effect on the depth of the lower moist layer of the trades which may be 6000 metres to the east of the trough as compared with 1500 metres to the west when such a wave reaches the Caribbean (Fig. 117). Disturbed convectional weather is probable within this deepened band for the trade-wind inversion is weakened or disrupted. The wave pattern thus involves the development of a poleward extension of the equatorial trough of low pressure, and within this extension a cyclonic circulation may be generated, especially at the middle levels where the wave development is most marked. It will be noticed that the linear pattern of disturbed weather thus generated is very similar to that predicated by Garbell's 'Temperate Fronts', but that the new model is much more elegant and more in accord with the observed facts. Nevertheless, all that is involved so far is no more than a short spell of disturbed weather, by no means uncommon towards the western and equatorward limits of the trades.

In the *upper* layer of the troposphere, between the 300- and 100-millibar levels, Riehl recognises an entirely different system of circulation. Although observations are fragmentary, he considers that the normal summer pattern involves the westward drift of a series of anticyclonic and cyclonic vortices which lie between the westerly circulation round the poles and the high-level easterlies near the equator (Fig. 53). Consecutive upper-level highs (or lows) in this series are often about 3000 miles apart and since they move westward at about 15 knots (17 mph) one can be expected overhead about once a week. Normally these upper air circulations have but a minor effect upon the underlying trades but when they bring air equatorwards over a low-level trough in which a cyclonic circulation has already been generated, the flow pattern is ripe for further deepening since convergence in the lower storey can be compensated for by divergence aloft. Furthermore the westward-moving train is occasionally interrupted by the equatorward swing of a 'long wave in the westerlies' during low index conditions. Since these waves move *eastwards* they first lower and then raise the upper air pressure pattern over the incipient cyclone, thus first deepening it and then providing for the evacuation of the rising air as the system springs into activity. Riehl (1954) does not make very clear which of these two processes he considers to be most productive of hurricanes but adds that though the situations thus described are 'very frequent . . . yet the number of storms is small'. He concludes (p. 334) that 'this suggests that the starting mechanism often misfires because other factors necessary for hurricane formation are not all present'.

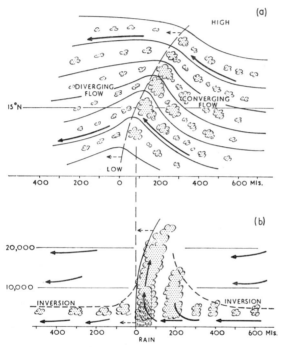

FIG. 117. An 'easterly wave' in the trades. (a) A general plan of the disturbance showing areas of convergent and divergent flow. (b) A cross-section of the feature at 15°N. Cloud widths are greatly exaggerated in both diagrams and should be interpreted as cloud groups of the appropriate height (after Malkus, 1957)

Riehl's observations are confined to the northern hemisphere. It is by no means clear how far these conditions apply south of the equator but the above description has been expressed in terms applicable to each hemisphere. It will be noted that the trigger mechanism he suggests is on the necessary sub-regional scale and that the rotary motion is regarded as developing most actively at first between the 3000-metre and 6000-metre levels, thereafter extending itself both upwards and downwards. This may prove to be a key to the formation of the 'eye' which Malkus (1957) regards as 'the point of no return': 'Without it the temperature of the core cannot reach the necessary levels and the winds cannot exceed about 45 miles per hour.' But, once an 'eye' has been formed, 'the closed central vortex takes over from the easterly wave as the dominant driving force of the air in its path'.

Riehl's (1954) broad conclusion that 'tropical storms move in the direction and with the speed of the steering current, which is defined as the pressure-weighted mean flow from the surface to 300 millibars over a band 8° latitude (560 miles) in width and centred on the storm' (p. 345) is not essentially different from Ferrel's but he does realise that this motion is not entirely passive. Vortices interact upon one another in proportion to their relative masses and a really large storm may cause an anticyclone on its poleward flank to move westwards with it, 'the pair advancing . . . as a couplet' (Riehl, 1954, 347). Recurvature is therefore by no means inevitable, indeed it seems to require the approach of a deep upper trough in the westerlies.

## Hurricane damage

Hurricanes at sea today bring discomfort rather than disaster for steam vessels can alter course to avoid the 'dangerous quadrant' forward and poleward of the approaching storm. It is thus where a hurricane makes a landfall that it is most to be feared. Danger arises from three quite distinct factors—the force of the wind, the reaction of the sea and the volume of the intense precipitation.

The effects of the direct impact of the wind are the most widespread but probably the least dangerous of the three. There may be extensive damage to growing crops (particularly to bananas and sugar-canes), the disruption of communications, and much superficial damage to structures but there is usually ample warning and precautions can be taken to minimise physical injuries. Tornadolike devastation occurs only locally as a result of small eddies. As a rule winds drop to well below hurricane force before the storm has passed more than about 50 miles inland.

Though much more local in its effects, the rising sea is far more lethal, especially on low and shallow coastlands. First comes a heavy swell which, having outrun the storm, may pound the coast for hours before the wind begins to rise, modifying the beach profile and undermining beach defences. Then, with the wind, comes the 'storm surge' which may raise the sea level some 15 ft above normal bringing huge waves inland over coastal settlements and later leaving vessels high and dry. In America the Galveston disaster of September 1900 cost over 6000 lives but, in recent years, the elaboration of an early-warning system based upon aircraft patrols, radar surveillance, satellite photographs and short-wave radio, has made possible the timely evacuation of the area at risk (see Dunn and Miller, 1964). The East Pakistan disaster of 12 November 1970 was sad evidence that the most vulnerable coast in the world is unquestionably the tidal delta of the Ganges river system. The storm had been under satellite observation for a week but the precise location of the danger area is rarely known more than 24 hours ahead. Evacuation of a densely peopled archipelago of sand and mud within this limited time would appear to be virtually impossible. Yet, as with the slopes of some great volcanoes, the land is far too rich to neglect.

The danger from flash floods following intense precipitation as a hurricane fills up over the land or encounters a relief obstacle is still more local but nonetheless severe. Much depends on the stream gradients and the extent to which valuable property has been allowed to encroach upon low terraces in the valley floors. Bridges are the first casualties and in severe cases a swath may be cut through riverside towns. Flooding is particularly characteristic of hurricanes which, having encountered a mid-latitude front, are about to convert themselves into an extratropical disturbance. The most notorious of these was 'Hazel' of October 1954 which, instead of decaying soon after it crossed the coast of North Carolina, actually revived along a cold front beyond the Appalachians and ran on to Ungava bringing devastation in its wake. Canada was shocked to receive some 7 inches of rain in 48 hours over an area of some 350 square miles north-west of Toronto, but this was no more than the last flick of the hurricane's tail (Boughner, 1955).

There is some evidence that hurricanes have been swinging across the eastern coasts of America with greater frequency in recent decades. American interest in hurricane research, however, has its purely scientific side. As Bergeron (1954) says, 'the tropical hurricane is one of Nature's magnificent geophysical experiments which . . . is continually repeated before our eyes—and nowadays also within reach of our instruments'. By studying these spectacular experiments it is hoped to gain a better understanding of all weather processes. Furthermore, hurricanes and typhoons may well play a significant role in the general circulation. Walker (1970) has suggested that a connection between the hurricane season and reduced frontal activity over western Europe is well worth further investigation.

It has been argued that by heavily seeding a nascent hurricane with silver iodide smoke its cloud masses might be glaciated, the object being to reduce fall-out or to rupture its symmetry. The process would certainly involve a continuous succession of plane sorties maintained over many hours. Such an expensive experiment was announced by the American authorities in 1967 but it was called off when the selected storm changed course. The effectiveness of the idea thus remains open to question. At best it would appear to offer no more than a local palliative for, since the basic forces remain unchanged, the storm might well regenerate elsewhere.

In conclusion we may note that intense tropical storms are not an unmitigated evil. Riehl (1965, 178) has pointed out that 'many areas, for instance south-eastern Asia and the west coast of Mexico, rely on tropical storms for much of their water-supply'. Furthermore, Weaver (1968), in an analysis of the economic effects of hurricane 'Janet' on the island of Grenada in 1955, has concluded that a 'disaster' may break a traditional dependence on outdated and uneconomic farming techniques and bring in a flood of fresh ideas and new capital as a result of aid from abroad.

# 13
# Local factors in climatology

## The problem of scale

It has already become apparent in the preceding discussion that the observation and analysis of the physical attributes of the air ocean raise significant problems of scale, both in space and in time.

Dynamic meteorology is concerned primarily with the 'free atmosphere', broadly speaking the air at least 1 kilometre (*c*. 3000 ft) above the earth's surface, where it may be presumed that terrestrial influences—particularly those arising from the diurnal cycle—if not entirely absent, will be so greatly weakened that they can be broadly ignored. The term 'air mass' itself carries some implication of the scale of the reasoning employed and the assumption that transformations occur 'adiabatically', although not entirely free from important qualifications, at least yields useful approximations. Yet, until the perfection of the radiosonde, this vast volume of air remained inaccessible to direct observation except in a most perfunctory fashion. Cloud type and cloud motion gave some clue to what was going on aloft, and precipitation was an end product which could be easily recorded even when the mechanisms involved were still wrapped in mystery. One thing, and one thing only, the earthbound meteorologist could do with precision: he could weigh the total load of the superincumbent atmosphere with his barometers and hence the overwhelming role of pressure in the development of weather science. Even today when balloon ascents are made at least once daily at some hundreds of stations (mostly in the northern hemisphere) the network is still widely meshed and pressure readings, now available at a variety of levels, remain of fundamental importance. This is partly because pressure is the best key to elevation but also because pressure discontinuities are for the most part unthinkable so that interpolation between spot readings is made much more reliably than for any other of the weather elements. Balloon observations have shown that the atmosphere is variously stratified and confirmed that vast sectors of it are involved in major systems of circulation. The most fascinating problem of the free atmosphere is how are these systems interlocked? Its thinking must thus be on a global scale.

The atmosphere familiar to man, and even more particularly to birds, from time immemorial is, on the other hand, dominated by terrestrial influences, though it also lies under the sway of events in the free atmosphere aloft. We take this lower layer of the atmosphere so much for granted that there is no really good name for it. Thus it is customary to speak of 'surface wind' and even of 'surface temperature' though what is really meant is weathercock wind and screen temperature, the Stevenson screen, not far from normal eye-level, having been specifically designed to eliminate, as far as possible, most of the surface effects in which the earth abounds. Climatology is primarily concerned with the weather-station atmosphere and draws its data from no less than 2000 official stations scattered unevenly over the world. Here measurements of temperature and humidity are paramount and because the analysis is to be on a broad regional basis much care is taken to secure a representative 'exposure' for each land station. In contrast to upper air data, land based records can now be made continuous in time, but they remain spot records and no really safe generalisation about temperature discontinuities is available to fill the gaps between them. The lowly rain gauge suffers heavily from this defect, even though the network is usually much closer than that of fully equipped stations. At sea the problems of instrumental exposure are increased tenfold and the 'stations', even when they are weather ships, do not stay in the same place. Sea temperature and wind direction present no great difficulties but there is little wonder that the barometer, proof against damp and erratic motion, has long been the seaman's pride and joy.

Yet if we are interested in climate and man and, even more so, in climate and the crops upon which man so heavily relies, it is self-evident that it is not the record from some ideally exposed site that we seek. Station data may supply an overall general picture but we must enrich this by an enquiry into the individuality of various kinds of sites. From this point of view 'exposure' is no longer seen as a kind of 'noise' obstructing the reception of the true signal, rather it forms the very stuff that demands analysis. Pressure ceases to be of much concern, except perhaps on high mountains, and what we chiefly desire to learn is the detailed texture of change, particularly of temperature, humidity and air motion, both from place to place and from time to time. This is the realm of local climatology but, here too, the problems to be faced present themselves over a wide spectrum of scales. We shall concern ourselves first with *mesoclimatology*, the local circulations often exhibiting a diurnal pattern, which are found in the neighbourhood of coasts and on mountain slopes. These are often of subregional relevance and thus clearly a part of standard climatology. There remains a much more specialised field, known variously as *microclimatology*, the climate near the ground, the climate within growing crops (including forests), or, more technically, the climate within boundary layers, with its own instrumental techniques and a growing body of theory. We must content ourselves here with no more than a summary of some of its conclusions. (See Barry, 1970, on the problem of scale.)

# Coastal zones

The single most obvious discontinuity which gives character to the face of the earth occurs along the 200 000 miles or more of ocean coastline where the comparatively dry and solid land surface meets that of the mobile sea. Along this geographical boundary two distinct economies of temperature and humidity are brought into close horizontal juxtaposition though the sharpness of the contrast must clearly depend upon latitude, the season, and the relief of the land on the one hand and on a complex series of interactions resulting from the general circulation, notably sea temperature, cloud amount and precipitation, on the other. The sharpest transition is to be expected along tropical desert coasts; the weakest where a frozen sea meets a snowclad land, though here the effective boundary is really shifted outwards to where fast ice meets the first extensive 'leads'. Within the continents a similar boundary follows the shores of the larger lakes and inland seas. On a microclimatic scale even rivers and pools have some effect, but these are matters of fine detail.

## Coastal drag

There is some evidence that, quite apart from the temperature and humidity budgets, even a very low coast may affect air flow via friction to a degree sufficient to produce significant climatic results. Thus Bergeron (1949) argued that a slight rainfall maximum recognised along the Dutch coast during rather stable conditions might be attributable to a check on surface air-flow rather than to a thermal effect from the coastal dunes (Fig. 118). However, it is not a simple matter to separate these two influences since the frictional drag over land surfaces results from turbulence which may be stimulated as much by thermal influences as by the presence of physical obstacles. Furthermore, since precipitation is initiated well above the earth's surface, any such effect must be a trigger action rather than the root cause of the effect observed.

Still more striking examples of the effect of coastal drag are found along trade-wind coasts where the wind direction is far less variable than it is in Holland. It is argued that, if the general set of the wind is parallel to the coastline, the difference in the frictional resistance over land and sea may be sufficient to cause local zones of convergence or divergence in the lowest 500 metres of the air stream. The consequent weakening or strengthening of the trade-wind inversion may then be reflected in rainfall amount. In the northern hemisphere this effect produces divergent flow, and hence subsidence aloft, strengthening the inversion over the land when the latter lies to the *left* of the direction of the prevailing wind. In the southern hemisphere a corresponding effect should be observed when the land is on the right. Lahey (1958)[1] has invoked this process to explain the three dry sectors of the northern

[1] See also reference in Trewartha (1961), 57–64.

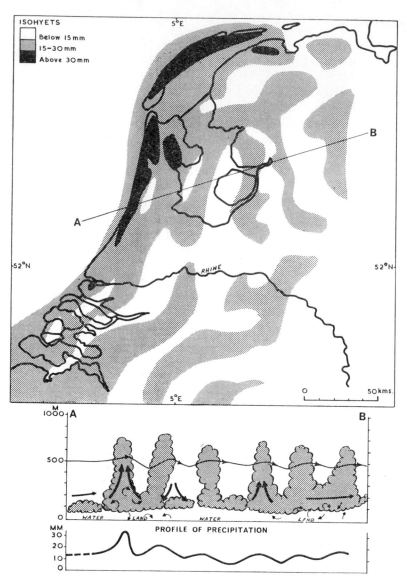

FIG. 118. Lee waves generating a belted distribution of rainfall over the Netherlands. The phenomenon was observed on 26 October 1945 (after Bergeron, 1960)

coast of Venezuela (west of each of the three towns, Cumana, La Guaira and Coro) where the shoreline runs closely parallel with and to the left flank of the ENE trade. Lahey was able to compute the extent of the divergent flow involved. It is possible that this effect is of equal importance along other trade-wind coasts and Fig. 119 indicates in diagrammatic form the nature of the relationships involved. Another possible example is the remarkable

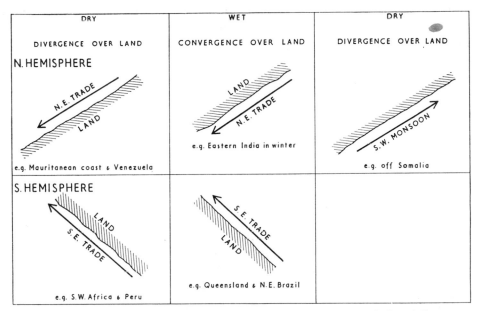

Fig. 119. A stylised interpretation of the effect of the run of the coast on induced divergence and convergence within the trades and other consistent tropical circulations

drought of the Somaliland coast on the left flank of the south-west monsoon. Clearly, however, such an effect is purely local and its importance must not be exaggerated. The overall stability or instability of the air current remains paramount and other local effects, such as land and sea breezes, must also be taken into account.

## Land and sea breezes: description

It has long been recognised that, given favourable circumstances, there is a general tendency at coastal locations for the development of a diurnal cycle of wind direction; landward winds, the 'sea breeze', predominating during the day and seaward winds, the 'land breeze', replacing it at night. Since both wind changes usually occur during daylight, that is 2 or 3 hours after the dawn and often an hour or two before sunset, this must long have appeared to the inshore fisherman, aware that the shoals move shorewards at night, as clearly the work of a beneficent providence.

The phenomenon is naturally best developed when the general wind is comparatively weak and we now know that it also requires strong insolation, i.e. fairly cloudless skies, and preferably the presence of hilly country not far from the shore. In the tropics the cycle thus persists throughout the year—it is recognised on 7–8 days out of 10 at Batavia (Djakarta)—but in temperate latitudes it is mainly a feature of the summer half-year (2–3 days out of 10 during western European summers). Almost everywhere the sea breeze is much the stronger of the two circulations though the land breeze is stimulated in the presence of coastal hills. In the trade-wind zones the general circulation is normally too strong to permit a reversal of wind direction but the effect may still be revealed by the superimposition of onshore and offshore components upon the general flow. The sea breeze may penetrate up to 50 miles inland under favourable conditions in tropical locations, bringing a welcome relief to the noonday heat, but much depends on the lie of the land. Its seaward root has been recognised at a comparable distance offshore. In temperate lands the range of the feature is much reduced though the author was surprised to find a definite diurnal variation in wind direction from the records of Ringway Airport, Manchester (Crowe, 1962, 18). When the sea breeze is fully developed, especially under tropical conditions, it may affect the lowest 1200 metres of the atmosphere though in temperate lands a depth of more than about 500 metres appears to be unusual. The land breeze is much more of a surface phenomenon.

Full comprehension of this interesting effect requires an integrated set of both landward and seaward stations all making balloon observations every hour or two both by day and night. Even when this has been attempted (see Fisher, 1960, in New England, and Moroz, 1967, on the eastern shore of Lake Michigan) it is an effort difficult to sustain over more than a few days. Our knowledge of the phenomenon is thus built up from a few studies in depth strictly relevant only to particular occasions, plus a great mass of observations of surface wind direction available from coastal stations all over the world at two, three and, occasionally, at four standard observation hours.

The classic account of the air motion over the landward sector of the circulation is still that given by van Bemmelen (1922) as the result of watching a number of free balloons (illuminated at night) over Batavia (Djakarta) in Java. He presented his results in the form of a time-section illustrated in Fig. 120. If we regard onshore and offshore components of less than two knots as within the range of observational error we must conclude that his analysis gives a convincing picture of the sea breeze by day but that the land breeze at night was scarcely recognisable. At its level of maximum development, broadly between 75 and 225 metres above sea level, the onset of the sea breeze is recorded at about 1030 hours, it strengthens rapidly, attaining speeds in excess of 12 knots from 1400 to 1700 hours, and then decays again by about 2000 hours. Nearer the ground the onset is delayed and the decay advanced by an hour or more and the maximum speed is rather less, doubtless owing to surface friction. Aloft the breeze develops in depth from about 450 metres at noon to at least 900 metres at sunset and, although weakening at all levels by that time, it persists aloft until 2100 hours or even later, having apparently been lifted bodily from the earth's surface, probably by the deepening of a stable layer near the ground as a result of night radiation.

Van Bemmelen's balloons also showed that this landward inflow was capped at levels between 1200 and 4000 metres by a complementary outflow. Over twice as deep as the sea breeze, this current attains maximum speeds (at the 1880–2200-metre level) barely half as great as those recorded nearer the ground. Although the evidence is rather inconclusive, there is some suggestion that this upper return current lags an hour or so behind the surface flow in onset, maximum development and eventual decay. The diurnal disturbance would appear to fade out completely above 4000 metres (13000 feet), a remarkable elevation nevertheless for the penetration of even the indirect effects of surface heating.

As we have already noted, the night circulation has virtually to be read into the diagram. Nevertheless van Bemmelen regarded it as a real current

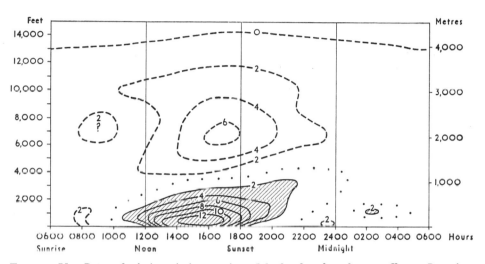

Fig. 120. Van Bemmelen's (1922) time section of the land and sea breeze effect at Batavia (Djakarta), Java. On-shore and off-shore components are shown by isopleths in knots and the on-shore sea breeze above 2 knots is emphasised by shading. Note the contemporaneous off-shore flow aloft. It is much deeper but of lower velocity than the sea breeze

though less than 300 metres deep and attaining 2 knots only shortly before midnight and again an hour or two after sunrise. The latter anomaly he explained as the arrival of a wave of cold air from the uplands some thirty miles to the south.

Observations of surface wind direction at coastal stations, particularly when these have been averaged over a considerable period of time, provide a much less vivid picture of the land and sea breeze effect, of less value perhaps to the meteorologist who is looking for causes and processes but of real importance to the descriptive climatologist. Under these circumstances the phenomenon has to make itself apparent in the face of the vagaries of the general circulation and within limits rigidly prescribed by local features of site exposure. Figures 121, 122 and 123 present a series of wind-roses for a few selected stations ranging from Finland to southern Africa and Table 47 presents winter and summer data for three British stations.

Helsinki and Tallinn provide an interesting first example because they lie, at a distance apart of some 50 miles, on opposite sides of the Gulf of Finland under a favourite track of mid-latitude depressions which should bring winds from all points of the compass. A land-and-sea breeze effect here is thus operating under a major handicap yet, in the summer months (May to August) there can be no doubt about its reality despite the latitude of approximately 60° N. At *Helsinki* winds from the south-west, south and south-east, the sea breeze directions, increase in frequency from 37.5 to 69 per cent of occasions between 0700 and 1500 hours whilst at *Tallinn* where the sea breeze direction is from north-west, north and north-east, a comparable increase from 32 to 56

TABLE 47. *Diurnal variation in wind direction in eastern and southern Britain. Percentage of all winds from various directions in winter (November, December, January, February) and summer (May, June, July, August) (10-year means)*

| LOCATION | N | NE | E | SE | S | SW | W | NW | CALMS |
|---|---|---|---|---|---|---|---|---|---|
| **ABERDEEN** | | | | | | | | | |
| Winter | | | | | | | | | |
| 0700 | 3 | 1 | 3 | 10 | 18.5 | 22 | 15.5 | 21 | 6 |
| 1300 | 4 | 1.5 | 5 | 10 | 22.5 | 18.5 | 18 | 16.5 | 4 |
| Summer | | | | | | | | | |
| 0700 | 9 | 5 | 4 | 8.5 | 17 | 14 | 8.5 | 25 | 9 |
| 1300 | 9.5 | 9 | **13.5** | **19** | 23.5 | 5.5 | 5 | 14.5 | 0.5 |
| **GREAT YARMOUTH** | | | | | | | | | |
| Winter | | | | | | | | | |
| 0700 | 5 | 7 | 9 | 7 | 15 | 22 | 21 | 11 | 3 |
| 1300 | 6 | 8.5 | 8 | 7.5 | 16.5 | 19.5 | 18 | 13 | 3 |
| Summer | | | | | | | | | |
| 0700 | 11 | 9 | 5 | 6 | 11 | 18 | 21 | 13 | 6 |
| 1300 | 16 | **14.5** | 7 | **15** | 12.5 | 12.5 | 11.5 | 8 | 3 |
| **CALSHOT** | | | | | | | | | |
| Winter | | | | | | | | | |
| 0700 | 13.5 | 11 | 6.5 | 6.5 | 12.5 | 17 | 16 | 15 | 2 |
| 1300 | 12.5 | 12 | 8.5 | 8 | 11 | 18.5 | 14 | 12 | 3.5 |
| Summer | | | | | | | | | |
| 0700 | 16.5 | 8.5 | 9.5 | 4 | 7 | 18.5 | 17 | 16 | 3 |
| 1300 | 8 | 4 | 10 | **12** | **13** | **34** | 8 | 9 | 2 |

per cent is to be noted between 0700 and 1300 hours. It is unfortunate that the observation hours are not identical at the two stations and that there is apparently some local interference with northerly winds at Tallinn, but the evidence is clear enough, the two sets of roses being virtually mirror images of each other. From November to February (not plotted) on the other hand, the corresponding frequencies are 48–49.5 per cent at Helsinki and 21.5–24 per cent at Tallinn, differences which are well within the range of chance over a 10-year period. At both stations throughout the year calms are less frequent in the afternoon than at the morning hour of observation.

In Britain, with its more maritime summers and more frequently disturbed weather situations, the effect upon mean wind data might be expected to be weaker yet, at Aberdeen, winds from the north-east, east and south-east increase from 17.5 to 41.5 per cent of occasions from 0700 to 1300 hours in summer whilst at Calshot the increase of south-east, south and south-west winds is from 29.5 to 59 per cent over the same time interval. Again, this feature is completely absent from the winter data. Baxendell (1935) has recognised a similar effect upon the west coast at Southport in a careful survey

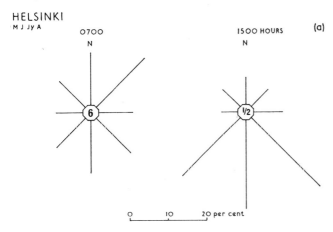

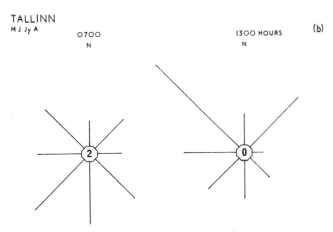

FIG. 121. The diurnal variation of the surface wind on the opposite shores of the Gulf of Finland in summer

covering 35 years and the present author has suggested that a land-sea circulation provides a useful ventilation to the whole of Lancashire (Crowe, 1962, 17–46). At Aberdeen and in Lancashire the land and sea breezes are probably supplemented by a hill-and-valley circulation to be discussed later.

At Oporto, Fig. 122, the coastal relief is much more impressive and a strong diurnal variation of wind direction is reported at all seasons. The winter roses certainly suggest that here the down-valley wind is the predominant partner although a morning hour of observation as late as 0900 hours might well be thought not to show it at its best. In summer, on the other hand, both the frequency and strength of the afternoon sea winds—north-west and west winds increasing in frequency from 22.5 per cent at 0900 hours to 75.5 per cent at 1500 hours—suggest the active cooperation of sea breeze and up-valley effects.

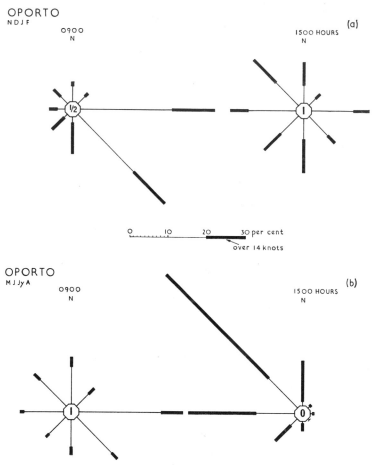

FIG. 122. The diurnal variation of wind direction at Oporto

Under fully tropical conditions where day and night remain of approximately equal length throughout the year and where daytime insolation is often intense, land and sea breezes become a regular and by no means unimportant part of coastal climates. Here too the effect observed depends upon local relief as well as upon the relation between the run of the coast and the prevailing trade-wind direction. Table 48 gives data for three different types of exposure in the north-eastern trades of the Atlantic Ocean as well as annual means for two stations in the Gulf of Guinea where the general circulation is weak throughout the year. *Funchal* (32°38′N, 16°54′W) lies on the southern coast of a lofty island in virtually complete shelter from the trades, the sea breeze thus comes from a direction diametrically opposed to the trades and

TABLE 48. *Diurnal variation of wind direction* (a) *on the Old World flank of the north-east trades and* (b) *at two equatorial stations. Percentage of all winds from various directions in winter* (*November, December, January, February*) *and summer* (*May, June, July, August*)

| LOCATION | N | NE | E | SE | S | SW | W | NW | CALMS |
|---|---|---|---|---|---|---|---|---|---|
| (a) *Trade-wind Stations* (*10 year means*) | | | | | | | | | |
| FUNCHAL (Madeira) | | | | | | | | | |
| Winter | | | | | | | | | |
| 0900 | 13 | 7 | 7.5 | 10 | 6 | 8 | 7 | 5.5 | 36 |
| 1500 | 1 | 0.5 | 2 | 14.5 | 30.5 | 25.5 | 3.5 | 0.5 | 22 |
| 2100 | 34 | 9 | 6 | 5 | 2 | 5 | 7 | 9.5 | 22.5 |
| Summer | | | | | | | | | |
| 0900 | 4.5 | 5.5 | 9.5 | 20 | 18 | 19.5 | 11 | 2.5 | 9.5 |
| 1500 | 1 | 0.5 | 1.5 | 10 | 39 | 37.5 | 4.5 | 0.5 | 5.5 |
| 2100 | 16 | 3 | 1 | 2.5 | 5.5 | 9 | 4.5 | 3 | 55.5 |
| SANTA CRUZ DE TENERIFE | | | | | | | | | |
| Winter | | | | | | | | | |
| 0600 | 19 | 8.5 | 7.5 | 1 | 1 | 2 | 29 | 21 | 11 |
| 1200 | 14 | 21 | 26 | 8 | 9 | 4 | 2 | 15 | 1 |
| Summer | | | | | | | | | |
| 0600 | 15 | 12.5 | 1.5 | 0.5 | 2.5 | 4.5 | 12.5 | 25.5 | 25.5 |
| 1200 | 13.5 | 26 | 28 | 4.5 | 6.5 | 4 | 3 | 12.5 | 2 |
| PORT ETIENNE | | | | | | | | | |
| Winter | | | | | | | | | |
| 0700 | 42 | 24.5 | 20.5 | 1 | 0 | 0 | 1 | 6 | 5 |
| 1700 | 34 | 13 | 9.5 | 2 | 1.5 | 1 | 10.5 | 21 | 7.5 |
| Summer | | | | | | | | | |
| 0700 | 55.5 | 23.5 | 4 | 0 | 0.5 | 0.5 | 2.5 | 10.5 | 3 |
| 1700 | 31 | 10.5 | 0.5 | 0 | 0.5 | 1.5 | 14 | 40.5 | 1.5 |
| (b) *Equatorial stations* (*means for whole year*) | | | | | | | | | |
| ACCRA | | | | | | | | | |
| 0700 | 1 | 1 | 0 | 0 | 1 | 7 | 55 | 14 | 21 |
| 1400 | 0.5 | 0.5 | 0.5 | 4.5 | 62 | 29.5 | 1.5 | 0.5 | 0.5 |
| DOUALA | | | | | | | | | |
| 0700 | 4 | 11 | 20 | 6 | 5 | 4 | 6 | 2.5 | 41.5 |
| 1400 | 1 | 1 | 1 | 0.5 | 3 | 40 | 43.5 | 3.5 | 6.5 |

there is strong evidence, particularly in winter, of a mountain wind effect at 2100 hours. *Santa Cruz de Tenerife* (28°28′N, 16°16′W), a popular oil-bunkering port, has a south-eastern exposure so that the trades are almost parallel to the lofty and rocky coast. The effect of the sea breeze is to divert the wind towards a more easterly direction during the day whilst the land wind blows from the

hills to the north-west. At *Port Etienne* (20°56′N, 17°93′W) on the comparatively open coast of the Sahara, the dominance of the trades is never lost
though they swing from west of north during the day to east of north during

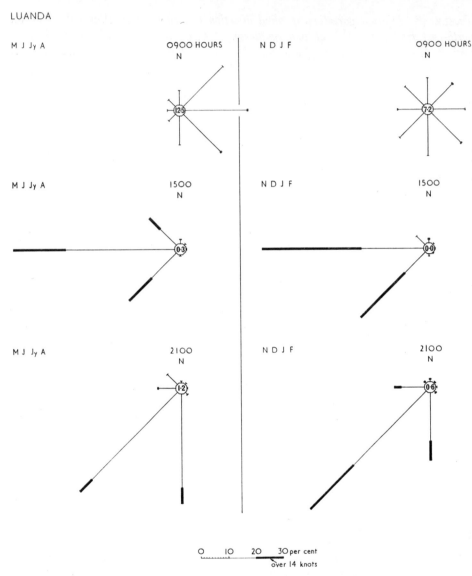

LUANDA

Fig. 123. The diurnal variation of wind direction at Luanda, Angola

the night. At all these stations the diurnal sequence is recognisable throughout
the year, but is better developed in summer than in winter.

At *Accra* (5°33′N, 0°12′W) (M.O. 492, 1949, 156 and 204) and particularly
at *Douala* (4°03′N, 9°41′E) (*ibid*, 162) there is so little seasonal variation in

the incidence of land and sea breezes that annual data will suffice. At these two almost equatorial stations this local circulation becomes, in fact, the most significant fact in surface air motion.

We have plotted the data for *Luanda* (8°49′S, 13°13′E) (*ibid*, 167, 207–8), a port with a north-western exposure on the coast of Angola, because it illustrates several features of land and sea breezes with remarkable clarity (Fig. 123). The station lies on the eastern flank of the south-east trades but these show little development within a hundred miles of the African shore in this latitude and thus do little to complicate the picture. Despite the low latitude some slight seasonal variation is apparent. Thus in the southern winter (May to August) a weak land wind is still in evidence at 0900 hours though in summer at the same time the wind-rose has become almost symmetrical. By 1500 hours the main seasonal contrast is in the greater strength of the summer sea breeze and this remains true of the 2100 hour record. The main feature of the transition from midday to evening, however, is the general swing of the sea breeze from west-south-west to south-south-west so that by 2100 hours it is blowing almost parallel to the general trend of the African coast. Theorists have long maintained that as the sea breeze develops, it should swing progressively so as to bring the presumed low pressure over the land on to its left flank in the northern hemisphere and to its right flank in the southern hemisphere in accordance with Buys Ballot's Law. It is rare to find this tendency so clearly illustrated by mean data.

## Land and sea breezes: the processes at work

It is an observed fact that, under comparable conditions of insolation, the diurnal variation of screen air temperature is greater over the land than over the sea. Since land and sea breezes follow a diurnal cycle it was very natural to associate them with this fact. Hence the initial popularity of a simple

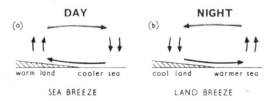

FIG. 124. The traditional view of land and sea breezes

convectional model, often illustrated as in Fig. 124 as a virtually closed circulation, reversing its direction at night. If this were even a first approximation to reality then, given clear skies and a weak general circulation, two consequences should follow: (1) the phenomenon should occur with almost clockwork regularity, and (2) since, by definition, the night minimum over land must fall as far below the general mean over land and sea as the day maximum rises above it, there would seem to be no reason why the night circulation (*b*) should not be the mirror image of the day circulation (*a*), if not precisely in strength and depth, at least in the general volume of air transferred. Although detailed night observation seems rarely to have been attempted (the difficulties are obvious), there is little indication that this is at all in close accord with observation. Even in Britain, Bilham (1938, 69) reminds us of 'stifling summer afternoons when the sea breeze has been conspicuously absent' whilst van Bemmelen's careful work has suggested that land and sea breezes are of quite different orders of magnitude. Fisher's (1960) work on the New England coast (see Fig. 125) leaves the night situation to be largely

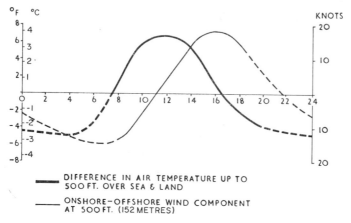

DIFFERENCE IN AIR TEMPERATURE UP TO
500 FT. OVER SEA & LAND

ONSHORE–OFFSHORE WIND COMPONENT
AT 500 FT. (152 METRES)

Fɪɢ. 125. The diurnal cycle of temperature and wind direction on the south coast of New England as observed by Fisher (1960). The data refer to 5 August 1959 when the general air motion was gently southward, i.e. seaward. Full lines cover the interval actually observed and broken lines indicate the inferred course of the curves during the rest of the 24-hour period. The phase lag was probably due in part to the general north wind but some degree of lag is to be anticipated

inferred and, furthermore, he deliberately chose situations with a general northerly, i.e. seaward, air stream. His data for 5 August 1959, however, remind us of a further difficulty, by no means confined to that area, namely the important phase-lag between the temperature and wind direction cycles.

Indeed, far from being an elementary matter, the generation of local land and sea breezes raises many of the problems encountered in the study of the general circulation. It too has proved a stimulating field for the exercise of mathematical ingenuity and a fully acceptable theory still evades us. It will suffice here to mention in a qualitative sense a few of the factors concerned.

1. The different response of land and sea surfaces to incoming solar radiation depends, firstly, on the proportion immediately lost by reflection and, secondly, on the amount of the energy absorbed but promptly converted into latent heat

through evaporation. Since, with a high sun, the albedo of the sea is con-siderably less than that of most land surfaces, the scales are tipped initially in the sea's favour. Evaporative losses are notoriously difficult to estimate but much of the existing evidence suggests that losses from lush vegetation are of much the same order of magnitude as those from open water. Only when the land surface is dry, or perhaps when the sea surface is considerably disturbed, does the balance swing back clearly in favour of the land. The real key to land and sea contrasts thus lies under a third heading, the disposal of this balance. At sea, owing to ray penetration and, above all, to eddy diffusion, the daily solar income is spread through a large volume of a material with a notoriously high capacity for heat. On land it is fed into a comparatively shallow layer ranging from the topmost foot or so of the soil to a complex amalgam of leaves, twigs, branches and air perhaps 20 to 30 ft thick at the top of dense woodland. It is thus basically a difference in heat *penetration* that is responsible for the observed temperature contrasts.[1]

2. Though heat is transferred by conduction, eddy diffusion and radiation from the various terrestrial surfaces into the overlying atmosphere, it is clear that to understand even a local circulation set up by temperature contrasts we must known more than is available from surface (screen) data. Indeed, once the circulation is initiated it is so directed that it tends to reduce the temperature contrast to which, according to the convection theory, it owes its origin. It then becomes a matter of assessing *the rate at which sensible heat is fed into a moving current* rather than of registering surface temperatures at two different localities. Here a clear distinction between the day and night situations over land becomes evident. Air warmed at its base is readily stimulated into upward eddies, the heat thus penetrates the atmosphere in depth and the rise in the screen maximum is proportionately reduced. Air chilled at its base becomes stable and inert, resisting upward displacement, so that the chilled layer tends to be shallow and the chill therefore comparatively intense. As triggers of a circulation that is to be operative over distances of tens of miles these situations are in no sense equal and opposite. We have already presented ample evidence that the sea breeze is by far the most ready response to land and sea contrasts; it is triggered off as soon as an isentropic layer of critical depth is built up over the land. The land breeze is a relative sluggard and usually requires the assistance of a relief gradient to set it in motion.

3. We have already seen (p. 235) that, since air is a compressible fluid, all convective processes within it must operate within an initial framework

[1] The above discussion ignores the loss of insolation from cloud tops which would become significant if an inequality of cloud cover over land and sea, latitude for latitude, were clearly established. A long series of satellite photographs may eventually provide convincing evidence one way or the other. In passing let us note that the hoary myth that 'clouds come from the sea' can certainly not be sustained. In western Europe the sunshine records of coastal stations reinforce the visual impression of the air traveller, particularly in summer, that much cloud is a landward phenomenon. Was not New Zealand recognised by the first Maori settlers as 'the long cloud'?

supplied by the vertical stratification of the atmosphere at the appropriate time and place and the relation of that stratification to the adiabatic gas laws for rising or subsiding air parcels. Where large broadly uniform surfaces are exposed to insolation, relief to growing instability is achieved by bubble or cellular overturning, i.e. by alternating cells of uplift and subsidence. We saw too (*a*) that this may be inhibited by a stable stratification aloft, (*b*) that to occur on any considerable scale it must be prefaced by the build-up of a surface isentropic layer deep enough to provide a reservoir for the rising

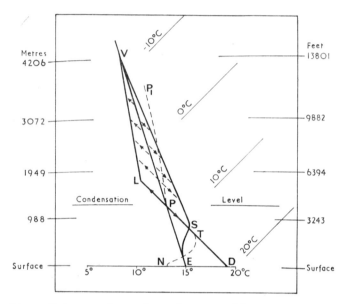

Fig. 126. An experimental exercise on a tephigram to illustrate the operation of land and sea breeze effects. See text.

bubbles, and (*c*) that, in estimating the penetrative power of the bubbles, suitable adjustments must be made both for the possibility of condensation and for the loss of lift entailed by entrainment of environmental air en route. There is no reason whatever for ignoring any of these points in relation to land and sea breezes. The only real difference is that, in the near presence of a contrasting underlying surface, the overturning is rather differently organised. Away from the coastal zone in both directions the normal bubble process must continue to hold sway, at least in the absence of major weather systems with which we are not at present concerned.

Figure 126 is an attempt to illustrate in an elementary fashion the general nature of the changes in vertical stratification induced by the sea breeze effect. We start with an environment curve *EV* which is to be regarded as representing the atmosphere over both land and sea before any surface temperature differential has developed. It is in fact the curve for the 'standard atmosphere' with a surface temperature of 15°C (59°F) and a lapse rate of 0.65°C/100 m and is thus rather stable, potential temperature increasing steadily with height. It is convenient to follow the events which ensue stage by stage though we must not forget that we are in fact dealing with a continuously evolving process.

*Daytime.* During the day insolation raises surface temperatures over the land from $E$ towards $D$ and the local overturning this produces gradually builds up an isentropic layer, i.e. a layer with a profile which corresponds with the DALR which we illustrate by $DP$. Note that this is a joint product of both local uplift and local subsidence within the air stratum between the surface and the level of $P$. Since the environment is a stable one the layers above $P$ remain immune from this turnover and, at least as a first approximation, the curve $DPV$ may be taken as representing the new environment established over the land whilst, at sea, where diurnal heating is almost negligible, the stratification illustrated by $EPV$ remains little changed. An air mass contrast has thus been generated across a very narrow transition zone following the outline of the coast.

As long as the layer between the surface and $P$ remains shallow there is some evidence that the inherent inertia of the atmosphere may restrain the development of a circulation in response to this contrast. As it deepens a whole complex train of events follows. The expansion of the heated layer $DP$ takes place simultaneously in both a vertical and a horizontal direction. The vertical expansion raises all the superincumbent layers (or pressure surfaces) above $P$ to a point where a slight upper-air outflow from land to sea is initiated. The horizontal expansion at the same time reduces the load per unit area imposed by the layer below $P$, i.e. in the densest layers of the atmosphere. Both of these effects combine to lower surface pressure over the land whilst the transference of air aloft raises surface pressure over the sea, the most readily available destination for the air removed. A sea-to-land circulation of surface air is thus initiated, the cooler sea air undercutting the landward air and lifting it bodily throughout a zone whose width will depend upon the distance achieved by sea air penetration. Since the volume of sea air involved has been extracted from the lower layers over the sea it is evident that uplift over the land will involve subsidence over the sea. Although the area over which subsidence occurs is not rigidly defined there appears to be considerable evidence that, in the land and sea breeze effect, the areas of uplift and subsidence are broadly of the same order of magnitude.

It is evident that the landward invasion of sea air must tend to eliminate the pressure differential which gave rise to it. To persist, if only for a few hours, the circulation must be *maintained*. To illustrate this as simply as possible we have assumed that, at least during the heat of the day, the invading sea air is warmed sufficiently by contact with the land to maintain the lower part of the landward curve at $DP$. It is necessary now to consider events above $P$. If both uplift and subsidence occur at the dry adiabatic lapse rate, points on the landward curve above $P$ will cool and points on the seaward curve will be warmed in the direction illustrated by the small arrowheads on Fig. 126. To make this clearer we have indicated the situation when the base of the land column has been lifted some 50 millibars, the uplift decreasing to zero at point $V$ (about 4200 m) beyond which it is assumed that the atmosphere remains undisturbed by the events below. The new landward curve is thus of the form $DLV$ whilst the seaward curve approaches the form $ESV$ where $S$ is a point, rather vaguely defined, which depends upon the depth of the surface flow. Now, to the extent to which surface pressure is affected by the mean temperature of the overlying air column, these dry-adiabatic changes will tend to exert a braking effect on the sea-to-land circulation which would be absent in an incompressible medium. Yet, in the middle layers, the density contrasts at corresponding levels in $DLV$ and $ESV$ will tend to accelerate the return flow from land to sea. The mass transfer involved will thus revive the flagging surface-pressure differential and the sea breeze will continue under a stimulus now derived mainly from aloft. Quantitative estimates of these conflicting forces involves mathematics beyond our scope, but it does appear that the direction of events has now shifted to the middle layers where mass-exchange takes time. A phase-lag of 2 or 3 hours between the sea breeze and the land-to-sea temperature contrast then becomes much more intelligible.

Now if a condensation level at about $P$ is introduced the curve for the landward environment is profoundly modified. Certainly the area $PLV$ disappears and a curve to the right of, but not far from, $PV$ is probable. Much will depend on the effect on the humidity of the layers between $P$ and $V$ of events during the previous days and on the extent to which mixing takes place during uplift. To keep the diagram simple we show only the limiting case, the saturation adiabat $PP_1$ appropriate for rising air with a condensation level at $P$. The landward curve thus becomes approximately $DPP_1$, a column that now contains cloud. Yet over the sea the introduction of the humidity factor will have no such profound effect. It is still safe to presume that subsidence will occur broadly at the DALR, modified only to the degree to which scattered clouds may be consumed in the process; the seaward curve $ESV$ is thus still valid. While the braking process has thus been reduced, so also has the temperature differential (and hence the pressure gradient) in the levels above $S$ which supply the return current. The role of the middle layers is thus substantially reduced and, if the clouds intercept much of the insolation over the land, the exchange may come to a premature conclusion.

*Night-time.* As overland surface temperatures fall with the approach of evening from $D$ towards $E$ and beyond it, the results which follow may be usefully considered in two stages. At first, night radiation chills only the superficial layers of the atmosphere so that the land curve below $P$ takes on a form represented diagrammatically by $NTP$. The thermal impetus to the sea breeze is thus (*a*) reduced and (*b*) lifted above the surface, but it does not entirely disappear until about the time when point $T$ finds itself on the curve $ESV$. The flagging sea breeze is thus lifted from the surface as van Bemmelen has shown.

Secondly, as the uplift weakens, the upper air contrast between the landward and seaward environment curves can no longer be *maintained* against the seaward circulation which it has stimulated aloft so that, above *P*, the two curves begin to approach one another once again. That this process may take time seems entirely reasonable in view of the distance involved. Whilst this transfer of atmospheric load is being effected in the upper air the landward air flow should continue in the lower layers until equilibrium is re-established

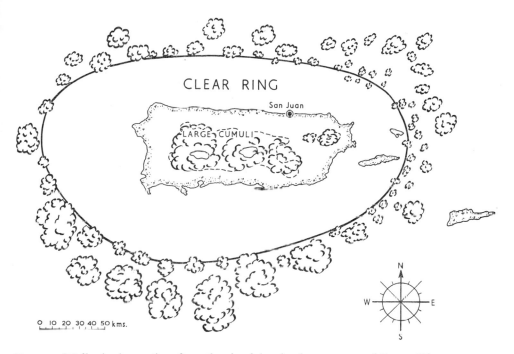

FIG. 127. Malkus's observations from the air of the cloud pattern round Puerto Rico at about noon. The flight was made on 25 June 1952. The sea breeze has helped to generate towering cumulus clouds over the lofty centre of the island but a corresponding zone of subsidence over the sea has consumed the normal trade-wind cumuli throughout a wide ring in all directions

at all levels. A land-to-sea breeze at night may well contribute to this end but its cooperation scarcely seems essential. Only a shallow layer of air is normally involved and the repercussions aloft are likely to be so slight that they could scarcely be represented upon our diagram. At first, particularly when cloud formation has been presumed to take place, the new curve at which equilibrium is reached is unlikely to coincide precisely with *EPV*, where the process began. A return to the *status quo* demands (*a*) that there should have been no precipitation, (*b*) that the daily cloud output should be consumed by nocturnal subsidence, and (*c*) that the heat transferred into the upper air should be lost to space by outgoing radiation.

This exercise will have served its purpose if it reminds us of the two-sided nature of the convectional process. Far too often in the literature we are presented with only the updraught side of the balance, thus perpetuating the myth that 'warm air rises' and leaving the descent side of the balance as vague as the song about an arrow which fell, 'I know not where'! All the evidence suggests that the sea breeze circulation is a relatively closed one, it occurs in a long and narrow ribbon astride the coast, there being insufficient time for the vast pools of continental and oceanic air on either flank to have much impact upon the proceedings. This view finds confirmation in the remarkable observations on cloud patterns made by Malkus (1955) round the island of

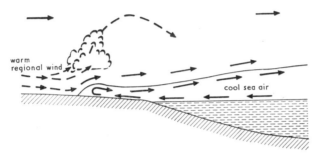

FIG. 128. A sea-breeze 'front' (after Estoque, 1961)

Puerto Rico. Figure 127, derived from this paper, shows that, even in the presence of a general trade-wind air stream, the updraught over the mountainous core of the island was able to produce a ring of subsidence extending some 25 to 45 miles (40–70 km) from the coast, and sufficiently active to suppress completely the normal considerable cloud pattern of the western trades.

Even though it is reasonable to expect that the sea breeze should develop initially in close proximity to the coast, its advance inland is often neither immediate nor regular. Time is required for the build-up of the critical isentropic layer and the breeze may have to overcome the resistance of a regional wind from the land. If this exceeds about 15 knots the sea breeze may be suppressed altogether. Against weaker resistance the sea breeze may emerge triumphant, though delayed. Often it then advances with some violence and the effect has been compared to a miniature cold front (Fig. 128). Given favourable circumstances this has been known to trigger off a line of thunderstorms in southern England, a few miles inland from the Channel coast and even when the updraught is weaker it has been observed to give a useful lift to glider pilots (J. E. Simpson, 1964, 1967). There is evidence also for more or less periodic surges in the strength of the breeze, a fact which will cause no surprise if the resistances encountered are kept in mind.

Leopold (1949, 320) has shown that the interaction between the sea breeze and the trade wind in the Hawaiian Islands depends upon the physical form of different islands in the group. Figure 129 gives a schematic representation of his findings.

From theoretical considerations it can be argued that as soon as the sea

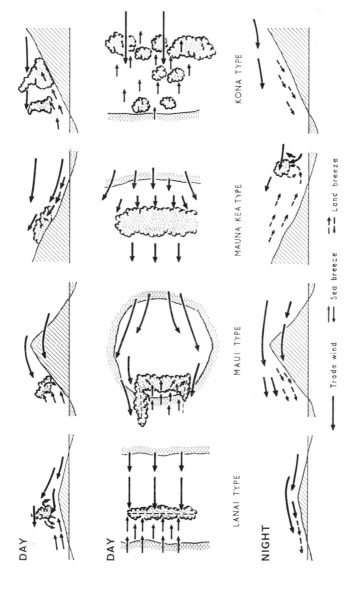

Fig. 129. The interaction of the sea and land breezes with the trades on various islands of the Hawaiian group (after Leopold, 1949)

breeze circulation develops in fetch, its direction should show a progressive change, the wind veering in the northern hemisphere and backing in the southern hemisphere under the influence of the earth's rotation. In its simplest form this means that, as lower pressures develop over the land, the air flow changes so as to keep the low pressure more and more to its left in the northern hemisphere and to its right in the southern, in accordance with Buys Ballot's Law. This tendency is easily masked by local factors and though it should be strongest in extratropical latitudes, it must be remembered that the fetch of the

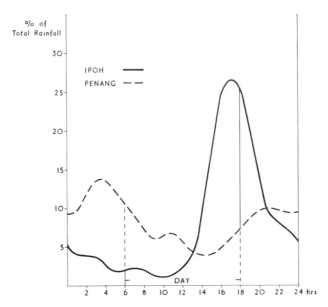

Fig. 130. The diurnal distribution of rainfall at Ipoh and Penang (Malaya). The two curves show the percentage of the mean annual total rainfall (1952–56) recorded over 2-hourly intervals of the day and night. Strictly such data should be presented by stepped histograms but continuous curves seemed preferable in a combined presentation

sea breeze is normally less there than between the tropics. Observational evidence for such a progressive change, taken from selected occasions, has been presented from Boston, Mass. (Haurwitz, 1947, 5) but it is remarkable that mean records yield a striking example at Luanda (Fig. 123) only 8°49′ S of the equator. Some scepticism that this may be attributable to the 'Coriolis force' is not out of place at this latitude.

At intertropical locations with a weak general circulation land-and-sea breeze effects can have notable results upon the diurnal distribution of rainfall particularly when the local circulation is reinforced by the presence of considerable relief. At many stations we have to be satisfied with subjective impressions but records at 2-hourly intervals are available from a number of aerodromes in Malaya. In Fig. 130 we present the proportion of the mean annual rainfall recorded at different times of the day at Ipoh and Penang airports (1952–56). The two stations are less than 80 miles apart and both recorded mean annual totals of the order of 100 to 105 inches (2500–2650 mm). The pattern at Ipoh is repeated at a number of stations on the mainland of Malaya though the afternoon maximum appears to be rather delayed, possibly because the station is in the comparatively sheltered Kinta valley. Penang is an offshore island and shows a fairly well developed night maximum associated with the land wind from the peninsula. Over the Malacca Straits very active night storms (sumatras) are thought to be stimulated by converging land breezes from the uplands of Malaya and Sumatra.

Lumb (1970) has noted a similar diurnal distribution of thunderstorm rain at Kisimu and Entebbe.

## Mountain obstacles and other relief effects

It has long been recognised that the presence of massive relief features on the surface of the continents is a matter of the highest climatological importance, but the full implications of their effect are still far from clearly understood. Where a mountain system maintains elevations of the order of 18000 feet (5500 m) over distances of many hundreds of miles, as in the Himalayas and Andes, it presents a solid barrier extending throughout the whole of the lower half of the mass of the atmosphere. Under such conditions, since any system of circulation must have both an upper and a lower limb, mass exchange between the lower halves of the atmosphere on either flank would appear to be largely inhibited. Transverse flow aloft will still be possible but it must find its compensation in the lower levels indirectly and circuitously by some modification of the surface circulation. Such mountain systems must therefore provide a permanent air mass 'wall', the barrier effect being supplemented, where the system is wide, by the generation over the heights of a local air mass (superior air) differing in character from the free air at comparable elevations on either flank. Since the general trend of the Andes is meridional whilst that of the Pamir–Himalayas is broadly zonal, both massive systems are unique features of the earth's surface and it is unlikely that useful comparisons could be drawn between them even if much more information were available.

Upland systems of lesser elevation than these two giants are much more widespread. Where they exceed 8000 feet (*c.* 2500 m) it should be noted that they still present a wall extending throughout the lowest, and most regionally variable, quarter of the total atmospheric load. A barrier effect is therefore by no means improbable where the general circulation is weak and, above all, where the system consists of broad plateaux as in western USA to Mexico, east-central Africa, Anatolia to Afghanistan, and eastern Mongolia to Siberia, each extensive enough to develop its own system of superior air. The broadly analogous Canadian Rockies and Patagonian Andes (and, on a smaller scale, the New Zealand Alps) are nevertheless regularly crossed by the jet stream with dynamic effects that are discussed below.

423

Of course, when it is a question of comparatively shallow outbreaks of very cold Arctic air, uplands of very modest elevation, e.g. those of the Crimea, may provide a high degree of protection. This serves to remind us that, even when an upland system is unable to present a formidable barrier to cross-wind circulation, there is no reason to imagine that *all* the air strata on the upwind side must necessarily share in the transit. As against a barrier effect, where an incident wind must escape laterally or upwards, it may be useful therefore to speak of a weir effect where the upper layers may be virtually skimmed off, leaving the lower layers on the windward side comparatively inert.

Owing to the complexity of most mountainous terrain and the wealth of relief-guided and eddy effects to be anticipated, at least between the peaks, observational data on cross-mountain flow tend to be fragmentary and even equivocal. Cloud forms may be most exciting but the nature and distribution of the clouds is one of the phenomena calling for explanation. This leaves the field open to a deductive approach and the problem has proved a stimulating challenge to the applied mathematician (see Scorer, 1958, 229–53). It appears that wind-tunnel experiments are usually misleading though interesting results emerged from a study of air motion around and over a model of the Rock of Gibraltar (Field, 1933 and M.O. 446b(1) 1944, 232, 234–5).

The difficulty of the deductive method arises from the number of the variables that have to be considered. Notable amongst these are: (*a*) the scale and general slope of the relief obstacle, (*b*) the strength and direction of airflow at different levels within the regional wind, and (*c*) the vertical stability conditions prevailing within the undisturbed air mass involved. Both (*b*) and (*c*) are likely to undergo considerable transformation in the presence of an obstacle and the degree of uplift which it involves whilst, although the physical characteristics of the barrier remain constant, this is by no means true either of its 'effective' height or of its effective width. Effective height may be materially increased by local updraughts along mountain slopes (see p. 439) during daily insolation and the effective width (together with the mean gradients of upward and downward motion on the upwind and downwind flanks) must clearly depend on the angle at which the obstacle is encountered. It is also sometimes materially modified by the generation within the airflow itself of eddies with a horizontal axis, usually known as 'bolsters' when they occur on the upwind side and 'lee eddies' when they occur downwind of the relief system (Fig. 131a). In most deductive models the mountain system is resolved into a 'ridge', often presumed to have a cross-section approximating a sine curve (Queney, Förchtgott, Scorer), the air stream is taken as meeting this obstacle at right-angles, and it is usually assumed that all the layers of the incident wind make an effective transit so that 'barrier' and 'weir' effects are at a minimum. At least as a first approximation it is then hoped to throw light on the degree to which the airflow is disturbed at various levels and to explain the generation of 'lee waves' which often produce interesting cloud forms and, occasionally, severe flying hazards. The results of these theoretical investigations have proved extremely stimulating but since the factors involved vary both with the time of day and with the nature of the prevailing weather-situation even R. S. Scorer, who has devoted much effort to this problem (1949–55) concedes that, in this field, theory has limited prognostic value.

When a mountain system is wide (a hundred miles or more across) and the transverse wind is both deep and persistent, as it is for instance where the

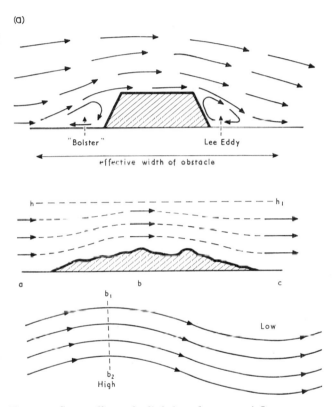

(a)

"Bolster"　　　　　Lee Eddy

ᵉffective width of obstacle

FIG. 131. Some effects of relief obstacles upon airflow.
(a) A 'bolster' and 'lee eddy' increasing the effective width
of an obstacle to air motion. (b) Section and plan to
illustrate the dynamic effect of a broad upland region upon
a strong transverse circulation

surface westerlies and the jet stream cross the Canadian Rockies and the Patagonian Andes, modern theory concludes that a significant distortion in the *direction* of the airflow is induced on a scale sufficient to affect the generation of weather patterns. The mathematical derivation of this effect is beyond the scope of this account but it carries the implicit assumption that vertical motion is damped with height so that at some level, represented on Fig. 131b by the line $hh_1$, the flow is undisturbed. It is possible from this diagram to gain at least some qualitative appreciation of what is involved. If a pressure

425

gradient between *a* and *c* is to be maintained (to support the transverse flow) the air stream must experience horizontal divergence between *a* and *b*; converging again as *c* is approached. The 'tubes' of cyclonic vorticity inherent in all major air masses on a rotating earth (see p. 164) are thus shortened during the mountain transit thus tending to induce a relatively anticyclonic curvature in the streamlines as shown in the lower part of the figure—drawn for the northern hemisphere. Considerable width in the system is required, since to produce a notable effect the deflection must be maintained over time. (A similar requirement is inherent in an alternative approach via the Coriolis parameter.) In the jet stream the most northerly point of the orographical deflection $b_1 b_2$ probably lies immediately above *b* but if the lower part of Fig. 131b is to be taken as giving a first approximation to the surface isobars there seems to be some justification for shifting $b_1 b_2$ a little upwind to remind us of yet another effect of a mountain barrier. This is the slightly increased pressure induced by the airflow on the upwind side of the obstacle, a feature which is independent of the width of the system. As we have seen (p. 215) the continued regeneration of horizontal wavelike patterns in the upper westerlies, triggered by orographic obstacles and then substantially reinforced by thermal contrasts, now plays an active part in meteorological theory, particularly as it affects the northern hemisphere. This is one of the factors at work in the generation of 'lee depressions' to be mentioned later.

Turning now to the apparently much more straightforward question of the nature of the *vertical* displacement of air crossing a hill or mountain system, it would seem a matter of simple common sense that updraught must prevail over windward slopes, downdraught over leeward slopes, and that the layers suffering maximum vertical displacement would be those nearest the surface. A very pertinent question—to what height will the disturbance extend?— meets with no such intuitive response. This may serve as a warning that there is more to this problem than meets the untrained eye.

Where the wind-speed is slight and slopes are gentle the intuitive picture is probably not far from the truth but with stronger winds and steeper slopes a number of dynamic reactions are set in train with quite remarkable results. On the theoretical plane Queney's (1949) calculations based on a dry air stream crossing a single ridge of unlimited length gave a wavelike distortion of flow extending to an unlimited height, crests and troughs succeeding one another in the vertical as elevation increased. Details of form were shown to vary with the width of the ridge (see his diagrams reproduced in Godske *et al.* (1957, 408–11)) but all the models show a 'tilt of wave troughs and wave crests against the wind' and a selective combination of descent with strong horizontal motion and of ascent with weak horizontal motion. Much more convincing are the firsthand observations of Manley (1945) in the neighbourhood of Crossfell Edge during strong easterly winds though the 'Helm wind' he describes is by no means of frequent occurrence.

Figure 132, based upon his account, shows a regional wind of 25 to 30 knots accelerating to about 40 knots as it reaches the crest of the escarpment and *maintaining that velocity* all the way down the leeward (south-western) slope. Events downwind are even more anomalous for, after traversing a gusty zone, one may reach a strip of the Eden Valley floor where a virtual calm prevails (occasionally even a light wind from the south-west) and the north-easterly airflow is not encountered again until some miles from the scarp foot. Evidence that the flow is broadly as indicated in the diagram is found, (*a*) in the fact that the roar of the wind aloft can often be heard from within the calm zone, (*b*) in an observed lift of a glider at 25 ft/sec (15 knots) within the upward current, and above all, (*c*) in the presence of a comparatively stationary roll

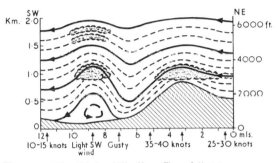

Fig. 132. The 'Helm Wind' at Crossfell Edge, Cumberland and Westmorland (after Manley, 1945)

of cloud—the 'Helm bar'—above the calm zone and at approximately the same height as the 'Helm cloud' which usually caps the escarpment itself. On some occasions the 'bar' is surmounted by lenticular clouds showing that the disturbance of the flow is of very considerable depth. A similar circulation on an even more magnificent scale is sometimes observed when strong westerly winds reach the great eastern escarpment of the Sierra Nevada in California (see Willett, 1959, 304). Desert dust has been observed to rise almost vertically in the ascending limb to elevations of the order of 15000 feet. In both of these cases it is evident that, from the glider-pilot's point of view, the best lift is to be found 2 or 3 miles to the *leeward* of the mountain obstacle rather than on its windward flank, a conclusion which it would be hard to reach from any in-tuitive approach.

Scorer's (1949-55) analysis has been devoted to cases where multiple 'bars' (a train of lee waves) are produced downwind from an obstacle. He shows 'that a stable layer in which the wind increases upwards is the circum-stance which most commonly makes lee waves possible'. In these circum-stances any initial upward or downward acceleration finds itself subjected to a restoring force generating a progression of vertical waves related in scale to the width of the obstacle and moving upcurrent often at the same speed as the wind. They are thus 'stable' waves more or less stationary in space and often produce lenticular clouds through which the wind blows without disturbing their position. In Fig. 133 we reproduce the essential features of his most realistic example (Scorer, 1953, 79). A long ridge about 3250 feet (1 km) high and some 6 to 12 miles (10–20 km) in width (a geographer might well feel

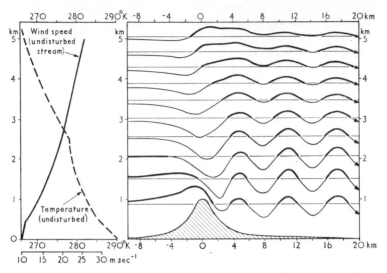

FIG. 133. Scorer's interpretation of the development of lee waves under specific conditions. The diagram expresses the result of a mathematical investigation but appears to be consistent with observation

inclined to measure this rather differently from a mathematician) is presumed. The cross-wind before it reaches the obstacle is given by the curve to the left, it is nearly 20 knots at the surface, 40 knots at 2500 metres and not far short of 60 knots at 5000 metres. The environment curve of the undisturbed current is also shown on the left and its general characteristics are tabulated below:

| | | Temperature | | Lapse rate per 100 m (°C) |
|---|---|---|---|---|
| Level (m) | | °K | °C | |
| | 5000 | 267 | −6 | 0.44 |
| | 2500 | 278 | 5 | 0.35 |
| | 500 | 285 | 12 | 1.00 |
| SFC | 0 | 290 | 17 | |

The lowest 500 metres is thus isentropic but above that layer the lapse rate is a good deal more stable than that presumed for the standard atmosphere.

For the rest the diagram may be left to speak for itself. Note that the streamlines are drawn more heavily where the air exceeds its initial elevation, thus showing positions where cloud *may* form when humidity conditions are appropriate. Note also the sharp *descent* of air in the immediate lee of the obstacle, a feature which recalls the Helm wind described above. Downstream this is succeeded by a train of stationary waves with a wavelength of some 4 miles and an amplitude of over 450 metres in the lower layers. The waves are still recognisable at an elevation four times that of the mountain chain and the total depth of the streamline disturbance may well be six or seven times the height of the relief obstacle. It must be recalled, however, that despite its intrinsic interest the figure represents a model situation only; all the implied transformations are dry-adiabatic and all frictional effects have been ignored.

The *forced descent* of air imposed by dyamical considerations such as have been described above can also occur within mountain systems of much more complex form, particularly where the flow is confined between the steep walls of a transverse valley which penetrates deep into a broad massif occasionally opposed to strong transverse currents. This is the origin of the foehn wind, named in the Alps but occurring with much the same features in many mountainous areas from the Canadian Rockies to New Zealand. Indeed, in the eastern Caucasus, Borisov (1965, 219) asserts that 'during anticyclones foehns occur simultaneously on both sides of the range', an observation which, if confirmed, would seem to throw most of the classical explanations of this phenomenon into the discard.

The foehn has attracted the attention of mountain peoples because it presents the curious anomaly of a warm dry wind apparently blowing down from cold and often snowclad passes. It is sudden in onset (often breaking a spell of cold calm weather), gusty in character, and often descends to the valley floor by fairly clearly marked stages or 'jumps'. Frequently it is prefaced by the development over the windward peaks of a flat-topped and forbidding cloud mass—the foehn wall—which may overlap the watershed but advances no farther downwind. Its flat top suggests marked stability aloft and its forward edge must surely be moulded by evaporation stimulated by widespread descent. The Helm cloud of Crossfell Edge would thus appear to be a foehn wall in miniature. Dynamic descent on an Alpine scale, however, produces much more impressive effects than can ever be experienced in Britain. Particularly in late winter and early spring, the warm dry air of the foehn may remove a heavy snow cover with almost miraculous speed.[1] The same is true of the 'chinook' or snow-eater to the east of the Canadian Rockies[2] which revived the winter-starved buffalo of the northern High Plains (Fig. 134). The 'Zonda' of west Argentina is similar in nature.

The classical studies of the foehn were made in the Reuss Valley between the St Gotthard Pass (2100 m) and Lake Lucerne (440 m) and the author vividly recollects the surprise of a field group of British students arriving on an

---

[1] Hann (1903, 346) quotes a case when 65 centimetres (2 ft) of snow was removed in 12 hours at Grindelwald.

[2] See Visher (1954, 153–4), and Brinkmann and Ashwell (1968).

Easter evening at Flüelen when informed that it was most strictly forbidden to smoke *out of doors*! The fire hazard presented by the foehn has burnt itself deep into Swiss consciousness and it was made very clear that this was one occasion when the traditional British defiance of authority (particularly of foreign authorities) could under no circumstances be excused or condoned. Surface observations were first supplemented by balloon ascents in this valley in March 1912 (Billwiller and Quervain, 1912). With a moderate foehn of about 12 knots it was observed that surface temperatures increased down-valley from −3°C at the pass to 12°C at Erstfeld (near Altdorf), an increase of 15°C during a descent of 1640 metres i.e. very nearly at the DALR. The

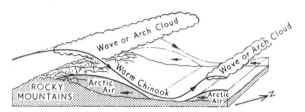

FIG. 134. A block diagram illustrating the Chinook (after Brinkmann and Ashwell, 1968)

balloon ascent followed the DALR with remarkable precision between 1000 and 2500 metres, recording free-air temperatures some 3°C higher than those measured on the ground at corresponding levels in the upper valley. The relative humidity decreased down valley from very nearly 100 per cent at the pass to only 37 per cent at Erstfeld. On this occasion the lower areas round the lake still lay under a mantle of much cooler air and the foehn left the surface near Erstfeld and streamed out across this mantle to produce a sharp inversion of temperature. This is quite a common feature for the inversion has often been recognised on the slopes of the Rigi (1787 m) some 15 miles from Erstfeld. A stronger foehn may sweep the whole lake as far as Lucerne and the remarkable fact is observed that the summit observatory on Pilatus, approximately at the same level as the St Gotthard Pass, then records temperatures from 7–10°C above those at the pass station. Clearly a fundamental transformation of the air mass has occurred.

The thermodynamic theory of the foehn was first propounded by J. Hann (1886). This is still to be found in the *Meteorological Glossary* (McIntosh, 1963). Briefly it compares wet-adiabatic ascent on one side of a ridge with dry-adiabatic descent on the leeward slope. Precipitation on the windward side is thus implied and the phenomenon thus seemed to provide an excellent illustration of the release of latent heat by condensation. The frequent presence of the foehn-wall cloud gave colour to this theory and quantatively it appeared to fit many of the observed facts. More recently the inevitability of windward precipitation has been questioned but the major problem posed by Hann's theory is why does the warmed air *descend*? Should it not continue its course *above* the undisturbed cooler air of the northern valleys? Indeed there is a growing body of evidence that on many occasions it does just that. A variety of ingenious ideas have been put forward to explain why the cool lee air is 'drawn away' but the gustiness and power of the foehn wind scarcely tallies with any view that it is no more than a passive partner. There seems to be no doubt that we are faced here with an example of forced descent in no way essentially different from the Helm wind though wavelike disturbances must be much less regular over such broken terrain as that presented by the northern flank of the Alps. Once this possibility is conceded, the upper layers of the regional wind, drier and with a considerably higher potential temperature than the layers beneath, then offer an almost inexhaustible source of foehn-wind air to which the lower layers may or may not make their own contribution. The greatly improved visibility brought by the foehn adds a pointer in this direction. On this view the mountain system is having what has been described above as a weir effect. Precipitation on the windward slope is no longer imperative and, indeed, the local circulation in the windward valleys may be nil or even gently down-valley against the regional wind.

In passing it may be noted that dynamically imposed descent of dry and potentially warm air may also occur in the absence of relief obstacles. Sawyer has recognised this feature aloft in research flights through temperate fronts (see p. 363) and Green (1965) has noted occasions when remarkably low humidities have been recorded on, of all places, lofty Scottish summits. The Russian sukhovei, paralleled it appears in the Great Plains of the USA, suggest that there is still a good deal to learn about this aspect of the circulation.

Sukhovei ('dry winds') have been defined more specifically as involving a relative humidity below 30 per cent, a wind velocity above 5 m/sec and air temperatures above 25°C (Sukachev and Dylis, 1964, 81), though naturally they vary in intensity. They are occasionally experienced over widely scattered areas of the USSR particularly in early and late summer. A 'sukhovei day' has been defined as one where the relative humidity does not exceed 50 per cent at any time during a period of 24 hours. Usually occurring on the southern flank of a high pressure cell situated over the continent, they blow characteristically from the east or south-east with no great strength.

References in Borisov (1965) cannot be built into a very consistent picture but, (*a*) simple advection of air from the south-eastern deserts appears to be ruled out; (*b*) local transformation, particularly the heating of dry Arctic air under sustained cloudless conditions over a dry soil, sometimes in the presence of suspended dust, seems to contribute much to their character; (*c*) some relation with subsidence, still rather obscure, appears to be probable.

The presence of high-pressure cells over a heated continent in these latitudes is something of an anomaly which itself suggests dynamic dictation from the jet stream aloft (see p. 368). Kaminskii regards them as areas of descending air which 'evidently reaches the land surface mainly at their limits' (Borisov, 1965, 165). The validity of this view has yet to be established but the term 'anticyclonic foehn' has already been coined to describe certain situations in the Austrian Alps which fail to conform to the wave theory outlined above.

## Eddy patterns

It is a matter of common experience that airflow past an object which is not streamlined produces eddies, i.e. the flow separates from a streamline and breaks into vortices. On a small scale the vortices most frequently observed have a roughly vertical axis and grit, leaves or snow is seen to be swept round to accumulate on the leeward side of the obstruction. As the scale increases to the size, for instance, of an office tower we become more and more conscious that eddies with a broadly horizontal axis are also involved, i.e. that the flow pattern may involve a forced upward or downward motion, sometimes with results that have given trouble to architects. Figure 135 is a rough sketch of the Senate House, University of London, in a brisk west wind showing the development of a 'bolster' at $A$ and of a lee eddy at $B$. In this building the bolster,

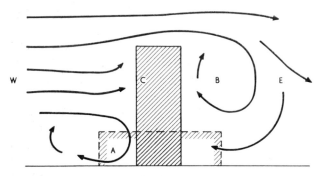

FIG. 135. The 'Senate House Wind', at London University. The diagram shows the effect on the local air circulation of a tall tower block. The most notable feature, the 'bolster' at $A$, is apparently contained and enhanced by the north and south wings indicated by broken lines

contained and probably exaggerated by the north and south wings (shown by broken lines) has surprised many a visitor since the structure was raised in the early 1930s. Entering from the car park on the western side he finds that he may have to fight his way against a steady gale, strong enough to be leaned upon, yet apparently blowing straight out from the closed doors of the building. Despite this object-lesson one reads of modern shopping precincts, e.g. in Boston, USA, where a similar nuisance can only be overcome by the installation of expensive roofing not envisaged in the original design. Since the Senate House tower is used entirely for library storage no trouble has been experienced from the lee eddy at $B$ but in office-blocks architects are now aware that at both $B$ and $C$ rain may come from *below* and windows should be designed accordingly. In very gusty conditions powerful suction effects may also be generated near $B$ and the author has seen a window carry part of its wooden frame outwards from the lee side of a new building when there was no apparent damage on the windward flank.

432

In nature such examples of sharp separation of flow are rare except near cliffs which should always be approached with caution in gusty weather. Nevertheless, even on the flanks of large mountain systems, bolsters and lee eddies sometimes occur giving local wind observations which are far from representative of the general flow. Their broad result is to produce an effective mountain cross-section of a more streamlined character. This is particularly true of deep valley trenches in a cross-wind. The consequences then vary according to the steepness of the slopes, the strength of the wind and, above all, the prevailing stability or instability of the atmosphere. In the English Lake District, for instance, the rain-bearing wind often continues to rise for

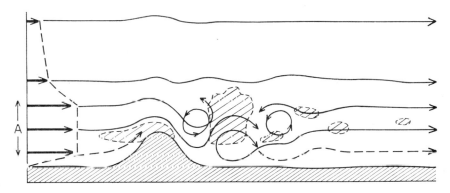

Fig. 136. Rotor streaming. The bold arrows on the left indicate the vertical wind profile and '*A*' marks the streaming layer (after Förchtgott, 1949)

some hundreds of feet after it has passed the ridge crest. The lee eddy then lies high over the lee flank and helps to sweep the rain downwind so that the catch on the lee side may equal or even exceed that on the exposed windward flank. With a more stable stratification, the eddy is likely to occupy a lower and more central location (see also López and Howell, 1967). A likely example is the Dead Sea trough in a westerly circulation. It is certainly wrong to make the facile assumption that the west wind will descend on the Palestinian flank and rise again in Jordan. The development of an independent eddy within the trough, isolating it to some degree from the humid Mediterranean air aloft, is certainly by no means improbable. This may not be without relevance to the aridity of the area. Certainly such eddies are usually fickle and unpredictable and for the sudden 'storms' (gusty squalls?) of the Sea of Galilee we have no less authority than that of Holy Writ.[1]

Figure 136, after Förchtgott (1949), illustrates another type of eddy effect. Förchtgott calls this 'rotor streaming' in contrast to standing eddy streaming illustrated in Fig. 131a and wave streaming illustrated in Fig. 133. In rotor

[1] The author is unaware of firsthand observations in this area under the conditions specified. A contributing cause of the aridity, and also of the relative aridity of such Alpine troughs as the Valais, would appear to be the evaporation of falling rain. Much too little is known about this although it can be *seen* to occur, for instance, in the Grand Canyon of the Colorado.

streaming lee eddies are generated but after attaining a critical scale they are carried away downwind and new growing eddies take their place. The flow pattern is thus unstable and the cloud effects are thus much more full of life.

Lee waves and lee eddies occasionally produce apparently anomalous results when winds reach gale force. A notable instance was the Sheffield gale of 16 February 1962 which came in from the west. The author woke in some alarm when a fierce squall struck his house near Manchester a hammer blow at about 4.30 a.m. and was astonished to find so little damage in the garden next morning. It was on the *lee side* of the Pennines that old trees were knocked over by the hundred around Leeds and new houses were stripped of their roofs in the Sheffield area. (See Aanensen, 1965.)

## Fallwinds

This is a slightly anglicised version of a German word for a feature virtually unknown in Britain, for it occurs where a cold and lofty continental interior abuts on a much warmer, usually maritime, lowland. The classical example is the *bora* of the northern Adriatic, a winter phenomenon, but around the icecaps of Greenland and Antarctica similar effects are observed throughout the year. The word has the great advantage of being self-explanatory for it refers to the descent of cold dense air under the force of gravity. Basically all that is necessary is a large catchment area generating cold air and a slope to set it in downward motion but to produce a wind worth a special name the upland must be offering some barrier effect to a circulation set in train by the general weather pattern.

The slow downward creep of a shallow layer of chilled air at night during calm weather is a universal phenomenon over sloping terrain (see pp. 441–43). Sir Napier Shaw coined the word 'katabatic' (downward-going) to cover all such circulations. He thought that a separate word was necessary since such downflows 'defy the laws of winds that apply to the free atmosphere' (*Manual II*, 1936, xxxi) in that (*a*) they are at least quasi-independent of the regional pressure pattern, and (*b*) they are not warmed dynamically at the DALR since they are continuously chilled by contact with the cold ground in transit. Indeed, as we have seen (p. 175), Shaw saw in this mechanism an important method of transferring air downward across his 'isentropic surfaces' to complete the general circulation. Given an appropriate orographic pattern, such as a group of converging valley-heads, it is conceivable that such a widespread downflow may be canalised into what is virtually a cold-air 'river' but a theoretical difficulty arises here in that as soon as the flow deepens much of its volume loses contact with the ground and the inexorable gas laws should again assume command. Since the lapse rate of environmental air rarely equals the DALR how is it possible for a *deep* downflow of air to continue?

It is clear that fallwinds must be a very special case of katabatic flow and indeed they often have an avalanchelike character that suggests the bursting of a dam or the opening of a sluicegate. Around the ice-clad plateaux of

Greenland and Antarctica the normal cold air 'rivers' of moderate depth and strength are often thrown into fearful spate by the arrival in the region of a temperate depression. This not only adds a regional pressure gradient favourable to air drainage from the heights as Shaw describes but also, and this is much more to the point, it introduces around the flanks of the ice-caps a volume of comparatively fresh 'tropical' air so deep and so warm that dynamic heating at the DALR is no longer able to bring the density of the downflowing air into equilibrium with this environment. Much the same

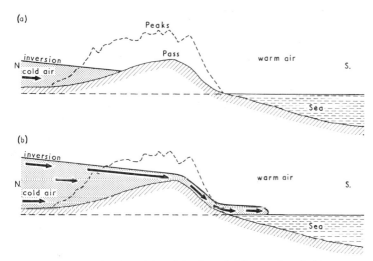

FIG. 137. The probable origin of the 'bora'. The sections indicate a pass through a broadly east–west mountain system in middle latitudes. (a) Here the mountains are ponding back an incursion of cold air and thus providing shelter to a 'riviera' coastland. (b) The bora breaks through across the pass. Dynamical heating consequent upon subsidence is presumed to be insufficient to eliminate the strong temperature contrast and the bora advances as a micro cold front

process appears to operate in winter along the coasts of Norway and Novaya Zemlya and similar events doubtless contribute to the violent blizzards of the Pacific fringe of Siberia.

The true bora of the northern Adriatic and of the Black Sea coast around Novorossiisk must have rather a different background since, despite the presence of mountains to the north of both Trieste (43°39′N) and Novorossiisk (44°43′N), they are neither high enough nor extensive enough to provide an ample reservoir of cold air which a fallwind demands. We are dealing here rather with a relief modification of what is basically a frontal effect since the origin of the cold air is a deepening outbreak from the Arctic. Fig. 137 gives a rather generalised illustration of the process. It shows the longitudinal profile of a pass leading southwards to a comparatively warm sea, the general altitude of the mountain barrier being indicated by the irregular broken line. Arctic outbreaks are normally shallow and produce an inversion with a gentle slope towards the south. In Fig. 137a the cold air is ponded back by the northward slope and a marked temperature contrast will be observed between the coastal plain and the continental interior. This is the very essence of the 'shelter' gained by the various 'Riviera' coasts of the Mediterranean. The reality of the inversion shown is amply confirmed by Borisov who describes vividly how upland areas throughout Eurasia rise like islands through the cold air wedge. In Figure 137b, however, the wedge has deepened, either by a revival of the Arctic outbreak or by radiation losses over the snowclad interior. Cold air may now flow southward through the pass and since its source lies in the upper strata of the cold pool there is no problem in accounting for its volume. Once released, the bora plummets downwards through the warmer maritime air in a highly turbulent flow, gusting to 100 knots or more. In the Adriatic, meteorologists distinguish between an 'anti-cyclonic bora' which comes with clear skies, is most frequent about sunrise and though of great violence does not extend far out to sea, and a 'cyclonic bora' when a depression over the Mediterranean draws northerly winds over most of the Adriatic under a canopy of precipitating cloud. Only the former is a genuine fallwind and its diurnal incidence suggests a delayed response to nocturnal radiation.

The association of fallwinds with polar outbreaks is even more clearly illustrated in the isthmus of Tehuantepec. Occasionally in winter a 'norther' sweeps right down the eastern flank of the Western Cordillera and spills over the 1500-foot pass into the virtually equatorial air of Tehuantepec Gulf with unexpected violence.

## Mountains and fronts

Since both warm and cold fronts have very low gradients (broadly about 1 in 100 and 1 in 50 respectively) it is to be anticipated that large-scale relief features with general gradients well in excess of these values will have a considerable effect upon their forward progress and development. The complex problems posed fall mainly within the sphere of the local meteorologist but a few comparatively large-scale effects deserve mention.

In many accounts the mistral of Provence and the Gulf of Lions is linked with the bora but, though both are cold and gusty, the mistral is scarcely a fallwind in the sense described above. It represents rather an orographical distortion of a cold front held back by the barriers imposed by the Pyrenees and the Alps but free to burst through into the Mediterranean via the 200-mile gap between them (Fig. 138). This gap is partially blocked by the Massif Central and the cold air flow is thus strongly channelled through the Gate of Carcassonne and the Rhone valley. Modified Arctic air is thus involved and since this is much more readily transformed over the Bay of Biscay than over west-central Europe it is the Rhone valley current to which the name is most usually applied. This bitter, dry wind is particularly feared in spring and tall cypress windbreaks, interspersed with light reed screens, form a characteristic element of Provençal market gardens. That the risk is much more wide-spread was brought home to the author when he noted one Easter that on farms near the now famous Costa Brava each individual seedling was pro-vided on its northern side with a tiny cardboard or plastic windscreen. Farmers do not take this kind of trouble without due cause.

Relief effects upon the advance of warm fronts have attracted much less notice since the usual result is little more than the prolongation of spells with

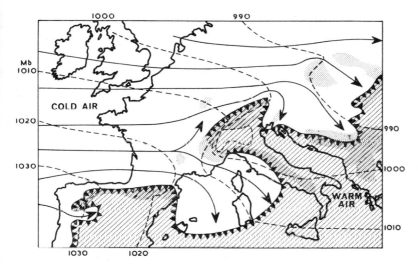

FIG. 138. A cold front being held in check by the mountain systems of western Europe but spreading as the mistral over the Gulf of Lions (after Bergeron, in Godske *et al.*, 1957)

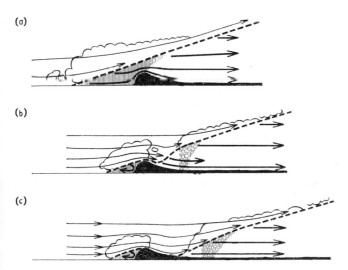

FIG. 139. A warm front encountering the uplands of southern Norway (after Bjerknes and Solberg, 1922)

cloudy skies and persistent rain long associated with the windward slopes of mountain areas before frontal theories were first propounded. Figure 139 shows an early interpretation of the facts made by Norwegian meteorologists. In (*a*) a warm front is shown approaching the Scandinavian uplands from the west or northwest. In (*b*) the lower part of the frontal surface has been intercepted by the relief obstacle whilst at greater elevations its forward progress continues unchecked. On the leeward side of the mountains a break in the great system of cloud sheets is seen to be developing though there is little change on the windward flank. In (*c*) the warm front inversion has stimulated foehn-like developments on the leeward flank but the warm front is regenerating again at a distance of some 150 to 300 miles from the mountain crest whilst cloud and rain persist on the windward flank. The diagram gives us a deeper insight into so-called 'orographic rain' and parallel conditions doubtless occur in British Columbia, the southern Andes,[1] the New Zealand Alps and elsewhere.

The distortion of frontal surfaces by considerable relief obstacles may also stimulate the generation of 'lee depressions', a common feature in the western Mediterranean. When the upper air flow is propitious these may grow into full-scale disturbances. For further details the reader is referred to meteorological texts.

### Upslope and downslope circulations

In our discussion of land and sea breezes it has been suggested that the general slope of the land is not without significance. Furthermore we have also seen that the *effective* height of a mountain obstacle partly depends on the nature of the local circulations to which it may give rise. When radiation is active under clear skies and when the regional wind is not so strong as to swamp all local patterns it is a common observation that an upslope circulation preponderates during the day, to be replaced by a downslope circulation at night. In so far as such circulations are free, i.e. not imposed by lee-eddy or similar effects, they must arise from local contrasts in air density. The key to the contrast in this instance lies in the different temperature regimes, level for level, of air in close contact with the valley side on the one hand, and of the comparatively 'free' air in mid-valley on the other.

Some idea of the magnitude of this effect may be gained from Fig. 140 which represents, with a true vertical scale, a cross-section of a valley with a flat floor a mile wide flanked by 1 in 3 (18°) slopes extending uniformly up to peaks some 6000 ft above valley-floor level. Starting from a presumed condition of equilibrium with an environmental lapse rate of 3°F/1000 ft (5°C in 3000 ft) and with a valley-floor temperature of $x$°C, the temperature at points $C$ and $E$, and at all points in the free air also at 3000 feet above the floor will be $(x-5)$°C and at the level $DF$ $(x-10)$°C. The presumed

[1] Indeed it can be seen here on December 4, 1967 in Fig. 69.

equilibrium thus entails characteristic but uniform air densities at each level and there is no local circulation.

*Daytime.* The response of the air in such a valley to insolation will clearly depend upon its general direction and the relationship of its various slopes to the observed diurnal path of the sun at a given latitude and season. Let us

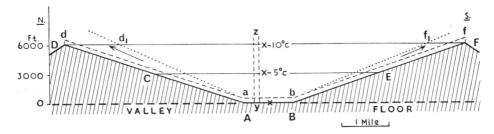

Fig. 140. Cross-section with a true vertical scale of a large valley with one-in-three slopes illustrating the operation of an upslope circulation. The valley is assumed to have an east–west trend and to lie in middle latitudes. See text

take the simple case when the direction is east–west and consider only a limited period around midday when the sun has a mean elevation above the horizon of 60°. (A fully integrated picture of the events from sunrise to sunset is clearly much more complex but the noonday situation broadly reflects the magnitudes appropriate to different latitudes and seasons.)

The angular incidence of the rays will thus be as follows:

|  |  | Sines | Ratios |
|---|---|---|---|
| Between $A$ and $B$ | 60° | 0.866 | 100 |
| Between $A$ and $D$ | 78° | 0.978 | 113 |
| Between $B$ and $F$ | 42° | 0.669 | 77 |

Insolation sufficient to raise the temperature of a shallow surface layer of air between $A$ and $B$ by, say, 5°C will thus raise that of a similar layer between $A$ and $D$ by 5.6°C and between $B$ and $F$ by 3.8°C. These superficial contrasts are doubtless interesting and, in certain circumstances, they may become important, but the significant feature about a slope is that it carries what is initially a superficial effect through a considerable range of altitudes. To assess the possible result of the insolation upon air circulation within the valley we have to compare the overall density of a *vertical* column of air ($yz$) above the valley floor with that of the two *sloping* columns ($AaDd$) and ($BbFf$) over a similar range in height. Initially the mean temperatures of all three columns over a range of 6000 feet was $(x-5)°C$, but whereas heat has been added only to the base of the vertical column, it has been fed into the sloping columns *throughout their whole length* owing to their close contact with the

439

ground. Taking the superficial layer as about 500 ft in depth, the mean temperatures and relative densities of the three columns will thus have become,

| Column | Mean temp. (°C) | If x = 15°C (288°K) (°K) | Density (ratios) |
|---|---|---|---|
| (yz) | (x − 4.6) | 283.4 | 100 |
| (AaDd) | (x + 0.6) | 288.6 | 98.2 |
| (BbFf) | (x − 1.2) | 286.8 | 98.8 |

Since a similar calculation is possible at any selected level it is clear that there will be a general tendency for the cooler mid-valley air to subside expelling the valleyside air upwards and outwards towards the peaks. The fact that, in nature, an initial condition of perfect equilibrium is unlikely and that the application of heat is continuous in no way invalidates this general conclusion. The vigour of the upslope circulation will clearly depend on the rate of input of solar energy and the extent to which this energy is absorbed by a variety of braking mechanisms that are brought into play as soon as the circulation is set in motion. Furthermore, since the upslope circulation is a cumulative process, in a dynamic model the upslope wind should increase in depth with increasing height as shown diagrammatically in Fig. 140 by the fine dotted lines $ad_1$ and $bf_1$.

Brakes on the system are supplied by various mechanisms. (a) Surface friction which depends on the roughness of the slope. On the great mountain wall north of Innsbruck, Defant (1951) measured upslope winds of about 7 knots between 15 and 45 metres from the ground though the speed fell off (rather gradually) at both lower and higher elevations. (b) Mixing between the rising warm airstream and the cooler free air above it reduces the buoyancy and disperses the flow. (c) Ascent entails falling pressure and hence cooling at the DALR. In a dynamic model, therefore, some of the heat supplied by the surface must be used to counter any difference between the observed environmental lapse rate and the DALR. (d) Similarly, especially as the system approaches maturity, subsidence in the free air will slowly raise its temperature, and thus diminish the density contrast which gave rise to the motion.

On the other hand, if the rising air reaches condensation level, a fresh source of energy is available in the release of latent heat though the picture will then be further complicated by the incidence of cloud shadow. Similarly, valley-shadow effects in the morning and evening add to the almost infinite variety of mountain climates (see Garnett, 1937).

*Night-time.* After sunset the effects of aspect and shadow disappear and condensation processes have a much more widespread distribution. Outgoing radiation now dominates the picture and heat losses are relatively greater from the peaks, partly owing to their wider angle of exposure to the sky and partly because the water vapour heat trap is less effective at greater elevations.

Temperatures in the two columns (*AaDd*) and (*BbFf*) will thus tend to become identical, level for level, but now the values will be *lower* than in the corresponding free air where the nocturnal fall in temperature is checked to some degree by the absorption of long-wave radiation from the valley sides. A broadly reversed circulation is thus set in train but it is usually much weaker than the upslope circulation whilst, again because of a cumulative effect, it increases in depth with *decreasing* elevation.

Braking systems operate but with rather a different emphasis than during the day. (*a*) Since nocturnal chill is very much a surface phenomenon a thin skin of very cold air is often held against gravity by vegetation. Above this lubricating layer, cold air may occasionally slide downwards like an avalanche. Downslope winds are therefore often gusty but, the supply of cold air being limited by the slow rate of diffusion of the surface chill, such gusts often require an interval of several minutes to regenerate. At the Innsbruck site Defant (1949) found the downslope wind considerably shallower than the upslope wind. It rarely exceeded 100 metres in depth and between 15 and 45 metres above ground level its characteristic speed was of the order of 4 knots. (*b*) Mixing is less effective at night owing to the more stable stratification. (*c*) Since descent involves rising pressure, downslope winds will tend to warm at the DALR. This braking mechanism is more effective at night since the lapse rate in the environmental air is normally lower at night than during the day. Shaw has pointed out that, at least on intermediate slopes, the mountain wind thus offers some compensation for the heat that would otherwise be lost through radiation. (*d*) Finally, as the downslope air floods across the valley floor and to the degree to which it may lie impounded there, the free air aloft must be lifted slowly, chilling at the DALR and thus gradually eroding the contrast in temperature which gave rise to the nocturnal circulation. Should it reach condensation level a stratus deck will be generated and the release of latent heat will then delay the attainment of equilibrium.

Since the alternation of upslope and downslope winds described above is a direct consequence of terrestrial radiation it must be profoundly affected by the state of the ground. Heat used in promoting evaporation is lost from the upslope wind process unless it is freed again by condensation aloft. A heavy snow mantle, reflecting incoming solar radiation, may well prolong the downslope wind throughout the day on the upper slopes even when an upslope circulation is generated upon the snowfree slopes below. The 'glacier wind' often experienced near the snouts of comparatively clean glaciers is merely a local example of the same effect. The strength of the glacier wind usually owes much to the channelling of the downflow by relief. All downslope winds, impelled by gravity, tend to seek ravines and mountainside farms on spurs or local promontories are thus comparatively free from their effects.

## Mountain and valley winds

It will be noted that the above discussion has been entirely in terms of the valley cross-section. If indeed there were no up-valley or down-valley motion and the circulation was 'closed', it is interesting to speculate how long it would take for the circulation to complete a full cycle, all the free air in the valley having passed at one time or another through the superficial layers. Defant's (1951) daytime estimate for the Inn valley was $4\frac{1}{2}$ hours though his cross-

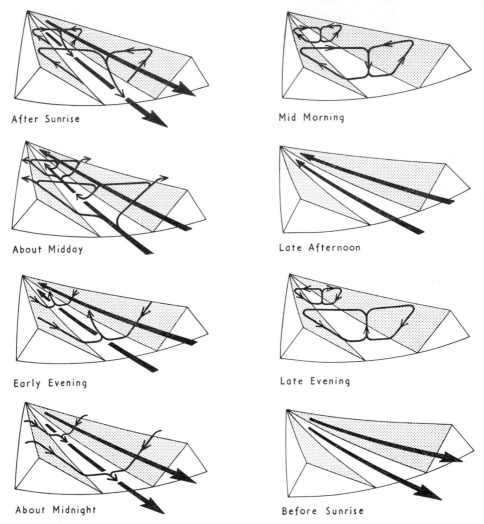

After Sunrise

Mid Morning

About Midday

Late Afternoon

Early Evening

Late Evening

About Midnight

Before Sunrise

Fig. 141. Defant's (1949) representation of the relationship between slope and valley circulations over a period of 24 hours

section seems open to question. In Fig. 140 the area of the cross-section over 500 feet above ground and below peak level, i.e. 6000 feet, is of the order of 125 million square feet; a section a foot wide would thus contain this number of cubic feet of air. Upslope winds with a mean speed of 6 knots (607 ft/min) throughout a depth of 300 feet on *both* sides of the valley within this section would deliver 364000 ft³/min, giving a complete turnover in less than 6 hours.

However artificial such a calculation may appear it serves to remind us that upslope winds tend to warm the whole valley. There is observational evidence that the free air in a valley exhibits a diurnal variation of temperature more than twice as great as that experienced at similar levels over open country. Density contrasts between valley and plain are thus generated, resulting in up-valley winds during the day and down-valley winds at night. The author recalls with some exasperation that during an August visit to the mountain hostel at Spiterstulen in Central Norway, the galley smoke invariably reversed its direction during the evening meal so that the actual moment of change was never observed. The feature was underlined by the fact that the wood smoke kept very near the ground and, in the pure mountain air, could be smelled if not seen at least a mile from the house.

Figure 141 (after Defant) shows the relationship in space and time of the slope and valley circulations. It can be left to speak for itself. Note, however, that the valley wind fills the whole trough and may reach 8 to 12 knots 1000 feet above the valley floor. It should also be remembered that in a great longitudinal depression like the Inn valley the time interval between the initiation of slope winds and the arrival of the corresponding valley circulation is likely to be greater than in a transverse valley with more direct access to the plains. The approach-run into Innsbruck airport is rarely easy but the valley circulation makes it possible to operate with a single east–west runway. Furthermore gorges, sharp bends and similar constrictions are very effective in checking the free evacuation of the cold mountain wind at night. Cold air flows 'like porridge or thick syrup' (quoted in Geiger, 1965, 412) and is thus readily ponded back to produce extensive 'frost hollow' effects of very considerable climatic importance.

On a still wider scale, yet partly as the cumulative effect of the processes described above, there is a broad tendency for pressure surfaces in the middle troposphere to sag over extensive uplands during the day and to take on a shallow domelike form during the night. A daily inflow of free air from over the surrounding lowlands may thus be observed at elevations well above peak level, followed by a corresponding outward tidal flow at night. The results on cloud formation and the incidence of precipitation may be considerable but much depends upon the character of the upland region. Extensive plateau surfaces under strong insolation will stimulate the inward tide whilst lofty, snowclad central peaks will enhance the nocturnal outward motion.

## Lakes

It has often been implied that the presence of a large body of inland water must modify the climate around its shores. There is evidence indeed of local land-and-lake circulations, of a local increase in air humidity, and above all of a notable reduction of the frost hazard. The latter may be important around quite small water-bodies but its effectiveness depends on the intensity of the cold, the depth of the water (which determines its capacity for heat storage and delays surface freezing), and above all on the activity and direction of the general circulation. Usually the protection offered is against some particular frost spell at a key period and the effect on mean minimum temperatures is not considerable. Whether evaporation from lake surfaces can have much effect in increasing local precipitation is still very much open to question. As we have seen, the precipitation mechanisms operate far above the earth's surface and depend much more upon atmospheric instability than upon its humidity content. Where there is high ground in the vicinity, as round the southern shores of the Caspian and Lake Baikal, a positive effect may be conceded. Furthermore the American Great Lakes do appear to affect the weight, and indeed the character, of winter snows. Yet it is a far cry from these facts to the assertion, sometimes made, that by artificially flooding large areas of desert or semidesert land, for example the Qattara depression or sectors of Russian Turkestan, a regional amelioration of climate might be induced. An object lesson in this sphere was offered by the natural flooding of Lake Eyre in 1949–50 when absolutely no increase in local precipitation was recorded by a team specially sent there to observe the facts (Peake Jones, 1955, 29). It was useful to have this negative result confirmed but the existence of desert *ocean* coasts on five continents made the outcome highly probable.

The basic fact, of course, is that, with the possible exception of the Caspian, no continental water-body is on a scale sufficient to affect the atmosphere materially *in depth*. They are not large enough to generate even a quasi-independent air mass so that their influence falls heavily under the sway of the regional circulation.

The *Caspian Sea*, with an area of some 170000 square miles (approximately that of the North Sea), is indeed a case apart, yet it is difficult to separate out local effects, due to the presence of the lake, from those arising from its range in latitude (some 10°) and the very different relief of its northern and southern shores. Suslov (1961, 437) writes: 'Neither the Caspian nor Aral Sea . . . appreciably influences the climate of Central Asia. Their influence is limited to only a narrow coastal strip.' Within such a strip one would expect some reduction in the severity of winters though the freezing of the northern section from about mid-November to mid-March must unquestionably weaken this effect. Similarly, the development of a sea breeze in summer, said to extend some 20 to 25 miles from the coast, must cause some reduction in maximum temperatures, though again one must not assume that this will be equally evident on all shores for the character of the hinterland is far from uniform.

Table 49 presents temperature data for five stations in close proximity to the Caspian and, for comparison, parallel data for stations some 250 miles or more to the west and east. Unfortunately both Stavropol and Tbilisi are at greater elevations than the rest but though a correction of $1°C/155$ m might be applied in summer, it would be a highly questionable adjustment during the Russian winter.

Taking the figures as they stand one has the broad impression of a general fall of both maximum and minimum temperatures from west to east across the northern sector of the Caspian in January–February. At this season the pre-

TABLE 49. *Temperatures in the Caspian region* $(^{\circ}C)$

| | WEST | | CASPIAN | | | | EAST | |
|---|---|---|---|---|---|---|---|---|
| NORTH | STAVROPOL (575 m) 45°02′N | | ASTRAKHAN (−14 m) 46°21′N | | GURIEV (22 m) 47°07′N | | KASALINSK (63 m) 45°46′N | |
| *January–February* | *Max.* | *Min.* | *Max.* | *Min.* | *Max.* | *Min.* | *Max.* | *Min.* |
| Mean daily | −2 | −7 | −4 | −9 | −7 | −14 | −7 | −15 |
| Mean monthly | 9 | −17 | 4 | −22 | 1 | −26 | 2 | −26 |
| Absolute extreme | 18 | −30 | 9 | −30 | 7 | −37 | 11 | −32 |
| *July–August* | | | | | | | | |
| Mean daily | 24 | 15 | 29 | 19 | 29 | 18 | 31 | 17 |
| Mean monthly | 31 | 10 | 34 | 15 | 36 | 13 | 38 | 12 |
| Absolute extreme | 35 | 5 | 37 | 11 | 41 | 7 | 41 | 8 |

| | TYUB-KARAGAN[1] (124 m) (44°36′N) | |
|---|---|---|
| *January–February* | *Max.* | *Min.* |
| Mean daily | −3 | −6 |
| Mean monthly | 6 | 16 |
| Absolute extreme | 11 | −24 |
| *July–August* | | |
| Mean daily | 27 | 21 |
| Mean monthly | 34 | 15 |
| Absolute extreme | 39 | 12 |

| | TBILISI (404 m) 41°43′N | | LENKORAN (19 m) 38°46′N | | KRASNOVODSK (−21 m) 40°00′N | | ASHKHABAD (226 m) 37°57′N | |
|---|---|---|---|---|---|---|---|---|
| SOUTH | | | | | | | | |
| *January–February* | *Max.* | *Min.* | *Max.* | *Min.* | *Max.* | *Min.* | *Max.* | *Min.* |
| Mean daily | 6 | −2 | 6 | 1 | 4 | 1 | 6 | −2 |
| Mean monthly | 13 | −9 | 11 | −5 | 11 | −7 | 18 | −11 |
| Absolute extreme | 20 | −14 | 15 | −12 | 19 | −17 | 29 | −26 |
| *July–August* | | | | | | | | |
| Mean daily | 28 | 18 | 28 | 20 | 32 | 24 | 36 | 21 |
| Mean monthly | 33 | 14 | 32 | 16 | 37 | 19 | 41 | 16 |
| Absolute extreme | 35 | 12 | 34 | 14 | 42 | 12 | 45 | 9 |

[1] On a peninsular site at the north-western extremity of the Mangyshlak range.
Data converted to metric scale from Met. Office Tables (MO 617c. and e), 1958.

vailing winds are from the north-east and the effect of the frozen water surface appears to be minimal. Yet there is a real difference between Guriev and Tyub-Karagan, some 200 miles downwind, involving a rise in the mean daily minimum of 8°C. The fact that this exceptionally 'maritime' station has

445

experienced an absolute minimum of −24°C in January shows, however, that at times very bitter cold waves can envelop the whole of the northern portion of the sea. Even far towards the south, where mean daily minima are above zero, it is quite clear that sharp frosts are by no means uncommon. Note, however, that in this sector the winter climate of the coastal stations is certainly less extreme than that to the west or east.

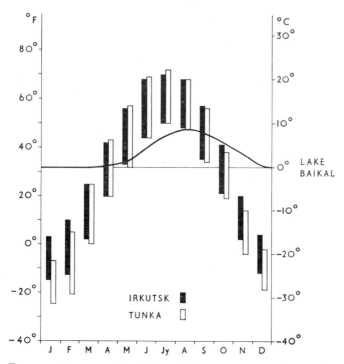

Fig. 142. The seasonal variation of the temperature of the surface waters of Lake Baikal compared with data for two stations some 40 miles from the lake shore. Mean monthly water temperatures are shown by the continuous curve. The level of the lake is at 1495 ft. For Irkutsk (1532 ft) and Tunka (2395 ft) monthly means of daily maxima and minima lie at the extremities of the vertical bars

In *July–August*, on the other hand, there is some general increase of temperature from west to east which the presence of the northern sector of the Caspian scarcely affects. Although a north-westerly circulation is then predominant even Tyub-Karagan shows only a slight reduction in mean daily maxima and extreme values remain very high, probably because they are brought by desert winds from the east. It appears, however, that the presence of the great expanse of warm water does reduce the fall of night minima. At this season there is little contrast in maximum temperatures between the northern and southern stations though Krasnovodsk is warmer than Lenkoran, again

446

suggesting that heat waves come from the east. Note the very high absolute maxima recorded at both Guriev (41°C; 105°F) and Krasnovodsk (42°C; 108°F). Of the five stations near the Caspian, Lenkoran alone records more than 10 inches (250 mm) of rainfall. Its mean annual total of nearly 50 inches (1250 mm) (mainly between September and April) appears to be representative of the whole of the Mazanderan coast which Blanchard (1929, 147) describes as 'une côte de Coromandel'. Here then the presence of the Caspian is paramount though the proximity of the Elburz mountains to the coast undoubtedly contributes much to the result.

*Lake Baikal*, with an area of some 12000 square miles, is in quite a different class yet Borisov declares (p. 133) that 'no other lake has such an influence on the temperature of the air around it'. This is partly a result of its exceptional depth (surface 456 m above sea-level, bottom to 1295 m below sea-level), and partly because, walled in by fault-line scarps rising from 600 to 1500 metres above lake level, its effects are concentrated upon a restricted area.

In Fig. 142 the mean temperatures of the open lake surface (after Suslov, 1961, 306) are compared with the mean daily ranges at both Irkutsk (467 m), some 40 miles down the Angara outlet, and Tunka (730 m), at about the same distance from the lake but in the south-westerly extension of the group of structural depressions to which it owes its origin. At this distance the climatic effect of the lake appears to be inconsiderable.

Suslov (p. 309) presents isotherm maps of the region for December and July but quotes mean monthly data for only one lakeside station, Listvyanka, at the Angara outlet. Below we compare his values with similar data for Irkutsk, 40 miles from the western shore, and Ulan Ude, 65 miles from the eastern shore on the lower Selenga river.

| | IRKUTSK | *Diff.* | *Mean monthly temperatures* (°C) LISTVYANKA | *Diff.* | ULAN UDE |
|---|---|---|---|---|---|
| December | −20 | +8.5 | −11.5 | +9.5 | −21 |
| January | −21 | +3.5 | −17.5 | +9.5 | −27 |
| April | −1 | 0 | −1 | −1.5 | 0.5 |
| July | 15.5 | −3.0 | 12.5 | −7.0 | 19.5 |
| October | −0.5 | +2.0 | 1.5 | +2.5 | −1 |

It is evident that the contrasts are considerably more marked than anything we were able to demonstrate around the shores of the Caspian despite the much shorter distances involved. The thermal effect of the lake is at its greatest in December when it is normally still unfrozen but even after a considerable proportion of its surface is ice-clad a moderating effect is still recognisable. Russian workers attribute this more to the release of the latent heat of fusion during cold spells than to the emission of heat from the water at

447

depth (Borisov, 1965, 134). It is claimed that the effect extends upwards to some 1000 metres above the lake surface. The presence of ice-floes retards the arrival of spring so that vegetation around the shore shows little sign of life until mid-June. This is particularly the case on the south-eastern coast towards which the broken ice is drifted by the prevailing north-westerly winds. The thermal effect in midsummer operates largely through the generation of coastal fog.

The steep slopes around Baikal are responsible for vigorous valley circulations, particularly in a downward direction. Notable amongst these is the 'sarma', named after a valley through the western scarp. It is a fallwind of bora type and occurs mainly from mid-August to November when cold waves from the north-west spill downwards towards the relatively warm lake surface, attaining speeds up to 80 knots and producing a heavy swell.

The effect of Lake Baikal on precipitation appears to be slight. Annual totals of 250 to 300 millimetres are characteristic of most of its coastlands but values up to 750 millimetres have been recorded in the uplands in the extreme south-west.

In view of their large total area of some 95 000 square miles, most of which remains unfrozen even during severe winters, it might be anticipated that the *Great Lakes of North America* would exert similar climatic effects. In fact, according to Visher (1954, 17) they raise mean January temperatures by some 5°F (2.8°C) and lower mean July temperatures by 3°F (1.7°C) as compared with their surroundings. They would thus appear to be less than half as effective as Lake Baikal. Clearly the influence of the water-body is more widely diffused, partly because the shores are low and wind round the three great 'peninsulas' of Upper Michigan, Lower Michigan and Ontario, and partly owing to the greater vigour of the general circulation. It is thus difficult to illustrate the effects of the lakes with much precision.

To avoid latitude effects we have selected a cross-section of Lake Michigan near the 43rd parallel, using records for the uniform period 1871–1930 at Madison, Milwaukee, Grand Haven and Lansing (Table 50) and employing a number of different criteria. Temperatures have been expressed in degrees Celsius to facilitate comparison with the data quoted above. Lines (a) and (b) are self-explanatory. For line (c) in winter we have used the five coldest Januarys during the period at Madison, namely 1875, 1885, 1912, 1918 and 1929; and in summer, the five warmest Julys, 1878, 1897, 1901, 1916 and 1921. The mean daily minimum and maximum values quoted in line (d) are for the same extreme years, as also are the averages of the absolute extremes given in line (e).

It is evident that the thermal effect of the lake is much greater in winter than in summer but that its scale is seriously obscured by the use of long-period means. During very cold spells the contrast between the two sides of the lake at Milwaukee and Grand Haven is quite remarkable for cold waves come mainly from the north-west. A parallel result is the delay of some 10 days or more in the arrival of the first autumn killing frosts along the eastern shore of

Lake Michigan and the southern shores of Lakes Erie and Ontario though the advantages of these areas from the point of view of the fruit-grower stem rather from a more subtle climatic effect not easily displayed quantitatively. Cool foggy air off the lakes delays the opening of buds in spring and thus reduces the *damage* done by late frosts.

In summer the most important effect of the cool air over the lakes would appear to be a reduction in the frequency and intensity of thunderstorms and hence of the hail hazard.[1]

TABLE 50. *Temperatures along 43° N across Lake Michigan (1871–1930), (°C)*

| | MADISON | DIFF. MIL–MAD. | MILWAUKEE | DIFF. G.H.–MIL. | GRAND HAVEN | DIFF. G.H.–L. | LANSING | DIFF. L.–MAD. |
|---|---|---|---|---|---|---|---|---|
| **WINTER** | | | | | | | | |
| (a) Mean December-February | −6.8 | +1.8 | −5.0 | +1.5 | −3.5 | +1.1 | −4.4 | +2.4 |
| (b) Mean January | −8.5 | +2.1 | −6.4 | +2.0 | 4.4 | +1.0 | −5.4 | +3.1 |
| (c) Mean 5 cold Januarys | −15.2 | +2.8 | −12.4 | +3.8 | − 8.6 | +2.0 | −10.6 | +4.6 |
| (d) Mean min. 5 cold Januarys | −19.4 | +2.4 | −17.0 | +5.2 | −11.8 | +3.8 | −15.6 | +3.8 |
| (e) Mean 5 Extremes | −30.6 | +2.3 | −28.3 | +6.6 | −21.7 | +4.4 | −26.1 | +4.5 |
| **SUMMER** | | | | | | | | |
| (a) Mean June-August | 20.9 | 1.1 | 19.8 | 0.6 | 19.2 | 1.0 | 20.2 | −0.7 |
| (b) Mean July | 22.3 | −1.1 | 21.2 | −0.7 | 20.5 | −1.0 | 21.5 | −0.8 |
| (c) Mean 5 warm Julys | 25.1 | −1.2 | 23.9 | −0.6 | 23.3 | −0.5 | 23.8 | −1.3 |
| (d) Mean max. 5 warm Julys | 30.2 | −1.9 | 28.3 | −0.3 | 28.0 | −2.3 | 30.7 | +0.5 |
| (e) Mean 5 extremes | 36.7 | 0 | 36.7 | −2.8 | 33.9 | −3.3 | 37.2 | +0.5 |

Since the lakes of *East Africa* lie in intertropical latitudes their thermal effects would seem to be confined to the generation of land-and-lake-breezes bringing some considerable modification of the diurnal range. This is particularly true of *Lake Victoria* in view of its great area of some 26000 square miles. The much steeper shores of the other lakes makes the comparison of lakeside and 'inland' stations virtually impossible.

Mean data for the two 3-month periods, December to February and June to August, at Kampala (0°20′N, 32°36′E, elev. 1312 m) and Mbale (1°06′N, 34°11′E, elev. 1220 m) are given overleaf. Mbale lies about 120 miles north-east of Kampala but is only 70 miles from the lake in a direct line. Their rainfall records are very similar and the difference in elevation is slight. Clearly the main effect is a reduction of midday temperatures at Kampala by

---

[1] Visher (1954) says that the lakes 'increase the average length of the frost-free season about 30 to 40 days on their eastern and southern sides'. This is difficult to substantiate either from his maps or from the data for the stations quoted above. The problem, of course, is to find two stations where all other conditions are approximately equal.

| | Average daily | | Average monthly | | Extreme | |
|---|---|---|---|---|---|---|
| | *Max.* | *Min.* | *Max.* | *Min.* | *Max.* | *Min.* |
| KAMPALA | | | | | | |
| December–February | 82 | 64 | 88 | 59 | 97 | 53 |
| June–August | 77 | 62 | 82 | 59 | 85 | 53 |
| MBALE | | | | | | |
| December–February | 89 | 61 | 95 | 56 | 101 | 52 |
| June–August | 82 | 61 | 87 | 57 | 94 | 52 |

*Temperatures in degrees Fahrenheit*

5–7° F, doubtless a welcome feature despite the greater humidity of the lakeside air since it is accompanied by a more active local circulation. The extent of the wind reversal at Entebbe is illustrated in Fig. 143. It appears to

ENTEBBE
Mean January and July

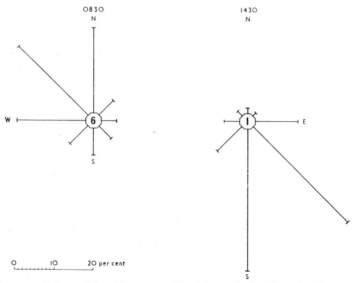

Fig. 143. Lake and land breezes at Entebbe on Lake Victoria. Since there is little or no evidence of a seasonal variation it is possible to combine the data quoted by Kendrew for January and July

persist with very little seasonal variation throughout the year. (See Lumb, 1970.)

The effect of Lake Victoria on precipitation is confined to its western shores, particularly between Entebbe and Bukoba where 60 to 80 inches is recorded on some 170 days. Elsewhere lakeside stations record annual totals very similar to the regional average, from 40 to 45 inches, though the presence of the lake does not prevent the development of a comparatively dry zone in the neighbourhood of Musoma (30 inches on 110 days) almost opposite Bukoma.

# Forests

Despite the great inroads made upon the forest lands of the world during the last two centuries it is estimated that some 20 to 22 per cent of the land surface of the globe (excluding Antarctica) is still forested, about half of that area being in tropical evergreens and a third in conifers particularly in sub-arctic and mountain regions. The deciduous woodlands have suffered most severely from the remorseless spread of cultivation but extensive patches remain as timber or game reserves, shelter belts or as shady retreats.

Western mythology, still preserved in our fairy tales, is deeply imbued with the feeling that the forest is a land apart, mysterious and sometimes forbidding. A parallel mythology may also be recognised in pseudo-scientific circles regarding the influence exerted by forests upon local climates. The pioneer farmer in northern lands is usually convinced that forest clearance will materially lengthen his all-too-brief growing season. The Men of the Trees, justifiably appalled by the consequences of deforestation in semi-arid lands, go so far as to claim that by tree-planting, and by tree-planting alone, man has at his hand a weapon with which he could 'conquer the desert'. The practical doctrines which emerge thus range from 'cut and burn' to 'conserve and replant' but little evidence is forthcoming to support either of these extreme views—it is usually implied that the facts are obvious.

One of the functions of science is to dissect the obvious and in a detailed analysis of the literature on this subject Geiger (1965, 298–368) makes the following points.

1. Any continuous cover of vegetation, whether it be short grass, a field of grain, macchia scrub or high forest, must have some effects on a micro-climatological scale because it changes the albedo of the earth's surface and provides some degree of shade and a restriction of air circulation at soil level.

2. The general effect of such a cover is to raise the 'active surface' at which the diurnal cycle of radiative and humidity exchanges occurs to the top of the closed canopy. This may be only a few inches above the ground over a meadow but may reach 100 feet or more above ground level in closed woodland.

3. Screen observations in the trunk area of a woodland are thus not comparable with screen observations over a lawn since they do not refer to the same horizon of reference. One set of readings is taken within the plant cover and the other above it.[1] A beetle entering a closed crop from a sun-drenched path has doubtless the same sense of relief, of reaching a realm of still, humid shade, as a gleaner who moves from the stubble to rest in a neighbouring wood. It is purely a matter of relative scale.

[1] The data presented by Zon (1941, 479) suffer from this defect. Ten-year means of daily maxima and minima are plotted for a station inside a birch-beech-maple forest at Dukes, Michigan and compared with similar data from a clearing. For the months June to September the maxima in the clearing are some 3° F higher than those in the forest whilst the minima are about 2.5°F lower; during the rest of the year the contrast is slight.

This tells us something about the climate *under* woodland but nothing about the climate *above* it.

4. Modern devices make it possible to measure climatic conditions within a growing crop. Just as these conditions vary according to the structure and density of the growth, so woodland climates differ according to tree type and tree density.

5. Geiger concludes that 'forests do indeed have a great influence on the nature of a country, but this is due only slightly to their effect on climate' (p. 361). He asserts that 'it is certain that any increase in precipitation due to the presence of a forest is bound to be very small' and that though forests may have 'a moderating influence on climatic extremes . . . the extent of this influence is even harder to assess' (p. 368). Nevertheless he concedes that 'there is no doubt that the influence of the forest is the greater the more inhospitable the nature of a territory is to the growth of trees'. Near the tree-line towards arid or arctic territories or on mountains and on other exposed sites he thinks that 'the beneficial effects of forests may become very great. There is here a very great and rewarding field open for scientific investigation'. Note, however, that these latter statements do not refer specifically to *climatic* advantages. Unless trimmed by fire as along their prairie or savannah margins, the natural limits of forests are usually fringed by a zone of open or stunted growth, the canopy is discontinuous and parallels must then be sought with comparatively open crops.

The general tenor of Geiger's argument is that forest climates certainly exist but that they are confined substantially to the zone within the canopy. Of unquestionable importance to all forest dwellers, to tree seedlings, herbaceous growth and soil processes, they are not necessarily of much relevance on a macro-climatic scale. This is where the proponents of extreme views are led astray. A forest climate is the climate within a growing crop and its recognisable features are paralleled within any close crop. What are those features? We must look particularly at the effects on insolation, temperature, humidity, air motion and the interception of precipitation.

(*a*) *Insolation and back-radiation.* During the day much of the incident short-wave radiation is intercepted by the tree crowns which respond to this effect like any other rough surface and radiate heat mainly to the free air above, thus reducing the heat input to the zone beneath. Hence the cool shade of a woodland environment. The extent to which direct insolation penetrates into a forest depends on the elevation of the sun as well as upon the density of the stand so that it has the effect of a much shortened as well as of a weakened day. It is estimated that only from 15 to 5 per cent of the short-wave radiation reaches the trunk space and still less attains the forest floor. Furthermore, except in direct sun-flecks,[1] there is a marked shift towards the red end of the spectrum, easily tolerated by the human eye but having a marked effect on woodland photography and on the energy available for photosynthesis.

[1] That these are circular 'pin-hole' images of the sun is beautifully illustrated during an eclipse when they become crescentic.

During the night, outgoing radiation also occurs most actively at the forest crown and air temperatures over woodland fall just as readily as they do over a meadow with, at times, the production of dew. Only within the canopy is the nocturnal fall reduced by back-radiation from leaves and branches so that heat loss from the forest floor is often delayed until after midnight.

This shift of the active layer from near the ground to the tree-crown area is best illustrated in tropical evergreen forests where flowers, fruits, insects and those that live upon them are to be found high aloft and the ground zone is left to a variety of waste-consumers or scavengers. It is thus possible for a man to die of starvation in the near presence of a riot of life.

(*b*) *Temperature.* The effect of these exchanges upon temperature must obviously be a marked reduction in the diurnal range, particularly on sunny days. An investigation by Geiger (1965, 318) in an old oak forest in Germany on a bright August day showed a rise in temperature from 9°C near sunrise to only 19°C in the afternoon at a height of 3 metres above the forest floor as compared with readings of 8° and 26–27°C within the crown area (23 m). The predominant effect is thus a lowering of the daily maxima and this must be reflected in the long-term summer means. Richards (1952, 164) plots some interesting data for two days in rain forest in Nigeria.

| | 11 May 1936 (wet season) | | | 9 March 1936 (dry season) | | |
|---|---|---|---|---|---|---|
| Level | 1300 (°C) | 0600 (°C) | Range (°) | 1300 (°C) | 0600 (°C) | Range (°) |
| Top of canopy (24 m) | 30 | 21 | 9 | 32 | 24 | 8 |
| Near the ground (0.7 m) | 27 | 23 | 4 | 29 | 24.5 | 4.5 |

Here the effect appears to be fairly evenly spread between day and night. Clearly this feature will vary with the density and seasonal persistence of the forest crown. In deciduous woodland in winter the effect is slight and it can be argued that the check on air motion induced by the trunks may well compensate for back-radiation from the branches during frosty nights.

(*c*) *Humidity.* Since all plants must lose moisture during transpiration it is readily appreciated that one of the chief ways in which a closed cover can make its own climate is by raising the humidity of the air within the stand. True to some degree of the level of *absolute* humidity, this feature becomes even more marked when the minimum levels of *relative* humidity are considered in consequence of the effect on the temperature regime noted above. In Nigerian evergreen forests Richards quotes the following data as characteristic of daily minimum values.

| | Wet season | | Dry season | |
|---|---|---|---|---|
| | *RH* (%) | *Sat. deficit* (mb) | *RH* (%) | *Sat. deficit* (mb) |
| Top of canopy | 67 | 14 | 64 | 17 |
| Near the ground | 92 | 3 | 77 | 9 |

Clearly evaporation is largely inhibited near the ground surface, particularly during the wet season, though it may be very active in the tree crowns during the heat of the day. During the night however a relative humidity of 95 to 100 per cent is characteristic of the trunk space.

Whether the total water loss by evapotranspiration from a close stand of vegetation exceeds that from an open water surface is still an issue open to warm debate (see p. 97). One school of thought points to the fact that even a meadow exposes some twenty to forty times the area to evaporative processes as compared with the ground on which it grows; the other stresses the fact that under all conditions the total loss by evaporation is strictly limited by the available heat supply. The physical arguments of the latter school are very hard to shake and the comparative irrelevance of leaf area becomes clearer when it is realised that vegetation must enhance the vapour-blanket effect recognisable even over open water when air motion is slight.

(*d*) *Air motion.* Any close cover of vegetation produces a sharp fall in wind speed within it as compared with the movement overhead. The very noticeable fall of wind speed in woodland is thus, once again, the expression of a general rule apparent only because of the greater scale in height. Winds in a large forest rarely exceed 2 to 4 knots and in tropical evergreen forest an almost permanent dead calm may prevail. Much depends on the leaf area and the proximity of an open border, rare in natural woodland which tends to seal its margins with more lowly vegetation. Above the crowns wind normally increases with height as over open country so that the general effect is merely to raise the effective surface above ground level.

It may be mentioned here that woodland shelterbelts narrow enough to permit considerable penetration of light may nevertheless produce a reduction of wind speed of some 50 per cent over a zone up to fifteen to twenty times as wide as the screen is high. In fact a permeable barrier has distinct advantages over an impermeable one (such as a high wall) since the flow through the obstruction damps out the formation of gusty lee eddies which do more harm than good. This effect may be puzzling to those unfamiliar with air motion but it should be realised that whereas light travels in straight lines and thus penetrates the screen wherever it does not encounter a solid object, air is thrown into innumerable eddies. These have the effect of increasing the general viscosity of the flow, the air itself thus generating its own invisible barriers. This fact, now well established by experiment, has important implications

454

on the horticultural scale. A screen is not necessarily more efficient because it *looks* substantial.

(*e*) *The interception of precipitation.* Whilst there is no certain evidence that the presence of forest can increase precipitation there is no doubt that a tree canopy intercepts a considerable proportion of the total which falls upon it, particularly in light showers. This moisture may then be re-evaporated without ever reaching the soil. The 'loss' thus entailed has been estimated at some 40 per cent of the total fall in tropical evergreen forests and figures of the same order of magnitude have been quoted for European beech forests during the summer. This failure of the rain to reach a standard gauge is put into proper perspective if we recall the equally significant losses suffered by raindrops during their fall path (see p. 117). In neither case has the moisture completely gone to waste since it has increased the humidity of the lower layers of the atmosphere and thus provided some check upon further evaporative losses. In fact there would appear to be some relation between these processes since light rains can only fall through humid air. In more arid lands losses by interception are less serious since the showers that reach the ground are necessarily more torrential.

On balance the interception of precipitation involves no serious loss to the water economy. Note that it occurs on any plant cover though since the more lowly plants are below gauge level it may pass unnoticed. Certainly most water supply authorities preserve or plant woodland on their catchment areas since the increased regularity of flow saves flood wastage and reduces silting hazards. Where there is heavy winter snow, as in Norway or the Alps, forests preserve a considerable proportion of the cover into late spring and thus provide an even more effective water storage. Though this may be at the expense of some retardation of spring temperatures the general weight of the evidence is that this is a good thing wherever late frosts present a hazard. It is not the forest which generates the frost despite widespread beliefs to the contrary.

## Urban climates

The most self-evident effect of a large town upon the atmosphere is the production of a smoke pall, particularly in winter where conventional open coal fires are used for domestic heating. Varying in intensity with the weather situation, this is usually easy to recognise from a vantage point outside the town even when an observer beneath the screen is comparatively unconscious of its presence. Hence the rather shady vernacular term, 'the Smoke', for London and the more affectionate 'Auld Reekie' for Edinburgh. The situation in both these cities has shown some improvement in recent years but where there is a heavy concentration of industrial plants the pall persists throughout the year. The author noted in his diary that on 8 June 1926, Ben Lomond, Goat Fell, Tinto Hill and, most surprising of all, the Pentland Hills were all visible from the tower of Glasgow University. The weather though fine was not exceptional, this was a direct effect of the General Strike.

An interesting feature, particularly of the domestic pall, is that it *clings to the city* like a veil, shifting only slightly with a change in wind direction and showing 'no great progressive accumulation of smoke in surface air as it blows across a town' (Chandler, 1965, 98). A parallel observation is that there is no obvious correlation between intensity of pollution and the size of the town. Anyone who has watched the behaviour of smoke from a point source, a factory chimney, a ship at sea, or even from a bonfire in open country, particularly in a fresh breeze, will find himself instinctively reluctant to accept this—clearly the smoke streams downwind. Of course there is some

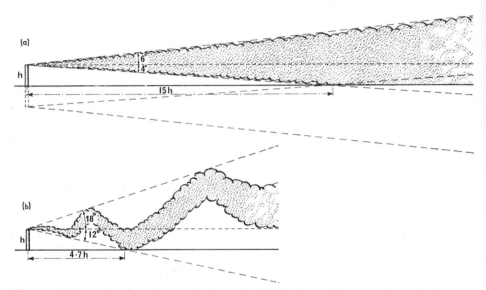

FIG. 144. The smoke plume from a single stack. (a) Under conditions of almost laminar airflow. Note the 'mirror image' useful in estimating smoke concentration at distances over 15 times the stack height. (b) Under much more turbulent conditions. The smoke travels in variable waves which swing between the limits indicated. The angle subtended by the plume may vary widely and the values quoted are no more than representative

export of industrial filth aloft, the sheep on the mid-Pennine moors make that only too evident, but the great bulk of solid pollutants hangs in a low dome over a city until it falls out by gravity or is washed out by rain. It follows as a corollary that 'smokeless zones' of quite modest dimensions, a few acres, may produce quite remarkable effects. The author was sceptical about this until he found that the small zone in central Manchester could be 'seen' from a skyscraper. Pollution readings within the square mile covered by Hyde Park and Kensington Gardens also substantiate this effect.

The basic reason for this feature is that household smoke issues from a great multitude of sources scattered more or less evenly over a large area. The problem of diffusion from a point source has proved fascinating to applied mathematicians though it is not capable of very precise formulation (see Sutton, 1962, 159–67 and Scorer, 1965, 186–217). Here we simply present Fig. 144 to show the spread of smoke from a tall stack under different types of air motion. As can be checked by direct observation, the smoke forms a conical plume so that pollution density decreases as the square of the distance travelled downwind. In (*a*) we have a fairly normal pattern with a cone angle of 10° and in (*b*) the effective cone angle is trebled as a result of large-scale turbulence; a notable gain in the rate of mean diffusion is then accompanied by a shortening of the distance at which the effluent may reach the

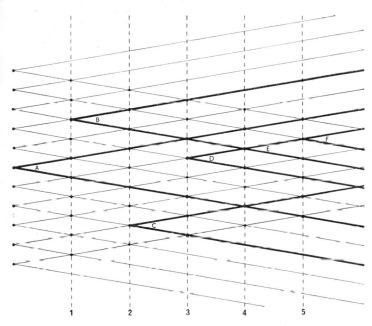

FIG. 145. A view in plan of a group of interlocking half-cones from a regular array of equal smoke sources. This is a model situation from which numerical results are derived in the text. One source in each successive row is indicated by the letters

surface, if only at irregular intervals. The question of multiple-point emission requires a rather different treatment and in Fig. 145 we have approached it in ground-plan via a simple graphical exercise. Smoke sources near the ground will tend to form flattish half-cones but the inverse square law can still be used as a first approximation even if the width of the spread tends to increase rather more rapidly than the height. For graphical simplicity we have employed an angle of 19°, the cones then merge downwind at a distance three times the regular distance assumed between the sources.

For a single smoke source, *A*, if the plume is sampled at regular intervals downwind $(1, 2, 3, \ldots n)$, the inverse square law gives successive concentrations in the series:

$$100, 25, 11, 6, 4, \ldots 100/n^2.$$

But if there is a uniform row of sources at $A$ then a pattern like that shown emerges. Cross-section 1 is so placed that the concentration from a line of sources is still 100, but every point along line 2 now lies within the range of two cones, on line 3 within three and so on. The series produced by a single line of sources is thus of the form, in round figures:

$$100, 50, 33, 25, 20, 17, 14, \ldots 100\ n/n^2\ (= 100/n)$$

We keep the figure comparatively simple if we now add further lines of sources along the cross-sections, that is, groups of which cones $B$, $C$, $D$, etc. are representative individuals. An estimate of the increase in smoke concentration downwind can then be arrived at by a process of cumulative addition.

To give three examples,

$$
\begin{aligned}
\text{Concentration at section} \quad 5 &= 100 + 50 + 33 + 25 + 20 & &= 228 \\
\text{at section } 10 &= (228) + 17 + 14\tfrac{1}{2} + 12\tfrac{1}{2} + 11 + 10 & &= 293 \\
\text{at section } 15 &= (293) + 9 + 8\tfrac{1}{2} + 8 + 7 + 6\tfrac{1}{2} & &= 332
\end{aligned}
$$

These values have been plotted (curve $C$) in Fig. 146. It will be noted that an increase to some three times the value registered where the cones first interlock occurs quite rapidly but thereafter the proportionate increase is relatively slight. By section 50 the value reaches 450 and it is probably unwise to push the model much farther than that.

This exercise helps to explain some of the features described above and, before we leave it, there may be some interest in introducing a small open space or smokeless zone into the pattern. In Fig. 147 the area is only five units in width and runs for only five section-intervals downwind, yet the effect produced is quite remarkable. Curve $D$ in Fig. 146 shows the fall and recovery of concentration values downwind through the middle of this area. We have not attempted to give a scale to this diagram but if the smoke sources are interpreted as chimney clusters in terraced housing the smoke-free zone or open space could be as small as two to three acres.

Although other pollutants, such as sulphur dioxide from industrial plants and petrol or diesel fumes from road traffic, can have much more serious effects on urban health, it is smoke which produces the chief climatic effects since it intercepts a considerable proportion of the incoming solar radiation, particularly when the sun is relatively low. There is therefore a sharp reduction in the number of hours of bright sunshine. The Oldham Road station, east of Manchester, now closed, achieved the unenviable mean annual total of 970 hours (of a possible duration of 4460 hours). This helped to give that city a reputation which it now scarcely deserves, yet a more modern figure of the order of 1100 hours is still some 15 per cent less than that recorded at Ringway Airport only eight miles from the city centre. Central London yields a mean annual total of 1315 hours as compared with 1580 hours in open country to the north or south, a reduction of some 17 per cent, though the range in distance is greater. During the summer months this means that both cities lose about half an hour of bright sunshine each day, a considerable loss in terms of amenity.

That there is no consequent fall of temperature in large towns arises from a variety of factors. (*a*) The intercepted radiation warms the air aloft. (*b*) Rapidly drying surfaces of roads, walls and roofing respond readily to the balance of radiation received; losses via evaporation are thus confined to a short period after showers. (*c*) Air motion within a town is considerably

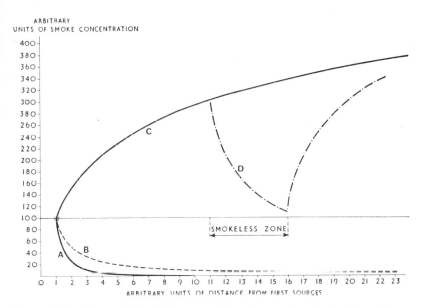

FIG. 146. The downwind variation in smoke concentration derived from the model illustrated in Fig. 145. A value of 100 is assigned to the concentration where the cones of the first row of sources interlock—at row 1. Curve A indicates the progressive decrease in the concentration from a *single* source such as A in Fig. 145. Curve B plots the decreasing concentration downwind from a *line* of equidistant sources all similar to A. Curve C shows the progressive build-up from a multiple *array* of line sources as in Fig. 145. Curve D is a downwind cross-section through the middle of the very limited smokeless zone introduced into the pattern at the eleventh row as in Fig. 147

impeded by the structures. (*d*) The smoke or haze pall also intercepts back radiation into space and thus checks the fall of night minima. All these factors depend to some degree on the prevailing weather situation but the net result is that, in the absence of relief effects, towns produce 'heat islands' of varying intensity both by day and by night. Attempts to demonstrate this by the use of mean daily maxima and minima have usually proved disappointing since the former are rarely as much as 1°C higher in town than in country though the latter may differ by up to 2°C. It is rather on particular occasions that the heat island becomes noteworthy but even so it is much more of a night-time than a daytime feature. By combining the evidence from a network of stations with traverses made in an instrumented vehicle, Chandler (1965, 150) has greatly increased the precision with which this feature can be analysed. His best daytime example (3 June 1969, *op. cit.*, 158) shows maxima in central London 2.2°C (4°F) higher than at equivalent levels east or west of the city. At night however in May (p. 164), June (p. 168) and October (p. 170) he found occasions when minima near the city centre were some 6°C (11°F) above those recorded in the environs. It must be emphasised however

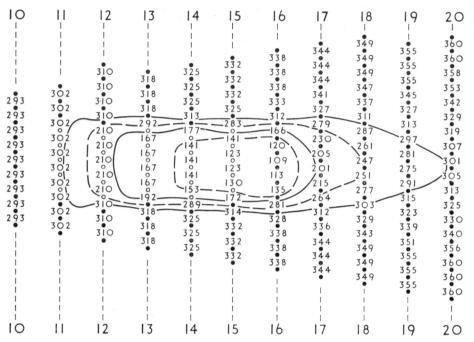

FIG. 147. A view in plan of the effect of the introduction of a smokeless zone only five units wide and five rows deep into a regular pattern of smoke sources. It is assumed that the wind is constant in direction. The figures give the relative smoke concentration as estimated in the text. Isopleths have been drawn from relative concentrations of 150, 200, 250 and 300. Smoke sources are indicated by the black dots and open circles represent sources closed down by the regulations. The effect is dramatic

that such readings are exceptional and the frequency group of daily minima (differences between Kensington and Wisley over the period 1951–60) given by Chandler (p. 150) has the form

| | $D_9$ | $Q_3$ | Median | $Q_1$ | $D_1$ | |
|---|---|---|---|---|---|---|
| | 5.6 | 3.9 | 2.1 | 0.8 | −0.2 | °F |
| | 3.1 | 2.2 | 1.2 | 0.4 | −0.1 | °C |

Thus on 66 nights out of 365, town temperatures were either lower than country temperatures or the difference was negligible: only on 36 nights was the margin in favour of the town station in excess of 3.1°C. Chandler notes that the heat island effect is most marked during periods of light winds and cloudless skies. Winds in excess of 23 knots eliminate it but on calm nights it usually has 'a steep (often cliff-like) margin' and may extend upwards for about 500 feet. Over pockets of open land, and particularly along the Thames, cooler air may descend from above thus providing some ventilation for the 750 square miles covered by the city. Similar effects over other large towns await more detailed investigation. (See Garnett, 1967, on Sheffield.)

A pertinent question is how far does the heat from domestic and industrial appliances affect the temperature regime of towns? An estimate quoted by Geiger (1965, 489) puts this at 40 cal/cm²/day, that is, of the same order as that received from total sun and sky radiation in midwinter in north-central Europe. However, we must not allow this figure to lead us to rash conclusions. Midwinter temperatures in Europe owe much more to advection of heat from the west than to direct radiation. The fact that Chandler found his best examples of the heat island effect on summer nights suggests that the contribution made by domestic heating is not decisive.

It has been suggested that the thermal effect of towns may help to trigger off local rainstorms when the general stratification of the atmosphere is favourable. Specific examples from Göteborg, Mannheim, Munich and Budapest are quoted in Geiger, and Parry (1956) has analysed a parallel event over Reading. There is nothing inherently improbable about this but a better case for town-generated thermals would be based on cloud amount and particularly on cloud height. This would require careful and prolonged observation from at least two points, one within and one outside a town in a region where relief contrasts were minimal. The author is unaware that such a task has been undertaken. Atkinson (1968, 1969) has indeed approached the question of cloud depth over London by an indirect route, i.e. through a careful and elaborate study of 'thunder rainfall'. He argues that this is 'formed in clouds predominantly caused by convective motions' and his maps indicate a distinct maximum over the conurbation, statistically significant perhaps only over its most central portion. On the evidence presented it is difficult to be sure that this activity is thermally induced, i.e. that it is a direct response to higher daytime maxima. The relative weakness of daytime urban heat island effects suggests that thermals would have little penetrating power. He offers as an alternative the play of mechanical turbulence arising from the greater 'roughness' of the city centre but wind-speed has played no part in the investigation. These two factors must clearly interact, the one upon the other, but Atkinson's case is the best yet made for an urban effect on the type, if not the amount, of precipitation. It appears to be strongest in summer during storms associated with warm fronts.

A very puzzling fact has been brought to light by Barrett (1964, 1966), namely that both summer and winter rainfall at four stations on the north-eastern fringe of Manchester have shown a remarkable upward trend from 1890 to 1960. Running 10-year means increase with astonishing consistency during this period from 29–32 inches to 38–42 inches, a feature not shared by other stations in the vicinity. The scale of this phenomenon, both in time and degree, is very different from an occasional shower. The suggestion that it may be a lee-effect associated with the growth of the conurbation meets with two difficulties: (*a*) that the spread has been of a comparatively open suburbia, and (*b*) that the city is normally so cloudy that insolation effects are far from powerful. The answer more probably lies in some quirk in the local circulation within the Manchester embayment.

Snowfall on low ground in Britain depends on such a delicate balance of forces that forecasters are reluctant to mention it until it has been reported from some station. The view that falls over towns are less both in frequency and quantity than over the surrounding country is thus not inherently improbable though the effect of relief must never be overlooked. Even mean data for number of days with snow lying (covering at least half of the ground) are largely vitiated by the wide variation from year to year. Chandler (1965, 240) was able to find little evidence of any such effect around London. For the Manchester area the following figures for two sharply contrasting winters give some support to the view that there is indeed a real contrast.

|  | Winter | |
| --- | --- | --- |
| *Snow lying* | *1946–7 (days)* | *1949–50 (days)* |
| Whitworth Park | 35 | 4 |
| Ringway Airport | 49 | 1 |

On the Continent and in America, where winters are more severe and snow is often of a dry powdery consistency, it is more likely to accumulate in city eddies, notably in courtyards and at road junctions, than over open country.

City fogs are notoriously unpleasant because of pollution but the notion that cities generate fog is largely fallacious; they generate smoke which turns a beautiful opalescent mist into a stinking abomination. To this they add days with smoke haze when mist is slight or absent. In both cases the essential requirement is a wind speed below 10 knots and the development of a low-level inversion, particularly over a valley site. Data quoted for the London area by Chandler (1965, 212) make it evident that 'first define your fog' is a necessary rule for clear thinking on this murky question.

*Estimated fog frequencies, London 1947–56*

| *Visibility (yards)* | *< 44* | *< 220* | *< 440* | *< 1100* |
| --- | --- | --- | --- | --- |
|  | (*hours per year*) | | | |
| Kingsway | 19 | 126 | 230 | 940 |
| Heathrow Airport | 46 | 209 | 304 | 562 |
| Southeast England | 20 | 177 | 261 | 494 |

To the pilot of an aircraft it is the righthand column that is significant but the driver of a private car has more interest in that on the left. Note that he is most likely to encounter real trouble in the suburbs. Memorable occasions occur when visibility drops to below 10 yards but a brightly illuminated (shallow) white fog then presents much the same problem for movement as the dark and filthy 'London particular'.[1]

# Climate in boundary layers: Microclimatology

Although the air actually penetrates into the pore spaces of rocks, soil and snow unless it is ousted by water, it is the outer or upper limits of these substances which forms the climatically 'active surface' wherever this is not displaced upwards by a close cover of vegetation. A similar surface overlies all open sheets of water though the reactions are there modified by the mobility and relative transparency of the medium. As we have seen, the activities involved include complex radiational exchanges, the molecular diffusion of water, the provision of fine aerosols[2] and the imposition of frictional drag upon air motion. Such an active surface thus underlies the whole of the air-ocean and in its immediate neighbourhood the air acquires a fine stratification or lamination which cannot be recognised by traditional screen observations. This is the field of enquiry of microclimatology.[3]

That the term 'lamination' is not too strong when applied to *air temperature* is evident from mirage effects, notably the sheen over a smooth road on any warm sunny day. Geiger (1965) quotes 40°C (104°F) as a characteristic surface temperature on natural ground during a sunny July afternoon at Leipzig though the mean screen maximum was only 23°C (74°F) and the absolute screen maximum 36°C (97°F). In lower latitudes ground temperatures of 60°C (140°F) are by no means uncommon and rocks are almost too hot to touch. Seeking some relief from the burdens of office, congressmen have been known to fry eggs on the marble steps of the Capitol in Washington. Since air is a notoriously poor conductor and since air motion in the lowest few centimetres is usually slight, a thin layer of air soon takes on similar characteristics. A contrary effect at night is less readily observed except where it leads to the deposition of hoar frost. As Sutton (1953, xi) puts it, 'the layers of air within a fraction of an inch of the ground may experience both tropical heat and icy cold in the course of a single day'. Lowly creatures seek refuge

[1] On one afternoon during the infamous London 'smog' of December 1952 the author found it impossible in the middle of a seven-yard pavement to see either the shop-fronts to his left or the curb to his right. The eerie sensation of stepping into a dimensionless world whilst still in a busy and familiar street was quite unforgettable.

[2] Aerosols are very fine particles such as smokes, dust, spores or spray which remain suspended for so long that they can be regarded as virtually a component of the air itself. They form only a tiny proportion of its total mass.

[3] Because similar techniques are employed this title has also been given to studies of climate *within* crops and soils. There is a real distinction here which would be clearer if this were called Bioclimatology. Obviously the two fields are related.

from these extremes either under vegetation or below ground. The general effect on air and soil temperatures on five spring days at Seabrook, N.J. (39°34′N, 75°13′E) is illustrated in Fig. 148. The diurnal range at screen level was 11°C, at 10 cm 15°C and at ground level probably over 20°C. A few centimetres below ground it became almost negligible.

Investigations over a number of years at stations in southern England covering the lowest 350 feet above the surface yield the following seasonal values of the lapse-rate expressed in degrees Celsius per 100 metres (Table

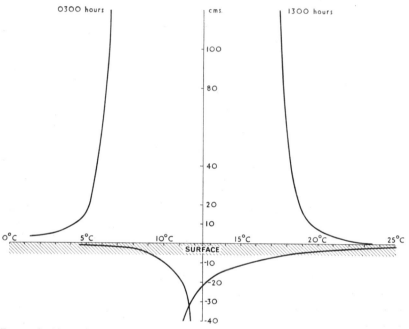

FIG. 148. Air and soil temperatures near the ground by day and by night. The data were obtained at Seabrook, N.J. on five bright days in spring but the contrast between near-surface and screen-level temperatures is normal

51). If some of the figures look surprising it must be remembered that, particularly in the lower ranges, the rate indicated is maintained only over a very limited vertical distance. The height ranges were in fact as follows:

|  | Elevation | | Range in metres |
|---|---|---|---|
|  | Feet | Metres |  |
| E | 350–155 | 106.7–47.2 | 59.5 |
| D | 155–50 | 47.2–15.2 | 32.0 |
| C | 50–3.5 | 15.2–1.1 | 14.1 |
| B | 4–1 | 1.2–0.3 | 0.9 |
| A | (12–1 inches) | 0.3–0.025 | 0.275 |

Data for height ranges A and B are given in detail in *Geophysical Memoir* 65 (Best, 1935) and those for levels C, D, and E in *Geophysical Memoir* 89 (Best *et al.*, 1952). Neither the stations employed nor the years covered by the two investigations were identical but the records in this form may be usefully combined. It should be recalled that, in the free atmosphere, the DALR is $-1.0°C/100$ m and the normal environmental lapse rate about $-0.65°C/100$ m. It is evident at once that both by day and by night conditions near the ground are highly abnormal. If all positive lapse rates are described as 'inversions', all negative lapse rates beyond $-1°C/100$ m called 'superadiabatic', and rates between $-1$ and $0°C$ (isothermal conditions) are regarded as 'normal', then the total mass of the observations yields the following frequencies. (The data for layer B are actually derived from monthly means quoted in Geiger, 1965, 86, but appear to be representative. Layer A yields very similar proportions.)

|  | Superadiabatic (%) | Normal (%) | Inversions (%) |
|---|---|---|---|
| Layer E | 9 | 59 | 32 |
| Layer D | 33 | 28 | 39 |
| Layer C | 36 | 10 | 54 |
| Layer B | 39 | 1 | 60 |

Hence, in close proximity to the ground, the 'normal' fall of temperature with height is only experienced for a brief period during the changeover from night-time inversions to daytime superadiabatic conditions. Occasionally indeed an inversion may persist for more than 24 hours.

During the day, nevertheless, even in cloudy and windy Britain, superadiabatic lapse rates are the rule rather than the exception though in winter they occur only between 1000 and 1400 hours and extend only a few feet from the surface. In summer, on the other hand, they normally prevail from 0600 to 1800 hours and the lowest 155 feet of the atmosphere is commonly affected. Under very sunny conditions values far in excess of those quoted in Table 51 have been observed; thus at Rye in layer C the lapse rate has reached $-20.5°C$ in May and $-9.8°C$ even in December. Under these exceptional conditions layer E was also affected. In our treatment of convection (p. 247) we followed Shaw in stressing the significance of the progressive growth of an 'isentropic layer' (a layer of uniform potential temperature) in the lowest few hundred metres of the atmosphere. It is this layer which forms the *reservoir* of warm air on which active convection depends. It will be recollected that the lapse rate within such a layer (the DALR) is represented by a sloping straight line when a lineal scale of temperature is plotted against a logarithmic scale of pressure or a lineal scale of height. With great ingenuity K. Brocks (quoted Geiger, 1965, 77–82) has demonstrated that near the ground a new rule prevails. To produce a straight line graph it is here necessary to use double-log paper and plot log temperature against log height. Brocks called this zone the 'unstable sublayer'. This begins to grow when the elevation of the sun reaches some 10° above the horizon and deepens until it reaches 30°. Thereafter the unstable sublayer does not thicken but additional heat is passed upwards to deepen the isentropic layer which succeeds it aloft. Under strong

TABLE 51. *Mean lapse rates °C/100 m over southern England at five levels*

| SEASON | 0400 hrs | 0800 hrs | 1200 hrs | 1600 hrs | 2000 hrs | 2400 hrs | HEIGHT RANGE |
|---|---|---|---|---|---|---|---|
| *Winter (Dec., Jan., Feb.)* | | | | | | | |
| E | 0.27 | 0.13 | −0.33 | −0.25 | 0.06 | 0.19 | 59.5 m |
| D | 0.83 | 0.66 | −0.95 | −0.35 | 0.52 | 0.75 | 32.0 m |
| C | 4.15 | 3.25 | −0.45 | 2.50 | 4.50 | 4.85 | 14.1 m |
| B | 25 | 15 | −13 | 17 | 30 | 27 | 0.9 m |
| A | 104 | 77 | −129 | 110 | 148 | 142 | 27.5 cm |
| *Spring (March, Apr., May)* | | | | | | | |
| E | 0.63 | −0.23 | −0.89 | −0.69 | 0.22 | 0.63 | 59.5 m |
| D | 2.17 | −0.97 | −1.60 | −1.27 | 0.83 | 1.63 | 32.0 m |
| C | 7.10 | −1.75 | −5.35 | −2.70 | 4.80 | 7.10 | 14.1 m |
| B | 33 | −26 | −62 | −27 | 41 | 40 | 0.9 m |
| A | 111 | −239 | −440 | −136 | 170 | 152 | 27.5 cm |
| *Summer (June, July, Aug.)* | | | | | | | |
| E | 0.84 | −0.30 | −0.53 | −0.38 | 0.28 | 0.86 | 59.5 m |
| D | 1.65 | −1.60 | −2.10 | −1.61 | 0.62 | 1.42 | 32.0 m |
| C | 5.65 | −3.45 | −5.95 | −3.25 | 3.80 | 6.25 | 14.1 m |
| B | 20 | −32 | −61 | −36 | 26 | 30 | 0.9 m |
| A | 52 | −375 | −559 | −258 | 122 | 104 | 27.5 cm |
| *Autumn (Sept., Oct., Nov.)* | | | | | | | |
| E | 0.64 | 0.51 | −0.43 | −0.25 | 0.35 | 0.63 | 59.5 m |
| D | 2.84 | 1.10 | −1.16 | −0.45 | 1.72 | 2.34 | 32.0 m |
| C | 5.35 | 0.30 | −3.10 | −0.05 | 6.70 | 6.05 | 14.1 m |
| B | 26 | −2 | −28 | 4 | 35 | 32 | 0.9 m |
| A | 94 | −51 | −233 | 35 | 144 | 117 | 27.5 cm |

*Source:* Geophysical Memoirs, 65, 89.

insolation Brocks considers that the unstable sublayer may reach some 40 metres in thickness in summer. Note however that even this rule does not apply to air in immediate contact with (i.e. within 10 cm of) the ground.

At night positive lapse rates (inversions) occur at every level given in Table 51 and there is little variation in their intensity throughout the year. Also built up from below, as is clearly shown by the figures for 1600 hours in both autumn and winter, this feature results in the production of an 'inversion sublayer' closely analogous to the unstable sublayer but with a temperature gradient in the reverse direction. This rarely exceeds 20 metres in depth, after which the chill spreads through an increasing height until sunrise. It will be noticed that the strongest positive gradients at level D occur at around 0400 hours in all four seasons, i.e. at a time when the gradients at lower levels are beginning to weaken as a result of the thickening of the cold layer. Again it must be emphasised that occasionally the inversion is exceptionally strong, extreme values observed at Rye reached $53.5°C/100$ m for layer C and $13.5°C/100$ m even as high as layer E. The latter record was in the presence of a snow cover, but did not persist for long.

In general terms the changeover from a predominantly outgoing to a predominantly incoming radiation pattern, and vice versa, occurs about an hour after sunrise and before sunset respectively though the time taken by the sun to move through the lowest ten degrees of elevation above the horizon must clearly vary with both season and latitude (see Fig. 3).

Evidently the almost universal occurrence of inversions at night must have a marked effect upon the incidence of 'ground frost'. The difference between screen minima and minimum values recorded by a thermometer exposed at a height of 5 centimetres above short grass (grass minima) has been carefully analysed by F. Witterstein for a station in the Rhineland. His frequency groups for dull and clear nights with a wind speed of less than 9 knots (Geiger, 1965, 100) may be conveniently summarised as follows:

*Difference between screen minima and grass minima* $(°C)$

|  | Dull nights | Clear nights |
| --- | --- | --- |
| Least difference | $-1.4$[1] | 0 |
| $Q_1$ | 0.25 | 2.0 |
| Median | 0.75 | 2.6 |
| $Q_3$ | 1.3 | 3.25 |
| Greatest difference | 4.0 | 6.5 |

[1] Grass minimum above screen minimum.

It was noted that some of the greatest differences occurred on clear nights in April and May, just when the frost hazard is greatest for small plants. A similar investigation at Cranwell on nights with less than four-tenths cloud cover showed grass minima to be $7°F$ ($3.9°C$) lower than screen minima when the wind at 13 metres was less than 7 knots though the difference was halved when the wind was above 13 knots.

It may at first appear rather surprising that there is a similar stratification in water vapour content in the lower layers of the atmosphere and that here too we find both negative (humidity decreasing with height) and positive gradients (humidity increasing with height). Table 52 presents a summary of the observations made at Rye (Best *et al.*, 1952), again in gradient form, for the three levels C, D, and E. It may be noted that an overall analysis of the data showed negative gradients for only 62 per cent of occasions (positive values 38 per cent) at *all three levels*. This fact is rather obscured by the process of averaging, particularly in level E. Since evaporation is a process of removing water vapour *upwards*, negative gradients might well be regarded as normal and the high proportion of positive values at this station may well be a result of the close proximity of the sea.

Negative values are particularly large during the day in layer C ($3\frac{1}{2}$–50 ft) both in spring and summer. Note that they rise to an early maximum a few hours after sunrise when insolation is stimulating evaporation whilst eddy diffusion is still comparatively weak. Subsequently the gradient is gradually weakened as moisture is more freely transferred upwards until towards sunset the evaporative process is itself slowed down. On occasion during the day values far in excess of those given have been observed, the maximum gradient recorded in layer C was $-40.8$ gm³/100 m and in layer E, $-7.5$ gm³. The low values in winter should be noted.

Positive gradients at night develop in the reverse fashion. They first appear in rather a weak form soon after sunset and become strongest shortly before dawn. This humidity 'inversion' may be responsible for dew though a deposit may occur when no such feature can be recognised, the moisture coming more or less directly from the soil. It will be noticed that in autumn, the 'season of mists', the 'inversion' is normally deeper than at any other season of the year. Extreme gradients reaching 22.2 gm³ in layer C and 11.3 gm³/100 m in layer D have been observed at this time.[1]

We conclude this section with a brief reference to the variation of *wind speed* with height in close proximity to the earth's surface. The essential truth has been expressed by Sutton (1962, 136) in a striking phrase, 'even a gale is brought to rest on the surface of the ground'. It is thus a fair presumption, well attested by observation, that wind speed normally increases with height though the range over which that increase extends must depend upon the circumstances. In the case of a shallow fall-wind or katabatic downflow, for instance, the rule applies within the bulk of the active cold air whilst the atmosphere above may remain comparatively inert.

A question of obvious importance follows: at what *rate* does the increase of wind speed occur?—or, put more technically: what is the *form* of the velocity profile? Here difficulties begin to multiply. Over a flat plate in the controlled conditions within a wind tunnel the rate of increase has been shown to be exponential, the mean speed increasing as the one-seventh power of height. It may be inferred that wind profiles over natural surfaces belong broadly to a similar family of curves but two new factors are introduced; (*a*) turbulent flow mechanically induced in varying degree by the roughness of natural surfaces, and (*b*) the marked influence on the development of vertical eddies exerted by the thermal stratification of the lower layers of the atmosphere. Vertical eddies transport momentum thus decreasing the exponent by bringing a more active circulation towards the lower layers. They are particularly

---

[1] Geiger (1965) gives an estimate of 30–40 mm as the annual total of dewfall in western Europe. This is about a tenth of the water lost by evaporation. The upward and downward transports of water-vapour in the atmosphere are so far out of balance because the vertical circulation at night is much weaker than by day. Precipitation from far aloft is required to square the account.

characteristic of the daytime situation. Low-level inversions at night produce a more laminar flow, the motion of the lower layers is then more independent of the wind aloft and the exponent must therefore be increased.

In view of the great importance of low-level air motion in the diffusion of heat, humidity, smoke and airborne insects and seeds a great deal of mathematical ingenuity has been employed in the endeavour to find ex-

TABLE 52. *Mean lapse rates of humidity* $(gm^3/100\ m)$ *over southern England*

| SEASON | 0400 hrs | 0800 hrs | 1200 hrs | 1600 hrs | 2000 hrs | 2400 hrs |
|---|---|---|---|---|---|---|
| *Winter (Dec., Jan., Feb.)* | | | | | | |
| E | −0.01 | −0.03 | −0.16 | −0.11 | −0.06 | −0.03 |
| D | −0.02 | −0.01 | −0.31 | −0.12 | −0.09 | −0.02 |
| C | 0.66 | 0.31 | −0.74 | 0 | 0.31 | 0.64 |
| *Spring (Mar., Apr., May)* | | | | | | |
| E | 0.04 | −0.09 | 0.01 | −0.01 | 0.01 | 0.01 |
| D | 0.12 | −0.51 | −0.25 | −0.19 | −0.25 | −0.13 |
| C | 1.33 | −2.36 | −2.12 | −1.59 | 0.52 | 1.31 |
| *Summer (June, July, Aug.)* | | | | | | |
| E | −0.12 | −0.21 | −0.20 | −0.15 | −0.19 | 0.17 |
| D | −0.34 | −1.00 | −1.20 | −1.04 | −1.04 | −0.68 |
| C | 1.28 | −4.87 | −4.49 | −3.92 | −1.59 | 0.74 |
| *Autumn (Sept., Oct., Nov.)* | | | | | | |
| E | −0.18 | −0.20 | −0.44 | −0.37 | −0.36 | −0.30 |
| D | +0.59 | −0.20 | −0.44 | −0.02 | 0.06 | 0.23 |
| C | 1.16 | −1.05 | −2.37 | −1.45 | 0.91 | 1.12 |

Positive values indicate an *increase* in humidity with height.
Negative values indicate a *decrease* in humidity with height.

pressions to fit varying conditions of surface roughness and static stability. Here we shall simply draw attention to some comparative observations made at 5 and 400 feet (from a captive balloon) over level grassland (Frost, 1947). The rule which emerged was of the form

$$\bar{u} = \bar{u}_1 \left(\frac{z}{z_1}\right)^p$$

where $\bar{u}_1$ is the mean wind speed at the ground-based anemometer at height $z_1 = 5$ feet and $\bar{u}$ is the mean speed at some other level, $z$. The exponent $p$ was observed to vary with the lapse-rate.

Under adiabatic conditions (a fall of temperature of $1.2°$C over the 120 metres range between the 5- and 400-feet levels) $p = 0.15 = 1/7$, so the rule corresponds very closely to that derived from wind tunnel experiments. When the layer was virtually isothermal (lapse rate $0°$C) $p = 0.3$ proved more appropriate. With a moderate inversion (a positive lapse rate of $2.5°$C)

$p = 0.45$ and when the inversion reached $5°C/120$ m, $p = 0.6$. The general effect of this variation in the exponent is readily appreciated if the heights selected are successively doubled, for $(z/z_1)$ then equals 2. Since

$$(2)^{0.15} = 1.11; \qquad (2)^{0.3} = 1.23; \qquad (2)^{0.45} = 1.37; \qquad (2)^{0.6} = 1.52$$

we can then build up the following geometric series, taking the wind speed at 5 feet at any convenient figure. We select *5 knots* for the adiabatic and isothermal cases but for inversions *2 knots* is considered more appropriate. In both instances it is the proportionate increase of wind speed with height that is really significant.

TABLE 53. *Estimated wind speeds at selected heights under different lapse rate conditions (knots)*

| | | | INVERSIONS | |
|---|---|---|---|---|
| LEVEL (ft) | ADIABATIC $p = 0.15$ | ISOTHERMAL $p = 0.3$ | *Moderate* $p = 0.45$ | *Strong* $p = 0.6$ |
| 5 | 5.0 | 5.0 | 2.0 | 2.0 |
| 10 | 5.5 | 6.1 | 2.7 | 3.1 |
| 20 | 6.2 | 7.6 | 3.8 | 4.6 |
| 40 | 6.8 | 9.3 | 5.1 | 7.0 |
| 80 | 7.6 | 11.4 | 7.1 | 10.7 |
| 160 | 8.4 | 14.0 | 9.7 | 16.3 |
| 320 | 9.3 | 17.3 | 13.3 | 24.7 |
| 640 | 10.3 | 21.3 | 18.2 | 37.5 |

Since the 'standard' exposure for anemometers is 10 metres (33 ft) above ground level the wind speed record is thus considerably above that experienced at face-level, particularly during inversions.

# 14
# Climatic change

We now reach a number of baffling problems which have stimulated a vast literature touching, at some point or other, almost every branch of scientific enquiry. Does climate change? Are there bounds beyond which change is improbable, or indeed impossible? What rational grounds can be found to explain apparent changes for which the evidence to hand, gathered from a multitude of specialisms, seems quite irrefutable? The questions are straightforward enough, but the answers still largely elude us. Nevertheless the speculation involved has played an interesting part in the history of the ideas man has entertained about the planet on which he lives and a critical survey of these notions will provide a useful check on the present state of knowledge.

It is necessary first to clarify some of the concepts involved in this discussion; for instance, the nature of 'climate' itself. In the throes of a recent examination a student coined the aphorism, 'Climate is the invention of Man'. He could have sharpened the point by adding 'though weather is real'. Meteorological processes are energy exchanges which operate through time; short-period change is thus of their very essence. Some of these changes, such as those associated with the diurnal march of the sun or the rather less regular march of the seasons, are so clearly periodic in character as to suggest mechanical repetition. Others contain so great a random element that, in the western world, the weathercock is a recognised symbol of inconstancy. In 'inventing' climate, that is, in building up a mental concept apparently valid over a considerable range of space and time, we draw on both tradition and experience to assure ourselves that only a limited sector of the total spectrum of possible weather events is likely to occur. We are thus assessing probabilities and these are best expressed in terms of a likely range of variation. Change is still with us but we try to locate its bounds. Do these bounds change through time?

The question raises two others, over what space and over what time? As comparatively sedentary creatures we associate climate with *place* but weather events occur in a very mobile medium and often at considerable height. The connection between the factors of macroclimate and the underlying ground is thus often quite tenuous. Air masses are not firmly anchored in position, they can migrate, within limits it is true, but nevertheless with a considerable degree of freedom. It follows that, if the displacement is at all prolonged or frequent, climatic patterns may migrate with them. True, this is one aspect of 'climatic change' but it is a special aspect which does not call for any very fundamental changes in the terrestrial heat budget nor indeed for any highly exceptional distortion of the systems of circulation. British summers may be 'Mediterranean' (1947, 1959) or 'Icelandic' (1954)—the contrast is remarkable—yet both are comprehended within the same worldwide system of processes. The idea of climatic migration has been frequently canvassed in the past but far too often it has been conceived as a rather rigid and mechanical shift of 'zones' which might well require some extraterrestrial or astronomical foundation. It is pertinent therefore to emphasise that the climatic zones are no less an 'invention of man', particularly of our classical ancestors. The tradition dies hard but modern thinking lays less and less stress on purely zonal concepts.

Whether we are thinking of worldwide climatic change or only of some degree of climatic displacement, the time scale is of paramount importance. In this sphere the rule 'the present is the key to the past' operates only within rigid limitations. Any rational theory of climatic change based upon present-day knowledge must concern itself with a world not radically different in physical structure from the world we know. Geological evidence from before Tertiary times is thus largely irrelevant and nothing but confusion can result from the attempt to cover all geological time. This was true even when a place was considered to have geographical coordinates which were fixed and immutable though oceans could come and go. The attempt to reconstruct oceanic circulations and to assess varying degrees of continentality was then highly speculative though not entirely irrational. With the introduction of the concepts of 'continental drift' and of 'polar wandering' the whole system of geographical reference is disoriented and, as is evident from the literature, nothing but the crudest of zonal climatic concepts can be applied. All that a climatologist can contribute in this sphere is the warning that evidence must not be taken at face value since the general circulation under such conditions is completely unknown. As we shall see, he has problems enough during the Quaternary.

It is now realised that climatic distributions are the result of a balance between stupendous forces, often opposed and subject to complex interactions within the atmosphere itself. To these are added feedback processes from the earth beneath—heat and humidity from the oceans, the radiative effects of persistent snow—which often involve a considerable lag in time. Climate is thus just as much a matter of residuals as the economic health of a modern state. Little wonder that the balance swings this way and that, and that sometimes the change seems to betray a persistent trend. Residuals are always difficult to evaluate and thus it is rarely possible to be sure that a trend is real and not illusory. The basic problem of climatic change is thus the recognition and explanation of truly secular trends, changes maintained over such a time and to such a degree that an immediate return to the *status quo* is unthinkable. Faced with the evidence of the Ice Ages it is impossible to question that such changes do occur. Most of the historical evidence is much more questionable since parallel effects might well arise from well-known processes of physical geography often exacerbated by careless husbandry or overstocking.

# The evidence for climatic change

## The glacial period

The evidence of surface geology and the form of the land indicates that on three or four successive occasions in comparatively recent times continental ice caps were built up on the Baltic and Laurentian shields, i.e. in latitudes where, although modern snowfall is considerable, it never persists throughout the summer on low ground. The absolute time-scale of these Pleistocene events is disputed, estimates vary from 300000 years (Flint, 1957) to 1500000 years (Ericson and Wollin, 1966), but the relative scale is now generally agreed. Each glacial invasion is estimated to have endured for at least one-twentieth of this period and the remarkable facts emerge that (*a*) they were broadly concentric, (*b*) they were of approximately similar intensity, and (*c*) the relative positions of their southernmost margins in the two continents (51°N in Europe and 40°N in America) correspond with reasonable accuracy to the relative location of modern midwinter isotherms. True, their limits towards the drier continental interiors were more varied but this is hardly surprising as no one expects that the surface contours of the ice-caps would have been identical. It is not known conclusively whether the ice-caps completely disappeared between each successive incursion but the apparent length of the interglacial phases and the evidence that they were at least as warm as at present certainly suggests that they did so. Whether these events on the two sides of the North Atlantic were strictly contemporaneous seems to be incapable of proof, the general presumption by geologists is that broadly they were but this is little more than an inspired guess based on the parallel nature of the histories revealed. The climatologist could make much more of this problem if he knew which ice-cap developed first; he may suspect it was the American but positive knowledge is denied him. The dating of the final phases is much more accurately known, it appears to have been broadly similar in the two continents with America lagging perhaps by some thousands of years. This does not prove much since the climatological problems posed by the 'retreat' of the ice are different from, and even more intractable than, those presented by its initial accumulation. Postglacial time may be variously defined but few estimates exceed 18000 years. It is thus possible that we are now in the early stages of yet another interglacial period. This gives the subject much more than purely academic interest. Some comfort may be drawn from the facts, (*a*) that the cycle is not regular (see Fig. 149) and (b) that no evidence of a similar fluctuation can be found during the 65–70 million years of Tertiary time. The series of events described may thus be termed 'unique'. It is true that geologists have found convincing evidence of much earlier glaciations but, since these occurred when terrestrial geography was fundamentally different from its present pattern, it would appear that the attempt to embrace those events within a single comprehensive theory is doomed to generate much more confusion than light.

Parallel with the development of these remarkable ice-shields on two sectors of the earth's crust noted for their tectonic stability there is ample evidence of a changing snowline in many unstable and now lofty regions where mountain-building cannot be ruled out as a contributory cause. Notable examples include the Alaskan–Canadian cordillera and the Sierra Nevada in America and range from the Atlas, through the Alps to the Caucasus, Tian Shan and beyond in the Old World. Most of these mountains still carry a permanent snow cover of more restricted dimensions. From a climatological point of view it seems wise to distinguish these 'local' glaciations quite clearly from the 'regional' glaciations described above. Powerful reasons for such a distinction, not always made in the literature, include:

1. The fact that, with the possible exception of the Alaskan–Canadian dome, the area of ice involved, even at its maximum, was unlikely to have been large enough to have had much effect upon the general circulation.
2. These local glaciations might well have been no more than secondary effects generated by the presence of the regional ice-shields.
3. Though snow accumulation everywhere involves a complex and rather delicate balance between snowfall, ice movement and ablation, the relative weight of these various factors will not be the same in regions of high relief or lower latitude as in the economy of a comparatively low-level ice-shield of subcontinental proportions. The question whether glacial episodes are initiated primarily by a fall in temperature or rather by more active precipitation (particularly in winter) cannot be regarded as closed. Rightly or wrongly, the general weight of opinion favours the former but the point made here is that the evidence of mountain snowlines is, at best, equivocal. Here more active precipitation, at any season of the year, could have had much the same effect. Once established, a continental ice-sheet is perfectly capable of lowering temperatures in its vicinity, but this does not prove beyond doubt that the fall in temperature was worldwide.

It was perhaps a little unfortunate that comprehensive glaciological observations were first made in the Alps, a local system but near enough to the continental ice-sheet, when fully developed, for both temperature and precipitation effects to have been experienced together. It was here that Penck and Brückner (1909) first demonstrated the multiple nature of the events, attaching the names of four Danube tributaries, Günz, Mindel, Riss and Würm to the successive stages they recognised. They also provided a relative time scale for the series based upon the degree of weathering of the successive foreland deposits. Although Flint (1957) considers such evidence unreliable it would appear that their story is broadly confirmed by the testimony of ocean cores (see Fig. 149).

An intelligent appreciation of Ice Age climatology demands factual information drawn from a much wider range of latitudes. What happened in the southern hemisphere during the Pleistocene and, above all, what is the testimony of the Tropics?

Antarctica remains covered by the only great ice-shield in the southern hemisphere and it is highly improbable that it ever had a serious competitor. Its age may never be determined but there is no reason for doubting that its history would take us well back into the Pleistocene. Goldthwait (1966, in discussion p. 229) has called it 'potentially the longest weather record in the world', but it is certainly not ready of access. There is some evidence that its depth was formerly greater but its limits so frequently reach the sea that methods of interpretation so carefully evolved in Europe and America cannot be applied. Evidence from the ocean floor suggests that pack-ice was formerly much more extensive. The glaciers of Chile and New Zealand are of the local or mountain type which we have learned to interpret with caution. In Chile the mountain ice formerly reached the sea in latitude 42°S as compared with 46°40′S today (compare with the corresponding figures of 48 and 57°N in western Canada). Both here and in the South Island the lower snowline appears to have been broadly contemporaneous with the last glaciation in the northern hemisphere but, in the absence of proven interglacials, the parallel can be drawn no further. Perhaps this is just as well, too close a fit between such very different parts of the world would invite more suspicion than trust.

The great bulk of the solar income received by the globe is intercepted between 40°N and 40°S latitude. Surely it is here rather than towards the poles that we should seek evidence of any overall change in world climate? The search is being made but so far the evidence seems either doubtful or contradictory. There are several reasons for this.

1. There are good reasons for believing that the equatorial rain-forest is of great age. This would imply that temperature variations have been kept within a limited range. Against this it can be argued that beneath the larger trees the forest maintains its own climate to a degree that could be proof against changes sufficient to have important repercussions in more marginal regions.

2. There is evidence that the snowline on lofty intertropical mountains was formerly lower by some 600 to 750 metres (2000–2500 ft). Unfortunately when this occurred and whether it was repeated on more than one occasion is invariably obscure. To leap to the conclusion that this is proof positive of a fall in mean temperature throughout the tropical girdle of the order of 6–8°C is climatologically naïve. It would imply that neither cloudiness nor precipitation had been affected by the change. Though widely scattered, the areas involved form a tiny fraction of the vast spread of the intertropical zone and the same result might be expected from a slight increase in atmospheric instability. This could arise from a number of causes amongst which an actual *increase* of sea-level temperatures by a degree or two cannot be ruled out. More persistent daytime cloud and an increase of precipitation on and around the

peaks could then easily invalidate conclusions based upon the normal fall of temperature with height. An alternative approach via the vertical zonation of vegetation is no more reliable (Morrison, 1966).

3. It has been argued with more conviction than insight that the glacial episodes in middle latitudes must have been accompanied by 'pluvial' periods in lower latitudes. Too often indeed the argument rests on a purely deductive approach, e.g. the glacial anticyclones over the Laurentian and Fenno-Scandinavian ice-domes must have displaced the belt of cyclonic storms southward 'through a distance of about 15° of latitude from the position it occupies today' (Flint, 1957, 244). This is altogether too rigid a view. It is conceivable that a great ice-cap over north-western Europe might well affect the course of the jet stream over a limited range of longitude, particularly in summer, but in winter the cold pads of Arctic air over Canada and the USSR, though invisible, are just as effective on the steering pattern as if they were composed of some tangible material. One hazards the guess, and it can be little more, that the major effect of any such displacement would be confined to the Pacific coast of America and the Atlantic coast of Mauritania. In any event 'glacial anticyclones' now run at a discount (see p. 220) for they fail to answer the question, 'whence the snow supply?'

For positive evidence of pluvial periods one is compelled to turn to the archaeological record, evidence of the former levels of existing or now extinct lakes and, above all, to signs of migration of the arid border. Vital to this issue, since the problem is there faced on a grand scale, is the question of a more humid Sahara. Here we are faced with a paradox. Geomorphological and geochemical evidence points firmly towards the great age of vast expanses of this desert. Rock paintings of elephants, rhinoceros, and giraffes in sites many hundreds of miles from where such beasts could today satisfy their huge appetites seem to imply a catastrophic decline in the vegetative cover. The key to the paradox may rest with the great wadi systems which radiate from the central highlands. Though these have carried water at some time, it follows neither that there was much rainfall on the interfluves nor that the flow was other than intermittent. Some increase in the frequency and perhaps the violence of torrential cloudbursts, concentrated for the most part over the mountains, could easily raise the water-table in the wadi floors to a level which might support a gallery forest. If this is what is meant by a pluvial period, the term may be accepted but let us not imagine that large areas of the desert ever blossomed as the rose. Butzer (1966, 73) has estimated that mean annual rainfall in the central highlands may have reached 50 to 200 millimetres as compared with the 5 to 75 millimetres recorded to day. In arid lands mean annual totals obscure more than they reveal; such totals could be easily achieved in a very few days. Towards its southern limits the Sahara presents a complex story. Wide belts of ancient dunes, now weathered and stable, suggest a long period of greater aridity but these are interspersed with old lake basins, notably that of the 'Mega-Chad' which rose at least once some 140 feet (42 m) above its present level. Dating was once obscure but Grove and

Warren (1968) now attribute Mega-Chad entirely to the postglacial period (10000–5000 BP). It should also be noted that everywhere along this border the source of run-off was the belt of equatorial rains. The link with glacial episodes is thus tenuous in the extreme.

Rain, like snow, leaves little evidence except where it accumulates but, unlike snow, it may accumulate many hundreds of miles from where it falls, forming a lake in perhaps an entirely different climatic province. Furthermore, wherever a lake could possibly have been reached by meltwater from continental or mountain ice, a rise in lake level is by no means positive proof of increased precipitation. Even where no such connection can be established (the examples are not numerous) changes in level may be the result of minor earth movements, the erosion or blocking of outlets and even the growth or decline of aquatic vegetation. In regard to Lakes Bonneville and Lahontan it has been argued that the high stages were essentially synchronous with glacial phases—'High-water lake phases occurred only slightly after glacial maxima' (Flint, 1957, 223). In the Caspian it has been established that high stages corresponded to low stages of the Black Sea, presumably the result of ocean water being locked away as continental ice. Though these observations may serve to prove that high stages were not interglacial, they do not resolve our difficulty. We must be careful in our use of terms. From a strict geological point of view the end of a glacial phase may be reasonably dated at the virtual disappearance of land ice but, from a hydrological angle, the important date comes much earlier, it is when the ice *begins to wane*. It seems the most natural thing in the world that meltwater-fed lakes like the early Caspian should rise rapidly until they overflowed, the level then falling as outlets were eroded and as evaporation overtook a depleted supply. All these events might well occur within a 'glacial' phase, as geologically defined, and the response of ocean-level would be much later. What rainfall had to do with such changes in lake-level demands a precision of timing not yet achieved.

4. Ocean cores. The evidence of climatic oscillations in intertropical lands having thus proved both fragmentary and equivocal, scientists turned with high hopes to the record locked away, apparently in complete security, beneath the ocean floor. Cores about four feet in depth obtained by the 'Meteor' from the tropical Atlantic in 1925–7 showed relics of existing types of temperature-sensitive planktonic foraminifera only in the top fifth of the record; below there was an abrupt change to types now found only in middle and high latitudes. Was this the result of the last Ice Age and, if so, what would be revealed if much longer cores could be secured?

The technique of deep-sea core sampling has made great strides since 1947 and the work of one organisation, the Lamont Geological Observatory of Columbia University, has been vividly described by Ericson and Wollin (1966). It has become evident that the ocean deeps are not as still as had been imagined and that processes of erosion, transport and redeposition by 'turbidity currents' may truncate a record in one place and intercalate graded

477

material in the sequence in another. The new geology is thus almost as complex as the record of the normal sedimentary series. With great ingenuity and patience the Lamont school has fitted together the records of 26 cores obtained from the Caribbean and tropical Atlantic to reveal a presumed

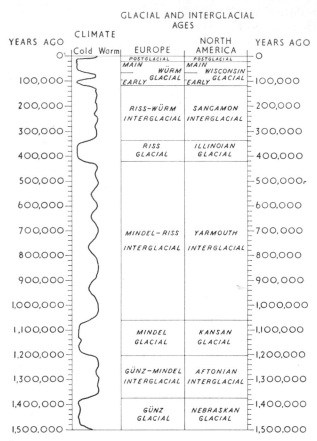

Fig. 149. Ericson and Wollin's interpretation of climatic change during the Pleistocene as indicated by a number of deep-sea core samples from the tropical Atlantic and the Caribbean

history which, at least in a qualitative sense, bears a remarkable resemblance to the conclusions reached by Penck and Brückner (1909) (Fig. 149). Two noteworthy features emerge: (*a*) the faunal changes both at the beginning and the end of each glacial epoch were abrupt. It is therefore inferred that 'long intermediate intervals of slowly rising or falling temperatures do not seem to have occurred' (Ericson and Wollin, 1966, 136). (*b*) Eight cores, regarded as beginning in recent times, indicate that the last glaciation (the Würm or Wisconsin) had a two-stage character, a short early phase being separated from a later or main phase by an interval of a virtually interglacial nature.

478

The interpretation of this evidence is certainly open to debate. Leaving the time factor to the geologists, can anything more positive than the adjectives 'warm' and 'cold' be inferred about surface-water temperatures?

Emiliani (1955), using the ratio of $O^{18}$ to $O^{16}$ in the calcium carbonate of shells, has inferred that the mean temperature of the equatorial Atlantic changed by some 6°C between warm and cool periods. Since the presentday conditions are 'warm', Flint has pointed out that ice could have persisted in polar regions during interglacials. The oxygen-isotope technique is now under question (see Ericson and Wollin, 1966, 108–11) but even if we accept Emiliani's figure for the Atlantic it would be rash to conclude that it applies to tropical temperatures as a whole. The North Atlantic is particularly open to the receipt of both glacial meltwater and icebergs and at its maximum the ice reached the sea over a very wide frontage indeed. Ice accumulation in no way implies a *cessation* of wastage, particularly along an ocean front. Here ice motion, not temperature, is the dominant control and nobody has ever suggested that the caps ground to a halt at their maximum stage. Wastage at an accelerated rate would thus be characteristic of glacial maxima and the author is not aware that the probable effects on ocean temperatures have ever been computed. That they might be considerable is revealed by the great distance which cool water of low salinity can travel without intermixing with warmer water of greater salinity beneath. This is one way in which the side-effects of glaciation could be transmitted to lower latitudes but it would be a great mistake to confuse effect with cause.

Cores from the Pacific and particularly from the Indian Ocean may eventually clinch this question though there the possible effects of a larger Antarctic ice-cap would have to be borne in mind. The evidence today is meagre but the presence of a temperature anomaly over the eastern third of the equatorial Pacific at the present time is a clear indication of the great power of the oceanic circulation.

It is thus possible for the Atlantic floor to provide an effective time-scale for the events of the Pleistocene without bringing us appreciably nearer to an answer to the vexed question of ultimate causes.

## The postglacial period

It is now generally agreed that the termination of the Würm glaciation in Europe can be conveniently placed at about 11 500 years before the present time (written 11 500 B.P.—where greater accuracy is justified the 'present' is taken as 1950). This date did not see the complete disappearance of Scandinavian ice and a considerable volume still hung about northern Canada for another 5000 or 6000 years.

The climatic history of this period is being worked out with great assiduity along various avenues of approach, notably varve analysis, pollen analysis, radio-carbon dating, the archaeological and historical record and, for the most recent centuries, the direct measurement of both temperature and precipitation. Many of the conclusions reached depend on inferences which are open to challenge but the variety of techniques employed does provide some basis for cross-checking and confirmation. The attractiveness of this pursuit arises largely from its relevance to the early history of modern man but it is also hoped that some key may be found to the more massive climatological events of the preceding epoch.

## POSTGLACIAL CLIMATIC CHANGE

| | YEARS B.P. | CLIMATE | PERIOD | BALTIC | POLLEN ZONE | VEGETATION W. EUROPE | MAN |
|---|---|---|---|---|---|---|---|
| POST GLACIAL | | SLIGHT DETERIOR-ATION | SUB-ATLANTIC | MYA | VIII | *BEECH, OAK* *ALDER, ASH* *IVY, HOLLY* | |
| | 2,500 | | | LIMNAEA | | | IRON AGE |
| | | | SUB-BOREAL | | b | *OAK, ASH* *LINDEN, IVY* | BRONZE AGE |
| | | CLIMATIC OPTIMUM | | LITTORINA | VII | | NEOLITHIC |
| | 5,000 | | ATLANTIC | | a | *OAK, ELM* *LINDEN, IVY* | |
| | 7,000 | | | | | | MESO-LITHIC |
| | | RAPID AMELIOR-ATION | BOREAL | ANCYLUS | VI | *PINE, HAZEL* *BIRCH* | |
| | 9,000 | | | | V | | |
| | 10,250 | | PRE-BOREAL | YOLDIA | IV | *BIRCH, PINE* | |
| LATE GLACIAL | 11,000 | COLD | YOUNGER DRYAS | ICE LAKE | III | *BIRCH PARK TUNDRA* | |
| | 12,000 | MILDER | ALLERØD | | II | *BIRCH PINE, WILLOW* | |
| | 13,000 | COLD | OLDER DRYAS | | c | *TUNDRA* | UPPER PALAEO-LITHIC |
| | 13,750 | SLIGHTLY MILDER | BÖLLING | ICE CAP SHRINKING | I b | *BIRCH PARK TUNDRA* | |
| | 15,000 | COLD | LOWER DRYAS | | a | *TUNDRA* | |

Fig. 150. A tabular presentation of postglacial climatic change as derived from various terrestrial sources

The difficulty of the task must not be underestimated for, as Faegri (1950, 194) has pointed out, it is often true that 'those who can judge the evidence (biologists, geologists) cannot judge the conclusions, and those who can judge the conclusions (meteorologists), cannot judge the evidence'. Yet a general picture has emerged of an aperiodic oscillation between 'warm' and 'cool' interludes, the amplitude of the contrast increasing with the length of the period of reference. (Lamb *et al.*, 1966, 177–8). Though Faegri considers this to be largely an illusion, Lamb suggests 'that on each of these time-scales different causative influences are at work'. If this is true the near-present must offer an indifferent key to the distant past.

In western Europe we now have a remarkable record of the relative frequencies of pollen grains preserved in peat and similar terrestrial deposits. Many sites have been investigated and the data give us a history of the vegetation going back to the last interglacial period. At first the time-scale could only be relative but the discovery, about 1950, of radio-carbon ($C^{14}/C^{12}$) techniques has raised palynology to the status of an almost independent science. What has this history to offer to the climatologist? Godwin (1966, 5) has claimed that the record 'is neither more nor less than the most complete and the most realistic register of climatic fluctuations throughout the past which we now have at our disposal' though, in the same volume, Frenzel (1966, 101-2) has outlined reasons why the climatic interpretation of past changes in vegetation may often be both difficult and ambiguous. For a broad view we rely on the general pattern of plant associations revealed but plant succession is affected by other than purely climatic factors. To achieve greater precision stress is laid on the distribution of known temperature-sensitive species, such as ivy, holly and mistletoe, but here we may run into trouble with the seasonal factor (see Faegri, 1950, 189-90).

It is sufficient here to sketch only the broad outlines of the picture, dealing first with the inferred changes in temperature (Fig. 150).

*Temperature.* The climatic history of western Europe over the last 15 000 years falls broadly into two periods, a 'late-glacial' phase to about 10 000 B.P. followed by a postglacial phase of some complexity which is regarded as continuing to the present time. During the former period the continental ice was in general retreat and the newly exposed surface of the ground was being colonised by tundra. The glaciological evidence suggests that the retreat was not continuous; it was marked by periods of stillstand and even local readvance. Pollen study confirms this by indicating that at two periods, broadly between 14 000 and 13 000 B.P. (the Bölling) and between 12 000 and 11 000 B.P. (the Alleröd) forests, first of birch and later of birch, willow and pine, advanced into the tundra only to suffer a later decline. It is argued that summer temperatures during the greater part of late-glacial time were some 5 °C below those prevailing at present though they are thought to have attained almost modern values during the Alleröd amelioration. How far this sequence of events was paralleled in other glaciated areas is debated but Heusser (1966, 124-41) considers that he has found evidence of the Alleröd phase in both western Canada and southern Chile.

Postglacial climatic history has been characterised as a sequence of 'increasing warmth, culminating warmth and then decreasing warmth' but this is a very general view. Birch and pine return and are then overtaken by hazel, oak, elm, linden and ash which, for a time, apparently flourished well beyond their present northern limits. In both Britain and Scandinavia there is some evidence of tree growth about 300 metres above the existing limits. This was the so-called 'climatic optimum' which lasted broadly from 8000 to 4000 B.P. and it is inferred that temperatures must have been 2-3 °C above those now prevailing. Heusser finds a parallel feature in western Canada and southern Chile though the tree species involved were different in the two regions and utterly foreign to those of Europe. Furthermore, from the rate of carbonate sedimentation in ocean cores, Wiseman (1966) has inferred that 'the main thermal maximum in the equatorial Atlantic was about 3000 B.C.' (4950 B.P.). The chief anomaly during this warm period was the Cochrane halt stage or readvance in eastern Canada. It occurred about 7000 B.P. but the effect appears to have been local.

The climatic optimum apparently came to a rather abrupt end and in Britain there was a fall in the treeline and the development of extensive areas of blanket peat. The detailed sequence of events has been analysed in numerous papers by both Manley and Lamb. The decline in temperature was punctuated by a succession of particularly cool spells, notably from 3000 to 2400 B.P. and from 400 to 250 B.P. (A.D. 1550–1700), thus once again the duration of the spells apparently decreases as we approach modern times. Manley (1964, 164) has pointed out that these changes 'all lie within the range of behaviour shown by extreme years and extreme seasons at the present day. The main problem to be solved by the meteorologist is that of "persistence of type" over a considerable number of years.'

*Humidity.* The interpretation of the pollen record in terms of water-balance is even more difficult. The distribution of precipitation in both space and time is inherently much more irregular than that of temperature, and ground-water conditions are not determined exclusively by meteorological factors. In western Europe the great bulk of evaporation occurs in summer (see Fig. 37) and it is hard to see how a 'humid' phase could possibly be combined with high summer temperatures and hence presumably cloud-free skies. Surplus winter rainfall runs off for soil-storage is limited. The botanical sub-divisions of the climatic optimum (Boreal, Atlantic and Sub-boreal) are thus not only unhappily named, they also imply meteorological assumptions that are difficult to substantiate. Certainly 'recurrence surfaces' in peat, formerly regarded as evidence of drier intervals, have now been shown to be much too transgressive in time for any such conclusion to be relied upon. Raikes (1966) has indeed roundly declared that 'in European and northern American latitudes . . . it can be said with confidence that bogs and water-logging are primarily dependent on drainage conditions and hardly dependent at all on variations of rainfall'. He would thus accept an 'Atlantic' phase only in an ecological sense, relating it to a rise in water-table consequent upon marine transgression.

Turning to more recent times, it has been argued that the erosion of 'moss hags' in upland peat in Britain is a sign that the deposit is past its prime and that therefore conditions must be drier now than formerly. Closer examination suggests that this is a perfectly normal erosional effect wherever water motion is organised. Similar peat cliffs surround each of the innumerable lakes of the Outer Hebrides but active growth continues on the low domed surfaces between them. Certainly anyone who has tried to maintain a lawn in north-western England during recent years is entitled to have serious doubts whether the era of peat is past.

# Theories of climatic change

During the last century a wide variety of hypotheses have been proposed as possible explanations of as much of the above evidence as was known at the time. When wholly or in part refuted, many of these notions have shown a singular power of recovery; they have been revived in modified form and again become the subject of current debate. A review of these ideas is therefore not merely of historical interest; the question is still very open. The chronological sequence being confused, it is wise to group the theories into types.

## Astronomical or orbital theories

The seasonal distribution of solar radiation on the earth is affected by three periodic aberrations in the relative position of earth and sun which occur concurrently but with different periodicities. The most notable of these is the 'precession of the equinoxes' with a period of some 21 000 years. Thus about A.D. 1200 (750 B.P.) the earth was nearest to the sun (perihelion) on 21 December but about 9300 B.C. (11 250 B.P.) perihelion occurred on 21 June. Although the effect of solar distance is partially offset by the greater angular velocity of the earth during perihelion—it passes through the 180° from equinox to equinox in a few days less than half a year—the net effect in the northern hemisphere, it is argued, was to produce mild short winters and long cool summers around 22 000 B.P., 43 000 B.P., etc. and cooler and longer winters with shorter though warmer summers around the years 11 000 B.P., 32 000 B.P. etc. However, as the sequence is carried backward through time, two other factors come into play. They are a variation of the order of ± 1.3° in the angle between the earth's axis and the ecliptic (now 66.55°) with a period of some 41 000 years, and a slow change in the eccentricity of the earth's orbit with a period of some 92 000 years. The former involves a corresponding displacement of the Tropics and Polar Circles and its major effect would be registered in high latitudes; the latter must modify the relative lengths of the intervals between the equinoxes. Computations of the presumed effect of these astronomical factors upon terrestrial temperatures are complex (see Woerkom, 1953), but it is alleged that they could have caused cold spells in latitude 65°N at rather irregular intervals, notably around 25 000 B.P., 70 000 B.P., 115 000 B.P. (severe), 185 000 B.P., and 230 000 B.P. (severe), to mention only the most recent. All these periods were comparatively short-lived.

This material has been interpreted in various ways. Croll (1875) argued that a succession of cool, long winters would favour the accumulation of snow and ice to such a degree as would neutralise the effects of the concomitant warm, short summers. He thus laid stress on a winter aphelion. Most of the later work in this field has taken the opposite view, long cool summers with short mild winters are regarded as most favourable for glaciation and hence great stress has been laid upon estimates of *summer* radiation. Despite this rather basic difference of opinion the astronomical theory dies hard and Zeuner (1958, 137) has roundly declared that though 'it cannot answer the question why glacial phenomena abound in the Pleistocene', 'it can and does explain the fluctuations of the Pleistocene climate'. It is thus now relegated to a comparatively secondary role.

Criticisms of the astronomical theory are fairly obvious.

1. It produces a cyclical series of events with a time interval which is too long to be relevant to postglacial fluctuations of climate and too short to throw much light upon the glacial epochs. Furthermore there is every presumption that these aberrations have persisted throughout geological time, why then so little evidence of climatic fluctuation in the Tertiary?

2. The computed variation in insolation is never more than a very few per cent. Is this sufficient to produce the results inferred? When this view is pressed, recourse is made to the argument that, whilst astronomical events may initiate a change, the total outcome is affected by a number of multiplying factors, notably changes in albedo and in the circulation pattern. Unless these factors can be very precisely specified this savours strongly of special pleading.

3. The variations apply to solar income at the *upper limits* of the atmosphere. What the response near the surface would be is still a matter of open debate (see later, p. 503). Certainly they would be complex and the case for a damping of solar effects, via evaporation, cloud amount, and albedo, is at least as strong as the case for 'multipliers'.

4. The presumed effect would be slight within the tropics and most marked towards the poles where north and south could not agree in phase. These last objections now carry little weight. Such evidence for climatic change within the tropics as we have been able to find begins to look very much like a side-effect of events in the north and this may even have affected the rhythm of the far south.

Quite a different astronomical hypothesis suggests the random passage of the solar system through presumed clouds of galactic dust or other widely diffused matter of a density sufficient to intercept a proportion of solar radiation. This is too conjectural to deserve the title of a theory. If solar income could be thus reduced the effect would be the same as a reduction of the solar constant, to be discussed later. Even the basic assumption has been challenged, some astronomers believing that the sun's gravitational attraction on such matter would actually *increase* its luminosity.

## Theories involving the displacement of the continents with respect to the poles or to each other

These have been reviewed at length by Schwarzbach (1963) but they scarcely concern us here since no material displacement has occurred since Tertiary times. The idea of 'continental drift', initiated as a wild surmise by Wegener (1924), himself a meteorologist, has been given a considerable degree of respectability by recent work on geomagnetism. The theory appears to throw much light on the known distribution of old coral reefs, dune sands, coals and evaporites, but it can be thanked chiefly for taking the vexing problems of the Permo-Carboniferous glaciation out of the field of climatology. The terrestrial pattern that it presumes for that period is so different from modern geography that any attempt to draw analogies with the existing general circulation must be rash indeed.

## Theories involving a change in the composition of the atmosphere

This group postulates changes of a semi-catastrophic nature in the opacity of the atmosphere to either solar or terrestrial radiation. Likely agencies are thought to be a marked change in the carbon dioxide content or the widespread distribution of volcanic dust. Again, these theories are most appropriate to climatic change on a geological time scale but, since the processes invoked are of a strictly meteorological nature, they deserve some consideration.

*Carbon dioxide.* This gas is now present in the atmosphere in the low proportion of 0.3 parts per thousand by volume, but it is known to absorb radiation at terrestrial wavelengths so that it contributes to the 'greenhouse effect' of the atmosphere due mainly to the presence of water vapour. Geiger (1965, 19) puts its present contribution to counter-radiation at one-sixth of the total. At various times it has been suggested, notably by Arrhenius (1896, 237), Callendar (1939), and Plass (1956), that an increase in the proportion of this gas in the atmosphere could have a notable effect upon mean surface temperature. Plass argues that, with half-clouded skies, doubling the proportion would raise surface temperatures by $2.5°C$, whilst halving the present figure would lower them by $2.7°C$. It has been claimed furthermore that, as a result of industrial combustion, the proportion of $CO_2$ increased by 10 per cent during the first half of the present century, a period of rising temperatures in northern latitudes, but the backbone of this argument has been weakened, if not removed, by the generally downward trend which has supervened since 1940.

There are numerous difficulties facing this interesting hypothesis:

1. Perhaps the most significant is how can the $CO_2$ content of the atmosphere in past ages be known? It can only be inferred, and that hazily, so we are not far advanced from a position where climatic change is frankly admitted to be due to 'causes unknown'.
2. The ocean holds a vast reserve of dissolved $CO_2$ and apparently acts as an atmospheric regulator. In an elaborate argument Plass has deduced a periodicity of some 50000 years if the air-ocean balance is initially disturbed but this involves the formation of land ice on a considerable scale. This might possibly have some relevance to the duration of ice ages but it is clearly not commensurate with postglacial oscillations.

485

3. The absorption bands of $CO_2$ lie mainly within the ranges 2.6–2.9, 4.1–4.5 and 13.8–15.4$\mu$. Although the second of these ranges falls within a clear water vapour 'window', only the third corresponds with terrestrial radiation at considerable intensity (see Fig. 17) and here it has commonly been argued that water vapour absorption is so strong as to mask most of the effect of variations in the $CO_2$ content. Plass (1962, 227) has challenged this view, arguing that, when the spectra are examined in detail, it is shown that carbon dioxide and water vapour are 'relatively independent of one another in their influence on the infrared absorption'. This is very much a matter for the specialist but, even if carbon dioxide is more influential than has been supposed, there is little doubt that water vapour plays the dominant role in producing the 'greenhouse effect'.

4. Conceding this, Plass turns to the upper air where water vapour is slight and where changes in $CO_2$ would be expected to have a greater relative effect. In so far as this view is accepted, direct temperature change at ground level recedes into the background and we are thrown more into the realm of atmospheric stratification. From the argument that a reduction in $CO_2$, by lowering temperatures aloft, would increase atmospheric instability, Plass reaches the conclusion that 'colder and wetter climates occur together'. This chain of reasoning appears to miss out so many middle terms that few meteorologists would follow him with confidence.

5. For large-scale changes in $CO_2$ content Plass has recourse to such geological processes as the formation of coals and the weathering of igneous rocks. This takes us into a time-scale of the order of millions of years and its relevance to climatic oscillations becomes extremely remote.

*Volcanic dust.* The suggestion that suspended volcanic dust may cut off enough direct solar radiation to have a significant effect upon surface temperatures seems, at first, to be eminently reasonable. Indeed, such a view was propounded by Benjamin Franklin to explain a cold spell in 1783–4. Short-term effects have been observed after explosive eruptions of unusual violence, but the experience is by no means universal and there is doubt whether surface temperatures for the world as a whole have ever fallen by more than 1°C, and then only for a limited period. The intercepted radiation is not all lost to space and, like cloud, suspended dust is effective in trapping terrestrial radiation which would otherwise escape through the water vapour 'window'. In the troposphere dust is readily removed by precipitation and a persistent veil could only be maintained by renewal at very frequent intervals. There is no geological evidence for such an almost continuous supply. Lodged in the stratosphere, it was formerly thought that dust could remain in suspension for many years but the study of radioactive fallout has indicated that mass-exchange between stratosphere and troposphere occurs on a scale far greater than was anticipated.

Humphreys (1940) revived and refined this theory. From optical effects observed after the Katmai explosion of 1912 he estimated that the mean diameter of relatively persistent particles was about $1.8\ \mu$. These, he argued, would effectively intercept or reflect solar radiation (wavelength mainly below $1.0\ \mu$, see Table 4) but would be bypassed by practically all terrestrial radiation (wavelength well in excess of $3.0\ \mu$). Thus the effect would be *selective* and a net fall in temperature becomes much more reasonable. Yet the Katmai material remained recognisable for no more than three years and the maintenance of such a screen over a much longer period is still open to question. To the geological argument that the sedimentary record shows no trace of a high degree of persistence in explosive eruptions Humphreys replies that an effective screen, even if maintained for 100000 years, would leave a deposit so thin that it would defy sedimentary analysis. This conclusion has the unfortunate effect that his theory can be neither proved nor disproved.

A similar increase in the earth's albedo could also conceivably arise from an increase in *cloudiness* but this too would leave no trace so again we are in the realm of pure supposition. Indeed, if loss from volcanic dust led to a reduction in cloud amount, a by no means improbable long-term development, the two effects would find themselves in opposition. It has been argued that both volcanic and meteoric dust can increase the supply of ice nuclei and thus stimulate cloud activity and precipitation. Even if it were known positively that the natural regeneration of freezing nuclei in the atmosphere often left it with a serious deficiency, it does not follow that there would necessarily be an increase in mean cloud amount. The reaction would be confined to areas where the other prerequisites for precipitation, notably the uplift of moist air, were already present. For the atmosphere as a whole the upward and downward mass exchanges must always be in balance. Though it does not follow that mean cloud amount is invariant, this balance would seem to imply that the range of variation moves within fairly strict limits.

## Geographical theories

All climatologists agree that the existing pattern of world climates points to the unquestionable significance of two groups of purely terrestrial factors, the relief of the earth's surface and the distribution of land and sea. An approach to the problem of climatic change by either of these routes can have no relevance to recent oscillations but, over a time range of the order of a million years, geology gives ample evidence that the pattern has been far from static. Nevertheless, acceptance of the principles of isostatic adjustment and of the permanency of the ocean basins sets firm limits to the range and location of the terrestrial distortion that can be invoked. Climatic changes arising from these factors are thus likely to be, initially at least, only subcontinental in scale. Although consequent reactions upon the circulation system may spread side-effects over a wider area, it is equally probable that there would be a good deal of regional compensation if the general level of solar income remained unchanged.

(*a*) *Elevation*. Since mean temperature falls by 1°F for every 280 feet of ascent (0.65°C/100 m), uplift of the land is a powerful coolant. Even though the relief of the earth in early Quaternary times can only be vaguely inferred, there is ample evidence of regional oscillation in level despite the fact that mountain-building, in its structural sense, had then come broadly to a close. Furthermore, since most precipitation in middle and high latitudes is initiated as snow, high ground intercepts this deposit before it has melted and subsequently preserves it because of low temperatures, particularly at night. During the day the cloud generated by high relief and the high albedo of fresh snow both contribute to the same end. Since total precipitation is also known to show a general increase with altitude, at least up to about 10000 feet (3000 m), little wonder that relief has exercised a strong appeal to those casting about for an initial cause of the development of continental ice-sheets. In the northern hemisphere the association of the Canadian Cordillera, the Labrador Heights, the Scandinavian Plateau and, to a smaller degree, the Western Highlands of Scotland, each with a known centre of ice dispersion during the Quaternary period, seems to give strong support to such a view. The evidence from other mountain regions appears to be equally convincing.

Yet it is one thing to claim that uplands provide the initial snow-trap, but quite another to explain either the enormous spread of the ice in both North America and Europe or the waxing and waning phases so characteristic of each. A reasonable case for the former can be made on the ground that snow accumulation on a massive scale must eventually generate its own relief feature (an ice dome) which, in its meteorological repercussions, will in no way differ from a vast snowclad plateau. Plastic because of its crystalline structure, such a dome can be expected to spread under gravity. In fact such domes now exist in both Greenland and Antarctica. But if elevation alone is the prime cause of glaciation such a dome, once established, should be self-perpetuating and it is extraordinarily difficult to see why it should ever disappear. The weight of the ice load is now known to have depressed the surface of the earth beneath it, but (*a*) isostatic motion is slow and delayed— recovery is still proceeding; and (*b*) density considerations rule out the possibility that such a motion could ever have exceeded the rate of snow renewal.

(*b*) *Continentality*. Similar difficulties are encountered when emphasis is laid upon a changed distribution of land and sea. The present pattern is known to produce a strong azonal anomalies of both temperature and precipitation, particularly in winter in upper-middle latitudes. Here continental winters generate vast pads of dense Arctic air (Fig. 83) and repercussions on the circulation of both air and sea spread side-effects far and wide. Can the results of reasonable changes in this pattern be assessed?

A serious attempt to estimate the effect of continentality upon temperature was made by C. E. P. Brooks in two early papers (1917 and 1918), and the results are summarised in Chapter 8 of his *Climate through the Ages*, 1949. The discussion is still worth following though Brooks's attempt to apply the argument to the whole of geological time might now be regarded as outmoded. Recognising that 'the effect of land in low latitudes is to raise the temperature and in high latitudes to lower the temperature' (1949, 133) and that 'when water freezes and loses its mobility, it takes on some of the properties of land' (*ibid*, 32), he sought to discover from existing climatic data the scale of the changes that would follow if an island arose in mid-ocean at 60°N latitude

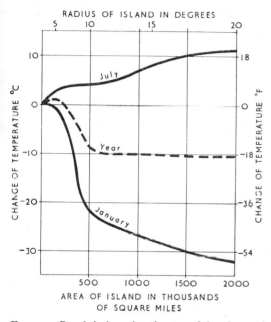

FIG. 151. Brooks's (1949) estimates of the change in temperature at a coastal station which would result from the progressive growth of an oceanic island until it assumed continental proportions. The latitude is assumed to be about 60°N

and grew progressively in size. The anticipated variation from mean zonal January and July temperatures at a near-coastal location is indicated in Fig. 151 (*ibid*, 145).

From this exercise Brooks was led to the interesting concept that there is a *critical size* in the influence of a land mass upon temperature in middle latitudes. The effect of a small island is negligible for it is swept by sea winds. A low island with a radius of some 5° of arc, i.e. with diameter 700 miles and area 375000 ml², will raise summer temperatures more than it lowers winter temperatures but the net effect is still slight. But when an island, now a continent, reaches a radius of 10°, i.e. diameter 1400 miles and area 1500000 ml², winter temperatures are lowered to such an extent that the mean annual temperature near its margin can be expected to be some 10°C below the oceanic mean for the latitude. The critical size thus occurs at about radius seven degrees of arc, i.e. diameter 950 miles and area 725 000 ml$^2$. The reason adduced for this rapid transition is that, at this scale, the land is able to develop its own winter air mass so that a new system of circulation supplements the effects of radiation balance.

Turning to the problem of the growth of a floating ice cover over a previously open Arctic, Brooks (1949) develops the parallel concept of a *critical temperature*. With mean January temperatures of the order of $-3°$ C over an open Arctic, he considers that ice would be unlikely to survive the summer. If winter temperatures are gradually lowered for some reason he thinks that, at first, an ice-cap would still have considerable difficulty in establishing itself but that 'once it has reached a certain size, the temperature on its edge will be below the freezing point of sea water ($-2°$ C) and it will continue to expand owing to the lowering of temperature which the ice itself introduces' (p. 35). This growth will continue until an equilibrium has been reached at the ice edge with the normal increase in sea temperature at lower latitudes which he takes as $5°$ C per $10°$ from the pole towards the equator. Assuming an initial fall in temperature at the pole of only $0.35°$ C below the freezing point of sea water, he thus deduces the following remarkable result.

| Mean latitude of ice edge (Lat.) | $90°$ | $85°$ | $80°$ | $75°$ | $70°$ | $65°$ |
|---|---|---|---|---|---|---|
| Normal temperature rise from pole ($°$ C) | 0 | +2.5 | +5.0 | +7.5 | +10 | +12.5 |
| Cooling at ice edge ($°$ C) | −0.35 | −6.6 | −10.1 | −11.3 | −11.8 | −12.3 |
| Net effect ($°$ C) | −0.35 | −4.1 | −5.1 | −3.8 | −1.8 | +0.2 |

'Hence an initial winter cooling to $0.6°$ F ($0.35°$ C) below the freezing point will result in the formation of a floating ice-cap with a radius of nearly twenty-five degrees' (p. 38). When this stage has been reached Brooks estimates that, at the pole itself, winter temperatures may well have fallen by no less than $25°$ C, a slight initial fall in temperature has thus been magnified many times by self-acting processes. Brooks's treatment of the summer melt is rather less convincing, for reducing such a vast amount of ice must draw heavily on supplies of heat. As long as the ice persists it will still exert a chill and the above table shows this to be greatest between latitudes 78 and 84° N. Within this range the ice will grow most rapidly in winter and retreat most slowly in summer, so slowly indeed that it is likely to be overtaken by the succeeding winter. This appears to be broadly the position at the present time.

Similar considerations apply to the growth of a continental ice-sheet, though new factors such as elevation, precipitation and air-drainage compli-cate the picture. Particularly because of the first two factors there is ample reason why such ice caps may form in localities many miles from the poles. Having reached a certain critical scale they too will tend to distort the atmospheric circulation in such manner as to favour their growth and self-preservation. Ice recession is thus again the most puzzling aspect of the question.

(c) *The form of the North Atlantic and Arctic basins.* In recent years increasing stress has been placed on the view that shield glaciation in the northern hemisphere was an *Atlantic* rather than a polar phenomenon. This change of emphasis still leaves many questions open but it at least has the advantage that the argument is expressed in terms of *existing* geography. The North Atlantic is thus seen as a theatre where a comparatively slight change in climatic pattern (cause unknown) could have the most momentous consequencies.

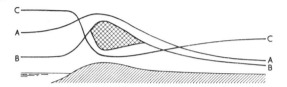

FIG. 152. Leighly's interpretation of the factors affecting snow accumulation. The section is to be taken as extending from an open ocean across a system of coastal uplands towards a continental interior.
$A - A$ = Total annual precipitation.
$B - B$ = Total annual snowfall.
$C - C$ = Heat available to melt snow.
All are to be regarded as plotted on a comparable scale expressed in terms of water equivalents. The hatched area then represents net snow accumulation during the year

Leighly (1949) argues this case with great acumen. Pointing to 'the nearly symmetrical spread of the ice sheets inland from near the shores of the North Atlantic', he associates this with the 'unique injection of warm air and water into high latitudes' which still occurs over the Norwegian Sea, particularly in winter. That a closely comparable moist air stream was the source of the continental snow is indicated by: (*a*) the reasonably presumed profiles of both ice domes (Fig. 152), (*b*) the apparent freedom of St Kilda from ice accumulation, and above all, by (*c*) 'the swing of the edges of the two ice sheets northward' as one moves away from the Atlantic in both continents. The edge of the continental ice reached the Arctic coast at about 130–140°W in America and 110–120°E in Siberia and between these longitudes, in both Alaska and Yakutia, the shores of the inner-Arctic remained free of land ice. Given a geography much like the present, Leighly argues that a comparatively slight change in the snow accumulation-ablation balance would produce ice-caps from Scandinavia to Novaya Zemlya on the one flank and from Labrador to Ellesmere Island on the other. Such caps have only to grow in a logically comprehensible manner to produce the results observed. For this growth he invokes the process of autostimulation outlined by Brooks, but points out an interesting contrast between the two continents, which helps to explain why the American ice-sheet attained an area three times greater than that of its European equivalent. 'Any tendency toward enhanced anticyclonic circulation over the north-European ice-sheet would hamper rather than aid its growth into the interior' but 'an ice sheet already in existence ... would create precisely the conditions that at the present favour abundant snow in the interior of North America: flow of air from some direction between south and east' (Leighly, 1949, 144). Put rather differently, this means that Arctic outbreaks from Fennoscandia would occur most frequently over eastern Europe whilst from Laurentia, as from Greenland today, they would occur most freely over the open sea, chilling the coast to at least the latitude of New York. On the other hand, the return flow of warm moist air which took place off Europe over the open sea would be directed in America towards the heart of the continent. Such a circulation would help to explain the close proximity of the treeline to the ice front in the interior of America. It is also not inconsistent with the presence of ice on high ground from the Alps to the Atlas whilst the southern Appalachians remained ice-free. These ideas make the nature of a glaciation much more comprehensible but we are no nearer to a root cause and the waning stage is still wrapped in mystery.

Ewing and Donn (1962) have proposed a new geographical theory which, although highly conjectural, does squarely face the problem of waxing and waning development. In their view this arises from an automatic variation in the exchange of Arctic and Atlantic waters across the Wyville-Thomson sill, rarely more than 400 metres deep. Arguing that an open Arctic is necessary for glacial accumulation on the surrounding land masses, they associate this with a period of free water exchange. As land-ice develops, the sea-level falls, exchange is reduced and a freezing Arctic depletes the snow supply. A 'gradual wastage of the continental glaciers' is held to follow and the restored sea-level renews the exchange and paves the way for a new glacial period. Repercussions on the temperature of the Atlantic and, indeed, on the albedo of the whole earth are regarded as enhancing the effect. Thus 'when the Arctic Ocean is ice-covered, surface temperatures in the Atlantic increase and continental glaciers decline; when the Arctic is open, surface temperatures in the Atlantic decrease and continental glaciers develop'.

The ingenuity of this argument cannot be denied but it involves such a bewildering array of assumptions that one scarcely knows where to begin.

1. Is the inferred variation in the rate of exchange of Atlantic and Arctic water across the Wyville-Thomson sill quantitatively sufficient to produce such widespread results? The authors present no computations but it will be noted that the maximum fall in sea-level (taken as 100 m) leaves the straits far from closed. Warm water enters the Norwegian Sea as a surface, wind-induced flow between Iceland and Norway and the cross-section (Ewing and Donn, 1962, 209) shows the straits reduced by no more than 20 per cent. Furthermore, the chill of Arctic water depends largely upon its ice burden and

it is hard to see how an ice-free East Greenland current could possibly chill the North Atlantic more than it does at present.

2. The motion of the atmosphere is free from these restraints and the authors are on stronger ground when they argue that an open Arctic might well increase cyclonic activity and precipitation in high latitudes. How far this would affect the nourishment of ice-caps centred outside the Arctic Circle is open to doubt: surely the major source of moisture must always remain the south. Even today winter depressions sweep far into the Barents Sea, and further penetration would carry their effect into the inner Arctic, i.e. just where the evidence of glaciation is least.

3. In existing summer conditions an at least partially open Arctic raises no problem, indeed we may well be moving towards such a phase. With ice-clad continents and a cooler Atlantic such a possibility seems extremely remote. At least a third of the Arctic is shallow and thus very sensitive to winter temperatures. It is not well designed by nature to play the part of a polar Mediterranean.

4. That the development of continental ice-caps of considerable area might help to freeze the Arctic is by no means improbable, it may even be conceded that some reduction in snow supply might follow; but that such an event would necessarily lead to the 'gradual wastage' of massive ice-domes already in existence—relief features in their own right and open to all the winds that blow—seems to be asking for results that are entirely out of scale. We are therefore, not really nearer to an answer to the problem of ice recession unless we accept the theory in its entirety.

5. The Arctic Ocean occupies rather less than 5 per cent of the surface of the globe. Is it not inherently improbable that the freezing and thawing of this surface should have had major repercussions over the whole world? In estimating the effect upon albedo the authors see fit to treble the cloud amount over 'about 12 million square miles of arid regions' which 'were well watered (pluviated) during glacial stages'. Such an estimate goes far beyond the evidence and, if it were true, would it not suggest that the authors were looking for the cause of glaciation in the wrong latitude?

It may be remarked that interest in the Wyville-Thomson ridge as a climatic factor is not new. Previous workers had speculated on the probable consequences of the complete or partial closure of the straits by crustal movements and it was generally conceded that a colder Arctic would follow. However, no one has ever imagined that the inferred motion was rhythmical. Despite its faults the new theory thus gives an old idea a completely new twist. To explain the absence of glaciation over long periods of geological time the authors turn to polar wandering though the past history of the sill would not be irrelevant. More recently stress has been placed on the rather delicate balance prevailing in the Arctic Ocean where winter freezing occurs more readily owing to the relatively low salinity (20 parts per 1000) of the surface waters. This appears to be due to the massive input (in summer) of fresh water from the Ob, Yenesi, Lena and Mackenzie rivers. It forms a fragile layer of no great depth covering the denser saline waters beneath (Lamb, 1970). What would be the repercussions of the onset of a glacial period upon this layer? During the waxing stage it is a not unreasonable presumption that the run-off might be reduced but the supply of fresh water would surely be increased during the waning stage. The results of this variation upon the possibility of an 'open Arctic' are thus by no means obvious and the problem is further complicated by Koerner's paradoxical conclusion that the present ice supply is substantially augmented by open leads (Koerner, 1970). This effect might well diminish as the proportion of open water increased but what is the limiting value?

## Theories involving a change in solar radiation

When dealing with the astronomical theories of climatic change it was assumed that the output of solar energy was constant, but what if the sun is in fact a variable star? It must be emphasised that in regard to total solar radiation there is little positive evidence for such a supposition, the records to hand suggest a variation of less than 0.2 per cent, but the period of reliable observations is short and greater changes may have occurred in the past. A near-constancy in the total emission also does not rule out some degree of change in the relative contributions of different wavebands.

(*a*) *Sunspots*. The most clearly observed change in the face of the sun is the occurrence of sunspots. Occasionally these are large enough to be seen with the naked eye through smoky haze or, preferably, through a piece of smoked glass.[1] Good records of sunspot frequency are available since 1749, the figure being expressed as a relative number where 100 represents spots covering one five-hundredth of the visible disc. Mean annual sunspot numbers since 1749 have been plotted in Fig. 153. They will be seen to vary from 0 in 1810 to 190 in 1957. Descriptive accounts suggest that the numbers were still higher in A.D. 1077–9 and in 1372. A notable feature is the periodicity of the 'sunspot cycle', usually stated as having a mean duration of $11\frac{1}{2}$ years though the intervals between annual minimum values in Fig. 153 will be observed to range from 9 to 13 years. This makes it a little difficult to establish the mean form of the cycle but a reasonable arrangement of the data for nineteen complete periods from 1755 to 1963 inclusive yields the slightly asymmetrical pattern shown in Fig. 154. Willett (1949) has laid considerable stress on the alternation of 'major' and 'minor' sunspot maxima but, though such a 'double sunspot cycle' can be recognised from 1848 to 1937, it is less evident from the record as a whole. What does emerge, however, is the occurrence of groups of high maxima at intervals of 80 to 90 years.

[1] Although the spots appear black in comparison with the sun's brilliant surface their temperature is estimated at about 4000° C

A great deal of ingenuity has been displayed in attempts to relate sunspot numbers to various climatic indicators. At first it was thought that the spots must involve a reduction in solar output but later it became clear that, if anything, they are a sign of increased activity. Huntington and Visher (1922) sought to show that sunspot maxima coincided with periods of increased storminess, greater precipitation, and hence ice accumulation. Rather more reasonably, Brooks (1934) claimed to show a direct relation with thunder-

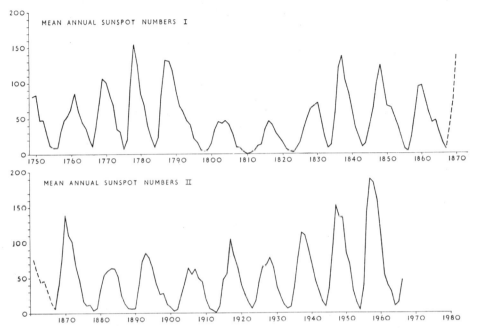

FIG. 153. Mean annual sunspot numbers from 1749 to 1966. It will be observed that the alternation of 'major' and 'minor' sunspot maxima is not very persistent but that groups of high maxima appear to recur at an interval of 80 to 90 years

storm frequency. At that time the analysis of periodicities was very much in fashion, but such exercises are full of statistical pitfalls particularly when, as Brooks discovered, the inferred terrestrial changes are not proportionate to sunspot number, the relative effect falling off rapidly as the number increases.

Yet a popular theory rarely dies and more recently Willett (1949) gave this approach a new lease of life. He was convinced that a more glacial weather pattern must arise from 'increased activity of the general circulation and of the condensation cycle'. Seeing no reason why an overall reduction of the solar constant should have this effect, he wondered whether the atmosphere might not respond to 'irregular emissive activity of the highly selective sort that accompanies sunspot disturbances'. i.e. to variations which leave the great bulk of solar income substantially unchanged. Willett admits that such a theory is 'highly controversial in character' and reasons are not far to seek.

495

1. Emphasis is shifted from the known and measurable aspects of solar radiation to the fields of ultraviolet and corpuscular emission which cannot be explored by conventional surface observations. Although these are now thought to be liable to wide variations, direct measurements are very recent, and Willett is compelled to employ the indirect key provided by sunspot numbers. It may well be unreliable.

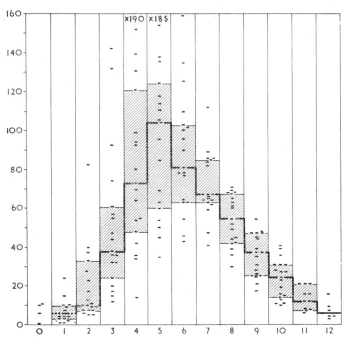

FIG. 154. An attempt to express the form of the sunspot cycle as a frequency group. The diagram includes nineteen complete cycles from 1755 to 1963 inclusive. The cycles are delimited from the occurrence of minimum values. In the four cases when the cycle occupied thirteen years the plot begins in column '0'; in the two cases where it occupied only 9 years the plot begins in column '2'; all other cycles begin in column '1'. A fairly consistent form emerges

2. The evidence offered for a relation between sunspots and the distribution of mean seasonal pressure, temperature and precipitation does not appear to have been subjected to statistical tests. The sharp distinction between the effects of 'major' and 'minor' maxima shown by Willett's maps is open to challenge on this ground, particularly since a rational basis for such a difference is hard to find. Wexler (1953), in a parallel series of maps, makes no such distinction and, indeed, he frankly admits (p. 84) that a random selection of years produced similar patterns.

3. It is claimed that such special types of radiation, entirely absorbed

high in the atmosphere, must exert their inferred effect by producing a rearrangement of the heat and cold sources at high levels 'without any significant change of the overall temperature of the lower atmosphere' (Willett, 1949, 308). How this is achieved is not explained. At best this is asking a very light tail to wag a heavy dog but the real difficulty with stratospheric heating by ultraviolet radiation is that its distribution would be relatively uniform; how then could it help to develop strong geographical contrasts? Corpuscular radiation, associated with the aurora, is certainly directed towards the polar regions by the earth's magnetic field but outbursts are short in duration and the energy is absorbed at such a high level that climatic implications are highly conjectural.

(*b*) *Variation in the solar constant.* A ready source of worldwide climatic change is offered by the possibility of variations in the total output of thermal energy by the sun. If we start from first principles such changes may be conceived as (1) progressive, (2) periodic, (3) irregular, or (4) some combination of two or more of these. Unfortunately, in the absence of positive direct evidence of such changes, we are thrown back on the indirect evidence that terrestrial climates have changed and the reasoning thus becomes completely circular. The geological record, as far as it goes, appears to rule out any marked progressive change in either direction but assumptions regarding rhythmical or irregular changes appear to be both open-ended and incapable of proof. Such theories thus do not work within the rigid limits which curtail the speculations of the astronomical school.

A further difficulty is that the climatic consequences of any postulated change in total solar radiation are by no means self-evident. Catastrophic change would undoubtedly produce catastrophic results, but it does not necessarily follow that a minor increase or decrease in solar output must involve a general rise or fall in terrestrial temperatures. The atmosphere contains far too many self-regulating mechanisms and its circulation depends far too strongly on temperature differentials near the earth's surface for its response to be either simple or automatic. As we have seen (Fig. 20) a high proportion of solar income is invested in the evaporation of water and the direct effects of cloud cover on the earth's albedo plus its indirect, and often opposing, effects on summer and winter temperatures, are factors which must not be ignored.

Sir George Simpson (1929) made these points with great force in his first paper on 'Past climates'. Though his glacial theory has been much quoted, this paper is still worth reading in its original form as it takes us back to first principles. Emphasising the fundamental part played by the atmosphere in maintaining terrestrial temperatures, he first shows that mean annual values at the poles and the equator are respectively some 70 and 50°C above those that would be expected from the radiation balance calculated with a solar constant reduced by 44 per cent to allow for the existing albedo. Zonal differences in radiative income also set the air in motion so that 'the temperature of any zone of latitude is not determined by the amount of solar radiation which falls on that zone, but by a number of interacting factors which make the temperature of every zone dependent on the temperature of every other zone' (p. 9). Of the 56 per cent of solar radiation which is absorbed by the earth and returned to space by long-wave emission, he estimates that '28 per cent is controlled by the temperature of the stratosphere, 17 per cent by the temperature of the ground and 11 per cent by the temperature and amount of cloud' (p. 17). A postulated change in solar output would presumably affect all three of these exchanges at once but, assuming that an increase of 10 per cent in solar radiation could be applied to each in turn, he concludes that radiative equilibrium would be re-established if (*a*) the mean temperature of the stratosphere were raised by 15°C, or (*b*) the mean temperature of the earth's surface were increased by 20°C, or (*c*) cloud cover increased by an additional one-tenth of the sky. The crux of his argument is that concentration of the effect at the surface as in (*b*) would inevitably so affect the activity of the circulation and the rate of evaporation that 'long before the temperature of the surface could change to the amount required to balance the increased radiation, the cloud amount would have increased sufficiently to effect the balance itself' (p. 18). Hence he sees 'no escape from the conclusion that changes in solar radiation are balanced mainly by changes in the cloud amount'.

Glaciation is thus viewed in terms of snowfall generation, snow accumulation, and snow-melt. Starting with a period of comparatively weak insolation, he associates this with low temperatures, a weak circulation and hence low precipitation. Although, in high latitudes, much of this might be in the form of snow, the rate of accumulation would be slow. With gradually increasing insolation the circulation will become more active and the evaporation–precipitation turnover will be increased. In high latitudes therefore, although the proportion of snowfall to total precipitation may fall, the total snow accumulation might reasonably be expected to increase to a stage when a glacial phase was initiated. But, carried still further, the increase in solar income must raise temperatures to a level where snow-melt overtakes accumulation and the ice would begin to retreat, introducing a 'warm interglacial' period characterised by heavy precipitation, particularly in the intertropical zone. A subsequent decline in solar output would cause terrestrial events to pass through the same series in the reverse order until conditions returned to the initial stage, a cool, dry interglacial. Accepting Penck's (1909) conclusion that the Pleistocene saw two pairs of glacial periods, Günz-Mindel and Riss-Würm, separated by a long dry interglacial, Simpson suggested that such a pattern would conveniently fit the passage of two complete solar cycles (Fig. 155). Interglacial periods (of different types) occur during both the troughs and peaks of this cycle whilst glacial phases of more or less uniform character are referred to the periods of *most rapid change* in solar output. The present time would thus be regarded as near the beginning of another long, cool interglacial.

498

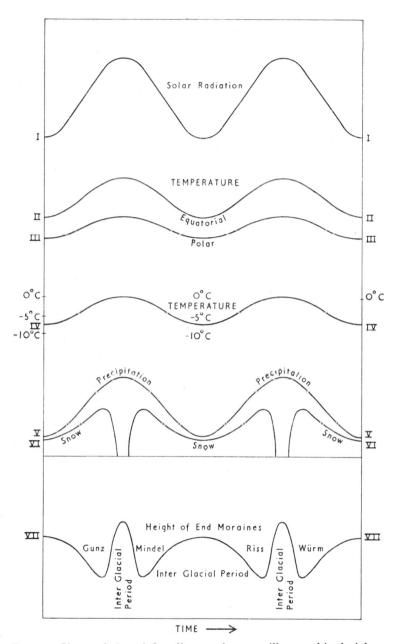

FIG. 155. Simpson's (1929) first diagram drawn to illustrate his glacial theory. He selects a place with a mean annual temperature of 0° C during the warmest epoch and assumes a temperature oscillation there of about 7° C (line IV). Curves V and VI then show the presumed variation in the proportion of snowfall to total precipitation in that locality. If the area is mountainous then curve VII indicates the variation to be expected in the elevation of end moraines

At first Simpson hesitated to attach a time scale to this sequence but in later papers (1957, 1959) when he was primarily concerned with the sequence of 'pluvial' stages, he inferred a solar period of the order of 380000 years and accepted the possibility of a third, earlier cycle of similar magnitude. Three pluvial periods each with a duration of some 140000 years and extending from one glacial maximum to the next during the times of maximum solar activity were thus fitted to the scheme. He also now tries to assess the scale of the temperature changes involved and concludes 'that the greatest change in the zonal temperature at the maximum of the glaciations compared with those of the present time was an increase of 6°C in all seasons at the equator and in all latitudes during January; and the greatest decrease was 12°C during July at the margin of the ice-cap in lat. 55°N' (1957, 479). The maximum range envisaged is thus not more than 18°C spread over a period of some 190000 years, a rate of change so small that it would be 'unrecognisable in the whole of historical time'.

This theory has now held the field for 40 years and, though it has not passed unchallenged, it must be admitted that, if some extraterrestrial cause must be found for climatic change, no other theory rivals it in its comprehensive sweep. This does not mean that it is *proved*, for long-period changes in solar radiation are apparently incapable of proof, but it certainly presents a number of salient points which no alternative theory can safely ignore. The major difficulty is the absence of clear evidence of still earlier solar pulsations of similar intensity and period, for such a solar characteristic is unlikely to be new. Possible escape routes from this difficulty may be found in (*a*) the fragmentary nature of the earlier geological record, or (*b*) the possibility that polar wandering and continental drift may have produced a terrestrial pattern less appropriate to ice accumulation.

On more recent evidence the implied alternation both of warm and cold interglacials and of glacial and pluvial phases has also been questioned. To counter this difficulty Bell (1953) has suggested a modification which doubles the number of solar pulses. Arguing that a cool period which lowers ocean temperatures and increases the extent of sea ice is a necessary prerequisite for the development of land-ice, she regards glaciation as the first result of an increase in solar output which stimulates precipitation under conditions when a thermal response in high latitudes is inhibited by the presence of the sea ice and the nature of the circulation system it engenders. As solar radiation increases still further 'summer melting will eventually predominate'. In her view no glaciation will accompany the decreasing phase because 'by the time the polar seas . . . are permanently frozen the precipitation has fallen to too low a level for large-scale glaciation' (Bell, 1953, 130). This variant makes some useful points but the timelag in sea *surface* temperatures is open to question and it calls for a strange behaviour on the part of the sun.

It should be remarked in conclusion that Simpson did not rule out either the astronomical factors or minor short-period solar variations as providing an embroidery to his grand design.

# Conclusion

It is evident from the above summary that existing theories of climatic change are many and various.[1] Why is this so? Can we discern any common ground amidst these varied approaches which may serve to indicate the path ahead?

Broadly the solutions suggested fall into two main categories, those which place major stress on extraterrestrial factors such as the nature of the sun or the orbital path of the earth and those concerned more specifically with terrestrial patterns such as the opacity of the atmosphere, the physique of the continents and the nature of the general circulation. Although these approaches lead back to different ultimate causes they do meet on common ground when we turn to the *processes* by which climatic change is made manifest. As both Brooks and Simpson have emphasised, postulated changes of an extraterrestrial nature must be interpreted via their presumed effect on the circulation of the atmosphere in three dimensions plus the activity of the oceans which is, to a large extent, wind-induced. The problem of climatic change on an earth with broadly its existing physical features thus resolves itself into the question under what conditions and to what degree could that circulation be conceivably modified and what would be the probable results.[2] Since the existing pattern of the general circulation is not yet fully understood this may not carry us very far forward on our quest but some progress has been made if we conclude that atmospheric behaviour is paramount.

If this is accepted we immediately encounter two groups of complex difficulties. (*a*) There can be no such thing as 'world climate' since the circula-

---

[1] One is reminded of the medical maxim that where manifold treatments for some condition are recommended no single one of them is likely to be really effective.

[2] Evidently the results would differ widely with latitude. Ångström's (1949) computations are of interest here. Presuming a change in the eddy conductivity of the atmosphere 'through a change in the average wind speed as well as through changes in the size of the circulation elements' (the zonal radiation-balance remaining unmodified) he reaches conclusions which may be summarised briefly as follows:

| Conductivity of air circulation (%) | (*Index*) | Mean zonal temperature | |
| --- | --- | --- | --- |
| | | 0–10°N (°C) | 80–90°N (°C) |
| Increased by 40 | (1.4) | 23 | − 9 |
| Increased by 20 | (1.2) | 24.5 | −12 |
| As *at present* | (1.0) | 27 | −17 |
| Decreased by 20 | (0.8) | 30.5 | −25 |
| Decreased by 40 | (0.6) | 36 | −37 |

In about latitude 35°N temperatures would remain unchanged. Of course this model has its limitations. On the one hand, the reaction upon the ocean should enhance the zonal response; on the other, the change in temperature distribution would be expected to react upon the circulation system itself.

tion system operates as a result of climatic differentials. Truly, climate is about *localities* and it is most unwise to jump to universal conclusions from evidence which, by its very nature, is of local relevance. This does not mean that climatic change can never be widespread (that would be unreasonable in the face of the glaciological evidence), but it does suggest that regional contrasts have always been considerable. (*b*) Climate is a complex, not a simple, variable. Many theories of climatic change involve gross over-simplification since they lay major stress on but one climatic element, usually surface temperature, thus yielding a twofold classification into 'warm' and 'cool'. In Simpson's theory humidity also pays a considerable role so that a fourfold possibility is opened up, 'warm and wet', 'warm and dry', 'cool and wet' and 'cool and dry'. Yet when stress is laid upon atmospheric behaviour as the essential key to climate it is evident that there are other elements of importance. To name only one, the strength of the surface circulation, is to open up the number of possible variants to *eight,* and so on in geometrical progression with the addition of every other element that may be judged to be significant, e.g. the temperature of the upper air or the strength of the upper air circulation. Of course, not all the possible variants are equally probable since there is some degree of correlation between the various elements of climate, yet the known relationships at any particular locality are usually far from perfect, so that a considerable proportion of the possible outcomes must be regarded as by no means unlikely. We are thus back to the basic difficulty of defining climate but it seems wise to spell out this purely arith-metical aspect of the problem since it is so frequently ignored in the theories outlined above.

A reason for this omission is not far to seek. It is to be found in the nature of the evidence for climatic change which has been reviewed above. Surface temperatures in the past may be inferred from the distribution of organic beings, from the rate of chemical processes, particularly in the sea, and, much less directly, from glaciological evidence. As we have seen, the humidity record is much more equivocal, but archaeological finds, tree rings, soil characteristics and the observed stages of inland water-bodies offer crude indicators. The strength of the surface circulation leaves no such trace though locally some indication of its prevailing direction may be written in sand. Clearly the paths of the upper winds, now regarded as playing a funda-mental part in the generation of climatic patterns, must remain entirely a matter for conjecture. At best therefore our appreciation of past climates must remain fragmentary.

A further very important consequence of focusing attention on the general circulation of the atmosphere is that it raises the question, to what extent is the system as we know it subject to restraints? That the system is inherently variable is generally conceded. Sheppard (1963) has made the point that the pattern 'now' implies a past evolution and a promise of further changes to come since the changing pattern carries 'along with it a corresponding change in the distribution and intensity of heat sources and sinks'. 'Every season is created anew out of an atmospheric structure at the beginning of the season which has not been quite realised before' (*ibid*, 331). Various feedback processes, particularly those associated with ocean temperature, may give to these changes a long-term trend; Lamb and Johnson (1959) and Lamb *et al.* (1966) have endeavoured to illustrate such changes in the circulation pattern since 6500 B.C. Yet a long-term trend in the circulation pattern in one part of the world is likely to involve corresponding changes in the reverse sense somewhere else in the system, and the innate variability of the general circulation does not necessarily imply that it is capable of massive distortion or reorganisation. (See Tullett, 1970.)

This view has been put forward with considerable force by Sawyer (1966). He seeks to distinguish between 'the sort of variations which, on the basis of present knowledge, seem to be likely as aberrations of the large-scale circulation of the atmosphere; and also those which seem unlikely or impossible'. Turning first to the alleged effects of corpuscular or short-wave radiation on the upper atmosphere, he says 'there is, as yet, little evidence, either theoretical or observational, that the 10 per cent of the atmosphere above 15 kilometres has an effect on the remaining 90 per cent out of proportion to its mass'. In any event, if changes at high levels are to affect the climate at ground level they must do so via a disturbance of the circulation in the troposphere. Within this system he recognises 'two essentially different mechanisms' of heat transfer from equator to pole. Broadly between 30°N and 30°S lie the traditional Hadley cells transporting a balance of potential energy polewards, mainly aloft and in a more or less zonal fashion. But polewards of these latitudes the circulation is 'quite different' for here 'an overriding transport of heat polewards' is effected by 'the day-to-day disturbances of the flow which carry relatively warm air poleward in one longitude and replace it by cool air transported equatorward in another longitude'. It is implied that much of this exchange occurs relatively near the surface and that it is inherently azonal.

Now there is evidence that 'the poleward limit of the meridional Hadley circulation of the tropics is primarily determined by geodetic parameters (the earth's dimensions, rotation and gravity)'. Hence it is hard to see how the mean latitude of the subtropical highs, the average equatorward limit of the westerlies or the general location of the subtropical jet stream at 12 kilometres could be substantially changed by 'variations in solar radiation or other factors which might change the terms in the heat budget of the atmosphere by a few per cent'. Evidence is found in favour of this view in the slight displacement (some 6° of latitude) which actually occurs today between winter and summer although the poleward transport of heat 'varies by more than 7 to 1' between these seasons. This emphasis on the relative immobility of the intertropical wind belts seems to imply that climatic change within these latitudes can only occur on a minor scale, though Sawyer does not specifically commit himself to any such statement nor does he give any indication of what degree of change could be regarded as reasonable. The author's work on the trade winds (Crowe, 1949, 1950, 1951) may give some useful hints here.

It strongly supports Sawyer's view that the poleward limits of trade-wind systems show little seasonal shift in latitude over the open ocean but makes three other points which are surely not irrelevant. As between late summer and early autumn, when latitudinal temperature contrasts are least, and late winter and early spring, when such contrasts are strongest, each trade-wind system was shown to double in vigour, to spread westward as a comparatively stable air stream across the whole width of each ocean, and to lose its poleward 'root' in its eastern sector. It seems not improbable that the response of the trades during a 'glacial' phase would be broadly in the same direction. 'Pluvial' conditions would be reasonable along the whole extent of the equatorial convergence zone though this might be narrower than at present if similar developments were occurring in the southern hemisphere; the summer drought of 'mediterranean' lands would be reduced in intensity and duration; but, in the absence of strong relief, the trade-wind zones themselves could be expected to be *drier* than at present and hurricane generation would be damped down, if not completely suppressed. Under such conditions it seems very doubtful whether the total area of cloud in the intertropical girdle would undergo much change. Conversely, during interglacial periods, one would expect more active and more widespread hurricanes and, perhaps, a prolongation of the hurricane season[1] whilst, in their eastern sectors, the trades with their inbuilt inversion would remain supreme and the 'mediterranean' fringe might well be more arid than at present (see Kraus, 1960, 10). Whatever the value of these speculations, it does appear that longitudinal contrasts should not be overlooked even within the intertropical zone.

It is when he turns to the polar caps north of 30°N and south of 30°S that Sawyer (1966) himself stresses the longitudinal factor. Assessing the thermal effect of an increase in solar output, of such changes in the distribution of solar energy as might follow from the astronomical factors described above, and even of the reduced albedo of the polar regions as might follow from an ice-free Arctic, he concludes that the net results on the general distribution of surface temperature (as approached from the radiation angle alone) would be slight. Furthermore, whilst 'quite conceivable changes in heat transport by the ocean or atmosphere (100 per cent change for the ocean transport or less than 10 per cent for the atmosphere) could lead to equilibrium conditions with an ice-free Arctic ocean', the present Arctic inversion is so strong that 'the temperatures of the middle and upper troposphere' (above the 700-mb level) 'would not be affected by more than a few degrees and the direct effect on the general circulation of the change in the poleward temperature gradient would probably be of a minor character'. He is thus compelled to seek the key to climatic change in *longitudinal* differences in the general circulation such as can be observed for a spell at the present time. The crux of the problem thus becomes the origin of *persistent* anomalies in the circulation pattern. Sawyer freely confesses that the causes of persistent anomalies in the circulation are not

---

[1] With possible repercussions within the temperate zone (Walker, 1970).

known but concludes that interpretations of climatic change must be sought 'in terms of shifts in position and changes in amplitude of the favoured forms of the waves in the westerlies of the temperate belt'. These might result 'from the longitudinal differences in the heat budget of the atmosphere' arising 'from quite small variations in ocean temperature as well as from other causes.'

With some justification the reader may feel that the general tenor of Sawyer's paper is rather negative but at least he has restated the problem of climatic change in terms which are acceptable to modern meteorological science and he has opened a line of enquiry where the present, or near-present, may offer a key to the past. We still have much to learn about 'waves in the westerlies' and when ocean temperature and 'variations in snow cover in middle latitudes where the sun is relatively high in the sky in autumn and spring' are invoked, we are clearly involved in complex feedback processes which require further investigation. Sawyer does not attempt to make a frontal attack on the problem of the glacial epochs but he does not think that their heat economy can be grasped 'without an understanding of longitudinal differences'. This tallies with the view, already expressed, that in origin continental ice-sheets are *local* developments, however widespread their later side-effects may become. His approach has the further advantage that the 'waves' provide a series of channels by which a growing ice-sheet can be fed, a serious omission, as we have seen, in all theories based primarily on temperature considerations.

A satisfactory glacial theory which leans heavily on a distortion of the circulation pattern must cover three aspects of the problem:

1. The initiation of glacial phenomena in selected localities;
2. The development and sustenance of the ice-sheets when emplaced; and finally,
3. A stage of overdevelopment, presumably of the distortion, which will lead to regression and ultimate decay of the ice.

Sawyer's approach has no real difficulty with the first two points but its implications with regard to (3) still have to be worked out. If continental ice produces a widespread chill over ocean waters he must find it far from easy to steer the 'long waves' into a pattern which would promote recession. Perhaps we have too readily assumed that at a glacial maximum temperatures over the Arctic ocean must necessarily be lower than at present. Ewing and Donne do not believe this but possibly for the wrong reasons. Is it conceivable that the presence of large continental ice-caps in higher-middle latitudes might open or reopen the Arctic ocean by directing storm tracks right across the pole? And is this the first prerequisite for ice recession? The idea that the presence of ice-caps in the northern hemisphere must necessarily deflect westerly disturbances towards the *south* has held the field too long and, as we have seen, it has been challenged by Leighly. What would be the nature of the polar circulation if the two great pools of Arctic air (see Fig. 83) were largely replaced by masses of immobile ice with a rather constricted corridor between them? Transpolar circulation has been given little attention in the past for we have been obsessed with zonal concepts. Yet it is known to occur, in both hemispheres.

It is appropriate that this chapter should close with questions rather than assertions. A geographical question of great significance to glaciology is why does the Greenland ice-shield persist between the latitudes of 60° and 80°N in a hemisphere now substantially free of land-ice? Is it merely a decaying fossil remnant of past climates or does it follow from a condition of present balance? The economy of this ice-sheet is fundamental and it seems doubtful if investigations of smaller and hence more accessible glaciers can yield results of equal value. Judging from its presence, glaciation demands a unique combination of circumstances which has not always been recognised owing to a confusion between continental and mountain ice. They are (*a*) considerable relief, (*b*) a sea chilled, particularly in summer, by large quantities of floating ice, and (*c*) a very active circulation involving considerable precipitation and much cloud. It appears that the Antarctic ice-cap is sustained by much the same mechanism, particularly around its margins.

# Appendix 1
# Climatological statistics

Climatology derives most of its basic concepts from an appreciation of various series of meteorological observations taken over extended periods of time. To reduce a mass of data to an intelligible form it must be processed in some systematic manner. The subject thus inevitably involves statistics which may be described as the art of manipulating measurements so as to secure the maximum amount of ordered information from them. It is the object of this appendix to show that the processing of data is really a matter of plain common sense. Indeed, the simplest possible devices are often to be preferred.

This account falls into two sections. The first deals entirely with the statistics of description, particularly as applied to monthly data for temperature and precipitation. Although very elementary, the points made are of fundamental importance. The second section touches lightly on what might be called the statistics of comparison. This involves the introduction of some theoretical considerations which are handled at greater length in standard texts, here we are concerned only with their broad climatological implications.

## Description

Accurate description usually involves measurement, and all measurements, however refined, involve some degree of imprecision. In the exact sciences where the object measured is clearly defined this arises mainly from the design of the instrument used. The range of possible 'error' is usually slight and it is a common practice to repeat the reading several times and strike an average of the results. It must be noted that 'error' in this context means inherent uncertainty, something quite different from a 'mistake'. A mistake can be rectified but statistical error is always with us. This is particularly true when we seek to measure the characteristics of 'groups' rather than of single objects. In weighing boys at their fifth birthday we know we shall obtain a cluster of answers and the 'error' here stems from the intrinsic nature of the group rather than from any fault in the scales. If therefore we infer an 'average boy' we are taking a step from the world of reality to the world of ideas. The transition may be fruitful but it has its pitfalls. It is a sound maxim that the

507

'average boy' should never be left on his own. He should always be accompanied by some statement as to the distribution of his fellows.

In climatology we are also concerned with the measurement of phenomena which are known to be variable; the statistical problems encountered are thus analogous to those met in biology though the range of 'error' may be more extreme. Sometimes the individuals composing the relevant group have a fair degree of concrete reality, e.g. the total rainfall accumulated at a given site during a month. Often our concepts are more abstract; the incessant periodic and aperiodic variation of both temperature and pressure make some kind of processing necessary before the climatologist is ready to take over. The mean temperature for a given day or month is thus already a statistic but in the discussion which follows we shall regard them as 'given data' or 'observations' and focus attention on the implications which have to be faced if these are employed to develop concepts valid over a very much longer period of time. To date by far the most popular way of combining records from different years has been by striking an arithmetic average. We seek to examine whether the method which works well for instrumental imprecision is necessarily the best approach to errors of the other category.

## Temperature

The basic type of question which arises here is of the form, 'How warm is it at Greenwich in July?' An answer may be required for some specific purpose, such as the design of an air-conditioning plant, or merely as the basis of a comparison, of July with another month at Greenwich or of July at Greenwich with the same month elsewhere. Whatever the object, it is in answers to questions like this that we find the bricks with which climatological structures are erected. The question may seem simple enough but how do we set about finding an answer? Ignoring the complexities of diurnal and day-to-day variation it may seem that the mean temperatures for 120 Julys at Greenwich given in *World Weather Records* should offer the necessary clue. We give them, arranged in order of magnitude, in Table A.1.1.

It is evident at once that, even in the comparatively equable climate of Britain, there is a remarkable range of possible answers to our question. The highest mean monthly value over this period was 68.9°F and the lowest, 57.5°F, a range of no less than 11.4°F. That 1859 and 1919 were nevertheless not freak years is suggested by the comparison of the second highest value with the second lowest, the third highest with the third lowest, and so on. It is clear from inspection that there can be no question of a serious mistake (or misprint) and that all the values fall into a single comprehensible group. How should we set about trying to generalise about it? As we have seen the traditional approach is to compute the arithmetic average. This comes out as 62.71°F. Actually only the years 1895 and 1956 gave July values within 0.1°F of this value, nevertheless it is clear that in this case the average is a pretty fair representation of the general magnitude of the group as a whole. The example has been selected with this aim in view.

Let us look a little more critically at this conclusion. Sixty-five of the 120 Julys recorded *less* than the average though the highest extreme was 6.2°F above it and the lowest extreme was only 5.2°F below it. Furthermore if we sum up all this information in no more than a single figure we are throwing away a great deal of useful and perhaps relevant information. Clearly an

TABLE A.1.1. *Mean monthly temperature in July at Greenwich, 1841–1960 (°F)*

| | | | | | | | |
|---|---|---|---|---|---|---|---|
| 68.9 | (1859) | 64.9 | (1951) | 62.3 | (1848) | 61.0 | (1927) |
| 68.1 | (1868) | 64.8 | (1869) | 62.3 | (1864) | 60.9 | (1902) |
| 67.6 | (1921) | 64.8 | (1901) | 62.3 | (1908) | 60.8 | (1843) |
| 67.3 | (1911) | 64.7 | (1846) | 62.2 | (1849) | 60.8 | (1936) |
| 67.0 | (1852) | 64.6 | (1865) | 62.2 | (1850) | 60.7 | (1940) |
| 66.7 | (1876) | 64.5 | (1897) | 62.2 | (1917) | 60.6 | (1915) |
| 66.7 | (1959) | 64.1 | (1925) | 62.0 | (1871) | 60.4 | (1954) |
| 66.6 | (1900) | 64.0 | (1873) | 61.9 | (1866) | 60.3 | (1851) |
| 66.5 | (1887) | 64.0 | (1945) | 61.9 | (1894) | 60.3 | (1882) |
| 66.3 | (1923) | 63.6 | (1885) | 61.9 | (1898) | 60.1 | (1842) |
| 66.3 | (1933) | 63.6 | (1946) | 61.7 | (1844) | 60.1 | (1867) |
| 66.3 | (1934) $-D_9$ | 63.5 | (1948) | 61.7 | (1942) | 60.1 | (1891) |
| 66.0 | (1870) | 63.5 | (1958) | 61.7 | (1948) | 60.0 | (1909) |
| 66.0 | (1905) | 63.4 | (1906) | 61.6 | (1856) | 59.9 | (1845) |
| 66.0 | (1935) | 63.4 | (1929) | 61.6 | (1880) | 59.9 | (1875) |
| 66.0 | (1941) | 63.3 | (1912) | 61.6 | (1903) | 59.8 | (1883) |
| 66.0 | (1949) | 63.2 | (1878) | 61.6 | (1924) | 59.8 | (1916) |
| 65.8 | (1899) | 63.2 | (1884) | 61.6 | (1930) | 59.6 | (1862) $-D_1$ |
| 65.7 | (1928) | 63.1 | (1886) | 61.5 | (1861) | 59.6 | (1890) |
| 65.5 | (1872) | 63.1 | (1926) | 61.5 | (1877) | 59.5 | (1892) |
| 65.5 | (1881) | 63.1 | (1937) | 61.4 | (1858) | 59.4 | (1920) |
| 65.5 | (1904) | 63.0 | (1944) | 61.4 | (1863) | 58.6 | (1907) |
| 65.5 | (1952) | 63.0 | (1950) | 61.4 | (1938) | 58.6 | (1922) |
| 65.3 | (1847) | 62.9 | (1893) | 61.3 | (1918) | 58.5 | (1913) |
| 65.3 | (1955) | 62.8 | (1956) | 61.3 | (1930) | 58.3 | (1860) |
| 65.3 | (1957) | 62.7 | (1895) | 61.3 | (1960) | 58.2 | (1879) |
| 65.2 | (1896) | 62.6 | (1855) | 61.2 | (1939) | 58.1 | (1910) |
| 65.2 | (1947) | 62.5 | (1914) | 61.0 | (1853) | 58.0 | (1888) |
| 65.1 | (1857) | 62.4 | (1932) | 61.0 | (1854) | 57.7 | (1841) |
| 64.9 | (1874) | 62.4 | (1953) | 61.0 | (1889) | 57.5 | (1919) |
| | $-Q_3$ | | $-M$ | | $-Q_1$ | | |

expression of the mean scatter of the data is highly desirable. This can also be obtained arithmetically by working out the 'mean deviation' of the group (the average difference between the actual values and the general average taken irrespective of sign). The reader may care to work this out for himself. A still more sophisticated procedure is to work out the 'standard deviation'. Here the deviations are squared to bring them all to the same sign but it is clear that with the data in their present form this involves embarking on a major arithmetical exercise. We shall return to this matter in the next section.

We now come to a rather revolutionary thought. A student can become 'numerate', i.e. he can learn to appreciate the significance of numbers, without necessarily doing sums. When we work out an average we are really seeking a middle value; clearly Table A.1.1. gives us this without more ado. Evidently 60 Julys gave values in excess of 62.35°F and the other 60 gave values that were less. This is known as the 'median' of the group and it is no less valid than the arithmetic average as a general expression of the scale of the group as a whole. In this instance the difference between the average and the median is slight, but in some of the cases which follow it is considerable and a case can then be made for preferring the median as a more representative middle value. But having once embarked upon this non-arithmetical exercise, there is no logical reason why we should not continue it. Clearly a quarter of the values lie above 64.9°F and a quarter fall below 61.0°F. These positions are known as the upper $(Q_3)$ and lower $(Q_1)$ 'quartiles' and they present us with a perfectly reasonable statement of the mean scatter or spread of the data. Even the labour of constructing Table A.1.1. is avoidable for the data can be plotted on a line graph and the entries counted from each extremity or, in a shorter series, the positions of the median and the two quartiles can be located by simple inspection. An expression of the 'mean extremes' of the group can also be obtained by striking off the upper and lower ten per cent of the instances. These positions are known as the upper $(D_9)$ and lower $(D_1)$ 'deciles'. They are of most value when the absolute extremes are suspected of having something of a 'freak' character. In the table before us the deciles lie at about 66.15°F and 59.6°F.

The mass of information given by Table A.1.1. can thus be succinctly summarised as follows:

*July temperatures at Greenwich* $(°F)$

|  | $D_9$ | $Q_3$ | $M$ | $Q_1$ | $D_1$ |
|---|---|---|---|---|---|
| 1841–1960 | 66.15 | 64.9 | 62.35 | 61.0 | 59.6 |
| 1841–1900 | 66.25 | 64.75 | 62.1 | 60.55 | 59.6 |
| 1901–1960 | 66.2 | 65.25 | 62.85 | 61.25 | 59.6 |

In the lower half of this table the two 60-year periods, 1841–1900 and 1901–1960 have been analysed separately. Does it yield evidence of a secular rise in July temperatures? The verdict is equivocal. This gives a first warning that where individual values are spread over a considerable range the middle point may be partly the product of mere chance. In the above case even the first place of decimals is clearly suspect.

Table A.1.1. is undoubtedly cumbersome and the information it contains can be expressed much more briefly and with little loss of accuracy if it is put into a classified or 'graded' form as in Table A.1.2. A difficulty here is that the choice of classes or 'grades' is entirely arbitrary. We illustrate this by grouping the data in two versions of two-degree intervals (Table A.1.2).

TABLE A.1.2. *July temperatures at Greenwich 1841–1960 (°F)*

| (a) | Instances (years) | (b) | Instances (years) |
|---|---|---|---|
| Below 58 | 2 | Below 59 | 9 |
| 58–59.9 | 15 | 59–60.9 | 20 |
| 60–61.9 | 36 | 61–62.9 | 38 |
| 62–63.9 | 28 | 63–64.9 | 23 |
| 64–65.9 | 23 | 65–66.9 | 25 |
| 66–67.9 | 14 | Over 67 | 5 |
| Over 68 | 2 | | |
| | 120 | | 120 |

It is usually wise to experiment a little with the class interval and with the starting-point of the classification, the object being to secure a relatively simple and smooth 'curve'. The latter word reminds us that visual appreciation of the data is then possible through the construction of a histogram as in Fig. 156. Alternative solutions are shown but it must not be forgotten that both diagrams represent the same distribution. This introduces us to a third type of middle value, the grade at which there are most instances. This is known as the 'mode' in the sense that it is the most fashionable category and it is in this sense that the word 'average' is often employed in common speech. Yet the mode is not a very popular statistic because it is a little indeterminate. It depends in the first instance on an intelligent choice of a class interval and even then it has to be 'felt for'. Thus it is evident from Table A.1.2. that 61–62.9°F with 38 instances is preferable to 60–61.9°F with only 36 but reference back to Table A.1.1. will show that there are also 38 instances in the interval 60.5–62.4°F. The fact that the mode also covers a *range* of values is also a little off-putting to those in search of precision but this is a point in its favour. It is the object of this appendix to demonstrate that the apparent precision of the arithmetic average is very largely fallacious.

These three expressions of a mean value, the mode, the median and the arithmetic average, respond differently to the addition of one exceptionally high (or low) value to the series. The mode remains unchanged since, by definition, the new entry will fall outside the modal class. The median may shift slightly but only to the same degree as it would be affected by any other new entry above (or below) the median position. The arithmetic average, on the other hand, responds according to the *weight* of the new record (the magnitude of its deviation from the average); in a short series the shift may be quite considerable. The type of mean we employ for descriptive purposes thus largely depends on the significance we attach to the exceptional occurrence.

In conclusion let us note that a long series of July temperatures at Greenwich thus presents a compact, closely grouped and relatively symmetrical distribution. To entertain the notion of a 'normal' temperature is therefore

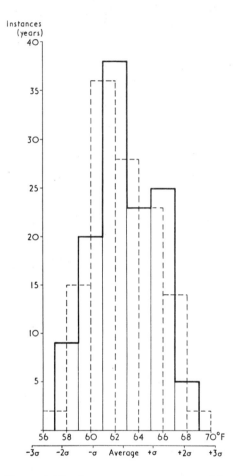

Fig. 156. The frequency group presented by the mean monthly temperatures of 120 Julys at Greenwich (1841–1960). The data are grouped in two-degree (F) classes or grades and alternative solutions are illustrated. The bar beneath the histogram shows the horizontal scale transformed to units of the standard deviation ($\sigma$) from the arithmetic average. See text

not unreasonable and all three expressions of it fall within a degree or so of one another. The extent of the scatter may have surprised us but it will be shown later than this is much what could be expected from the operation of 'pure chance'.

To illustrate a case of much greater temperature variability we have selected January temperatures at Winnipeg over the period 1873–1960 (88 years) (Table A.1.3.). The data are summarised in classified form.

TABLE A.1.3. *January temperatures at Winnipeg 1873–1960 (°F)*

| (a) | Instances (Years) | (b) | Instances (Years) |
|---|---|---|---|
| Below −16 | 1 | | |
| −16 to −14.1 | 4 | Below −14 | 5 |
| −14 to −12.1 | 4 | | |
| −12 to −10.1 | 5 | −14 to −10.1 | 9 |
| −10 to −8.1 | 5 | | |
| −8 to −6.1 | 10 | −10 to −6.1 | 15 |
| −6 to −4.1 | 7 | | |
| −4 to −2.1 | 10 | −6 to −2.1 | 17 |
| −2 to −0.1 | 4 | | |
| 0 to 1.9 | 7 | −2 to 1.9 | 11 |
| 2 to 3.9 | 9 | | |
| 4 to 5.9 | 5 | 2 to 5.9 | 14 |
| 6 to 7.9 | 10 | | |
| 8 to 9.9 | 3 | 6 to 9.9 | 13 |
| 10 to 11.9 | 3 | | |
| Over 12 | 1 | Over 10 | 4 |
| | 88 | | 88 |

It is evident that a two-degree class interval is quite inadequate to iron out the irregularities of this distribution and that even a four-degree interval fails to indicate any very obvious mode. Indeed there is some suggestion that Januarys at Winnipeg tend to alternate between two situations, the very cold with temperatures around −6°F (−21°C) and the not-so-cold with temperatures of 6°F (−14°C). This may well arise from whether one or more spells of intensely cold weather occur during the month. At all events it is clear that, in these circumstances, the notion of a monthly 'normal' can scarcely apply and that any statement of a mean value must be fairly unreliable.

A summary of the record by the method outlined above yields the following result:

*January temperatures at Winnipeg (°F)*

| $D_9$ | $Q_3$ | M | $Q_1$ | $D_1$ |
|---|---|---|---|---|
| 7.7 | 4.15 | −2.45 | −7.6 | −12.4 |

The arithmetic average is −2.0°F. The mean scatter, given by $(Q_3 - Q_1)$, is thus 11.7°F at Winnipeg as compared with only 3.9°F in the Greenwich case cited above. Such values are usually quoted in the form ±5.85 and ±1.95 degrees F respectively. This statistic is known as the quartile deviation or, more picturesquely, as the 'probable error', the implication being that there is an

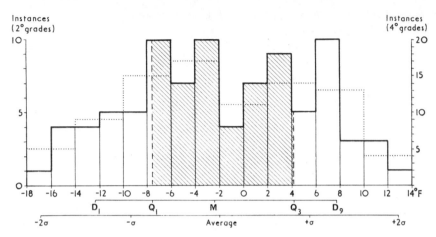

Fig. 157. The frequency distribution of the mean monthly temperatures of 88 Januarys at Winnipeg (1873–1960). The data are presented in both two-degree (F) and four-degree class intervals or grades. Other groupings are possible but no really convincing modal value emerges. The position on the horizontal scale of the median (M), the quartiles ($Q_1, Q_3$) and the deciles ($D_1, D_9$) is indicated and the interquartile range is cross-hatched. A scale in terms of the standard deviation from the average is appended. Like Fig. 154, the distribution is broadly symmetrical but the values are not closely concentrated about a central point

equal chance that any single record will fall within this range or outside it. In other words it encloses the central 50 per cent zone, a concept used in naval gunnery-practice.[1] Obviously the tighter this range the more accurate the shooting. Clearly Januarys at Winnipeg are rarely on target (Fig. 157). This affects the degree to which we can make accurate generalisations about them.

## Rainfall

Still more serious problems arise when we face the problem of generalisation about rainfall. It is often optimistically assumed that if we have a good long record (some forty years or more) we can handle this climatic element with some confidence. Yet the road is full of pitfalls for the unwary. The record for Milan covers no less than 197 years from 1764 to 1960 and we select the month of October, the 'wettest' month of the year.

Our basic question is now of the form, 'How wet is it at Milan in October?' Until we have some kind of answer to this we cannot compare October with any other month at Milan or Milan with any station in some other part of the

---

[1] To hit and demolish the target is to suspend operations.

world. Here, at least, we are dealing with a simple monthly total and not with a statistic as complex as mean monthly temperature. Yet is the answer any more straightforward? Apply the previous drill. The largest total recorded over the period was 376 millimetres (14.8 in) in 1872; the smallest was nil in 1921. The second and third largest were 319 millimetres (12.6 in) and 315 millimetres (12.4 in) in 1953 and 1907; the second and third lowest were 4.8. millimetres (0.19 in) and 5.5 millimetres (0.22 in) in 1871 and 1774. Despite the very wide range involved there is therefore no obvious case for regarding any element in the series as a freak occurrence though the 1872 total was certainly exceptional. In the light of this range the question certainly arises, how much reliance can be placed upon the 197-year average which comes out as 117 millimetres (4.6. in)? In fact 114 years recorded totals less than the average and only 83 years (42 per cent) recorded more. Furthermore as Fig. 158 shows, the concentration of the records about the average was anything but strong. Clearly, as we have already mentioned, the average is supersensitive to the occasional very high values.

A more balanced view of the group as a whole is yielded by the method we have described. The results which emerge are tabulated below together with details for subdivisions of this extremely long record.

*October rainfall at Milan (cm)*

|  | $D_9$ | $Q_3$ | $M$ | $Q_1$ | $D_1$ | *Average* |
|---|---|---|---|---|---|---|
| 1746–1960 | 23.4 | 16.4 | 9.8 | 5.8 | 3.3 | 11.7 |
| 1911–1960 | 22.1 | 14.8 | 10.0 | 6.0 | 3.3 | 11.0 |
| 1861–1910 | 26.3 | 17.0 | 9.4 | 4.7 | 3.3 | 12.1 |
| 1811–1860 | 25.3 | 20.0 | 12.1 | 7.4 | 4.8 | 13.8 |
| 1746–1810 | 18.1 | 15.3 | 8.6 | 4.8 | 2.4 | 9.9 |

It is clear that the spread of the values has varied considerably from period to period though there is little reason to challenge the general consistency of the record, except perhaps before 1810 when it is possible that the exposure of the gauge was not ideal. Since it is sometimes argued that the median is less stable under sampling than the arithmetic average we have quoted the appropriate averages for comparison. The only serious discrepancy is in the period 1861–1910 when, as the upper decile shows, there was an unusual number of very heavy falls. This has raised the average whilst the median has responded to the considerable number of very moderate falls signalled by the lower quartile. Since rainfall is in no sense cumulative over periods measured in decades this can scarcely be regarded as a fault.

The full record is summarised in classified or graded form in Table A.1.4 and the story is told clearly in Figure 158. Note that the mode of this group is as low as 4 to 6 centimetres though an alternative interpretation as 4 to 8 centimetres is probably preferable; no less than 49 Octobers fell into this category. When there is a large number of instances in a group, as in this case, statisticians are sometimes tempted to fit a smooth curve to the histogram and to speak of the mode as at its maximum point. This will clearly fall very near 6 centimetres. The rule for positively skew curves that the

$$\text{MODE} = \text{AVERAGE} - 3(\text{AVERAGE} - \text{MEDIAN})$$

is thus beautifully illustrated by this example. This is mentioned only in passing as, in the author's opinion, the fact that the mode normally draws our

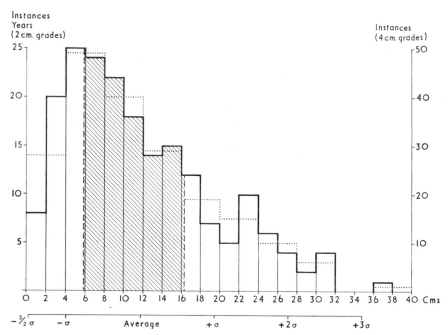

FIG. 158. The frequency distribution of October rainfall totals for 197 years at Milan (1764–1960). The data are grouped into both 2-centimetre and 4-centimetre grades, the latter yielding a smoother 'curve'. The inter-quartile range is cross-hatched. Note the concentration towards relatively low values. Most rainfall records show a 'positively skew' distribution of this nature. The arithmetic average thus lies well above the major concentration of values. The scale in terms of the standard deviation of the group shows that the record for one year (1872) may be regarded as highly exceptional

attention to a *range* of values is a distinct advantage when we are thinking in descriptive terms. However, it may interest the reader to note that these three types of mean values, reached by such different routes, can be shown to be related. Most rainfall data give a skew curve of this character.

The arithmetic average is even more deceptive when a considerable number of very low records occur at a station which nevertheless occasionally experiences quite heavy falls. We select as an example the July record for Karachi over the period 1856 to 1960 (Table A.1.5.). It should be borne in mind that this is at the height of the south-west monsoon and that July has a

TABLE A.1.4. *October rainfall at Milan 1764–1960 (cm)*

| cm | Instances (years) |
|---|---|
| 0–2 | 8 |
| 2–4 | 20 |
| 4–6 | 25 |
| 6–8 | 24 |
| 8–10 | 22 |
| 10–12 | 18 |
| 12–14 | 14 |
| 14–16 | 15 |
| 16–18 | 12 |
| 18–20 | 7 |
| 20–22 | 5 |
| 22–24 | 10 |
| 24–26 | 6 |
| 26–28 | 4 |
| 28–30 | 2 |
| 30–32 | 4 |
| 36–38 | 1 |
| | 197 |

claim to be considered the wettest month of the year at this station. To emphasise the nature of the problem which faces climatological description it is necessary to present the data in half-inch classes though this produces a rather ragged diagram in Fig. 159.

TABLE A.1.5. *July rainfall at Karachi 1856–1960*

| Inches | Instances (years) | Inches | Instances (years) |
|---|---|---|---|
| 0–0.5 | 38 | 8.0–8.5 | 3 |
| 0.5–1.0 | 6 | 8.5–9.0 | 1 |
| 1.0–1.5 | 6 | 9.0–9.5 | 0 |
| 1.5–2.0 | 8 | 9.5–10.0 | 0 |
| 2.0–2.5 | 4 | 10.0–10.5 | 1 |
| 2.5–3.0 | 6 | 10.5–11.0 | 2 |
| 3.0–3.5 | 2 | 11.0–11.5 | 3 |
| 3.5–4.0 | 3 | 11.5–12.0 | 0 |
| 4.0–4.5 | 3 | 12.0–12.5 | 0 |
| 4.5–5.0 | 5 | 12.5–13.0 | 1 |
| 5.0–5.5 | 4 | 13.0–13.5 | 1 |
| 5.5–6.0 | 2 | 13.5–14.0 | 0 |
| 6.0–6.5 | 4 | 14.0–14.5 | 0 |
| 6.5–7.0 | 0 | 14.5–15.0 | 0 |
| 7.0–7.5 | 0 | 15.0–15.5 | 1 |
| 7.5–8.0 | 0 | 18.5–19.0 | 1 |
| | | | 105 |

The problem which confronts us is also revealed if we turn to the original record and look for the extreme values. The highest fall recorded was 18.6 inches in 1894, the second highest 15.4 inches in 1933, and the third 13.4

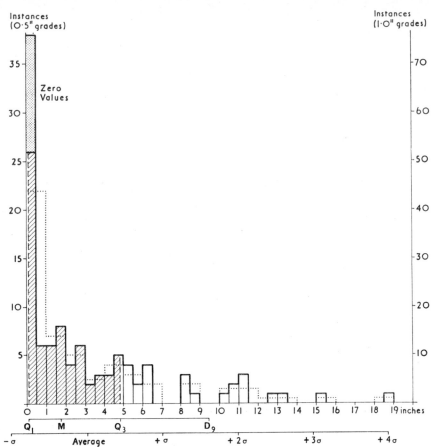

Fig. 159. The remarkable frequency distribution of the July rainfall totals during 105 years at Karachi (1856–1960). The data are grouped into both ½-inch and 1-inch grades but neither method gives a continuous curve. Twelve years when the record gave no measurable precipitation face us with a problem in graphical representation. With such a skew distribution the arithmetic average ceases to be representative of the group as a whole. The curve could be made more 'normal' in form if the horizontal scale were expressed in logarithmic terms but the treatment of the zero values would still present difficulty. The standard deviation is less appropriate a measure of scatter in a group of this form but it does indicate that the records of two years (1894 and 1933) were highly exceptional. The interquartile range is cross-hatched.

inches in 1930. Yet the first, second and third lowest values were all zero and no less than twelve years of the 105 covered by the record registered no rainfall whatsoever. These are very strange bedfellows which will defy any system of generalisation which demands a single mean. The arithmetic average comes out at 3.1 inches but only 37 years (35 per cent of the series) recorded more than this as compared with 68 years which recorded less. A summary statement in the form we have proposed above is certainly much more revealing. It is given below both for the whole period and for the last two half-centuries.

*July rainfall at Karachi (in)*

|  | $D_9$ | $Q_3$ | $M$ | $Q_1$ | $D_1$ | *Average* |
|---|---|---|---|---|---|---|
| 1856–1960 | 9.4 | 4.85 | 1.75 | 0.10 | 0 | 3.1 |
| 1911–1960 | 10.3 | 4.85 | 1.90 | 0.45 | 0.05 | 3.3 |
| 1861–1910 | 8.6 | 4.75 | 1.25 | 0.05 | 0 | 3.0 |

It is true that the average is thus more stable under sampling than the median but should we not count the cost? It is in fact largely achieved by spreading the single heavy fall of 1894 over the nine years in 1861–1910 when the record was nil. This is quite impossible in any practical or physical sense.

It is evident from Fig. 159 that the modal rainfall in July at Karachi is 0–0.5 inches, or alternatively 0–1.0 inches. These class intervals contain respectively 38 (36 per cent) and 44 (42 per cent) of the 105 instances. Clearly if the practical question arises, 'How much water can be obtained from run-off at Karachi even during the wettest season of the year?' the only sensible answer is, 'In most years very little, especially with the mean temperature running in the neighbourhood of 85°F'. The water-supply engineer would thus turn his attention to ground-water or Indus-river sources whilst the drainage engineer would face the problem of the disposal of over 9 inches of rain during 10 per cent of the years. A refugee encampment on the floor of a dry watercourse within the city is wiped out at intervals.

## Comparison and correlation

When we seek to institute comparisons between climatic data or to demonstrate the relationships which may be suspected between them, it is no longer possible to avoid statistical theory. Even from the 'kindergarten' approach outlined above it should be apparent, however, purely on commonsense grounds, that differences of mean temperature of less than 1°F or of mean rainfall of less than one inch may often be entirely meaningless. They may arise purely from the chance factor involved in the choice of the period of reference.

The reader is referred to statistical texts for a full statement of these problems but a brief introduction may serve to whet his appetite for more and to demonstrate some of the implications of the above non-mathematical section.

The fact is that the products of 'pure chance' are often calculable. Without this bookmakers would face ruin and the Premium Bond fund might run dry!

A pin-board available as a parlour game and usually present in some form or other in any amusement arcade forms the best introduction to those not too happy about their grasp of the binomial theorem. A ball passes successive rows of pins so that, at each encounter, it meets the pin dead-centre. On each occasion therefore there is a fifty-fifty chance that the ball will pass to the right or to the left. A single ball thus zig-zags down the board and we are apt to watch it with an empty mind. But what if we have a number of balls and can accumulate them in boxes at the base? A rule then emerges which is given by 'Pascal's triangle' as follows:

| Rows | | | | | | | | | | | | | Totals |
|------|---|---|---|---|---|---|---|---|---|---|---|---|--------|
| | | | | | | | 1 | | | | | | 2 |
| 1 | | | | | | 1 | | 1 | | | | | $4 = 2^2$ |
| 2 | | | | | 1 | | 2 | | 1 | | | | $8 = 2^3$ |
| 3 | | | | 1 | | 3 | | 3 | | 1 | | | $16 = 2^4$ |
| 4 | | | 1 | | 4 | | 6 | | 4 | | 1 | | $32 = 2^5$ |
| 5 | | 1 | | 5 | | 10 | | 10 | | 5 | | 1 | $64 = 2^6$ |
| 6 | 1 | | 6 | | 15 | | 20 | | 15 | | 6 | | 1 $128 = 2^7$ |
| 7 | 1 | 7 | | 21 | | 35 | | 35 | | 21 | | 7 | |

This can be extended at leisure by adding pairs of values in the manner indicated.

The conclusions which follow from this 'game' are of great importance.

(*a*) An ordered symmetrical pattern is seen to arise from the apparently chaotic behaviour of a considerable number of independent individuals. The only rule imposed is that, at every pin it meets, each ball should be equally free to go one way or the other.

(*b*) The values along any row may be plotted to give one version of the infinite but related forms taken by the 'normal curve of error'. This is thus relieved of its considerable mathematical terrors. Figure 160 has been constructed in this manner from the 17th series. Anyone can do this and it will surely be conceded that the curves are not without an austere beauty.

(*c*) The rules of chance are not inexorable laws, they are only statements of tendencies. Using 128 balls down 7 rows on any single trial we must not therefore expect to get precisely the totals indicated but repeating the experiment several times and averaging the results we should approach these figures more and more closely. It should be apparent nevertheless that, in these circumstances, only the central eight of the boxes can hope to make a capture. Those beyond this range could only be reached as a result of bounce which would indicate a fault in design. The total *range* of the distribution is thus quite rigidly circumscribed.

(*d*) The model also presents a useful rule about 'grading' data and thus we get back to climatology. If we suspect that temperature variation involves a considerable element of 'chance' and we have a 32-item record it would be unwise to classify the data into more than six groups; with 128 items we might profitably increase the number to eight. It so happens that we have already grouped 120 July temperatures at Greenwich into both six and seven class intervals and we know that the curve is at least roughly symmetrical. A

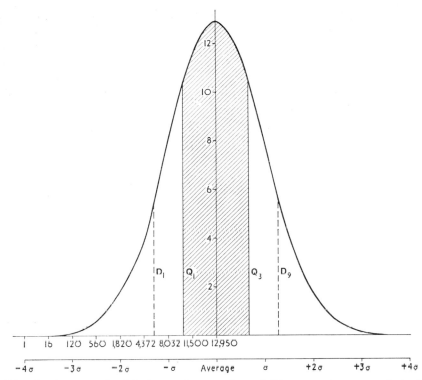

FIG. 160. An example of the 'normal curve of error'. The curve shown was derived from the seventeenth row of Pascal's triangle ($2^{17}$ instances). The vertical scale is thus in thousands. This is thus one member of a whole family of curves which can be expressed by a single mathematical formula to be found in statistical texts. Because it is perfectly symmetrical the arithmetic average, the median and the mode all coincide. The interquartile range, covering the central half of the area enclosed by the curve, is cross-hatched. Ten per cent of the area lies to the left of $D_1$ and another 10 per cent lies to the right of $D_9$. A horizontal scale in terms of the standard deviation of the group is also provided. It will be observed that the great bulk of the values lie within $\pm 3\,\sigma$ of the average. The text explains how this type of curve is derived from the operation of pure chance. It provides a standard which lies at the very foundation of all statistical theory

rather crude comparison with random production may thus be revealing. It can be achieved as follows:

|        |   |    |    |    |    |    |    |   | *Total* |
|--------|---|----|----|----|----|----|----|---|---------|
| *Row 7* | 1 | 7 | 21 | 35 | 35 | 21 | 7 | 1 | 128 |
| *Group (b)* | — | 9 | 20 | 38 | 23 | 25 | 5 | — | 120 |
| *Group (a)* | 2 | 15 | 36 | 28 | 23 | 14 | 2 | | 120 |

The correspondence is certainly imperfect but it does appear that 'chance' has played a considerable role in moulding the Greenwich record.

As might be expected from Fig. 157 even the four-degree classes (b) at Winnipeg are much too weak in the central values to fit any reasonable interpretation of chance production. There is evidently some real process encouraging dispersion here. It may also be noted in passing that in group (*a*) we clearly had too many class intervals. The effect of this can also be appreciated from the model. Thus if only 16 balls are dropped through the seven rows, even if they achieve the ideal distribution 1, 4, 6, 4, 1, as they pass row 4, this will be progressively broken down as they strive to reach 8 boxes. The mean chance of their reaching each of these is given by the figures in row 7 divided by 8 but, since a fractional chance still represents a *possible* success, a distribution such as 1, 1, 2, 4, 4, 3, 0, 1 is not to be ruled out and farther down the scale all trace of a concentration might well disappear. In these circumstances the chances of random production are not being fully exploited and we are entering on a pure gamble. Beware!

Since rainfall distributions are usually skew there is little point in attempting to try out this kind of exercise on them.

The great advantage of the normal curve of error is that, since it can be expressed mathematically, one can make precise calculations about it. As it is a symmetrical curve the arithmetic average, median and mode are all at the same central point. In Fig. 160 it will be seen that the horizontal scale is expressed outwards from that point in positive or negative units of $\sigma$ (sigma). This symbol expresses the standard deviation of the whole group, i.e. the square root of the average of squares of the deviations measured from the arithmetic average. It is then possible to make the following important statements about the various proportions of the total area of the curve enclosed within given limits and hence of the mean chance that any single record, taken at random, will fall within the limits specified.

| Range ($\sigma$) | Percentage of records within range | Chance | |
|:---:|:---:|:---:|:---:|
| ±0.675 | 50.0 | 1 : 1 | (*Q*) |
| ±1.0 | 68.3 | 2.1 : 1 | |
| ±1.28 | 80.0 | 4 : 1 | (*D*) |
| ±1.5 | 86.6 | 6.5 : 1 | |
| ±2.0 | 95.45 | 21 : 1 | |
| ±2.5 | 98.76 | 80 : 1 | |
| ±3.0 | 99.73 | 370 : 1 | |
| ±3.5 | 99.954 | 2175 : 1 | |
| ±4.0 | 99.99 | 10 000 : 1 | |

Note that (*Q*) represents the interquartile range and that (*D*) represents the interdecile range with which we are now familiar; also that rather less than 5 per cent of the values lie outside a range of ±2$\sigma$ from the arithmetic average.

This interesting distribution is sometimes used to weed out extraneous entries from a group of data. In climatology the greatest danger is from misprints, but these can usually be spotted if one is alert. Much more important is the contribution it makes towards the decision whether the observed difference between the averages of two groups of data is 'statistically significant', that is, whether the distinction is real or more probably merely the result of chance. We have already seen reason to doubt the validity of very small differences in average temperature or rainfall; the method outlined below provides a positive check. Other methods are given in statistical texts but they all stem from the same principles. This matter is of quite outstanding importance. To go to some length to 'explain' an observed difference and then to have it demonstrated that it is statistically insignificant is to develop a very red face indeed.

The matter is not easy on first encounter for two reasons. In the first place it involves the concept that there is an inherent 'scatter', not only in the individual items of a group, but also in the mean values which may be derived from it. The former has surely been demonstrated beyond doubt but the latter is by no means so obvious. Certainly we have shown that the averages of October rainfall at Milan for different 50-year periods are not the same, and reasons have been given for suspecting that they are affected by the odd chance of one or two very high readings but how can the average for the whole period of 197 years be at fault? The answer lies in the appreciation that this too is in fact only another random sample. Surely it has been raining over the site of Milan for a very long time indeed. In the second place, when all the calculations have been done, this matter still lies in a twilight zone between mathematical precision and critical judgment. The answers emerge in terms of *probabilities* and different interpretations are therefore possible. This aspect of the matter often troubles students who are used to answers to mathematical problems being either right or wrong.

The theory involves two basic assumptions:

1. That all the items involved in a comparison are entirely independent of one another. In regard to the type of data with which we have been dealing, this appears to be generally true. In comparing records, for instance, between two adjacent months we may sometimes find that two warm or wet months may succeed each other but this is usually balanced by a considerable number of warm-cool or wet-dry transitions.
2. That even when the initial groups have a skew form, the inferred scatter of mean values and the observed or inferred scatter of the differences between successive pairs of values from each group have both a much more normal form.

Two useful formulae then emerge:

$$\text{(i)} \quad \sigma_{\text{mean}} = \frac{\sigma_{\text{group}}}{\sqrt{n}}.$$

where $n$ = number in group.

$$\text{(ii)} \quad \sigma^2_{\substack{(u+v) \\ (u-v)}} = \sigma^2_u + \sigma^2_v.$$

where $u$ and $v$ represent two groups under comparison.

The first formula is unavoidable. It states quite simply that the degree of 'scatter' of the mean of nine observations is one-third of the scatter of the initial group; that of the mean of 100 observations is only a tenth of the scatter of the group. Hence the advantage of long-period means. The second formula can be sidestepped if we are willing to undertake the labour, for instance, of actually subtracting November from October rains at Milan over a considerable period of years and building up a new group from the differences, but it offers a short cut for multiple comparisons. The fact that scatter is *increased* by this process may be surprising, but think of two pendulums oscillating independently, the distance between the two bobs will vary more widely than the swing of either of them. Note that the addition of pairs of months has the same effect. In this case however all the values in the new group are greater than before and the *percentage* scatter about the new mean is reduced. Formula (i) says that this will be by $1/\sqrt{2}$ if the two records move independently. Many years ago the author was a little taken aback when he found he was illustrating this rule by combining pairs of months for the rainfall records of stations in the dry zone of north-eastern Brazil. Here it certainly looked as if wet-wet and dry-dry combinations were sufficiently frequent to vitiate the assumption of complete independence.

We must illustrate this matter very briefly to give some idea of the orders of magnitude involved.

For the two 60-year periods, 1841–1900 and 1901–1960 at Greenwich the average July temperatures were 62.56°F and 62.87°F, a difference of 0.31°F. Are we then to conclude that Julys have been getting a little warmer? We have already seen reason to doubt this. The standard deviations of the two groups are 2.60°F and 2.48°F and, dividing by $\sqrt{60}$, we find that the standard deviations of the two means are 0.336 and 0.320°F.

By formula (ii)

$$\sigma^2{}_{\text{diff.}} = (0.336)^2 + (0.320)^2 = 0.214°\text{F}$$

therefore

$$\sigma_{\text{diff.}} = 0.463°\text{F}$$

The observed difference between the means is thus considerably less than its own standard deviation and can be confidently regarded as a product of pure chance. To have been significant the difference should have been at least $2\sigma = 0.93°\text{F}$; $2.5\sigma = 1.16°\text{F}$; or $3\sigma = 1.39°\text{F}$. We give alternative solutions because this is where subjective judgment comes in. The author prefers the more severe test. It can thus be asserted with considerable confidence that in Britain mean monthly temperatures are not reliable within a range less than 1°F and that decimal values are spurious. January values are likely to show a still wider range than July and the reader is invited to explore a short series.

Naturally the range of error is much greater at Winnipeg in January. The mean of the group presented above has been given as −2°F and its standard deviation comes out as 7.44°F; the standard deviation of the mean of 88 years is thus $7.44/\sqrt{88} = 0.794°\text{F}$. Let us make the not unreasonable assumption that either December or February values would show much the same degree of scatter. The standard deviation of the difference between two adjacent winter months is then found by

$$\sigma^2_{\text{diff.}} = (0.794)^2 + (0.794)^2 = 1.26°F$$

therefore

$$\sigma_{\text{diff.}} = 1.12°F$$

The values $2\sigma = 2.24°F$, $2.5\sigma = 2.80°F$ and $3\sigma = 3.36°F$ thus suggest that, at this station in winter, differences in mean values of up to some $3°F$ are likely to be suspect.

It is true that even this degree of error does not trouble us unduly in climatology since at most stations the seasonal rhythm of temperature is so strong that doubts arise only about the location of the turning points in midsummer and midwinter. It is quite otherwise with respect to precipitation where the random factor is much stronger. Taking the very long record for Milan for instance, the overall mean for October is 117.5 millimetres and that for November is 100.7 millimetres. To what extent does the difference of some 17 millimetres represent a real contrast? The standard deviations of the two groups of data are very large, namely 76.5 and 69.6 millimetres respectively. By formula (*a*) we divide these by $\sqrt{197} = 14.04$ for the SDs of the two mean values and obtain 5.45 and 4.96 millimetres. Then

$$\sigma^2_{\text{diff.}} = (5.45)^2 + (4.96)^2 = 54.3$$

and

$$\sigma_{\text{diff.}} = 7.37 \text{ mm}$$

so that $2\sigma = 14.74$; $2.5\sigma = 18.4$; $3\sigma = 22.1$ mm. Some statisticians might therefore regard the result as meaningful but others would treat it with reserve. An interesting comment on this difference of opinion is offered by a direct comparison of the two months, year by year. Over the full period, November actually recorded higher totals than October on no less than 93 occasions. Does 104 : 93 represent a convincing majority on the part of October? It is not very considerable.

Records as long as this are very rare and a further commentary upon the series may be obtained by breaking it down into four approximately equal periods. The relevant facts are given below:

*October and November rainfall at Milan (mm)*

| | October | | | November | | | Relative count | |
|---|---|---|---|---|---|---|---|---|
| | Mean | $\sigma_{\text{gp}}$ | $\sigma_{\text{m}}$ | Mean | $\sigma_{\text{gp}}$ | $\sigma_{\text{m}}$ | Oct. gtr | Nov. gtr |
| A. 1764–1810 | 99 | 60 | 8.8 | 115 | 70 | 10.2 | 19 | 28 |
| B. 1811–1860 | 138 | 76 | 10.8 | 106 | 71 | 10.0 | 34 | 16 |
| C. 1861–1910 | 121 | 91 | 12.9 | 106 | 66 | 9.3 | 26 | 24 |
| D. 1911–1960 | 110 | 69 | 9.7 | 101 | 70 | 9.9 | 25 | 25 |
| | | | | | | | 104 | 93 |

Note the following four points:

1. The mean scatter within these samples is much the same as for the whole record; hence the reliability of the average falls off, by formula (i), as the number in the sample is decreased.

2. In the case in question, the relative predominance of October stems largely from events in 1811–60. Inspection of the record reveals that this included two anomalous 'runs', in 1811–24 and again in 1846–59. Are these no more than the familiar 'run of luck'? Certainly during the last century the two months have been running neck and neck. Strange features like this are encountered in rainfall records and the author has illustrated a run of dry Septembers in eastern Britain from 1881 to 1930 (Crowe, 1940).

3. Applying the significance test, we find that the difference between the means for October and November is actually less than its standard deviation in periods C and D and only $1.2\sigma$ in period A when November apparently had the edge. All these apparent contrasts can therefore be dismissed as fortuitous.

For period B the calculation runs as follows:

Difference between means $= 32.5$ mm

$$\sigma^2{}_{\text{diff.}} = (10.8)^2 + (10.0)^2 = 216.6$$

and

$$\sigma_{\text{diff.}} = 14.72 \text{ mm}$$

so that $2\sigma = 29.4$; $2.5\sigma = 36.7$; $3\sigma = 44.2$. Significant by the 5 per cent test, the difference is thus questionable if higher standards are demanded. In the author's opinion the 5 per cent test often leads to the search for 'causes' in circumstances where random production is more probable. In ordinary life we rarely regard a 20 to 1 chance as a remarkable coincidence.

4. Turning finally to the internal consistency of the two groups, we can make similar comparisons between the various means for each month. All the November values are well within the range of pure chance. In October a comparison between period A with periods C and D both lead to the verdict 'not significant' thus suggesting that there was nothing wrong with the early exposure of the gauge after all. (The November record has confirmed this.) The October record in period B is thus the 'odd man out'. The difference between the means for A and B is 39 mm with a standard deviation of 13.9 mm; this is the nearest approach to a significant contrast presented by the data.

If we may generalise from a single example it is thus not unreasonable to conclude that when quite long-period means of monthly rainfall are of the order of 10 to 14 cm (4.0 to 5.5 in) it is a wise precaution to question the significance of differences of less than some 4 cm or 1.5 in. Many distinctions between areas with June and July maxima in continental locations would not stand up to this test.

It is, however, when only short records are to hand that this matter achieves supreme importance. We quote a very extreme case. Consider the following two records of rainfall given to the nearest inch.

| $X$ | $d_{(23)}$ | $d^2$ | $Y$ | $d_{(14)}$ | $d^2$ |
|---|---|---|---|---|---|
| 33 | 10 | 100 . | 9 | 5 | 25 |
| 18 | 5 | 25 . | 5 | 9 | 81 |
| 31 | 8 | 64 . | 17 | 3 | 9 |
| 5 | 18 | 324 . | 11 | 3 | 9 |
| 30 | 7 | 49 . | 28 | 14 | 196 |
| 5)117 | | 5)562 | 5)70 | | 5)320 |
| 23.4 | | 112.4 | 14.0 | | 64.0 |

Variance (i) $= 112.4 - (0.4)^2$     Variance (i) $= 64.0$
          $= 112.2 = (10.6)^2$              $= (8.0)^2$
S.D. (i) $= 10.6$ in          S.D. (i) $= 8.0$ in

Variance (ii) $= 5/4 \, (112.2) = (11.8)^2$     Variance (ii) $= 5/4 \, (64.0) = (8.9)^2$
S.D. (ii) $= 11.8$ in             S.D. (ii) $= 8.9$ in

We quote alternative interpretations of the variance and standard deviations here because many statisticians are of the opinion that, for small groups, the division should be by $(n-1)$ rather than by $(n)$. Clearly for larger groups the distinction is slight. The difference between the means of $X$ and $Y$ is 9.4 inches. Could such a contrast fail to be 'significant'? Doubts may be entertained owing to the remarkable degree of scatter shown by both sets of values. Applying our test, using the first interpretation, we have

$$\text{for } X, \sigma_{\text{mean}} = 10.6/\sqrt{5} = 4.7; \quad \text{for } Y, \sigma_{\text{mean}} = 8.0/\sqrt{5} = 3.6 \text{ in}$$

Since formula (ii) involves the addition of the *squares* of these quantities, a stage in the arithmetic is saved if we express these as follows:

$$\text{for } X, \sigma^2_{\text{mean}} = \frac{112.2}{5}; \quad \text{for } Y, \sigma^2_{\text{mean}} = \frac{64.0}{5}$$

hence

$$\sigma^2_{\text{diff.}} = \frac{112.2 + 64.0}{5} = 35.24 = (5.94)^2$$

so that $2\sigma = 11.9$; $2.5\sigma = 14.8$ and $3\sigma = 17.8$ in. On this showing, therefore, the significance of the apparent contrast is certainly dubious. The second interpretation of the standard deviations will be still more exacting.

It may now be revealed that the values quoted above are in fact the rainfall totals recorded at Bombay for June and August during the years 1936 to 1940. Here it is possible to extend the comparison over a much longer period, but at many tropical stations we may be faced with fragmentary information of this character and caution is clearly necessary.

We give below the means and variances for successive five-year periods at Bombay which may help to put the above exercise into perspective.

From Table A.1.6. it is possible to compare the months over any 10-, 15- or 20-year interval etc. of time by averaging both the means and the variances and applying the formulae given. It is a most instructive exercise despite the fact that, on the long-term view, when the pairs of months are compared

TABLE. A.1.6. *Bombay (rainfall in inches)*

|  | | JUNE | | AUGUST | | |
|---|---|---|---|---|---|---|
|  | | *Mean* | *Variance* | *Mean* | *Variance* | *Frequency count* |
| A. | 1911–1915 | 21.0 | 120.4 | 10.4 | 19.4 | 4 : 1 |
| B. | 1916–1920 | 14.6 | 25.8 | 14.6 | 105.8 | 4 : 1 |
| C. | 1921–1925 | 19.0 | 96.4 | 9.4 | 10.6 | 3 : 2 |
| D. | 1926–1930 | 23.8 | 86.2 | 11.8 | 42.2 | 4 : 1 |
| E. | 1931–1935 | 16.6 | 89.4 | 16.2 | 31.8 | 3 : 2 |
| F. | 1936–1940 | 23.4 | 112.2 | 14.0 | 64.0 | 4 : 1 |
| G. | 1941–1945 | 19.6 | 85.8 | 11.8 | 69.0 | 3 : 2 |
| H. | 1946–1950 | 12.8 | 41.4 | 14.2 | 36.2 | 3 : 2 |
| I. | 1951–1955 | 23.2 | 39.0 | 19.6 | 151.8 | 3 : 2 |
| J. | 1956–1960 | 27.4 | 95.0 | 23.4 | 197.0 | 3 : 2 |
|  | 1911–1960 | 20.1 | 79.2 | 14.5 | 72.8 | 34 : 16 |
|  | 1861–1910 | | | | | 33 : 17 |
|  | 1861–1960 | | | | | 67 : 33 |

irrespective of the *amount* of the difference, June comes out with a clear two-thirds majority. Of the ten 5-year periods quoted only A, C, and D satisfy even the moderate $2\sigma$ test and even in D the difference between the means (12.0 in) does not quite reach $2.5\sigma$ when this is computed by the less exacting method ($\sigma_{\text{diff.}} = 5.1$ in). For the 50-year period the difference is 5.6 in and

$$\sigma^2_{\text{diff.}} = \frac{79.2 + 72.8}{50} = 3.04 = (1.74)^2.$$

At this stage therefore the most stringent requirements are met. The student will have missed the whole point of this discussion if he leaps to the conclusion that we have been making much ado about nothing. What if only the record for 1946–50 had been at hand?

## Relationship

During most of this century geographers have talked rather glibly about 'correlation' and the author once found it necessary to remind his colleagues that this involves 'correlative variation in degree' and ought not to be applied simply to a crude correspondence between distributions on a map (Crowe, 1938, 8).

Statistical theory offers a mathematically elegant method of displaying such an inferred relationship and a brief comment may not be out of place here.

The standard formula has the form

$$R = \frac{\Sigma \; xy}{N \; \sigma_x \; \sigma_y}$$

The 'correlation coefficient', $R$, then always has a value between $\pm 1$ and values between $\pm 0.5$, some would say between $\pm 0.7$, are regarded as discouraging. It will be noted that the result depends upon the sum of the products of the related pairs, the number of such pairs ($N$), and the product of the two standard deviations.

In introducing students to this mystery it is usual to work out an example in some detail. This may have the unfortunate result of encouraging the belief that the quality of the result is in direct proportion to the labour involved. Nothing could be farther from the truth. There are a number of short cuts.

1. Since the individual deviations must be found, it is possible to work throughout in terms of them. They represent the two groups just as accurately as the original data.

2. Deviations may be written from any convenient origin as in the brief Bombay example given above (23 instead of 23.4) but in this case the *sign* of the deviation is vital. The variance and the sum of the products must then be adjusted later.

3. C. E. P. Brooks used to tell his junior colleagues that if they ever found themselves squaring numbers in excess of 20 they were indulging in gratuitous arithmetic (see his example in Brooks and Carruthers, 1953, 211). At least for simple correlation all numbers should be kept small. Anyone who feels that this is a very cavalier way of treating arithmetic should rework Gregory's (1963, 177) example using only values to the nearest 10. He may well be astounded at the outcome; it also yields the value for $R$ of $+0.92$. Such a reminder seems not out of place in a world where calculating machines are often available. They are marvellous toys but there is no reason for overworking them.

As a young enthusiast the author felt that Shaw's comment on such coefficients that 'for the most part they have only confirmed by a numerical expression what we knew already' indicated the depressing conservatism of an older generation. Experience has taught him otherwise. Widely employed both by Sir Gilbert Walker and C. E. P. Brooks, the device has, in fact, contributed little to climatology.[1] Why is this so? Several factors are worthy of note.

1. If the relationship is close enough to produce a significant coefficient it is is usually both obvious from first principles and self-evident from an inspection of the two series. Giving it numerical expression is then more of an aesthetic experience than a contribution to knowledge.

[1] Yet see the more advanced exercise in Gregory (1965).

2. Such a close relation can be convincingly displayed non-mathematically by plotting the data on a scatter diagram; $X$ against $Y$. It may be claimed that the visual impression conveyed by the resulting 'cluster' is more accurate and indeed more penetrating than any notion derived from a single coefficient. It is argued that the mathematical approach yields a 'regression line' of closest fit but in fact it produces *two* lines which intersect at the common average. The interpretation of the real message conveyed by these lines is by no means as obvious as is often assumed.

3. Calculation provides real assistance when the relationship is relatively obscure but moderate coefficients leave us with a serious problem of interpretation. Walker produced a number of quite unexpected 'relationships' but, sad to say, few if any of them have stood the test of time. Most alleged relationships between sunspot numbers and climatic data have shown a depressing habit of fading away as the record rolls forward in time.

4. A major difficulty is that the coefficient is sensitive to 'noise'. By this we mean random vagaries in the data or arising from the fact that we must perforce measure something other than the parameter we really need. We may entertain a fellow-feeling for economists here. Perhaps this is one reason why their predictions so often go astray. Early in 1940 the author was invited to attempt to forecast ice conditions in Denmark Strait. Rather crude records of the limits of the pack over the previous 40 years were available from Danish sources and for such factors as the mean run of the wind one was compelled to use mean pressure differences between distant stations. Needless to say, the degree of 'noise' proved overwhelming. Yet with modern satellite photographs and the data from weather-ships a reasonable forecast might now be possible.

5. Most statistical texts emphasise that even a good correlation is no 'proof' of an intimate relationship. Common sense must also be satisfied.

An alternative and rather simpler approach to correlation is offered by the 'rank' coefficient. The two series are replaced by numbers indicating whether each measurement is first, second or third, etc. in magnitude in its appropriate group and the differences between these rank numbers ($d$) replace the deviations used above. The formula used is then

$$R = 1 - \frac{6\,\Sigma(d^2)}{n\,(n^2 - 1)}$$

This method was originally designed to handle data, such as preferences or other subjective judgements, to which absolute values cannot be assigned and Brooks and Carruthers (1953, 236) comment that it 'is not likely to have much application in meteorology'. There is some evidence, however, that it may help to reduce the 'noise' factor.

In an example which the author has explored but which is not given here in full, the first approach gives a coefficient of +0.70 whilst the rank coefficient comes out as +0.79. The relationship, a rather surprising one, is between the rainfall totals from December to February at Ocean Island and Fanning Island, two tiny fragments of land no less than 2100 miles apart in the equatorial west Pacific. That strange events occur here may be deduced from a few sample values:

*December to February rainfall totals (inches)*

|  | Ocean Is. | Fanning Is. |
| --- | --- | --- |
| 1904–05 | 58 | 77 |
| 1916–17 | 1 | 1 |
| 1918–19 | 60 | 54 |
| 1926–27 | 2 | 2 |
| 1930–31 | 51 | 52 |
| 1949–50 | 0 | 1 |
| 1957–58 | 43 | 66 |

The correlation coefficient has the merit of placing these years in the context of the rest of the record but it is heavily weighted by them. That there is a real problem here awaiting solution is self-evident but a demonstrated correlation does nothing to solve it.

This appendix is in no sense a substitute for a statistical text. It is intended to illustrate a few of the challenges that face the budding climatologist and to put him on his guard. See Conrad and Pollak (1950), C. E. P. Brooks and Carruthers (1953), Gregory (1963), Cole and King (1968).

# Appendix 2
# A method for estimating mean monthly potential evaporation from a vegetated surface

(Source material required, (*a*) monthly values of mean daily maximum and minimum temperatures, and (*b*) mean monthly precipitation.)

**Stage 1**

Since evaporation occurs mainly during the day it has been suggested (*Geographical Studies* 4, 1957, 56–75) that a relationship with temperature will be more likely to emerge if mean daily *maximum* temperatures are used in preference to the means for the whole 24-hour period employed by Thornthwaite.

Also, since evaporation is greatest in summer, it was thought wise to explore this relationship, in the first instance, for the *warmest* month of the year at any given station.

The relationship which emerged from a number of stations in both hemispheres took the simple form:

$$E_{tr} = 7.3 \ (T_{mx} - 6)\degree C \text{ in mm}$$

or

$$E_{tr} = 0.16 \ (T_{mx} - 43)\degree F \text{ in inches}$$

Closer fit was obtained if the values were then adjusted according to Table A.2.1 which takes some account of the mean precipitation during the warmest month.

TABLE A.2.1. *Correction for precipitation during key month*

(Reduce values derived from the above formula by the following proportions)

| Precipitation | | Proportion |
|---|---|---|
| *In* | *mm* | |
| 0 | 0 | 100 |
| 1 | | 96 |
| 2 | 50 | 92 |
| 3 | | 88 |
| 4 | 100 | 84 |
| 5 | | 80 |
| 6 | 150 | 76 |
| 7 | | 72 |
| 7.5 or over | 190 or over | 70 |

For work in millimetres it would be convenient to draw a simple graph or re-state the table in approximate terms.

Adjustment to the nearest 1 per cent is effected by slide rule.

532

Stage 1 thus yields an absolute value as an estimate of potential evaporation of the warmest month.

## Stage 2

The appropriate values for the other 11 months are considered to be proportionate to this figure. In view of the number of factors which have to be taken into account, the proportion is derived from a number of indices. These involve:

(*a*) The difference between the saturation vapour pressure at mean daily maximum and minimum temperatures (Table A.2.2).

(*b*) A correction necessary because the mean maximum–mean minimum range is often well in excess of the true periodic diurnal range of temperature (Table A.2.3).

(*c*) A further adjustment for day-length like that proposed by Thornthwaite but carried logically beyond 50° N (Table A.2.4).

The method thus proceeds: Subtract the values appropriate to the mean daily minimum temperatures from those at the corresponding maxima in Table A.2.2 for each month of the year; multiply by the correction factors given in Table A.2.3 for selected latitudes; then multiply again by the appropriate factors for day-length given in Table A.2.4. Equate the index thus derived for the warmest month with the absolute value obtained in Stage 1 and read off the values for the other months in the proportions indicated.

The computation is readily effected by slide rule.

TABLE A.2.2. *Indices for monthly evapotranspiration*

| °C | | | | I. *Celsius scale* | | | | | |
|---|---|---|---|---|---|---|---|---|---|
| | −9 | −8 | −7 | −6 | −5 | −4 | −3 | −2 | −1 |
| −40 | 0.19 | 0.21 | 0.23 | 0.26 | 0.28 | 0.31 | 0.35 | 0.38 | 0.42 | 0.46 |
| −30 | 0.51 | 0.56 | 0.61 | 0.67 | 0.74 | 0.81 | 0.88 | 0.96 | 1.05 | 1.15 |
| −20 | 1.25 | 1.37 | 1.49 | 1.62 | 1.76 | 1.91 | 2.08 | 2.25 | 2.44 | 2.64 |
| −10 | 2.86 | 3.10 | 3.35 | 3.62 | 3.91 | 4.21 | 4.55 | 4.90 | 5.28 | 5.68 |
| °C | 0 | 1 | 2 | 3 | 4 | 5 | 6 | 7 | 8 | 9 |
| 0 | 6.11 | 6.57 | 7.05 | 7.58 | 8.13 | 8.72 | 9.35 | 10.01 | 10.72 | 11.47 |
| 10 | 12.27 | 13.12 | 14.02 | 14.97 | 15.98 | 17.04 | 18.17 | 19.37 | 20.63 | 21.96 |
| 20 | 23.4 | 24.9 | 26.4 | 28.1 | 29.8 | 31.7 | 33.6 | 35.6 | 37.8 | 40.1 |
| 30 | 42.4 | 44.9 | 47.5 | 50.3 | 53.2 | 56.2 | 59.4 | 62.8 | 66.3 | 69.9 |
| 40 | 73.8 | 77.8 | 82.0 | 86.4 | 91.0 | 95.8 | 100.9 | 106.2 | 111.7 | 117.4 |

(Since this table gives saturation vapour pressure over water in millibars it is of some value for general reference.)

**II.** *Fahrenheit scale*

| °F | | −9 | −8 | −7 | −6 | −5 | −4 | −3 | −2 | −1 |
|---|---|---|---|---|---|---|---|---|---|---|
| −40 | 0.5 | 0.5 | 0.5 | 0.6 | 0.6 | 0.6 | 0.7 | 0.7 | 0.7 | 0.8 |
| −30 | 0.8 | 0.9 | 0.9 | 1.0 | 1.1 | 1.1 | 1.2 | 1.3 | 1.3 | 1.4 |
| −20 | 1.5 | 1.6 | 1.6 | 1.7 | 1.8 | 1.9 | 2.1 | 2.2 | 2.3 | 2.4 |
| −10 | 2.5 | 2.7 | 2.8 | 3.0 | 3.1 | 3.3 | 3.5 | 3.7 | 3.8 | 4.1 |

| °F | 0 | 1 | 2 | 3 | 4 | 5 | 6 | 7 | 8 | 9 |
|---|---|---|---|---|---|---|---|---|---|---|
| 0 | 4.3 | 4.5 | 4.7 | 5 | 5 | 5 | 6 | 6 | 6 | 7 |
| 10 | 7 | 7 | 8 | 8 | 8 | 9 | 9 | 10 | 10 | 11 |
| 20 | 11 | 12 | 12 | 13 | 13 | 14 | 15 | 15 | 16 | 17 |
| 30 | 17 | 18 | 19 | 20 | 21 | 21 | 22 | 23 | 24 | 25 |
| 40 | 26 | 27 | 28 | 29 | 30 | 31 | 32 | 33 | 34 | 36 |
| 50 | 37 | 38 | 40 | 41 | 43 | 44 | 46 | 47 | 49 | 50 |
| 60 | 52 | 54 | 56 | 58 | 60 | 62 | 64 | 66 | 68 | 70 |
| 70 | 73 | 75 | 77 | 80 | 83 | 85 | 88 | 91 | 94 | 97 |
| 80 | 100 | 103 | 106 | 109 | 113 | 116 | 120 | 123 | 127 | 131 |
| 90 | 135 | 139 | 143 | 147 | 151 | 156 | 160 | 165 | 170 | 175 |
| 100 | 180 | 185 | 190 | 196 | 201 | 207 | 213 | 219 | 225 | 231 |
| 110 | 238 | 244 | 251 | 258 | 265 | 272 | 279 | 287 | 295 | 302 |

TABLE A.2.3. *Correction factors for variation in relative magnitudes of maximum–minimum range and the periodic diurnal range of temperature* (%)

| S. HEMIS.<br>N. HEMIS.<br>(°Lat.) | JULY<br>JAN. | AUG.<br>FEB. | SEPT.<br>MAR. | OCT.<br>APRIL | NOV.<br>MAY | DEC.<br>JUNE | JAN.<br>JULY | FEB.<br>AUG. | MAR.<br>SEPT. | APRIL<br>OCT. | MAY<br>NOV. | JUNE<br>DEC. |
|---|---|---|---|---|---|---|---|---|---|---|---|---|
| 72 | 0 | 5 | 20 | 70 | 100 | 100 | 100 | 100 | 75 | 20 | 0 | 0 |
| 68 | 0 | 25 | 65 | 90 | 100 | 100 | 100 | 100 | 85 | 55 | 10 | 0 |
| 64 | 20 | 45 | 80 | 95 | 100 | 100 | 100 | 100 | 90 | 70 | 30 | 15 |
| 60 | 35 | 65 | 90 | 95 | 100 | 100 | 100 | 100 | 90 | 80 | 45 | 25 |
| 56 | 50 | 70 | 90 | 95 | 100 | 100 | 100 | 100 | 95 | 85 | 60 | 40 |
| 52 | 60 | 75 | 90 | 95 | 100 | 100 | 100 | 100 | 95 | 85 | 70 | 55 |
| 48 | 65 | 80 | 90 | 95 | 100 | 100 | 100 | 100 | 100 | 90 | 75 | 65 |
| 44 | 80 | 85 | 90 | 95 | 100 | 100 | 100 | 100 | 100 | 95 | 85 | 75 |
| 40 | 85 | 90 | 90 | 95 | 100 | 100 | 100 | 100 | 100 | 95 | 90 | 85 |
| 36 | 90 | 90 | 95 | 95 | 100 | 100 | 100 | 100 | 100 | 95 | 95 | 90 |
| 32 | 90 | 95 | 95 | 100 | 100 | 100 | 100 | 100 | 100 | 95 | 95 | 90 |
| 28 | 100 | 100 | 100 | 100 | 100 | 100 | 100 | 100 | 100 | 100 | 100 | 100 |

Equatorward of about 30° N and S no systematic correction emerges.
Interpolate for intermediate latitudes.
Local correction factors may be obtained for any area near a station providing hourly temperature values but latitude appears to be the dominant control. Note that it is only seasonal changes in the *ratio* of the two ranges which affect the method.

Some comments on the general validity of this method are offered in the original paper together with a number of the actual results it yields. It must be stressed that it is designed for descriptive purposes only, that is, as a general indication of the effectiveness of monthly precipitation.

TABLE A.2.4. *Mean possible duration of sunlight expressed in units of 30 days of 12 hours each* (%)

|  | JAN. | FEB. | MAR. | APRIL | MAY | JUNE | JULY | AUG. | SEPT. | OCT. | NOV. | DEC. |
|---|---|---|---|---|---|---|---|---|---|---|---|---|
| °S | | | | | *Southern hemisphere* | | | | | | | |
| 50 | 137 | 113 | 108 | 89 | 77 | 67 | 74 | 88 | 99 | 119 | 129 | 141 |
| 40 | 127 | 107 | 107 | 93 | 86 | 78 | 84 | 92 | 100 | 115 | 120 | 129 |
| 30 | 120 | 104 | 106 | 95 | 92 | 85 | 90 | 96 | 100 | 112 | 114 | 121 |
| 20 | 114 | 101 | 105 | 97 | 96 | 91 | 95 | 99 | 100 | 108 | 109 | 115 |
| 10 | 108 | 98 | 105 | 99 | 101 | 96 | 100 | 101 | 100 | 106 | 105 | 110 |
| | | | | | *Equator* | | | | | | | |
| 0° | 104 | 95 | 104 | 101 | 104 | 101 | 104 | 104 | 101 | 104 | 101 | 104 |
| °N | | | | | *Northern hemisphere* | | | | | | | |
| 10 | 100 | 92 | 103 | 103 | 108 | 106 | 108 | 107 | 102 | 102 | 98 | 99 |
| 20 | 95 | 91 | 103 | 105 | 113 | 111 | 114 | 111 | 102 | 100 | 93 | 94 |
| 30 | 90 | 88 | 103 | 108 | 118 | 117 | 120 | 114 | 103 | 98 | 89 | 88 |
| 40 | 84 | 84 | 103 | 111 | 124 | 125 | 127 | 118 | 104 | 96 | 83 | 81 |
| 45 | 80 | 82 | 103 | 113 | 128 | 129 | 131 | 121 | 104 | 94 | 79 | 75 |
| 50 | 74 | 79 | 102 | 115 | 133 | 136 | 137 | 125 | 106 | 92 | 76 | 70 |
| 52 | 71 | 78 | 102 | 116 | 135 | 139 | 139 | 126 | 106 | 92 | 73 | 67 |
| 54 | 68 | 76 | 102 | 117 | 137 | 142 | 143 | 128 | 106 | 91 | 71 | 64 |
| 56 | 66 | 75 | 102 | 118 | 140 | 146 | 147 | 130 | 106 | 90 | 69 | 60 |
| 58 | 62 | 74 | 102 | 120 | 144 | 151 | 151 | 132 | 107 | 89 | 66 | 56 |
| 60 | 58 | 72 | 102 | 121 | 147 | 157 | 155 | 135 | 108 | 88 | 63 | 51 |
| 62 | 53 | 70 | 102 | 123 | 152 | 164 | 161 | 137 | 108 | 87 | 59 | 44 |
| 64 | 47 | 68 | 101 | 125 | 158 | 174 | 169 | 141 | 109 | 85 | 55 | 37 |
| 66 | 38 | 64 | 101 | 128 | 166 | 200 | 183 | 146 | 110 | 82 | 49 | 24 |
| 68 | 26 | 61 | 99 | 131 | 174 | 200 | 206 | 151 | 112 | 80 | 41 | 0 |
| 70 | 0 | 56 | 99 | 134 | 197 | 200 | 206 | 156 | 112 | 78 | 33 | 0 |
| 72 | 0 | 51 | 99 | 140 | 206 | 200 | 206 | 170 | 113 | 75 | 4 | 0 |
| 74 | 0 | 45 | 99 | 146 | 206 | 200 | 206 | 183 | 115 | 70 | 0 | 0 |
| 76 | 0 | 31 | 97 | 154 | 206 | 200 | 206 | 206 | 117 | 65 | 0 | 0 |
| 78 | 0 | 8 | 96 | 168 | 206 | 200 | 206 | 206 | 119 | 56 | 0 | 0 |

If the values obtained are compared with mean monthly rainfall a useful key to the objective classification of climates may be secured on the following basis:

| Humid months | $E < P$ |
|---|---|
| Rather dry months | $E \geqslant P$ but $< 3P$ |
| Dry months | $E \geqslant 3P$ but $< 6P$ |
| Really arid months | $E \geqslant 6P$ |

# Appendix 3

## Conversion tables

Except for temperature data, conversion from traditional to metric scales is readily made with the aid of a slide-rule. Where the task is a heavy one, tables such as those presented in *Smithsonian Meteorological Tables* (Smithsonian Miscellaneous Collections, Vol. 114, Washington 1951) may be preferred.

However, with a little practice, the following highly condensed presentation will be found to yield equivalents to a very considerable degree of accuracy. Care being taken in the placing of the decimal points, all that is required to cover any value likely to be encountered in climatology is a little simple addition (see Note on p. 539). The number of significant figures employed is, of course, entirely optional.

TEMPERATURE

*Fahrenheit to Celsius (Centigrade)*

$$C = 5/9\ (F - 32)$$

| °F | °C | °F | °C | °F | °C |
|----|----|----|----|----|----|
| 140 | 60.00 | 70 | 21.11 | 0 | −17.78 |
| 130 | 54.44 | 60 | 15.56 | −10 | −23.33 |
| 120 | 48.89 | 50 | 10.00 | −20 | −28.89 |
| 110 | 43.33 | 40 | 4.44 | −30 | −34.44 |
| 100 | 37.78 | 30 | −1.11 | −40 | −40.00 |
| 90 | 32.22 | 20 | −6.67 | −50 | −45.56 |
| 80 | 26.67 | 10 | −12.22 | −60 | −51.11 |

*Differences*

| | | | |
|---|---|---|---|
| 1 °F = 0.56 °C | | 6 °F = 3.33 °C | |
| 2 | = 1.11 | 7 | = 3.89 |
| 3 | = 1.67 | 8 | = 4.44 |
| 4 | = 2.22 | 9 | = 5.00 |
| 5 | = 2.78 | 10 | = 5.56 |

## Celsius to Fahrenheit

$$F = 9/5 \, C + 32$$

| °C | °F | °C | °F | °C | °F |
|----|----|----|----|----|----|
|    |    | 35 | 95 |    |    |
| 80 | 176 | 30 | 86 | −20 | −4 |
|    |    | 25 | 77 |    |    |
| 70 | 158 | 20 | 68 | −30 | −22 |
|    |    | 15 | 59 |    |    |
| 60 | 140 | 10 | 50 | −40 | −40 |
|    |    | 5 | 41 |    |    |
| 50 | 122 | 0 | 32 | −50 | −58 |
|    |    | −5 | 23 |    |    |
| 40 | 104 | −10 | 14 | −60 | −76 |
|    |    | −15 | 5 |    |    |

### Differences

| | | | |
|---|---|---|---|
| 1 °C = 1.8 °F | | 6 °C = 10.8 °F | |
| 2 | = 3.6 | 7 | = 12.6 |
| 3 | = 5.4 | 8 | = 14.4 |
| 4 | = 7.2 | 9 | = 16.2 |
| 5 | = 9.0 | 10 | = 18.0 |

## RAINFALL

### Inches to millimetres

1 inch — 25.400 millimetres

| Inches | Millimetres |
|--------|-------------|
| 10................ | 0 2 5 4 0 0 |
| 20................ | 0 5 0 8 0 0 |
| 30................ | 0 7 6 2 0 0 |
| 40................ | 1 0 1 6 0 0 |
| 50................ | 1 2 7 0 0 0 |
| 60................ | 1 5 2 4 0 0 |
| 70................ | 1 7 7 8 0 0 |
| 80................ | 2 0 3 2 0 0 |
| 90................ | 2 2 8 6 0 0 |

### Millimetres to inches

1 mm = 0.039370 inches

| Millimetres | Inches |
|-------------|--------|
| 100................ | 0 3 9 4 |
| 200................ | 0 7 8 7 |
| 300................ | 1 1 8 1 |
| 400................ | 1 5 7 5 |
| 500................ | 1 9 6 8 |
| 600................ | 2 3 6 2 |
| 700................ | 2 7 5 6 |
| 800................ | 3 1 5 0 |
| 900................ | 3 5 4 3 |

537

ELEVATION

*Feet to metres*

1 foot = 0.30480 metres

| Feet | | Metres | | |
|---|---|---|---|---|
| 100................ | 0 3 0 | 4 8 |
| 200................ | 0 6 0 | 9 6 |
| 300................ | 0 9 1 | 4 4 |
| 400................ | 1 2 1 | 9 2 |
| 500................ | 1 5 2 | 4 0 |
| 600................ | 1 8 2 | 8 8 |
| 700................ | 2 1 3 | 3 6 |
| 800................ | 2 4 3 | 8 4 |
| 900................ | 2 7 4 | 3 2 |

*Metres to feet*

1 metre = 3.280833 feet

| Metres | | Feet | | |
|---|---|---|---|---|
| 100................ | 0 3 2 8 | 0 8 3 |
| 200................ | 0 6 5 6 | 1 6 7 |
| 300................ | 0 9 8 4 | 2 5 0 |
| 400................ | 1 3 1 2 | 3 3 3 |
| 500................ | 1 6 4 0 | 4 1 6 |
| 600................ | 1 9 6 8 | 5 0 0 |
| 700................ | 2 2 9 6 | 5 8 3 |
| 800................ | 2 6 2 4 | 6 6 6 |
| 900................ | 2 9 5 2 | 7 5 0 |

DISTANCE

*Statute miles to kilometres*

1 mile = 1.609347 kilometres

| Miles | | Kilometres | | |
|---|---|---|---|---|
| 100................ | 0 1 6 0 | 9 3 5 |
| 200................ | 0 3 2 1 | 8 6 9 |
| 300................ | 0 4 8 2 | 8 0 4 |
| 400................ | 0 6 4 3 | 7 3 9 |
| 500................ | 0 8 0 4 | 6 7 3 |
| 600................ | 0 9 6 5 | 6 0 8 |
| 700................ | 1 1 2 6 | 5 4 3 |
| 800................ | 1 2 8 7 | 4 7 8 |
| 900................ | 1 4 4 8 | 4 1 2 |

### Kilometres to miles

1 km = 0.621370 miles

| Kilometres | | | | Miles | | |
|---|---|---|---|---|---|---|
| 100............... | 0 6 2 | | | 1 | 3 | 7 |
| 200............... | 1 2 4 | | | 2 | 7 | 4 |
| 300............... | 1 8 6 | | | 4 | 1 | 1 |
| 400............... | 2 4 8 | | | 5 | 4 | 8 |
| 500............... | 3 1 0 | | | 6 | 8 | 5 |
| 600............... | 3 7 2 | | | 8 | 2 | 2 |
| 700............... | 4 3 4 | | | 9 | 5 | 9 |
| 800............... | 4 7 9 | | | 0 | 9 | 6 |
| 900............... | 5 5 9 | | | 2 | 3 | 3 |

## AREAS

### Square miles to square kilometres

1 square mile = 2.5900 square kilometres

| Square miles | | | | Square kilometres | |
|---|---|---|---|---|---|
| 100............... | 0 2 5 9 | | | | 0 |
| 200............... | 0 5 1 8 | | | | 0 |
| 300............... | 0 7 7 7 | | | | 0 |
| 400............... | 1 0 3 6 | | | | 0 |
| 500............... | 1 2 9 5 | | | | 0 |
| 600............... | 1 5 5 4 | | | | 0 |
| 700............... | 1 8 1 3 | | | | 0 |
| 800............... | 2 0 7 2 | | | | 0 |
| 900............... | 2 3 3 1 | | | | 0 |

### Square kilometres to square miles

1 square kilometre = 0.38610 square miles

| Square kilometres | | | | Square miles | |
|---|---|---|---|---|---|
| 100............... | 0 3 8 | | | 6 | 1 |
| 200............... | 0 7 7 | | | 2 | 2 |
| 300............... | 1 1 5 | | | 8 | 3 |
| 400............... | 1 5 4 | | | 4 | 4 |
| 500............... | 1 9 3 | | | 0 | 5 |
| 600............... | 2 3 1 | | | 6 | 6 |
| 700............... | 2 7 0 | | | 2 | 7 |
| 800............... | 3 0 8 | | | 8 | 8 |
| 900............... | 3 4 7 | | | 4 | 9 |

*Note*

In the tables for rainfall, elevation, distance and area the vertical bar represents the position of the decimal point for the *values actually quoted*. Extension and interpolation from the table may be illustrated by the following example:

$$
\begin{aligned}
8{,}472 \text{ feet} = 8000 &= 2438.40 \\
400 &= \phantom{0}121.92 \\
70 &= \phantom{00}21.34 \\
2 &= \phantom{000}0.61 \\
\hline
&2582.27 \text{ metres}
\end{aligned}
$$

# References

## Sources of climatic data

### World data

*World Weather Records:* Smithsonian Misc. Collection (CLAYTON, H. H.), No. 79, *to 1920;* No. 90, *1921–30;* No. 105, *1931–40.* Smithsonian Institute, Washington, D.C. continued as:

*World Weather Records 1941–50,* Weather Bureau, U.S. Dept. Com., Washington, D.C.

*World Weather Records 1951–60,* 6 vols: 1. *North America;* 2. *Europe;* 3. *South America, Central America, West Indies, the Caribbean and Bermuda;* 4. *Asia;* 5. *Africa;* 6. *Antarctica, Australia, Oceanic Islands and Oceanic Weather Stations.* Environmental Sci. Services Admin., U.S. Dept. Com., Washington, D.C.

*Monthly Climatic Data for the World,* World Met. Organisation, Weather Bureau, U.S. Dept. Com., Asheville, N.C.

*Tables of Temperature, Relative Humidity and Precipitation for the World,* M.O. 617 (1958), Meteorological Office, H.M.S.O.

*Smithsonian Meteorological Tables,* 6th edn. (1951). Smithsonian Misc. Coll. vol. 114 (LIST, R. J.). Smithsonian Inst., Washington, D.C.

*Catalog of Meteorological Satellite Data—ESSA 3 and ESSA 5 Television Cloud Photography.* No. 5.316. Environmental. Data Service, U.S. Dept. Com., Washington, D.C.

   (The series began in 1961 and continues. The quality of recent mosaics is excellent.)

### Britain

*Monthly Weather Report of the Meteorological Office,* HMSO.

*British Rainfall.* Annual volumes since 1900. Met. Office, HMSO; formerly *Symon's British Rainfall,* Camden Square, London.

*Daily Weather Report of the British Meteorological Office,* Bracknell, Berkshire.

*Weather in Home Waters and the North Eastern Atlantic,* M.O. 446, Met. Office, HMSO, Vol. 2: Part 1. The Atlantic from the Azores to the African

Coast (1944); Part 2. The Bay of Biscay and the West Coast of Spain and Portugal (1944); Part 3. The English Channel (1940); Part 4. The Western Approaches to Great Britain and Ireland (1940); Part 5. The North Sea; Part 6. The Baltic Sea (1943); Part 7. The Norwegian and Barents Sea (1941).

## Other European countries

*Observações Meteorologicos*, Instituto Geofisco, Coimbra, Portugal.

SURING, R., *Ergebnisse der Met. Beobachtungen in Potsdam*, Veroff. des Preussischen Met Inst. n. 324.

*Bulletin Mensuel de l'Observatoire Met. de l'Université d'Uppsala*. Sweden.

ROLF, B. *Observations Met. à Abisko*

*Jahrbuch Det Norske Meteorologiske Inst.*, Oslo, Norway.

Årsbok. *Statens Meteorologisk-Hydrografiska Anstalt*, 27–28 (1945–46), Stockholm, publ. 1952–3.

*Nedbøren i Norge 1895–1943*, Det Norske Met. Inst., Oslo (1949).

*Nederbörden i Sverige 1901–30, 1921–50*, Statens Meteorologisk-Hydrografiska Anstalt, Stockholm; publ. 1951 and 1954.

*Anuário Climatologico de Portugal*, vol. 15, 1961, IV parte, Territórios Ultra-marinos (Observações de Altitude), Lisbon.

## Australia

*Yearbook of the Commonwealth of Australia*, 49 (1963) Comm. Bureau of Census and Statistics, Canberra.

## India and Malaysia

*Indian Daily Weather Report*, Delhi.

*Summary of Observations 1952–56*, Malayan Meteorological Service, Singapore.

## North America

*Climatic Summary of the United States*, Sections 49, 62, 63. U.S. Dept. Agric. Weather Bureau, Washington, D.C.

*Summaries of Radiosonde Observations in Canada 1954–1958*, Dept. of Transport, Met. Branch, Toronto.

## South Africa

*Climate of South Africa*, Part 1, *Climate Statistics* (1954). Weather Bureau, Pretoria.

# General

AANENSEN, C. J., ed. (1965) 'Gales in Yorkshire', February 1962, *Geophys. Mem. Met. Office* **108**.

AITKEN, J. (1881) 'On dust, fogs and clouds', *Trans. Roy. Soc. Edinburgh* **30**, reproduced with later papers of great interest in Aitken (1923).

AITKEN, J. (1923) *Collected Scientific Papers*, Cambridge University Press.

ANDERSON, C. E., ed. (1960) *Cumulus Dynamics*, Proc. Conf. on Cu. Convection, Portsmouth, N.H. 1959, Pergamon Press.

ÅNGSTRÖM, A. (1949) 'Atmospheric circulation, climatic variations and continentality of climate', *Geogr. Annlr.* **31**, 316–20. Stockholm.

APPLEMAN, H. S. (1960) 'An investigation into the formation of hail', in Anderson (1960, *q.v.*), 148–56.

ARRHENIUS, S. (1896) 'On the influence of carbonic acid in the air upon the temperature of the ground', *Phil. Mag.* **41**, 237–76.

ATKINSON, B. N. (1969) 'A further examination of the urban maximum of thunder rainfall in London, 1951–60', *Trans. Inst. Brit. Geogr.* **48**, 97–119.

ATKINSON, B. W. (1968) 'A preliminary examination of the possible effect of London's urban area on the distribution of thunder rainfall, 1951–60', *Trans. Inst. Br. Geogr.* **44**, 97–117.

ATLAS, D. (1966) 'The balance level in convective storms', *J. Atmos. Scis.* **23**, 635–51.

BAGNOULS, F. and GAUSSEN, H. (1953) 'Saison sèche et indice xérothermique', *Bull. Soc. d'Hist. Naturelle Toulouse* **88**, 193–239.

BALL, F. (1957) 'The katabatic winds of Adélie Land and King George V Land', *Tellus* **9**, 201–8.

BARRETT, E. C. (1964) 'Local variations in rainfall trends in the Manchester region', *Trans. Inst. Br. Geogr.* **35**, 55–71.

BARRETT, E. C. (1966) 'Regional variations of rainfall trends in northern England, 1900–59', *Trans. Inst. Br. Geogr.* **38**, 41–58.

BARRETT, E. C. (1967) *Viewing Weather from Space*, Longmans.

BARRETT, E. C. (1970) 'A contribution to the dynamic climatology of the equatorial eastern Pacific and Central America', *Trans. Inst. Br. Geogr.* **50**, 25–53.

BARRY, R. G. (1970) 'A framework for climatological research with particular reference to scale concepts', *Trans. Inst. Brit. Geogr.* **49**, 61–70.

BAUR, F. (1948) *Einführung in die Grosswetterkunde*, Wiesbaden.

BAUR, F. and PHILIPPS, H. (1934–5) 'Der Wärmehaushalt der Lufthülle der Nordhalbkugel in Januar und Juli und zur Zeit der Äquinoktien und Solstitien', *Gerlands Beitr. Geophys.* **42** and **45**, Pt i (1934) 160–207; Pt ii (1935) 82–132.

BAXENDELL, J. (1935) *Annual Report of the Fernley Observatory*, Southport.

BECKINSALE, R. P. (1965) 'Climatic change. A critique of modern theories', *Essays in Geography for Austin Miller*, ed. Whittow and Wood (*q.v.*) 1–38.

BELASCO, J. E. (1952) *Characteristics of Air Masses over the British Isles*, Geophys. Mem. 87. (M.O. 530 b), HMSO.

BELL, B. (1953) 'Solar variation as an explanation of climatic change', in Shapley (1953, *q.v.*), 123–36.

BEMMELEN, W. van (1922) 'Land und seebrise in Batavia', *Beitr. Phys. freien Atmos* **10**, 169–77.

BERGERON, T. (1928) 'Über die dreidimensional verknüpfende Wetteranalyse', *Geofys. Publ.* **5**, no. 6.

BERGERON, T. (1933) 'On the physics of cloud and precipitation', *Mem. Met. U.G.G.I. (Lisbon) Pt.* 2, 156–78.

BERGERON, T. (1949) 'The problem of artificial control of rainfall on the globe. I. General effects of ice-nuclei in clouds', *Tellus* **1**, no. 1, 32–43; 'The problem of artificial control of rainfall on the globe. II. The coastal orographic maximum of precipitation in autumn and winter', *Tellus* **1**, no. 3, 15–32.

BERGERON, T. (1954) 'The problem of tropical hurricanes', *Q. J. Roy. Met. Soc.* **80**, 131–64.

BERGERON, T. (1960) 'Problems and methods of rainfall investigation' in Weickmann, H. ed. (1960, *q.v.*), 5–30.

BERRY, F. A., BOLLAY, E. and BEERS, N. R., eds. (1945) *Handbook of Meteorology*, McGraw-Hill.

BEST, A. C. (1935) 'Transfer of heat and momentum in the lowest layers of the atmosphere', *Geophys Mem.* **65**, HMSO.

BEST, A. C. (1952) 'The evaporation of raindrops', *Q. J. Roy. Met. Soc.* **78**, 200–25.

BEST, A. C. (1957) *Physics in Meteorology*, Pitman.

BEST, A. C., KNIGHTING, E., PEDLOW, R. H. and STORMONTH, K. (1952) *Temperature and Humidity Gradients in the first 100 m over south-east England*, Geophys. Mem. 89 (M.O. 530 d), HMSO.

BILHAM, E. G. (1935) 'Classification of heavy falls in short periods', in *British Rainfall* 1935 (Met. Off. HMSO), 262–80.

BILHAM, E. G. (1938) *The Climate of the British Isles*, Macmillan.

BILLWILLER, R. and QUERVAIN, A. de (1912) 'Registrierballon aufsteige in einem Föhntale', *Met. Zeitschr.* **29**, 249–51.

BJERKNES, J. (1930) *Practical examples of polar front analysis over the British Isles* (1925–6), Geophys. Mem. 50 (M.O. 307 j), HMSO.

BJERKNES, J. (1951) 'Extratropical cyclones', in *Compendium of Meteorology*, ed. Maloney (*q.v.*), 577–98.

BJERKNES, J. (1954) 'The problem of cyclogenesis from Helmholtz's time to the present', *Proc. Toronto Met. Conf.* 1953, 133–8.

BJERKNES, J. and SOLBERG, H. (1921) 'Meteorological conditions for the formation of rain', *Geofys. Publ.* **2**, no. 3.

BJERKNES, J. and SOLBERG, H. (1922) 'The life cycle of cyclones and the Polar Front theory of atmospheric circulation', *Geofys. Publ.* **3**, no. 1.

BJERKNES, V. (1920) 'The structure of the atmosphere when rain is falling', *Q. J. Roy. Met. Soc.* **46**, 119–40.

BLANCHARD, R. (1929) 'Asie Occidentale', in *Géographie Universelle*, viii, pt. 1 (P. Vidal de la Blache et L. Gallois), Librairie Armand Colin, Paris.

BLÜTHGEN, J. (1964) *Allgemeine Klimageographie*, Berlin, De Gruyter.

BORISOV, A. A. (1965) *Climates of the USSR*, trans. R. A. Ledward, Oliver & Boyd.

BOUGHNER, C. C. (1955) 'Hurricane Hazel', *Weather* **10**, 200–5.

BOWEN, E. G. (1952) 'A new method of stimulating convective clouds to produce rain and hail', *Q. J. Roy. Met. Soc.* **78**, 37–45.

BRINKMAN, W. and ASHWELL, I. (1968) 'The structure and movement of the chinook in Alberta', *Atmosphere* (Bull. Canadian Met. Soc.) **6**, 1–10.

BRITISH ASSOCIATION (1962) *Manchester and its Region*, Manchester University Press.

BRITTON, G. P. and LAMB, H. H. (1956) 'A study of the general circulation over the far south', *Weather* **11**, 281–91, 339–54.

BROOKS, C. E. P. (1917) 'Continentality and temperature, I', *Q. J. Roy. Met. Soc.* **43**, 159–73.

BROOKS, C. E. P. (1918) 'Continentality and temperature, II', *Q. J. Roy. Met. Soc.*, **44**, 253–70.

BROOKS, C. E. P. (1932) 'The origin of anticyclones', *Q.J. Roy. Met. Soc.* **58**, 379–87.

BROOKS, C. E. P. (1934) 'The variation of the annual frequency of thunderstorms in relation to sunspots', *Q. J. Roy. Met. Soc.* **60**, 153–65.

BROOKS, C. E. P. (1949) *Climate through the Ages*, Benn.

BROOKS, C. E. P. (1954) *The English Climate*, English Universities Press, rev. by H. H. Lamb (1964), *q.v.*

BROOKS, C. E. P. and CARRUTHERS, N. *Handbook of Statistical Methods in Meteorology*, M.O. 538, HMSO.

BROOKS, E. M. (1951) 'Tornadoes and related phenomena', in *Compendium of Meteorology*, ed. Maloney (*q.v.*), 673–80.

BROWNING, K. A. (1968) 'The organisation of severe local storms', *Weather* **23**, 429–34.

BROWNING, K. A. and HARROLD, T. W. (1969) 'Air motion and precipitation-growth in a wave depression', *Q.J. Roy. Met. Soc.* **95**, 288–309.

BROWNING, K. A. and HARROLD, T. W. (1970) 'Air motion and precipitation-growth at a cold front', *Q.J. Roy. Met. Soc.* **96**, 369–89.

BROWNING, K. A. and LUDLAM, F. H. (1962) 'Airflow in convective storms', *Q.J. Roy. Met. Soc.* **88**, 117–35.

BRUNT, SIR DAVID (1941) *Physical and Dynamical Meteorology*, 2nd edn., Cambridge University Press.

BRUNT, SIR DAVID (1942) *Weather Study*, Nelson.

BUDYKO, M. I., ed. (1955) *Atlas of the Heat Balance*, Leningrad.

BUDYKO, M. I. (1958) *The Heat Balance of the Earth's Surface*, trans. N. Stepanova, U.S. Dept. of Commerce, Washington D.C.

BUTZER, K. W. (1966) 'Climatic changes in the arid zones of Africa during early to mid-Holocene times', in *World Climate from 8000 to 0 B.C.* (*q.v.*), 72–83.

BYERS, H. R. (1949) 'Principal results of a comprehensive investigation of the structure and dynamics of the thunderstorm', **1**, *Tellus*, no. 4, 6–17.

BYERS, H. R. (1951) 'Thunderstorms', in *Compendium of Meteorology*, ed. Maloney (*q.v.*), 681–93.

BYERS, H. R. (1959) *General Meteorology*, 3rd edn, McGraw-Hill.

BYERS, H. R. and BRAHAM, R. R. (1949) *The Thunderstorm*, U.S. Weather Bureau, Washington, D.C.

CALLENDAR, G. S. (1938) 'The artificial production of carbon dioxide and its influence on temperature', *Q.J. Roy. Met. Soc.* **64**, 223–40.

CALLENDAR, G. S. (1939) 'The composition of the atmosphere through the ages', *Met. Mag.* **74**, 33–9.

CARLSON, T. and LUDLAM, F. (1968) 'Conditions for the occurrence of severe local storms', *Tellus* **20**, 203–26.

CHANDLER, T. J. (1965) *The Climate of London*, Hutchinson.

*Climate and Man: Yearbook of Agriculture 1941*, U.S. Dept. Agric., Washington, D.C.

*Cloud Physics Conference*, Woods Hole, Mass. (1960), *Physics of Precipitation*, ed. Weickmann (*q.v.*), Geophys. Mon. 5. Amer. Geophys. Union, Washington, D.C.

COLE, J. P. and KING, C. A. M. (1968) *Quantitative Geography*, Wiley.

CONRAD, V. (1936) 'Die Klimatologischen Elemente und ihre Abhängigkeit von terrestrischen Einflüssen', in Köppen, W. und Geiger, R., eds., *Handbuch der Klimatologie, Band I, Teil B*.

CONRAD, V. and POLLAK, L. W. (1950) *Methods in Climatology*, 2nd edn., Harvard University Press.

COOKE, R. U. and HARRIS, D. R. (1970) 'Remote sensing of the terrestrial environment—principles and progress', *Trans. Inst. Brit. Geogr.* **50**, 1–23.

CORBY, G. A., ed. (1970) *The Global Circulation of the Atmosphere*, Royal Meteorological Society.

COULIER, P. J. (1875) 'Note sur une nouvelle propriété de l'air', *J. de Pharmacie et de Chimie*, **22**, 165–73.

CROLL, J. (1875) *Climate and Time in their Geological Relations*, Appleton, New York.

CROWE, P. R. (1938) 'On progress in geography', *Scottish Geog. Mag.* **54**. 1–19.

CROWE, P. R. (1940) 'A new approach to the study of the seasonal incidence of British rainfall', *Q.J. Roy. Met. Soc.* **66**, 285–316.

CROWE, P. R. (1949) 'The trade-wind circulation of the world', *Trans. Inst. Brit. Geogr.* **15**, 39–56.

CROWE, P. R. (1950) 'The seasonal variation in the strength of the trades', *Trans. Inst. Brit. Geogr.* **16**, 25–47.

CROWE, P. R. (1951) 'Wind and weather in the equatorial zone', *Trans. Inst. Brit. Geogr.* **17**, 23–76.

CROWE, P. R. (1954) 'The effectiveness of precipitation; a graphical analysis of Thornthwaite's climatic classifications', *Geogr. Studies* **1**, 44–62 (on p. 48, note (i)—for 'from' read 'by').

CROWE, P. R. (1957) 'Some further thoughts on evapotranspiration; a new estimate', *Geogr. Studies* **4**, 56–75 (on p. 69 for 'adequately' read 'inadequately').

CROWE, P. R. (1962) 'Climate', Chap. 2 in *Manchester and its Region*, Brit. Assoc. Handbook, Manchester University Press.

CROWE, P. R. (1965) 'The geographer and the atmosphere', *Trans. Inst. Br. Geogr.* **36**, 1–19 (on p. 17, line 37 for 'vertical motion' read 'vortical motion').

DALTON, J. (1793) *Meteorological Observations and Essays*, Richardson, London.

DEFANT, F. (1949) 'Zur Theorie der Hangwinde', *Arch. Met. Geophys. Biokl.* **1**, 421–50.

DEFANT, F. (1951) 'Local winds', in *Compendium of Meteorology*, ed. Maloney (*q.v.*), 655–72.

DEFANT, F. and TABA, H. (1957) 'The threefold structure of the atmosphere and the characteristics of the tropopause', *Tellus* **9**, 259–74.

DEPPERMANN, C. E. (1936) *Outlines of Philippine Frontology*, Bureau of Printing, Manila.

DOVE, H. W. (1862) *The Law of Storms considered in connexion with the ordinary movements of the Atmosphere*, 2nd edn., trans. R. H. Scott, Longmans.

DUNN, G. E. and MILLER, B. I. (1964) *Atlantic Hurricanes*, Louisiana State University Press.

DURAND-DASTÈS, F. (1961) 'Remarques sur les pluies d'été aux Indies'. *Ann. de Geogr.* **70**, 225–54.

DZERDZEEVSKII, B. L. (1958) 'On some climatological problems and microclimatological studies of arid and semi-arid regions in U.S.S.R.', in *Arid Zone Research*, Proc. Canberra Symp., Unesco, 315–25.

EADY, E. T. (1950) 'The cause of the general circulation of the atmosphere', in *Centenary Proc. Roy. Met. Soc.* 156–72.

EADY, E. T. (1953) 'The maintenance of the mean zonal surface currents', *Proc. Toronto Met. Conf.* 124–8.

EAST, T. (1957) 'An inherent precipitation mechanism in cumulus clouds', *Q. J. Roy. Met. Soc.* **83**, 61–76.

EMILIANI, C. (1955) 'Pleistocene temperatures', *Journal of Geology* **53**, 538–78.

ERICSON, D. B. and WOLLIN, G. (1966) *The Deep and the Past*, Cape.

ESTOQUE, M. A. (1961) 'A theoretical investigation of the sea breeze', *Q. J. Roy. Met. Soc.* **87**, 136–46.

EWING, M. and DONN, W. L. (1962) 'A theory of ice ages', in *Study of the Earth*, ed. White (*q.v.*), 203–24.

FAEGRI, K. (1950) 'On the value of palaeoclimatological evidence' in *Centenary Proc. Roy. Met. Soc.*, 188–95.

FERREL, W. (1889) *A Popular Treatise on the Winds*, Macmillan.

FICKER, H. (1936) 'Die Passatinversion', *Veröff. met. Inst. Univ. Berlin* **1**, Heft. 4.

FIELD, J. H. (1933) *A Survey of the Air Currents in the Bay of Gibraltar 1929–30*, Geophys Mem. 59, Met. Off., HMSO.

FINDLATER, J. (1969) 'A major low-level air current near the Indian Ocean during the northern summer', *Q.J. Roy. Met. Soc.* **95**, 362–80.

FISHER, E. L. (1960) 'An observational study of the sea breeze', *Journ. Met.* **17**, 645–60.

FLEAGLE, R. G. (1957) 'On the dynamics of the general circulation', *Q. J. Roy. Met. Soc.* **83**, 1–20.

FLEMING, R. (1930) *Wind Stresses in Buildings*, Wiley.

FLETCHER, N. H. (1962) *The Physics of Rainclouds*, Cambridge University Press.

FLETCHER, R. (1945) 'The general circulation of the tropical and equatorial atmosphere', *J. Met.* **2**, 167–74.

FLINT, R. F. (1957) *Glacial and Pleistocene Geology*, Wiley.

FLOHN, H. (1956) 'Investigations on the general atmospheric circulation, especially in lower latitudes', in *Sci. Proc. Intern. Assoc. of Met., Rome 1954* (*q.v.*), 431–41.

FÖRCHTGOTT, J. (1949) 'Wave currents on the leeward side of mountain crests', *Bull. Met. Tschécosl. Prague* **3**, 49, (fig. reprod. in *Met. Mag.* **94**, 1965, p. 11).

FÖRCHTGOTT, J. (1969) 'Evidence for mountain-sized lee eddies', *Weather* **24**, 255–60.

FOURNIER D'ALBE, E. M., LATEEF, A. M. A., RASOOL, S. I., and ZAIDI, I. H. (1955) 'The cloud-seeding trials in the central Punjab; July–Sept. 1954', *Q. J. Roy. Met. Soc.* **81**, 574–81.

FREEMAN, M. H. (1961) 'Fronts investigated by the Meteorological Research Flights', *Met. Mag.* **90**, 189–203.

FRENZEL, B. (1966) 'Climatic change in the Atlantic/sub-Boreal transition on the northern hemisphere: botanical evidence', in *World Climate from 8000 to 0 B.C.* (*q.v.*), 99–123.

FRISBY, E. M. (1963) 'Hailstorms of the upper Great Plains of U.S.A.', unpublished thesis, University of Reading.

FROST, R. (1947) 'The velocity profile in the lowest 400 feet', *Met. Mag.* **76**, 14–17.

FUKUI, E. (1954) 'Meridional distribution of radiant energy, its climatological and geographical significance', *Sc. Rep. Tokyo Kyoiku Daigaku* **3**, 1–22.

FULKS, J. R. (1951) 'The instability line', in *Compendium of Meteorology*, ed. Maloney (*q.v.*), 647–52.

FULTZ, D. (1951) Experimental analogies to atmospheric motions', in *Compendium of Meteorology*, ed. Maloney (*q.v.*), 1235–48.

FULTZ, D. (1961) 'Developments in controlled experiments on larger-scale geophysical problems', *Adv. Geophys.* **7**, 1.

FULTZ, D. and LONG, R. R. (1951) 'Two-dimensional flow around a circular barrier in a rotating spherical shell', *Tellus* **3**, 61–8.

GALTON, F. (1862–3) 'A development of the theory of cyclones', *Proc. Roy. Soc.* **12**, 385–6.

GAMOW, G. (1967) *A Star called the Sun*, Penguin Books.

GARBELL, M. A. (1947) *Tropical and Equatorial Meteorology*, Pitman.

GARNETT, A. (1937) *Insolation and Relief, Trans. Inst. Brit. Geogr.* **5**.

GARNETT, A. (1967) 'Some climatological problems in urban geography with reference to air pollution', *Trans. Inst. Brit. Geogr.* **42**, 21–43.

GARNIER, B. J. (1958) *The Climate of New Zealand*, Edward Arnold.

GEIGER, R. (1965) *Climate near the Ground*, 4th edn., trans. by Scripta Technica Inc., Harvard University Press.

GENTILLI, J. (1953) 'Une critique de la méthode de Thornthwaite pour la classification des climats', *Ann. de Géogr.* **62**, 180–5.

GENTILLI, J. (1969) 'Some regional aspects of southerly-buster phenomena', *Weather* **24**, 173–80.

GIBSON, T. T. and DOUGLAS, D. A. (1969) 'Simulation of antarctic circulations in a dishpan', *Weather* **24**, 309–18.

GILES, B. D. (1970) 'Extremely high atmospheric pressures', *Weather* **25**, 19–24.

GODSKE, C. L., BERGERON, T., BJERKNES, J. and BUNGAARD, R. C. (1957) *Dynamic Meteorology and Weather Forecasting*, Amer. Met. Soc., Boston and Carnegie Inst. Washington, D.C.

GODWIN, H. (1956) *History of the British Flora: a factual basis for phytogeography*, Cambridge University Press.

GODWIN, H. (1966) 'Introductory address', in *World Climate from 8000 to 0 B.C. (q.v.)*, 3–14.

GOLDTHWAIT, R. P. (1966) in *World Climate from 8000 to 0 B.C. (q.v.)*, discussion p. 229.

GORDON, A. H. (1962) *Elements of Dynamic Meteorology*, English Universities Press.

GREEN, F. H. W. (1965) 'The incidence of low relative humidity in the British Isles', *Met. Mag.* **94**, 81–8.

GREEN, J. S., LUDLAM, F. H. and MCILVEEN, J. F. (1966) 'Isentropic relative-flow analysis and the parcel theory', *Q.J. Roy. Met. Soc.* **92**, 210–19.

GREGORY, S. (1963) *Statistical Methods and the Geographer*, Longmans.

GREGORY, S. (1965) *Rainfall over Sierra Leone*, Dept. of Geogr. Research Paper 2. University of Liverpool.

GRINDLEY, J. (1953) 'Heaviest rainfall in 10, 20, 30, 60, 90 and 120 minutes on record for each county of the United Kingdom', *British Rainfall 1953*, 215–19, Met. Off., HMSO.

GROVE, A. T. and WARREN, A. (1968) 'Quaternary landforms and climate on the south side of the Sahara', *Geogrl. J.* **134**, 194–208.

HANN, J. VON (1886) 'Zur Frage über den Ursprung des Föhns', *Ztschr. Osterr. Gesch. f. Met.* **1**, 257–63.

HANN, J. VON (1897; trans. 1903) *Handbook of Climatology, Part I, General Climatology* trans. R. de C. Ward, Macmillan.

HARE, F. K. (1961) *The Restless Atmosphere*, Hutchinson.

HARE, F. K. (1968) 'The Arctic', *Q.J. Roy. Met. Soc.* **94**, 439–59.

HAURWITZ, B. (1947) 'Comments on the sea-breeze circulation', *J. Met.* **4**, 1–8.

HERBERTSON, A. J. (1905) 'The major natural regions of the world', *Geogrl. J.* **25**, 300–10.

HEUSSER, C. J. (1966) 'Polar hemisphere correlation: palynological evidence from Chile and the Pacific Northwest of America', in *World Climate from 8000 to 0 B.C.* (q.v.), 124–41.

HIDE, R. (1970) 'Some laboratory experiments on free thermal convection in a rotating fluid subject to a horizontal temperature gradient, and their relation to the theory of the global circulation', in Corby, G. A., ed. (1970, q.v.), 196–221.

HILDEBRANDSSON, H. H. and TEISSERENC DE BORT, L. (1898–1905) *Les Bases de la Météorologie Dynamique: Historique—État de nos Connaissances.* Paris, Gauthier-Villars et fils.

HOBBS, W. H. (1926) *The Glacial Anticyclones, the Poles of the Atmospheric Circulation.* Macmillan, New York.

HOBBS, W. H. (1935) 'The glacial anticyclones', in *Assoc. Met. U.G.G.I. Lisbon,* Paris, Paul Dupont, 1–11.

HOBBS, W. H. (1945) 'The Greenland glacial anticyclone', *J. Met.* **2**, 143–53.

HOUGHTON, H. G. (1954) 'On the annual heat balance of the northern hemisphere', *J. Met.* **11**, 1–9.

HUDSON, W. H. (1918) *Far Away and Long Ago.*

HUMPHREYS, W. J. (1940) *Physics of the Air,* 3rd edn, McGraw-Hill.

HUNTINGTON, E. and VISHER, S. S. (1922) *Climatic Changes, their Nature and Causes,* Yale University Press.

*International Association of Meteorology, Rome,* (1954) *Scientific Proceedings,* Butterworth.

*International Symposium at Imperial College, London* (1966), see *World Climate 8000 to 0 B.C.*

JOHN, I. G. (1949) '*The Properties of the Upper Air over Singapore,*' Mem. Malayan Met. Service, no. 4.

JOHNSON, D. H. (1970) 'The role of the tropics in the global circulation', in Corby, G. A. (1970, q.v.), 113–36.

JOHNSON, D. H. and DANIELS, S. M. (1954) 'Rainfall in relation to the jet stream', *Q.J. Roy. Met. Soc.* **80**, 212–7.

JOHNSON, N. K. (1929) *A Study of the Vertical Gradient of Temperature in the Atmosphere near the Ground,* Geophys. Mem. 46, HMSO.

JOHNSON, N. K. and HEYWOOD, G. S. P. (1938) *An Investigation of the Lapse-rate of Temperature in the Lowest Hundred Metres of the Atmosphere,* Geophys. Mem. 77, HMSO.

KENDREW, W. G. (1953) *The Climates of the Continents,* 4th edit. Oxford University Press.

KIDSON, J. W., VINCENT, D. G., and NEWELL, R. E. (1969) 'Observational studies of the general circulation of the tropics: long term mean values', *Q. J. Roy. Met. Soc.* **95**, 258–87.

KINGTON, J. A. (1969) 'A century of cloud classification', *Weather* **24**, 84–9.

KOERNER, R. M. (1970) 'Weather and ice observations of the British Transarctic Expedition, 1968–9', *Weather* **25**, 218–28.

KÖPPEN, W. (1922) 'Die Regenmenge an der Trockengrenze', *Met. Zeitsch.* **39**, 242–4.

KÖPPEN, W. (1931) *Grundriss der Klimakunde*, 2nd edn, Berlin, Walter de Gruyter.

KÖPPEN, W. and GEIGER, R. eds. (1932–9) *Handbuch der Klimatologie in fünf Banden*, Band 1. *Allgemeine Klimalehre* (1936); Band 2. *Amerika* (1938); Band 3. *Europa und Nordasien* (1932–9); Band 4. *Südasien und Australländer* (1938); Band 5. *Afrika, Ozeane und Register* (1933), Berlin, Gebrüder Borntraeger.

KOTESWARAM, P. (1958) 'The easterly jet stream in the tropics', *Tellus* **10**, 43–57.

KRAUS, E. B. (1960) 'Synoptic and dynamic aspects of climatic change', *Q.J. Roy. Met. Soc.* **86**, 1–15.

KUIPER, G. P. ed. (1949) *The Atmospheres of the Earth and Planets*, 50th Anniv. Sympos. of Yerkes Obs. 1947, University of Chicago Press.

LAHEY, J. F. (1958) *On the Origin of the Dry Climate in northern South America and the Southern Caribbean*, Dept. Met. Monograph, Univ. of Wisconsin.

LAMB, H. H. (1958) 'Differences in the meteorology of the northern and southern polar regions', *Met. Mag.* **87**, 364–79.

LAMB, H. H. (1959) 'The southern westerlies', *Q.J. Roy. Met. Soc.* **85**, 1–23.

LAMB, H. H. (1961) 'Atmospheric circulation, climate and climatic variations', *Geography* **46**, 208–22.

LAMB, H. H. (1964) *The English Climate*, English Universities Press.

LAMB, H. H. (1970) 'Climatic variation and our environment today and in the coming year', *Weather* **25**, 447–55.

LAMB, H. H. and JOHNSON, A. I. (1959) 'Climatic variation and observed changes in the general circulation', *Geografiska Annlr* **41**, 94–134.

LAMB, H. H., LEWIS, R. P. W., and WOODROFFE, A. (1966) 'Atmospheric circulation and the main climatic variables between 8000 and 0 B.C.: meteorological evidence', in *World Climate from 8000 to 0 B.C. (q.v.)* 174–217.

LANE, F. W. (1967) *The Elements Rage*, David & Charles.

LANG, R. (1920) *Verwitterung und Bodenbildung als Einführung in die Bodenkunde*, Stuttgart, E. Schweizerbartische Verlagsbhg.

LEIGHLY, J. (1949) 'On continentality and glaciation', *Geografiska Annlr.* **31**, 133–45.

LEOPOLD, L. B. (1949) 'The interaction of trade wind and sea breeze, Hawaii', *J. Met.* **6**, 312–20.

LIST, R. J., ed. (1951) *Smithsonian Meteorological Tables* see Sources of Climatic Data.

LOCKWOOD, J. G. (1962) 'Occurrence of Föhn winds in the British Isles', *Met. Mag.* **91**, 57–65.

LOCKWOOD, J. G. (1963) 'Winds at 200 mb. over the tropics', *Met. Mag.* **92**, 75–84.

LÓPEZ, M. E. and HOWELL, W. E. (1967) 'Katabatic winds in the equatorial Andes', *J. Atmos. Sci.* **24**, 29–35.

LORENZ, E. N. (1970) 'The nature of the global circulation of the atmosphere: a present view', in Corby, G. A., ed. (1970, q.v.), 2–23.

LUDLAM, F. H. (1951) 'The production of showers by the coalescence of cloud droplets', *Q.J. Roy. Met. Soc.* **77**, 402–17.

LUDLAM, F. H. (1955) 'Artificial snowfall from mountain clouds', *Tellus* **7**, 277–90.

LUMB, F. E. (1970) 'Topographic influences on thunderstorm activity near Lake Victoria', *Weather* **25**, 404–10.

MCINTOSH, D. H. (1963) *Meteorological Glossary* (M.O. 729) 4th cdn., HMSO.

MALKUS, J. S. (1952) 'Recent advances in the study of convective clouds and their interaction with the environment', *Tellus* **4**, 71–87.

MALKUS, J. S. (1955) 'The effects of a large island upon the trade-wind air stream', *Q.J. Roy. Met. Soc.* **81**, 538–50; also discussion **82**, 235–41.

MALKUS, J. S. (1957) 'The origin of hurricanes', *Sci. Amer.*, Aug., 3–9.

MALKUS, J. S. (1958a) 'On the structure and maintenance of the mature hurricane eye', *J. Met.* **15**, 337–49.

MALKUS, J. S. (1958b) 'On the structure of the trade-wind moist layer', Woods Hole Oceanog. Inst., *Papers in Physical Oceanography and Meteorology*, Vol. 13, no. 2.

MALKUS, J. S. and RIEHL, H. (1964) 'Cloud structures and distribution over the tropical Pacific Ocean', *Tellus* **16**, 275–87.

MALONEY, T. F., ed. (1951) *Compendium of Meteorology*, Amer. Met. Soc., Boston, Mass.

MANLEY, G. (1945) 'The helm wind of Crossfell 1937–9', *Q.J. Roy. Met. Soc.* **71**, 197–219.

MANLEY, G. (1964) 'The evolution of the climatic environment', in *The British Isles*, ed. J. Wreford Watson and J. B. Sissons, pp. 152–76.

MARTONNE, E. DE (1926) 'Aréisme et indice d'aridité', *C. R. Acad. Sci.*, 182.

MARTONNE, E. DE (1947) *Traité de Géographie Physique*, 2nd edn., Paris, A. Colin, 3 vols.

MASON, B. J. (1957) *The Physics of Clouds*, Oxford University Press.

MASON, B. J. (1962a) *Clouds, Rain and Rainmaking*, Cambridge University Press.

MASON, B. J. (1962b) 'Charge generation in thunderstorms', *Endeavour* **21**, 152–63.

MASON, B. J. (1969) 'Some outstanding problems in cloud physics—the interaction of microphysical and dynamical processes', *Q. J. Roy. Met. Soc.* **95**, 449–85.

MAURY, M. F. (1858) *The Physical Geography of the Sea*, 2nd edn., London, Sampson Low.

METEOROLOGICAL OFFICE (1963) *Meteorological Glossary* (M.O. 729), 4th edn., ed. D. H. McIntosh (*q.v.*).

METEOROLOGICAL OFFICE (1944) *Weather in Home Waters and the North-eastern Atlantic, Vol. II, Part 1, The Atlantic from the Azores to the African Coast* (M.O. 446 b.1.), HMSO.

METEOROLOGICAL OFFICE (1949) *Weather on the West Coast of Tropical Africa* (M.O. 492), HMSO.

METEOROLOGICAL OFFICE, *Geophysical Memoirs*, HMSO, 46, *see* Johnson, N. K. (1929); 50, *see* Bjerknes, J. (1930); 59, *see* Field, J. H. (1933); 65, *see* Best, A. C. (1935); 77, *see* Johnson, N. K. and Heywood, G. S. P. (1938); 87, *see* Belasco, J. E. (1952); 89, *see* Best, A. C. *et al.* (1952); 96, *see* Sawyer, J. S. (1955); 108, see Aanansen, C. J. (1965).

MIDDLETON, W. E. KNOWLES (1965) *A History of the Theories of Rain*, Oldbourne.

MILLER, A. AUSTIN (1947) *Climatology*, 5th edn, Methuen.

MONTEITH, J. L. (1962) 'Attenuation of solar radiation; a climatological study', *Q.J. Roy. Met. Soc.* **88**, 508–21.

MOROZ, W. J. (1967) 'A lake breeze on the eastern shore of Lake Michigan: observations and model', *J. Atmos. Sci.* **24**, 337–55.

MORRISON, M. E. S. (1966) 'Low latitude vegetation history with special reference to Africa', in *World Climate 8000 to 0 B.C.* (*q.v.*) 142–8.

MURRAY, R. and DANIELS, S. M. (1953) 'Transverse flow at entrance and exits to jet streams', *Q.J. Roy Met. Soc.* **79**, 236–41.

MURRAY, R. and JOHNSON, D. H. (1952) 'Structure of the upper westerlies', *Q.J. Roy. Met. Soc.* **78**, 186–99.

NAIRN, A. E. M., ed. (1963), *Problems in Palaeoclimatology*, Proc. N.A.T.O. Palaeoclimates Conf. at Newcastle upon Tyne, 1963, Wiley.

NAMIAS, J. (1958) 'Synoptic and climatological problems associated with the general circulation of the Arctic', *Trans. Amer. Geophys. Union*, **39**, 40–51.

NEIBURGER, M. (1960) 'The relation of air-mass structure to the field of motion over the eastern North Pacific Ocean in summer', *Tellus* **12**, 31–40.

NEWTON, C. W. (1966) 'Circulations in large sheared cumulonimbus', *Tellus* **18**, 699–712.

NEWTON, C. W. (1970) 'The role of extratropical disturbances in the global atmosphere', in Corby, G. A., ed. (1970, *q.v.*), 137–58.

NEWTON, C. W. and PERSSON, A. V. (1962) 'Structural characteristics of the sub-tropical jet stream and certain lower-stratospheric wind systems', *Tellus* **14**, 221–41.

NEWTON, C. W. and OMOTO, Y. (1965) 'Energy distribution near jet streams and associated wave-amplitude relations', *Tellus* **17**, 449–62.

PALMÉN, E. (1951a) 'The role of atmospheric disturbances in the general circulation', *Q.J. Roy. Met. Soc.* **77**, 337–54.

PALMÉN, E. (1951b) 'The aerology of extratropical disturbances', in *Compendium of Meteorology*, ed. Maloney (*q.v.*) 599–620.

PALMER, C. E. (1952) 'Tropical meteorology', *Q.J. Roy. Met. Soc.* **78**, 126–64.

PARRY, M. (1956) 'An "urban rainstorm" in the Reading area', *Weather* **11**, 41–8.

PEAKE-JONES, K., ed. (1955) *Lake Eyre, South Australia: the Great flooding of 1949–50*, Report of the Lake Eyre Committee, Roy. Geogr. Soc. of Australia, S. Austr. Branch, Adelaide.

PEDGLEY, D. E. (1962) *A Course in Elementary Meteorology* (M.O. 707), HMSO.

PEDGLEY, D. E. (1970) 'Heavy rainfalls over Snowdonia', *Weather* **25**, 340–50.

PENCK, A. and BRÜCKNER, E. (1909) *Die Alpen in Eiszeitalter*, Leipzig, Tauchnitz, 3 vols.

PENMAN, H. L. (1950) 'Evaporation over the British Isles', *Q. J. Roy. Met. Soc.* **76**, 372–83.

PENMAN, H. L. (1955) 'Evaporation from Lake Eyre', in *Report of the Lake Eyre Committee*, ed. Peake-Jones (*q.v.*) 57–61.

PENMAN, H. L. (1956) 'Evaporation: an introductory survey', *Netherlands J. Agric. Sci.* **4**, 9–29.

PENMAN, H. L. (1963) *Vegetation and Hydrology*, Tech. Comm. no. 53, Commonwealth Bureau of Soils, Harpenden.

PERLROTH, I. (1969) 'Effects of oceanographic media on equatorial Atlantic hurricanes', *Tellus* **21**, 230–34.

PETTERSSEN, S. (1941) *Introduction to Meteorology*, McGraw-Hill.

PETTERSSEN, S. (1956), *Weather Analysis and Forecasting*, 2nd edn, McGraw-Hill, 2 vols.

PIAZZI-SMYTH, C. (1858) 'An astronomical experiment on the Peak of Teneriffe', *Phil. Trans. Roy. Soc. London*, **148**, 465–534.

PLASS, G. N. (1956) 'The carbon dioxide theory of climatic change', *Tellus* **8**, 140–53.

PLASS, G. N. (1962) 'Carbon dioxide and the climate', in *Study of the Earth*, ed. White (*q.v.*), 224–38.

PRESCOTT, J. A., COLLINS, J. A., and SHIRPURKAR, G. R. (1952) 'The comparative climatology of Australia and Argentina', *Geogrl. Rev.* **42**, 118–33.

QUENEY, P. (1948) 'The problem of air flow over mountains', *Bull. Amer. Met. Soc.* **29**, 16–26.

RAIKES, R. (1966) in *World Climate from 8000 to 0 B.C.* (*q.v.*), discussion, p. 83.

RAMAGE, C. S. (1959) 'Hurricane development', *J. Met.* **16**, 227–37.

RAMAGE, C. S. (1966) 'The summer atmospheric circulation over the Arabian Sea', *J. Atmos. Sci.* **24**, 144–50.

RAMAMURTHY, K. (1948) 'Some aspects of the regional geography of Tamilnad; rainfall rhythm', *Indian Geogr. J.* **23**, no. 3, 34–64.

RAMAMURTHY, K. (1965) 'A Study of the Rainfall Regimes in India', unpublished thesis, University of Madras.

REITER, E. R. (1963) *Jet Stream Meteorology*, University of Chicago Press.

RICHARDS, P. W. (1952) *The Tropical Rain Forest*, Cambridge University Press.

RIEHL, H. (1948) 'A radiosonde observation in the eye of a hurricane', *Q. J. Roy. Met. Soc.* **74**, 194–6.

RIEHL, H. (1954) *Tropical Meteorology*, McGraw-Hill.

RIEHL, H. (1962) 'The tropical circulation', *Science* **135**, 13–22.

RIEHL, H. (1965) *Introduction to the Atmosphere*, McGraw-Hill.

RIEHL, H. (1969) 'On the role of the tropics in the general circulation of the atmosphere', *Weather* **24**, 288–308.

RIEHL, H. and FULTZ, D. (1957) 'Jet streams and long waves in a steady rotating-dishpan experiment: structure of the circulation', *Q. J. Roy. Met. Soc.* **83**, 215–31.

RIEHL, H. and FULTZ, D. (1958) 'The general circulation in a steady rotating-dishpan experiment', *Q. J. Roy. Met. Soc.* **84**, 389–417.

RIEHL, H., YEH, T. C., MALKUS, J. S. and LA SEUR, N. E. (1951) 'The north-east trade of the Pacific Ocean', *Q.J. Roy. Met. Soc.* **77**, 598–626.

RODDA, J. C. (1970) 'Rainfall excesses in the United Kingdom', *Trans. Inst. Brit. Geogr.* **49**, 49–60.

ROSSBY, C-G. (1941) 'The scientific basis of modern meteorology', in *Climate and Man* (*q.v.*), 599–655; first half reprinted in Berry *et al.* (1945) (*q.v.*).

ROSSBY, C-G. (1949) 'On the nature of the general circulation of the lower atmosphere', in *The Atmospheres of the Earth and Planets*, ed. Kuiper (*q.v.*), 16–48.

SAWYER, J. S. (1947) 'The structure of the intertropical front over north-west India during the south-west monsoon', *Q.J. Roy. Met. Soc.* **73**, 346–69.

SAWYER, J. S. (1955) *The Free Atmosphere in the Vicinity of Fronts*, Geophys. Mem. 96 (M.O. 584d), HMSO.

SAWYER, J. S. (1957) *The Ways of the Weather*, A. and C. Black.

SAWYER, J. S. (1966) 'Possible variation of the general circulation of the atmosphere', in *World Climate 8000 to 0 B.C.* (*q.v.*), 218–29.

SCHAEFER, V. J. (1946) 'The production of ice crystals in a cloud of supercooled water droplets', *Science* **104**, 457–9.

SCHAEFER, V. J. (1951) 'Snow and its relationship to experimental meteorology', in *Compendium of Meteorology*, ed. Maloney (*q.v.*), 221–34.

SCHWARZBACH, M. (1963) *Climates of the Past: an Introduction to Paleoclimatology*, trans. R. O. Muir, Van Nostrand.

SCORER, R. S. (1949) 'Theory of waves in the lee of mountains', *Q.J. Roy. Met. Soc.* **75**, 41–56.

SCORER, R. S. (1953) 'Theory of airflow over mountains, II. The flow over a ridge', *Q.J. Roy. Met. Soc.* **79**, 70–83.

SCORER, R. S. (1954) 'Theory of airflow over mountains, III. Airstream characteristics', *Q.J. Roy. Met. Soc.* **80**, 417–28.

SCORER, R. S. (1955) 'Theory of airflow over mountains, IV. Separation of flow from the surface', *Q.J. Roy. Met. Soc.* **81**, 340–50.

SCORER, R. S. (1958) *Natural Aerodynamics*, Pergamon.

SCORER, R. S. (1963) 'Cloud nomenclature', *Q.J. Roy. Met. Soc.* **89**, 248–53.

SCORER, R. S. (1965) *Vorticity in Nature*, Imp. College of Sci. and Tech., London.

SEILKOPF, H. (1939) *Handbuch der Fliegerwetterkunde*, Bd ii, Berlin, Radetzski.

SHAPLEY, H., ed. (1953) *Climatic Change; Evidence, Causes and Effects*, Harvard University Press.

SHAW, SIR NAPIER (W.N.) (1914, repr. 1965) 'Principia Atmospherica: a study of the circulation of the atmosphere', *Proc. R. Soc. Edinburgh* **34**. Repr. in *Selected Met. Papers of Sir Napier Shaw* (*q.v.*), 166–86.

SHAW, SIR NAPIER (1919–31) *Manual of Meteorology*, 4 vols.: I. *Meteorology in History* (1926); II. *Comparative Meteorology*, 2nd edn. (1936); III. *The Physical Processes of Weather*, (1930); IV. *Meteorological Calculus; Pressure and Wind*, rev. edn. (1931), Cambridge University Press.

SHAW, SIR NAPIER (1940) *The Drama of Weather*, 2nd edn, Cambridge University Press.

SHAW, SIR NAPIER (1955) *Selected Meteorological Papers of Sir Napier Shaw*, Macdonald.

SHAW, SIR NAPIER (W.N.) and LEMPFERT, R. G. K. (1906, repr. 1955) *Life History of Surface Air Currents, a study of the surface trajectories of moving air*, repr. in *Selected Met. Papers of Sir Napier Shaw* (*q.v.*), 15–131.

SHEPPARD, P. A. (1954) 'The maintenance of the wind systems of the world', *The Times Science Review*, 13–16.

SHEPPARD, P. A. (1958) 'The general circulation of the atmosphere', *Weather* **13**, 323–36.

SHEPPARD, P. A. (1963) 'Basic ideas on the general circulation of the atmosphere', in *Problems in Palaeoclimatology*, ed. Nairn (*q.v.*), 322–31.

SIBBONS, J. L. H. (1962) 'A contribution to the study of potential evapotranspiration', *Geografiska Annlr.* **44**, 279–92.

SIMPSON, SIR GEORGE C. (1929) 'Past climates', *Mem. and Proc. Manchester Lit. and Phil. Soc.* **74**, 1–34.

SIMPSON, SIR GEORGE C. (1941) 'On the formation of cloud and rain', *Q.J. Roy. Met. Soc.* **67**, 99–133.

SIMPSON, SIR GEORGE C. (1957) 'Further studies in world climate', *Q.J. Roy. Met. Soc.* **83**, 459–85.

SIMPSON, SIR GEORGE C. (1959) 'World temperatures during the Pleistocene', *Q.J. Roy. Met. Soc.* **85**, 332–49.

SIMPSON, J. (1967) 'An experimental approach to cumulus clouds and hurricanes', *Weather* **22**, 95–114.

SIMPSON, J., BRIER, G. W., and SIMPSON, R. H. (1967) '"Stormfury" cumulus seeding experiment 1965', *J. Atmos. Sci.* **24**, 508–21.

SIMPSON, J. E. (1964) 'Sea-breeze fronts in Hampshire', *Weather* **19**, 208–20.

SIMPSON, J. E. (1967) 'Aerial and radar observations at some sea-breeze fronts', *Weather* **22**, 306–16.

SIMPSON, R. H. (1952) 'Exploring the eye of the typhoon "Marge", 1951', *Bull. Amer. Met. Soc.* **33**, 286–98.

STARR, V. P. (1951) 'The physical basis for the general circulation', in *Compendium of Meteorology*, ed. Maloney (*q.v.*), 541–50.

STOUT, G. E., BLACKMER, R. H. and WILK, K. E. (1960) 'Hail studies in Illinois relating to cloud physics', in *Physics of Precipitation*, ed. Weickmann (*q.v.*), 369–83.

SUKACHEV, V. and DYLIS, N. (1968) *Fundamentals of Forest Biogeocoenology*, trans. J. M. Maclennan, Oliver & Boyd.

SUMNER, E. J. (1954) 'A study of blocking in the Atlantic-European sector of the northern hemisphere', *Q.J. Roy. Met. Soc.* **80**, 402–16.

SUSLOV, S. P. (1961) *Physical Geography of Asiatic Russia*, trans. N. D. Gershevsky, W. H. Freeman.

SUTCLIFFE, R. C. (1954) 'Cyclones and anticyclones; a comparative study', in *Proc. Toronto Met. Conf.* (*q.v.*), 139–43.

SUTTON, SIR GRAHAM (1953) *Micrometeorology*, McGraw-Hill.

SUTTON, SIR GRAHAM (1962) *The Challenge of the Atmosphere*, Hutchinson.

555

THOMPSON, B. W. (1951) 'An essay on the general circulation of the atmosphere over south-east Asia and the west Pacific', *Q.J. Roy. Met. Soc.* **77**, 569–97.

THORNTHWAITE, C. W. (1931) 'The climates of North America', *Geogrl. Rev.* **21**, 633–55.

THORNTHWAITE, C. W. (1948) 'An approach towards a rational classification of climate', *Geogrl. Rev.* **38**, 55–94.

THORNTHWAITE, C. W. and HOLZMAN, B. (1939) 'The determination of evaporation from land and water surfaces', *Monthly Weather Rev.* **67**, 1–8.

*Toronto Meteorological Conference, 1953* (1954) Amer. Met. Soc. and Roy. Met. Soc., *Proceedings*.

TREWARTHA, G. T. (1961) *The Earth's Problem Climates*, University of Wisconsin Press.

TRICKER, R. A. (1964) *Bores, Breakers, Waves and Wakes*, Mills and Boon, London.

TUCKER, G. (1965) 'The distribution and annual cycle of local heating rate throughout the troposphere in the northern hemisphere', *Met. Mag.* **94**, 205–14.

TULLETT, M. T. (1970) 'The nature of climatic change', *Weather* **25**, 465–71.

VISHER, S. S. (1954) *Climatic Atlas of the United States*, Harvard University Press.

VONNEGUT, B. (1949) 'Nucleation of supercooled water clouds by silver-iodide smokes', *Chem. Rev.* **44**, 277–89.

WALKER, J. M. (1970) 'After Bjerknes', *Weather* **25**, 175–9.

WALLINGTON, C. E. (1969) 'Depressions as moving vortices', *Weather* **24**, 42–52.

WATTS, I. E. M. (1955) *Equatorial Weather*, University of London Press.

WEAVER, D. (1968) 'The hurricane as an economic catalyst', *J. Trop. Geogr.* **27**, 66–71.

WEGENER, A. (1911) *Thermodynamik der Atmosphäre*, Leipzig, J. Barth.

WEGENER, A. (1924) *The Origin of Continents and Ocean Basins*, trans. J. G. Skerl, Methuen.

WEGENER, A. (1928) 'Beiträge zur Mechanik der Tromben und Tornados', *Met. Zeitschrift* **48**, 201–14.

WEICKMANN, H., ed. (1960) *Physics of Precipitation, Proc. Cloud Physics Conf.*, Woods Hole, Mass. (*q.v.*).

WENT, F. W. (1966) 'On the nature of Aitken condensation nuclei', *Tellus* **18**, 549–56.

WEXLER, H. (1953) 'Radiation balance of the earth as a factor in climatic change', in *Climatic Change*, ed. Shapley (*q.v.*), 73–105.

WEXLER, H. (1959) 'Seasonal and other temperature changes in the antarctic atmosphere', *Q.J. Roy. Met. Soc.* **85**, 196–208.

WHITE, J. F., ed. (1962) *Study of the Earth*, Prentice-Hall.

WHITTOW, J. B. and WOOD, P. D., eds. (1965) *Essays in Geography for Austin Miller*, University of Reading.

WILLETT, H. C. (1949) 'Solar variability as a factor in the fluctuations of climate during geological time', *Geografiska Annlr.* **31**, 295–315.

WILLETT, H. C. (1953) 'Atmospheric and oceanic circulation as factors in glacial-interglacial changes of climate', in *Climatic Change*, ed. Shapley (*q.v.*), 51–71.

WILLETT, H. C. and SANDERS, F. (1959) *Descriptive Meteorology*, 2nd edn., New York, Academic Press.

WISEMAN, J. D. H. (1966) 'Evidence for recent climatic changes in cores from the ocean bed', in *World Climate from 8000 to 0 B.C.* (*q.v.*), 84–98.

WOERKOM, A. J. J. VAN (1953) 'The astronomical theory of climatic changes', in *Climatic Change*, ed. Shapley (*q.v.*), 147–57.

WOODWARD, B. (1960) 'Penetrative convection in the sub-cloud regime', in Anderson, C. E. (1960, *q.v.*), 28–38.

*World Climate 8000 to 0 B.C.* (1966), *Proc. Internat. Symp. at Imperial College London*, 1966, Roy. Met. Soc.

ZEUNER, F. E. (1958) *Dating the Past*, 4th edn., Methuen.

ZON, R. (1941) 'Climate and the nation's forests', in *Climate and Man* (*q.v.*), 477–98.

# Index

# J

# N

# O